Principles and Practice of Analytical Techniques in Geosciences

RSC Detection Science Series

Editor-in-Chief:
Professor Michael Thompson, *University of Toronto, Canada*

Series Editors:
Dr Subrayal Reddy, *University of Surrey, Guildford, UK*
Dr Damien Arrigan, *Curtin University, Perth, Australia*

Titles in the Series:
1: Sensor Technology in Neuroscience
2: Detection Challenges in Clinical Diagnostics
3: Advanced Synthetic Materials in Detection Science
4: Principles and Practice of Analytical Techniques in Geosciences

How to obtain future titles on publication:
A standing order plan is available for this series. A standing order will bring delivery of each new volume immediately on publication.

For further information please contact:
Book Sales Department, Royal Society of Chemistry, Thomas Graham House, Science Park, Milton Road, Cambridge, CB4 0WF, UK
Telephone: +44 (0)1223 420066, Fax: +44 (0)1223 420247
Email: booksales@rsc.org

Visit our website at www.rsc.org/books

Principles and Practice of Analytical Techniques in Geosciences

Edited by

Kliti Grice
Curtin University, Perth, Australia
Email: k.grice@curtin.edu.au

THE QUEEN'S AWARDS
FOR ENTERPRISE:
INTERNATIONAL TRADE
2013

RSC Detection Science Series No. 4

Print ISBN: 978-1-84973-649-7
PDF eISBN: 978-1-78262-502-5
ISSN: 2052-3068

A catalogue record for this book is available from the British Library

Published by The Royal Society of Chemistry,
Thomas Graham House, Science Park, Milton Road,
Cambridge CB4 0WF, UK

Registered Charity Number 207890

For further information see our web site at www.rsc.org

Printed and bound by CPI Group (UK) Ltd, Croydon, CR0 4YY

Preface

The pace of revolution in analytical chemistry in the field of geosciences has been dramatic over the recent decades and includes fundamental developments that have become commonplace in many related and unrelated disciplines. Geoscientists have embraced all analytical tools related to the planet Earth, from ancient to modern environmental systems. The analytical tools used (nano- to macroscale, using stable or radioactive isotopes) have been applied to a wide range of applications from inorganic to organic geochemistry, biodiversity, and chronological tools, to build an understanding of how the Earth system evolved to its present state. Geoscientists apply new scientific approaches, often based on complex analytical data sets, to explain our forever changing climate, exploration, and production of our natural resources (water, energy) to explore sustainability and events related to human impact.

Presently there is no dedicated book that brings together frontline developments in analytical technologies applicable to geoscience research and ongoing global needs. This book is intended to fill that gap. The editor and authors are all recognised researchers in this field, and all chapters have been peer-reviewed to ensure high quality. The book is suitable for academics, industry personnel, and postgraduate students, and also as a teaching guide.

Kliti Grice
Curtin University, Perth, Australia

RSC Detection Science Series No. 4
Principles and Practice of Analytical Techniques in Geosciences
Edited by Kliti Grice
© The Royal Society of Chemistry 2015
Published by the Royal Society of Chemistry, www.rsc.org

Abbreviations

AFTT	apatite fission track thermochronology
AIT	ambient inclusion trails
ALF	airborne laser fluorescence
AMS	accelerator mass spectrometry
API	American Petroleum Institute
BC	black carbon
BCF	Barney Creek Formation
BP	before present
BSR	bacterial sulfate reduction
BT	benzothiophene
CAI	calcium–aluminium-rich inclusion
CCN	cloud condensation nuclei
CDT	Canyon Diablo Troilite
CSIA	compound-specific isotope analysis
DBT	dibenzothiophene
DCM	dichloromethane
DFA	downhole fluid analysis
DGDG	digalactosyldiacylglycerols
DIC	deconvoluted ion current, dissolved inorganic carbon
DMS	dimethylsulfide
DMSP	dimethylsulfoniopropionate
DOM	dissolved organic matter
DST	drill stem test
E&P	exploration and production
EA	elemental analysis
EEM	excitation–emission matrix
EI	electron impact, electron ionisation
EM	electron multiplier

RSC Detection Science Series No. 4
Principles and Practice of Analytical Techniques in Geosciences
Edited by Kliti Grice
© The Royal Society of Chemistry 2015
Published by the Royal Society of Chemistry, www.rsc.org

EMO	enhanced mineral oil
EPMA	electron microprobe analysis
EPS	extracellular polymeric substances
EqVR	equivalent vitrinite reflectance
FAMM	fluorescence alteration of multiple macerals
FC	Faraday cup
FID	flame ionisation detector
FISH	fluorescence in situ hybridisation
FLIM	fluorescence lifetime imaging microscopy
FWHM	full width at half maximum
FWL	free-water level
GC	gas chromatography
GDGT	glycerol dialkyl glycerol tetraether
GOC	gas–oil contact
GOR	gas-to-oil ratio
HI	hydrogen index
HISH	halogen-labelled *in situ* hybridisation
HPHT	high-pressure high-temperature
HPLC	high-performance liquid chromatography
HSE	highly siderophile elements
ICPMS	inductively coupled plasma mass spectrometry
IO	internal olefin
IR	infrared
IRMS	isotope ratio mass spectrometry
KCF	Kimmeridge Clay Formation
LAO	linear α-olefin
LC	liquid chromatography
LC-IRMS	liquid chromatography–isotope ratio mass spectrometry
LG	large geometry
LMIG	liquid metal ion gun
LSM	laser scanning microscopy
MALDI	matrix-assisted laser desorption/ionisation
MCI	molecular composition of inclusions
MC-ICPMS	multi-collector inductively coupled plasma mass spectrometry
MDR	methyldibenzothiophene
MGDG	monogalactosyldiacylglycerols
MPCA	multiway principal component analysis
MPI	methylphenanthrene
MRM	multiple reaction monitoring
MS	mass spectrometry
MSE	moderately siderophile element
MSSV	microscale sealed vessel
MSWD	mean square weighted deviation
NanoSIMS	nanoscale secondary ion mass spectrometry
NMR	nuclear magnetic resonance
OC	organic carbon

ODP	Ocean Drilling Program
OIB	Ocean Island basalts
OLED	organic light-emitting diode
OM	organic matter
OS	organic sulfur
OSC	organic sulfur compounds
OWC	oil–water contact
PAH	polycyclic aromatic hydrocarbons
PAR	photosynthetically active solar radiation
PCA	principal component analysis
PDB	Pee Dee Belemnite (standard)
PFA	perfluoroalkoxy alkane
PHD	peak height distribution
PIDD	primary ion dose density
PLS	partial least squares
POPC	palmitoyl-oleoyl-phosphatidylcholine
PVT	pressure–volume–temperature
QFT	quantitative fluorescence technology
QGF	quantitative grain fluorescence
QGF +	quantitative grain fluorescence plus
QGF-E	quantitative grain fluorescence on extracts
QSA	quasi-simultaneous arrivals
ROI	regions of interest
RSD	relative standard deviation
RT	retention time
SCD	sulfur chemiluminescence detector
SEM	scanning electron microscopy
SHRIMP	sensitive high-mass resolution ion microprobe
SIM	single-ion monitoring
SIMS	secondary ionisation mass spectrometry
SIR	selected ion recording
SLF	shipborne laser fluorescence
SOSB	South Oman Salt Basin
TEM	transmission electron microscopy
TIC	total ion current
TIMS	thermal ionisation mass spectrometry
TOF	time of flight
TR	transformation ratio
TRIFT	triple focusing time-of-flight
TSF	total scanning fluorescence
TSFS	total synchronous fluorescence scanning
TSR	thermochemical sulfur reduction
UCM	unresolved complex mixture
UV	ultraviolet
V-CDT	Vienna Canyon Diablo Troilite
VR	vitrinite reflectance

Contents

RSC Detection Science Series No. 4
Principles and Practice of Analytical Techniques in Geosciences
Edited by Kliti Grice
© The Royal Society of Chemistry 2015
Published by the Royal Society of Chemistry, www.rsc.org

**Chapter 3   Application of Radiogenic Isotopes in Geosciences:
Overview and Perspectives   49**
*Svetlana Tessalina, Fred Jourdan, Laurie Nunes,
Allen Kennedy, Steven Denyszyn, Igor Puchtel,
Mathieu Touboul, Robert Creaser, Maud Boyet,
Elena Belousova and Anne Trinquier*

**Chapter 4   Advances in Fluorescence Spectroscopy for Petroleum
Geosciences   94**
Keyu Liu, Neil Sherwood and Mengjun Zhao

Chapter 8   High-Precision MC-ICP-MS Measurements of $\delta^{11}B$: Matrix Effects in Direct Injection and Spray Chamber Sample Introduction Systems   251
Michael Holcomb, Kai Rankenburg and Malcolm McCulloch

Chapter 11 Applications of Liquid Chromatography–Isotope Ratio Mass Spectrometry in Geochemistry and Archaeological Science 313

Alison J. Blyth and Colin I. Smith

Chapter 12 Advances in Comprehensive Two-Dimensional Gas Chromatography (GC×GC) 324

Christiane Eiserbeck, Robert K. Nelson, Christopher M. Reddy and Kliti Grice

CHAPTER 1

Nanoscale Secondary Ion Mass Spectrometry (NanoSIMS) as an Analytical Tool in the Geosciences

MATT R. KILBURN*[a] AND DAVID WACEY[a,b]

[a] ARC Centre of Excellence for Core to Crust Fluid Systems, and Centre for Microscopy, Characterisation and Analysis, The University of Western Australia, 35 Stirling Highway, Crawley, WA 6009, Australia; [b] Department of Earth Science and Centre of Excellence in Geobiology, University of Bergen, Allegaten 41, N-5007, Bergen, Norway
*Email: Matt.Kilburn@uwa.edu.au

1.1 Introduction

1.1.1 Overview

The *in situ* chemical characterisation of rocks, minerals, and soils is fundamental to our understanding of the geological and environmental processes that have shaped our planet. Furthermore, we are becoming increasingly aware that evidence of large-scale phenomena, such as crustal evolution, mantle metasomatism, changes in atmospheric composition, or the emergence of life, is often only apparent at the micro- to nanoscale. Our ability to piece together clues about the Earth's evolution is therefore limited by the sensitivity and resolution of our analytical techniques. Over the past four decades, the development of microbeam technologies using electrons,

RSC Detection Science Series No. 4
Principles and Practice of Analytical Techniques in Geosciences
Edited by Kliti Grice
© The Royal Society of Chemistry 2015
Published by the Royal Society of Chemistry, www.rsc.org

ions, lasers, and X-rays has pushed to reduce the volume of material analysed while increasing sensitivity almost to the limits of counting statistics.

Of all the microbeam techniques available to geoscientists, secondary ion mass spectrometry (SIMS) is perhaps the most versatile. Combining *in situ* microbeam measurements with the high sensitivity and specificity of mass spectrometry, the technique has the ability to detect most of the elements in the periodic table with a high dynamic range (allowing both major and trace element analysis simultaneously), in a wide range of materials. SIMS is a well-established technique in the semiconductor industry, where it has been routinely used to measure the concentration of dopants and implants, and in forensic device failure analysis. The relative similarity between semiconductor materials and minerals has meant that the geological community has also employed SIMS to quantify trace elements in minerals at concentrations too low for detection by more typical methods such as electron microprobe analysis (EPMA). Furthermore, as a mass spectrometry technique, SIMS has proven itself to be highly effective in the analysis of stable and radiogenic isotopes in minerals, becoming the gold standard for uranium–lead geochronology. The lateral resolution, however, has typically been limited to tens of microns, so the demand for ever smaller length scales has necessitated the development of instruments that can produce an ion beam with orders-of-magnitude reductions in spot size while retaining ultra-high levels of sensitivity. In this chapter, we consider one such technological development, namely nanoscale secondary ion mass spectrometry (Nano-SIMS), and discuss its application within the geosciences.

1.1.2   Secondary Ion Mass Spectrometry

SIMS is a surface analysis technique that uses a highly energetic ion beam to 'sputter' material from a sample surface, which is then analysed in a mass spectrometer. The primary ion beam impacts the sample surface with energies ranging from a few hundred electron volts to more than 10 keV, causing a cascade-collision within the top few atomic layers which releases material from the surface. The sputtered material consists of atoms, atom clusters, molecular fragments, and backscattered primary ions, and the small proportion of this material that is ionised (secondary ions) is extracted into a mass spectrometer using electrostatic fields. The amount of ionised material sputtered is dependent on a number of factors, including the chemical structure of the substrate, the ionisation efficiency of the elements within the substrate, and the nature and energy of the primary ion used.

SIMS works in both 'static' and 'dynamic' regimes depending on the current density of the primary ion beam. In static SIMS, the primary ion dose is typically less than 10^{12} ions cm^{-2}, which is low enough that each impinging primary ion interacts with an essentially pristine area of the sample surface. This results in the sputtering of low-energy secondary ions from the topmost monolayers only. The surface is not actively eroded by the primary ion beam, hence the term 'static'. The secondary ions have velocities dependent on their mass, and by measuring the time of flight to reach the detector it is possible to differentiate between different elements, molecules and molecular fragments.[1] The pulsed primary ion beam, typically consisting of cluster projectiles such as C_{60} or Bi_{300}, can be rastered over the sample surface, producing images with lateral resolution up to 100 nm, which record the entire mass spectrum (up to tens of thousands of amu) on each pixel. The technique, however, must trade off lateral resolution for mass resolution and sensitivity, and even under optimum conditions cannot achieve the mass separation necessary to provide high-precision isotopic analyses.

Dynamic SIMS uses a high-energy beam that erodes the sample surface, producing a higher secondary ion flux, but destroying the molecular bonds between the atoms in the sample. This results in the detection of elemental ions only (although some non-stoichiometric ion clusters are formed in the sputtering process, as discussed below). Dynamic SIMS typically employs magnetic sector mass analysers, in which secondary ions are deflected in a magnetic field depending on their mass to charge ratio, m/z. Double-focusing mass spectrometers also use an electrostatic sector to filter the secondary ions by kinetic energy either before or after the magnetic sector. So-called sector-field mass analysers allow high mass resolution to be achieved with high transmission, providing optimal conditions for high-precision elemental and isotopic analyses. The development of dynamic SIMS over recent years has pushed towards higher sensitivity, for the measurement of high-precision isotope ratios, and high lateral resolution, for imaging.

1.1.3 Historical Use of SIMS in Geoscience

The use of SIMS in geoscience really began in earnest with the development of the sensitive high-resolution ion microprobe (SHRIMP) in the 1970s. With larger sector fields than had previously been employed in SIMS, its larger radius allowed high mass resolution to be routinely achieved. The ability to make spatially resolved measurements of uranium–lead isotopes within individual zircon grains revolutionised the field of geochronology, and the SHRIMP II is still the gold standard for uranium–lead dating in zircons and other minerals, today.[2] At the same time, smaller-radius instruments (namely, the Cameca IMS f-series) were beginning to gain popularity for the measurement of trace elements and stable isotopes. This was especially true among the cosmochemistry community, as SIMS had the ability to locate and measure tiny grains of isotopically unusual material (*i.e.* with pre-solar composition) against a background of 'normal' material.[3] This approach is also used by the nuclear safeguards community for identifying enriched uranium particles in environmental dust samples.[4] Cameca also developed a large-geometry ion probe in the 1990s, the IMS1270 (and later the IMS1280), based on the ion microscope concept of the f-series, but tailored towards the high-precision measurement of stable isotopes within minerals. The SHRIMP II and IMS1280 still dominate geoscience applications, with increasing degrees of overlap in their respective capabilities. The lateral resolution of these instruments, however, is still limited to microns or tens of microns. With the increasing need to reduce spot size, a new platform was sought that would allow the submicron elemental and isotopic characterisation of materials. An excellent history of the application of SIMS in geosciences is given in Stern.[5]

1.1.4   Development of NanoSIMS

The NanoSIMS 50, conceived by Georges Slodzian[6] and developed by CAMECA[7] (Gennevilleirs, France), was designed for imaging with submicron lateral resolution. This is achieved by positioning the primary probe-forming lens parallel and very close to the sample, allowing the beam to be focused to a very small diameter. The primary ion beam impacts the sample surface at $90°$, with the secondary ions extracted back through the same lens assembly. As the coaxial lens uses common focusing optics for both the primary and secondary ion beams, the polarity of the secondary ions must be opposite to that of the primary ions. Two primary ion sources are available for the NanoSIMS: a Cs^+ source for the generation of negative secondary ions, which can be focused to a sub-50 nm probe diameter; and a duoplasmatron source which can produce O^- and O_2^+ primary ions, with a smallest beam diameter of about 150 nm. Using the Cs^+ primary beam also allows secondary electrons, generated during the sputtering process, to be extracted to a photomultiplier, providing a simultaneous secondary electron image of the sample.

High mass resolution is achieved through the use of a double-focusing sector-field spectrometer, consisting of an electrostatic filter (to filter secondary ions according to their kinetic energy) and a magnetic sector (to separate ions by their mass). The geometry of the mass spectrometer has been optimised to give high transmission and high mass resolution at high lateral resolution, such that a $M/\Delta M$ mass resolution in excess of 9000 with a 100 nm beam diameter is achievable with only approximately 60% loss in transmission.

The secondary ions are focused along the plane of the magnetic sector where up to seven detectors can be positioned, allowing a large number of mass combinations to be measured. This multi-collection capability allows the same microvolume to be sampled on up to seven detectors simultaneously, minimising the loss of material and increasing throughput. Magnetic peak switching is also possible using a single detector. The latest generation of NanoSIMS instruments come equipped with both electron multipliers (EM) and Faraday cup (FC) detectors, and electronics that are housed under vacuum and regulated thermally. Imaging is achieved by scanning the focused primary ion beam across the sample, recording the number of incoming ions at the EM for each pixel of the scan (raster).

1.2 Technical

1.2.1 Instrument Specifications

1.2.1.1 Primary Ion Beam

The NanoSIMS is equipped with a Cs^+ microbeam source and an O^- duoplasmatron source. Due to the coaxial lens configuration, the polarity of the primary ion beam must be opposite to that of the secondary ions. Thus, the choice of a Cs^+ or O^- primary beam is determined by the polarity of the secondary ions to be analysed. The Cs^+ primary beam is generated by the thermal decomposition of $CsCO_3$, and the extraction of the resulting ionised Cs. The beam can be focused to a very fine probe diameter, depending on the beam current, which can be controlled through the use of an aperture. At each successively smaller aperture the beam current is decreased, resulting in a decrease in the secondary ion signal but producing a sharper image due to the smaller probe diameter.

As mentioned above, the impinging primary beam causes a cascade-collision in the top few atomic layers of the sample, implanting the primary ions into the sample surface. The sputtering of secondary ions begins once a sufficiently high dose of primary ions has been implanted—typically around 10^{16} ions cm^{-2}. Cs^+ enhances the generation of secondary ions by lowering the work function of the atoms in the sample, while O^- is known to enhance the ionisation probability.

1.2.1.2 Secondary Ions

The ionisation efficiency of a particular element is largely determined by its ionisation potential or electron affinity. Elements with a high ionisation potential will tend to form positive ions, while elements with a high electron affinity will tend to form negative secondary ions. Many elements, of course, form both positive and negative ions, but as a general rule, elements from the left side of the periodic table have strong positive ions yields, while elements from the right side have strong negative ion yields. Thus, elements such as carbon, nitrogen, oxygen and sulfur form negative secondary ions, which require the Cs^+ primary ion beam, while sodium, potassium, and calcium form positive secondary ions, necessitating the use of the O^- primary ion beam. Group 8 elements, however, do not ionise in the SIMS regime.

The sputtering process also produces molecular ions and complexes. This leads to one of the biggest problems in SIMS, namely isobaric interferences, where complexes of atoms add up to produce an ion species with the same mass as the element of interest thereby causing an interference on the desired signal (*e.g.* $^{12}C^1H^-$ interferes with $^{13}C^-$ at 13 amu). It can also, however, be advantageous: nitrogen, for example, does not ionise as a monoatomic ion species in SIMS, but it does form a complex with carbon, producing a strong negative ion signal at 26 amu ($^{12}C^{14}N^-$). Similarly, elements that nominally form positive ion species, such as calcium or iron, may be detected as negative ions in complexes with ubiquitous matrix elements, such as oxygen in silicate minerals (*e.g.* $^{40}Ca^{16}O^-$ at 56 amu), or sulfur in sulfides (*e.g.* $^{56}Fe^{32}S^-$ at 88 amu).

1.2.1.3 Multi-Collection Detector Setup

The multi-collection capability allows up to seven masses to be detected simultaneously, yet there is a physical limitation to how close the trolleys can be positioned, which is dependent on the mass range. For the standard NanoSIMS 50 (five detectors), one mass unit separation is only possible up to mass 30, making it possible to simultaneously detect the isotopes of silicon (^{28}Si, ^{29}Si, and ^{30}Si), but not ^{31}P and ^{32}S. On the NanoSIMS 50 L (seven detectors), it is possible to achieve one mass unit spacing up to mass 58, allowing the simultaneous detection of all the isotopes of iron. The Nano-SIMS, as with conventional single-detector SIMS instruments, also has the ability to switch the magnetic field during analysis to measure more masses than the multi-collection allows, and to allow one mass unit spacing where the physical separation is not possible. It is also possible to combine the multi-collection capability with magnetic field switching to cycle between two B-field settings, while using more than one detector. Of course, cycling the B-field adds significantly to the time taken for an analysis. In multi-collection mode, the B-field is regulated with an nuclear magnetic resonance (NMR) probe to ensure the peak does not drift over time.

The NanoSIMS 50 is equipped with EM and FC type detectors. EM detectors have a fast response time and are thus used for imaging, yet are susceptible to a number of intrinsic artefacts, including deadtime, ageing, and quasi-simultaneous arrivals (QSA), which are all discussed below. FC detectors have a higher sensitivity, but with an intrinsically high background can only be used to measure high secondary ion count rates (>1 million counts per second). The response time of the FC is too slow for imaging, yet they are far superior for quantitative ratio measurements at high count rate.

The geometry of the NanoSIMS has been optimised to give high secondary ion transmission at high mass resolution. It should be noted, however, that unlike Cameca's IMS-series instruments, the NanoSIMS is not an ion microscope (which projects an image of the sample onto the detector), but an ion microprobe, (where the detected signal is simply the total amount of ions sputtered from a single pixel within the primary ion beam raster). The image of the sample surface is digitally recreated by the software. High mass resolution is achieved through the use of slits; the mass spectrometer entrance slit (ES), the aperture slit (AS), and the energy slit (EnS). The ES is necessary to separate peaks, and should be used wherever there is the possibility of overlapping interfering peaks. The EnS is a continuous slit, with a variable width and position. It is used to reduce the amount of tailing from peaks that may overlap even though the peaks appear to be well resolved, which may be significant when the desired signal is small and the potentially interfering signal is large (abundance sensitivity). The EnS is positioned behind the electrostatic sector to allow the secondary ions to be filtered according to their energy. Positioning the slit to accept ions with higher energies can reduce interfering molecular ions and complexes, as the energy distribution of multi-atom ions is not as broad as for single ion species (energy filtering).

Each detector has its own exit slit and a set of deflectors to fine-tune the secondary ion signal. Applying small voltages to the deflector plates allows the positioning of the desired peak onto the conversion dynode (Figure 1.1).

1.2.2 Data Acquisition and Processing

1.2.2.1 Types of Analyses

Several types of analyses are possible with the NanoSIMS software: imaging, linescans, depth profiles, and element/isotope ratio measurements.

Images are acquired by rastering the primary beam over the sample surface, recording the number of secondary ion counts within the defined dwell time at each pixel, for each given ion species (Figure 1.2). The resolution of the image is generally determined by the number of pixels in the raster (*e.g.* 256×256, or 512×512) over the field of view—typically between 10 and 100 µm. A stack of images can be acquired consecutively to produce a 3D representation of the sample. Images are recorded with 32-bit data density, allowing quantitative data to be extracted using post-analysis processing tools, as detailed below.

Linescans can be acquired by two methods: stage control and beam control. In stage-control mode, linescans are performed by moving the stage under the stationary beam. The step size is defined by the user, but is limited to the minimum step size of the stage motors (∼1 µm). This method is ideal for coarse scans over large distances. In beam-control mode, linescans are acquired by moving the beam within a defined raster area. First, an image is acquired at the desired raster resolution. A line is then drawn onto the image using the software, which determines which pixels are to be measured. The line thickness can be modified by incorporating small rasters for each individual point (*e.g.* 3×3 pixels, centred on the middle pixel), to increase the signal-to-noise ratio and to reduce effects due to crater depth. The advantage of performing a linescan rather than acquiring an image and extracting the data along a similar line of pixels is that linescans can be performed much more quickly, allowing longer dwell times to increase the signal-to-noise ratio.

Depth profiles record the summed counts for individual raster scans acquired consecutively as the beam erodes the surface, producing a plot of intensity with depth/time.

Elemental and isotope ratios are acquired in the same manner as depth profiles, except that the counts from all of the individual scans are summed to increase the signal-to-noise ratio, and provide better counting statistics.

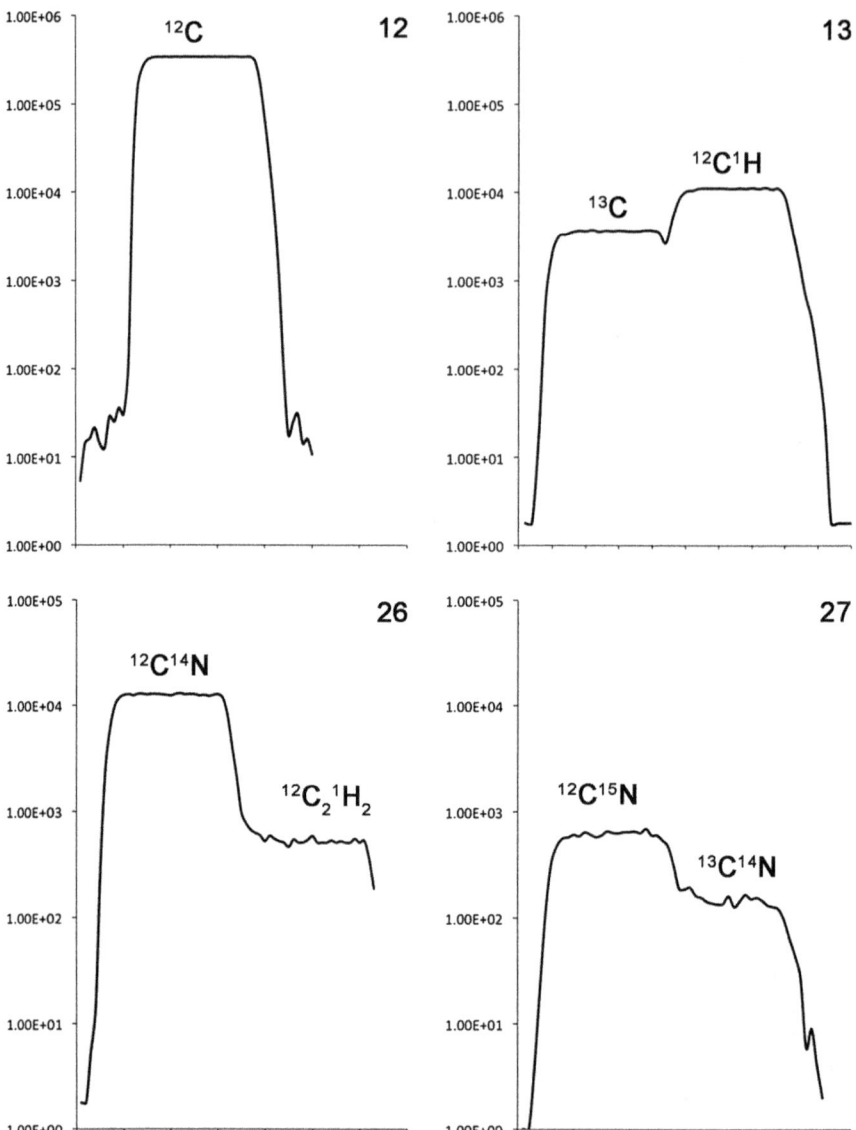

Figure 1.1 High mass resolution spectra of peaks on masses 12, 13, 26, and 27. On mass 12, there is only a single peak, $^{12}C^-$ ($m = 12.0000$). On mass 13, there are two peaks: $^{13}C^-$ ($m = 13.0034$) and $^{12}C^1H^-$ ($m = 13.0078$), which requires a $M/\Delta M$ of 2955 to resolve. The mass resolution is sufficiently high to see a small valley between the two peaks. On mass 26 the dominant peak is $^{12}C^{14}N^-$ ($m = 26.0031$), but there is a small interference from $^{12}C_2{}^1H_2{}^-$ ($m = 26.0156$). Mass 27 has two peaks: $^{12}C^{15}N^-$ ($m = 27.0001$) and $^{13}C^{14}N^-$ ($m = 27.0065$), which requires a $M/\Delta M$ of 4220 to resolve.

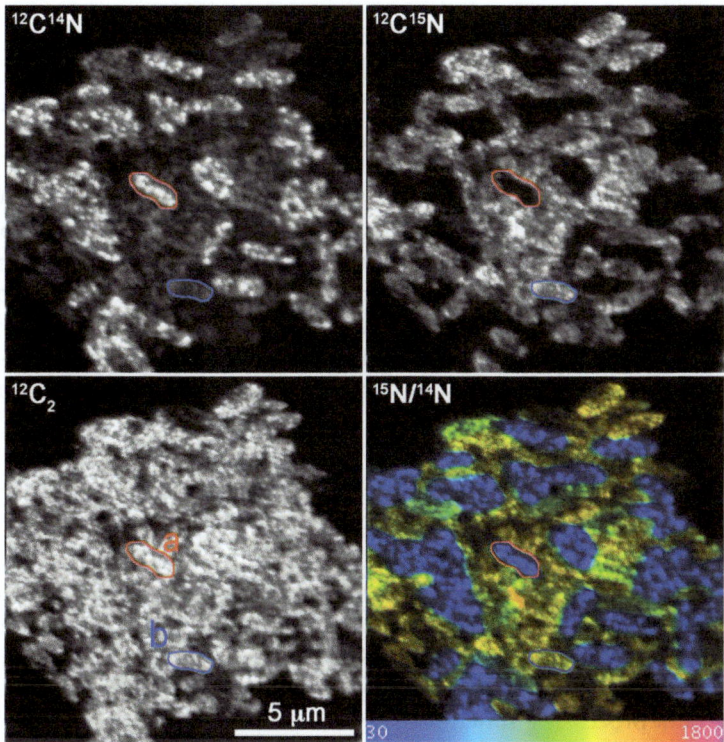

Figure 1.2 A mixed community of two bacterial populations, grown in ^{15}N-enriched medium, and mounted on a silicon wafer. The $^{12}C_2{}^-$ image does not give any indication that different bacteria are present, but the $^{12}C^{14}N^-$ and $^{12}C^{15}N^-$ images clearly reveal that one population is more enriched than the other. On the HSI colour scale (based on a ratio scale factor of 1000), the blue bacteria are enriched to about 4 at% ($^{15}N/^{14}N = 0.047$), while the yellow bacteria are enriched up to 60 at% ($^{15}N/^{14}N = 1.120$). The high lateral resolution allows data from regions of interest (ROI), here representing two individual bacteria, to be extracted.

A certain amount of consideration should be given to the size of the primary beam in relation to the size of the pixel, as the beam effectively digs a small crater at each pixel location. If the beam is much smaller than the pixel size (*i.e.* if the field of view is large and number of pixels low), then the beam will leave a gridwork of small craters on the sample surface. If the beam is much larger than the pixel, there will be a greater degree of pixel overlap, and the image/linescan will not be sharp. A good rule of thumb is to make the pixel size approximately equal to the size of the primary beam. Similarly, consideration should be given to the fact that the energy of the primary beam remains constant, regardless of the raster size, so the primary ion dose is much higher when the field of view is smaller. This results in the rapid erosion of the sample surface for small raster sizes, and conversely, very slow erosion rates when the raster size is large.

1.2.2.2   Corrections

EM detectors are subject to several intrinsic phenomena that must be accounted for in the data processing to avoid producing artefacts: deadtime, ageing, and QSA.

Deadtime is the time during which the detector is effectively blind to the secondary ion signal while the previous signal is being processed. When a secondary ion arrives at the conversion dynode it ejects an electron that triggers a cascade between the subsequent dynodes, resulting in an electronic pulse at the preamplifier electronics. During the time that it takes to record this pulse, the detector is blind to any other incoming ions arriving at the conversion dynode. This has the effect of undersampling the signal when the secondary ion count rate is high. A correction must therefore be made to determine the 'real' signal arriving at the detector. This is given by the expression:

$$N \approx \frac{N_m}{1 - N_m{}^\tau/T}$$

where N is the real number of secondary ions, N_m is the number of secondary ions detected, τ is the deadtime (in seconds), and T is the dwell time (in seconds) per pixel. The NanoSIMS EMs have a deadtime of 44 ns, set by the electronics, which must be applied to each pixel individually before any further data processing functions.

Detector ageing occurs with increasing cumulative counts due to the deposition of carbon on the last dynode. It manifests itself as a gradual change in the peak height distribution (PHD) that affects the response of the detector. To some extent this can be countered by checking the PHD regularly and adjusting the EM voltage. However, detector ageing can cause a drift in isotope ratios over the course of a single analytical session when count rates are very high (>300 000 counts per second). The PHD can be monitored during the analysis routine and this can then be used to determine the correction, which is typically linear with count rate.

QSA is a poorly understood artefact, which can have a significant effect on elemental and isotope ratios measured with EMs. The ionisation efficiency of some elements is so high that, for a given single impinging primary ion, two or more secondary ions may be ejected from the sample surface. These secondary ions pass through the mass spectrometer together and arrive at the conversion dynode of the EM more or less simultaneously. The EM is unable to differentiate between the ions, and records only a single pulse, regardless of the magnitude of the pulse. As with deadtime, the signal is undersampled. When measuring isotopes of a single element, the more abundant isotope will undersample more than the less abundant isotope, resulting in an inaccurate isotope ratio. The magnitude of this discrepancy increases linearly with secondary ion count rate. A theoretical model, based on Poisson statistics, was proposed by Slodzian *et al.*,[8] but found not to match the empirical data. A correction has been derived,[9] based on empirical observations of the linear change in the isotopic ratio with changing secondary to primary ion ratio:

$$R_{\text{true}} = R_{\text{meas}}(1 + \beta K)$$

where R_{true} is the real ratio, R_{meas} is the measured ratio, β is the slope of the linear regression, and K is the secondary to primary ion ratio. But as the actual value of K is not known (due to the undersampling), it must be determined experimentally, by:

$$K = \left(\frac{K_{\text{exp}}}{1 - K_{\text{exp}}/2}\right)$$

where K_{exp} is the measured secondary to primary ion ratio, based on the number of ions detected at the detector. It appears that different elements have their own intrinsic behaviour, and preliminary data suggest that this may be consistent across a range of materials: for carbon, sulfur, and silicon, the value of β has been found to be 1, 0.75, and 0.6, respectively.[10]

FC detectors are not subject to these issues as they record the continuous secondary ion current, and simultaneous arrivals are accounted for within the magnitude of the signal. It is imperative, however, to record the FC background noise during analysis, which is then subtracted from the signal.

1.2.3 Types of Data Processing

1.2.3.1 Images

Quantitative data can be extracted from the secondary ion images using one of the image processing software packages designed to work with NanoSIMS image files (*e.g.* ImageJ). The raw image files consist of 32-bit data, recording the counts and *xy*-coordinates for each pixel for each secondary ion species. The images should be corrected for EM deadtime and QSA (if possible) before extracting data. Data can be extracted by drawing regions of interest (ROI) around the pixels within the image that define certain features (Figure 1.2). The software then extracts the number of ions recorded for the pixel within the ROI, giving either a total or a mean. These data can be exported to spreadsheet software for further analysis.

1.2.3.2 Linescans/Depth Profiles

At present no software is available for processing NanoSIMS linescans and depth profiles. Raw data must be imported into spreadsheet software, and corrections performed manually.

1.2.3.3 Isotope Ratios

Whether ratio data has been acquired directly or extracted from images, the treatment is essentially the same. Natural isotopic fractionations in minerals are generally small, varying only in the third significant figure. It is therefore necessary to express the fractionation in units of per mil (‰) relative to a known standard using δ notation:

$$\delta^i x = \left(\frac{R_{\text{sample}}}{R_{\text{standard}}} - 1 \right) \times 1000$$

where i is the relevant isotope of element x (e.g. ^{34}S), and R is the measured ratio of isotope i to a reference isotope (e.g. $^{34}S/^{32}S$) for the sample and the standard, respectively. For a given analysis session, sample measurements must be bracketed by measurements of the standard reference material to determine the instrumental mass fractionation. This allows the measurement to be expressed relative to the international reference standard for that particular isotopic system (e.g. Canyon Diablo Troilite, or CDT, for sulfur), and to evaluate the stability of the instrument during the analysis (drift).

It should be noted that the small primary ion beam diameter used by the NanoSIMS precludes spot analyses. A focused beam would very quickly ablate a steep-sided crater in the sample, giving rise to potential artefacts caused by the different local extraction fields. Isotope data is generally acquired by rastering the beam across the sample surface, as in imaging, but without retaining any x-y spatial information.

1.2.4 Uncertainty Handling/Propagation of Errors

It is important to bear in mind that each measurement, data extraction protocol, and subsequent data processing function may generate an uncertainty that will be propagated to determine the final uncertainty in the measurement. With analytical measurements such as these, the most fundamental level of uncertainty is based on Poisson counting statistics. The minimum error associated with any given pixel, ROI, or total counts within an image frame, is determined by the number of counts recorded:

$$\sigma = \sqrt{N}$$

where σ is the uncertainty, and N the number of counts recorded. For a ratio of two different ion species, the expression becomes:

$$\sigma = \sqrt{\left(\frac{1}{N_1} \right) + \left(\frac{1}{N_2} \right)}$$

where N_1 is the number of counts for ion species 1 and N_2 the number of counts for ion species 2. For isotope ratio measurements, the level of precision required is largely dictated by the number of counts recorded on the least abundant isotope, such that a precision of 1‰ requires at least 1 million counts to be recorded on the least abundant signal. As mentioned above, the small primary beam diameter interacts with smaller volume of the sample than the larger beams used by SHRIMP or the IMS-series instruments, which means that to generate 1 million counts on the least abundant signal requires relatively long counting times. In fact, given the inherent problems with EM detector ageing, it is necessary to limit the count rate on the most abundant isotope to less than 500 000 counts per second, which increases the time required to achieve 1 000 000 total counts for the less abundant isotope. Given the small natural variations of isotopes in minerals (*e.g.* δ^{18}O in zircon), NanoSIMS struggles to achieve the precision necessary to provide useful data. But in some natural isotopic systems, such as δ^{34}S in sulfides, the fractionations are relatively large (*e.g.* >10‰), in which case NanoSIMS analysis is possible.[11] The latest generation of NanoSIMS 50L is equipped with multiple FC and more sensitive detector electronics, allowing higher precision than the older instruments.

Statistical uncertainty, such as the standard deviation of counts on pixels within an ROI, or between a number of ROIs, and the standard error of the mean of a number of measurements, should be propagated appropriately. In very general terms, the standard error of the mean (SE) of a number of measurements should approximate the counting statistics, such that the χ^2 value is close to 1:

$$\chi^2 = \left(\frac{\text{SE}}{\text{Poisson}}\right)^2$$

1.3   Examples of NanoSIMS Applications in the Geosciences

1.3.1   Biogeoscience

The capability of NanoSIMS to correlate nano- to microscale morphological features with chemical and isotopic signals characteristic of biology makes it a natural choice for studying microbe–mineral interactions in geological samples.

1.3.1.1   Iron Oxide Concretions

Concretions are preferentially cemented volumes within sedimentary rocks and are useful tracers of porewater chemistry during sediment diagenesis. The Jurassic Navajo Sandstone in Utah contains some of Earth's most famous concretions, ranging in size from small marbles to cannonballs, and consisting of a hard shell of iron oxide surrounding a softer sandy interior. Such concretions were thought to be the product of inorganic chemical re-actions involving the mixing of dissolved Fe(II) and oxic fluids without any structural precursor.[12] However, new research by Weber *et al.*[13] using NanoSIMS found that these Fe(III) oxide concretions formed from Fe(II) carbonate precursor material, and microbes were vital to their formation.

Weber *et al.*[13] showed that pre-existing Fe(II) carbonate concretions present in the sandstone were exposed to oxidising fluids flowing through an aquifer, establishing a redox boundary at the surface of the concretions. Indigenous Fe(II)-oxidising microbes used the Fe(II) carbonate as both a source of energy and a source of inorganic carbon to fix into organic carbon. This reaction produced acid that further dissolved the Fe(II) carbonate cement, and Fe(II) diffused from the interior to the exterior of the dissolving carbonate concretion with Fe(III) oxide biomineralisation occurring at the redox interface. This self-propagating process continued until Fe(II) carbonates were completely dissolved or microbial activity was terminated, ultimately leaving spherical concretions with hard Fe(III) oxide rinds and friable iron-poor sandy cores.

NanoSIMS was key to demonstrating the role of microbes in this process by enabling the correlation of nanoscale chemical features, namely enrichments in carbon, nitrogen, iron, and oxygen, with morphological remains of microbes (Figure 1.3). Supporting evidence for microbial mediation came from the carbon isotope signature of organic material extracted from the iron oxide rinds and from microbial morphologies imaged using scanning electron microscopy (SEM)) that resembled the Fe(II)-oxidising bacterium *Gallionella*.

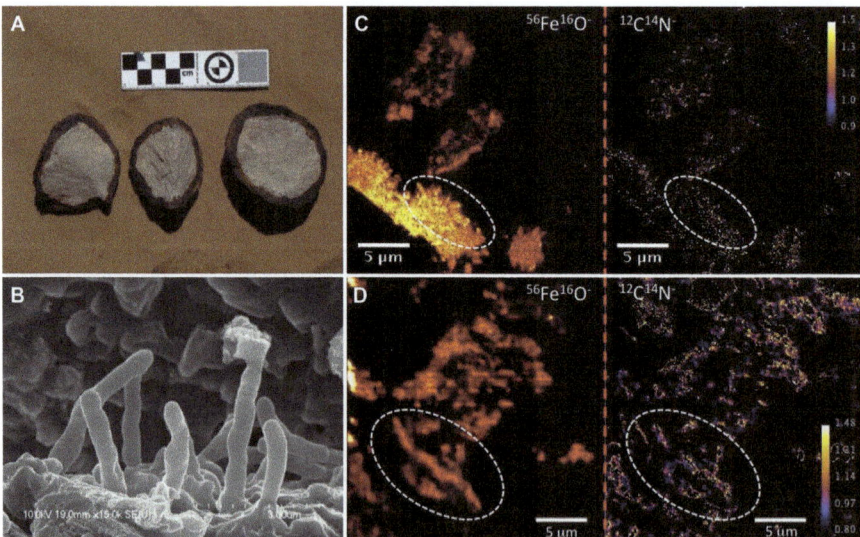

Figure 1.3   Biosignals in iron oxide concretions. (A) Iron oxide concretions from the Jurassic Navajo Sandstone of Utah (USA), consisting of a hard shell of iron oxide surrounding a softer sandy interior. (B) Filamentous microorganisms at the interface between the iron oxide shell and sandy interior. (C, D) NanoSIMS ion maps of the same interface showing organic material (CN$^-$) directly associated with iron oxides (FeO$^-$) that have morphologies resembling microbes (dashed circles).
Modified from Weber *et al.*[13]

The importance of this work is that it demonstrates that microbes can survive in rocky environments that are poor in organic carbon but rich in iron, and thus provides a new target in the search for life in unusual places. Of particular note are concretions with similar morphologies, nicknamed 'blueberries', identified by the Mars Opportunity Rover in 2004. The research of Weber *et al.*[13] provides the tantalising possibility that the Martian 'blueberries' may provide evidence of past microbial activity on the surface of Mars. It is hoped that the new Mars Rover, Curiosity, equipped with the ability to test for organic material, may find similar 'blueberries' in its new home of Gale Crater.

1.3.1.2 Ooids

Ooids are common components of carbonate successions in both modern and ancient environments. They are often used as palaeo-indicators of turbulent hydrodynamic conditions because a means of agitation was thought to be required for their formation.[14] An enigma exists, however, in that some high-energy, carbonate-supersaturated environments lack ooids, and some non-turbulent settings contain abundant ooids. New data from Pacton *et al.*[15] provides an alternative mechanism for ooid formation. These authors showed that photosynthetic microorganisms mediate freshwater ooid formation in western Lake Geneva, Switzerland. In this work, NanoSIMS was used to show the nanoscale chemical relationships of key components of the ooids, identifying a close spatial association of extracellular polymeric substances (EPS), bacteria, and magnesium silicate minerals in ooid cortices (Figure 1.4). This allowed the authors to advance a model of ooid cortex formation involving the adsorption of magnesium silicates onto cyanobacterial EPS, followed by transformation to the low magnesium carbonate phase typically found in fossil lacustrine ooids. These new findings cast doubt on the use of some ooids in palaeo-environmental reconstructions.

1.3.1.3 Microbialites

Microbialites are biosedimentary structures formed by the interaction of microbial communities with their environment. Microbialites in the rock record can often be identified by their laminated (stromatolites) or clotted (thrombolites) macrotexture, with the most ancient examples of stromatolites cited as some of the earliest evidence for life on our planet.[16] This has been questioned, however, because non-biological processes can also produce structures with similar macromorphologies to ancient stromatolites.[17] Recent work using NanoSIMS[18,19] has attempted to detect new biosignals at the micro- to nanoscale within modern microbialites to see if they may provide more robust evidence for the biogenicity and even community structure of ancient examples.

Wacey *et al.*[18] tested the preservation potential of cyanobacteria in microbialites using living and fossilised material (*c.*4000 years old) from

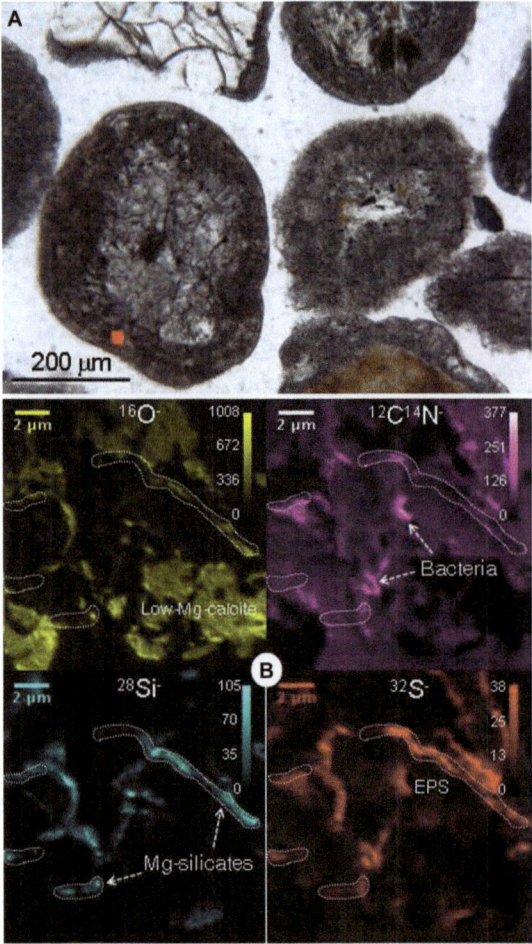

Figure 1.4   Biosignals in freshwater ooids. (A) Thin section photomicrograph of ooids
from western Lake Geneva (Switzerland). NanoSIMS analyses come from
area indicated by red box in the laminated organic ooid cortex. (B)
NanoSIMS ion images showing the variety and spatial association of
different components of the ooid cortex including magnesium silicate
permineralised EPS, bacteria, and low magnesium calcite.
Modified from Pacton *et al.*[15]

Lake Clifton in Western Australia. In the living material, NanoSIMS ion
maps ($^{12}C^-$, $^{26}CN^-$, and $^{32}S^-$) helped distinguish different components of
cyanobacteria (*e.g.* trichome, sheath, and EPS) and establish the nanoscale
spatial distribution of these in relation to magnesium silicate and aragonite
mineralising phases (Figure 1.5). In the fossil material it was found that
individual components of cyanobacteria were difficult to recognise but that
partially decomposed cyanobacterial filaments could still be detected by
$^{26}CN^-$ and $^{32}S^-$ NanoSIMS ion mapping. Similarly, the movement of magnesium from metastable magnesium silicates in the living material into

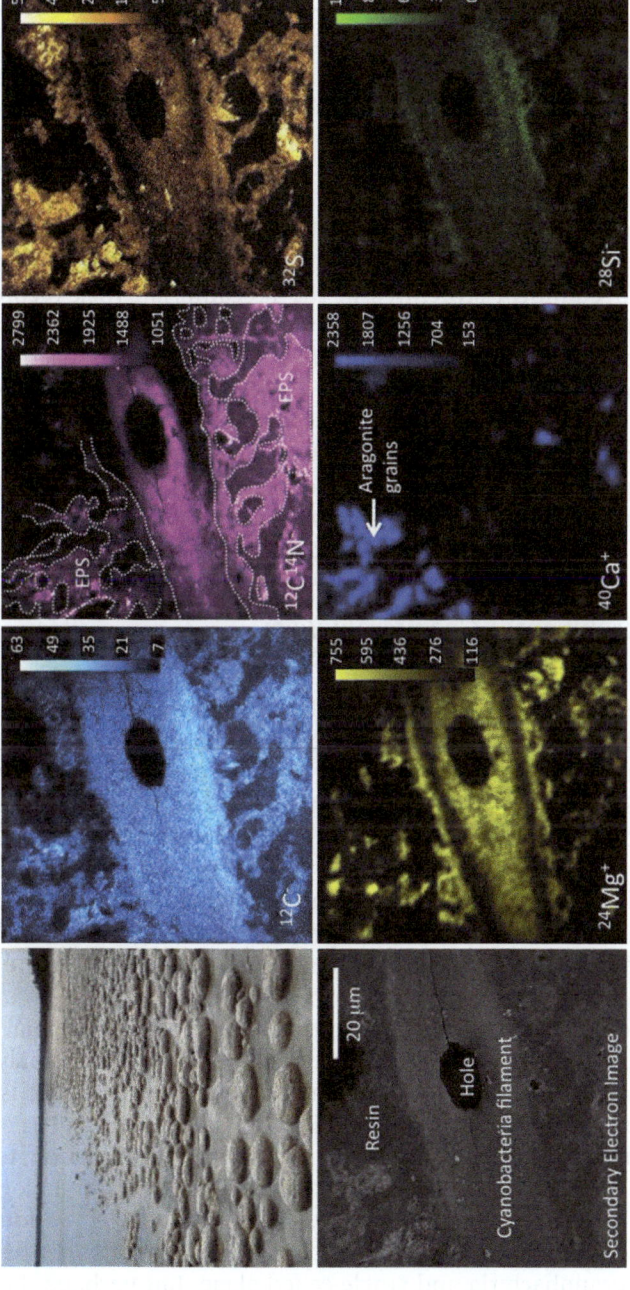

Figure 1.5 Biosignals in microbialites. NanoSIMS analysis of an actively growing microbialite from Lake Clifton, Western Australia. Sampling site is shown in the top left image and the NanoSIMS analysis area is shown in the bottom left 'secondary electron image'. NanoSIMS ion images show an organic-rich cyanobacterial sheath, surrounded by EPS enclosing aragonite grains. The sheath and EPS have been partially permineralised by magnesium silicate.

magnesium calcite in the fossilised material could also be visualised using NanoSIMS mapping of $^{24}Mg^+$.

Bosak *et al.*[19] used NanoSIMS as part of a suite of techniques to investigate the activity of cyanobacteria in conical stromatolites from Yellowstone National Park (USA). ^{13}C-Labelled bicarbonate was used as a carbon source and the incorporation of ^{13}C into biomass was traced using NanoSIMS imaging. Hence, active primary producers could be identified and their distributions and densities could be tracked in different parts of the stromatolites. This data combined with light-, epifluorescence-, and transmission electron microscopy (TEM), plus gene sequencing, showed that the growth of modern coniform stromatolites depends on a number of distinct cyanobacteria, and that most new biomass grows in the tips of the cones. This may help interpret similar coniform structures in the Precambrian rock record.[20] A similar study used NanoSIMS to look at the lateral and vertical variation of the abundance and isotopic composition of sulfide produced within a microbial mat.[21] NanoSIMS $\delta^{34}S$ data showed large variations (+10 to −30 per mil) over small spatial scales, plus repeated oscillations that corresponded to changes in sulfide concentrations and abundance of known sulfate-reducing microorganisms. This elegantly showed the spatial variability of sulfur cycling in a modern microbial mat and may likewise aid interpretations of sulfur cycling in ancient sediments.

1.3.1.4 *Environmental Geomicrobiology*

The ability to link metabolic function with phylogenetic identity has been a key goal of the environmental geomicrobiology community for many years.[22] The potential to simultaneously determine metabolic activity and identity of naturally occurring microorganisms has been demonstrated using conventional SIMS to analyse indigenous microbes in anoxic marine sediments.[23] Combining fluorescence *in situ* hybridisation (FISH) with SIMS measurements of carbon isotopes, Orphan *et al.*[23] found that cell aggregates binding a specific archeal probe were strongly depleted in ^{13}C, indicating a methane-based metabolism. Stable isotopes, such as ^{13}C or ^{15}N, can also be incorporated into an experimental study and used as tracers in the SIMS imaging of biological samples.[24,25] NanoSIMS, with its combination of high lateral resolution and high mass resolution, can image microbial communities at the scale of an individual bacterium (~ 1 μm) and analyse its isotopic composition (see Figure 1.2). Combining FISH with stable-isotope labelling allows us to determine the activity of specific microbes within mixed microbial communities.[26] Several variations on this approach, termed element-labelled fluorescent *in situ* hybridisation (EL-FISH) or halogen-labelled *in situ* hybridisation (HISH), have been reported.[27,28] Thompson *et al.*[29] employed HISH-SIMS to confirm symbiotic relationships between nitrogen-fixing cyanobacteria and single-celled algae. But perhaps the most promising approach is catalysed reporter deposition FISH (CARD-FISH), which deposits high concentrations of fluorine-containing fluorophores in target cells.[21,27] NanoSIMS can then be used to visualise the labelled cells by acquiring a signal for ^{19}F, which is not naturally present at high concentrations in cells. This technique has been used to identify active diazotrophism by cyanobacteria in microbial mats.[30]

The importance of microorganisms in determining nutrient flow within terrestrial ecosystems has also been demonstrated. Following studies by Herrmann *et al.*[31] that demonstrated the preservation potential of ^{15}N-labelled bacteria in soil, Clode *et al.*[32] went on to show how bacteria and plants compete for nutrients within the rhizosphere—the region of soil directly surrounding plant roots. ^{15}N-Ammonia was added to the soil, and at discrete time points the soil cores, including indigenous bacteria and wheat roots, were fixed, embedded in resin, and sectioned for NanoSIMS analysis. These data revealed that bacteria had assimilated up to 50 atom% ^{15}N within 30 min.

1.3.1.5 Biomineralisation and Palaeo-Environmental Indicators

Compositional variations within the biomineralised parts of marine organisms, such as coral, planktonic foraminifera, and shells, provide proxies for reconstructing past climate changes, palaeoceanography, and biomineralisation. Foraminiferal geochemistry has been used, in particular, for reconstructing past ocean seasonality, recording the amplitude of seasonal and interannual changes in seawater temperature and salinity reflecting the major characteristics of the ocean–climate system. Understanding the biological and environmental constraints on modern biomineralisation processes is key to unlocking the information stored in fossilised marine organisms that may reveal palaeoenivronmental conditions.[33] NanoSIMS again offers significant advantages over bulk analytical methods with its ability to determine compositional variations at the micron scale. Variations in magnesium/calcium ratios in the tests of foraminifera were first investigated by Sano *et al.*,[34] while Kunioka *et al.*[35] went on to illustrate the banding of magnesium, strontium, and barium within the calcite tests which appeared to be linked to corresponding organic bands (Figure 1.6). Seasonal trace element variations within coral skeletons are strongly overprinted at the micron scale by variations that are largely biologically controlled.[36–38] Incorporating isotopic labels into experiments with closely controlled environmental conditions further enhances our understanding of the biological response. Houlbreque *et al.*[39] and Brahmi *et al.*,[40,41] for example, used the distribution of an ^{86}Sr label to suggest that the growth front reflects heterogeneous cellular activity in the calicoblastic cell layer, where certain domains or compartments are actively forming skeleton while neighbouring cells are inactive. Marshall *et al.*[42] used ^{44}Ca label to illustrate the active transport of Ca^{2+} across the epithelia in coral polyp tissue to determine how calcium is directed to the calcifying cells.

1.3.1.6 Bioalteration of Volcanic Rocks

Seafloor and sub-seafloor basaltic glass represents a vast potential microbial substrate, and phylogenetic[43] and textural[44] studies confirm that microbes

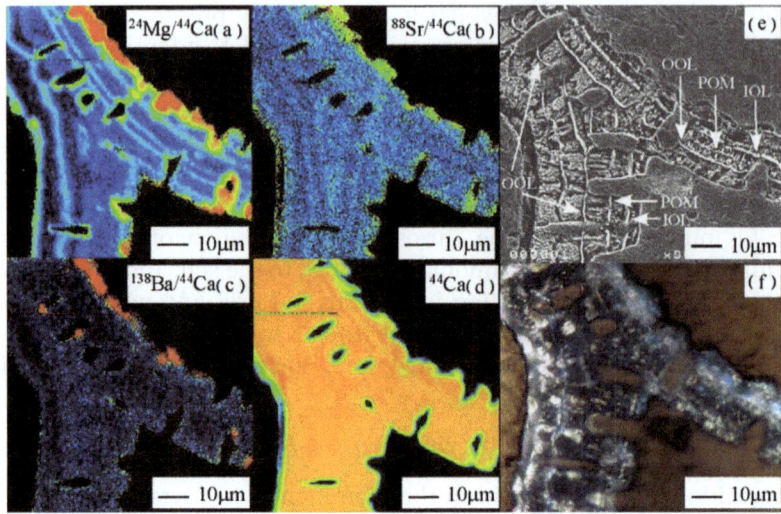

Figure 1.6   Elemental distributions within a biomineral. NanoSIMS ion images
showing the distribution of $^{24}Mg/^{44}Ca$, $^{88}Sr/^{44}Ca$, $^{138}Ba/^{44}Ca$, and ^{44}Ca
ions within the cross-section of a foraminifera test, with corresponding
SEM image (acid etched to expose the organic layers), and reflected light
image (stained to highlight proteins).
From Kunioka et al.[35]

inhabit such settings today. However, demonstrating this mode of life in
ancient basalts and determining the metabolisms that microbes use in such
settings is less straightforward. A recent study combining NanoSIMS with
TEM detected new nanoscale chemical biosignals in basaltic glass approximately 100 000 years old from the Arctic Mohns Ridge.[45] Here, fractures in
the basaltic glass were filled with palagonite that, in places, contained
bacteria-sized micropores. NanoSIMS mapping showed that these were frequently enriched in carbon and nitrogen and were surrounded by rings of
manganese (Figure 1.7). This led to the interpretation that the micropores
were casts of endolithic microorganisms (bacteriomorphs) that inhabited
the fractures in the glass, and they were likely manganese-oxidising bacteria.
This provided the first direct evidence for actual microbial microfossils
combined with a specific metabolism in such rocks; this can now be sear-
ched for as a biosignature in older samples.

1.3.2   Palaeontology: Origin and Evolution of Life

The timing of the origin of life on Earth and subsequent evolution of different microbial metabolic pathways are some of the biggest questions in
geoscience and some of the most difficult to resolve. NanoSIMS enables
isotopic measurements to be made from objects as small as 1–2 μm, and
brings with it the additional capability of elemental mapping on the naa-
nometre scale. This makes it ideally suited to the study of ancient putative

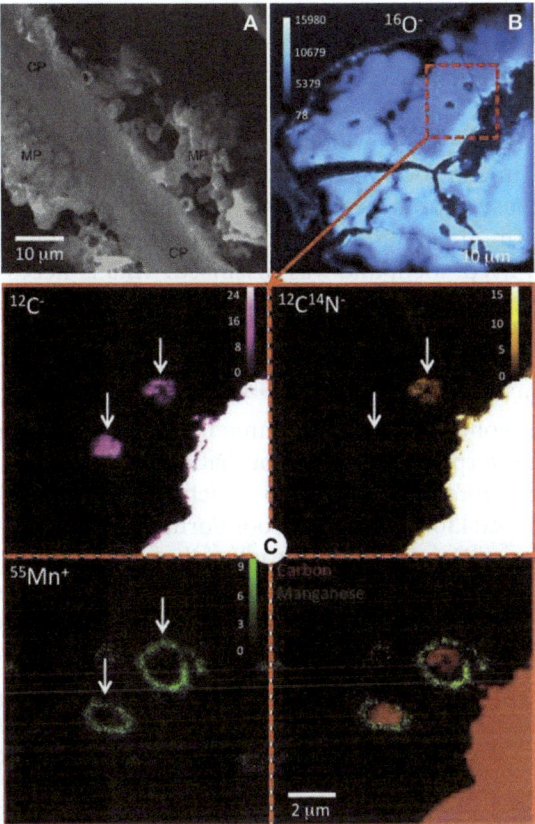

Figure 1.7 Bioalteration in volcanic glass. (A) Overview of a fracture in basaltic glass filled with compact (CP) and microporous (MP) palagonite, plus objects that resemble coccoid bacteria. (B) NanoSIMS oxygen ion image showing the fracture-filling palagonite with occasional subspherical pores (dashed square represents analysis area in C). (C) NanoSIMS ion images showing enrichments in carbon and (partially) nitrogen in the micropores, plus rims of manganese. This suggests the presence of degraded manganese-oxidising bacteria.
Modified from McLoughin *et al.*[45]

microfossils and associated biostructures and minerals. In recent years, NanoSIMS, in combination with other *in situ* and high-spatial-resolution techniques such as regular SIMS, TEM, laser Raman, and synchrotron radiation, has provided a better understanding of Earth's earliest life, with evidence collected from microfossils, trace fossils, stromatolites, and biominerals.

1.3.2.1 Microfossils and Trace Fossils

The first study of ancient microfossils using NanoSIMS was by Oehler *et al.*[46] These authors used material that was unquestionably biological in origin, from the 850 Ma Bitter Springs Chert of Australia, and showed that distinct cell wall and sheath-like structures were enriched in carbon, nitrogen, and sulfur. This correlation of multiple-element biological chemistry with morphological features characteristic of biology clearly observable under the microscope provides good evidence for the biogenicity of ancient microstructures in uncontroversial samples. For older, putatively biogenic microfossils, these morphological and chemical signals need to be integrated with a firm understanding of the depositional environment in which the microfossil lived, died, and was fossilised. This is necessary in order to reject non-biogenic formation mechanisms such as organic self-organised structures from hydrothermal systems.[47] Two further studies by this group extended their methodology to putative microfossils (*c.*3000 Ma) from the Farrel Quartzite of Western Australia.[48,49] Here, spheroidal and spindle-like microstructures contain similar enrichments of carbon, nitrogen, and sulfur and have been interpreted as highly probable microfossils.

The oldest microfossil material to which NanoSIMS has been applied comes from the *c.*3430 Ma Strelley Pool Formation of Western Australia.[50] These authors used NanoSIMS as one of a suite of six different techniques to demonstrate the biogenicity of spheroidal and tubular microfossils in this unit, with NanoSIMS again contributing high-spatial-resolution maps of carbon and nitrogen chemistry (Figure 1.8a). Uniquely, Wacey *et al.*[50] were

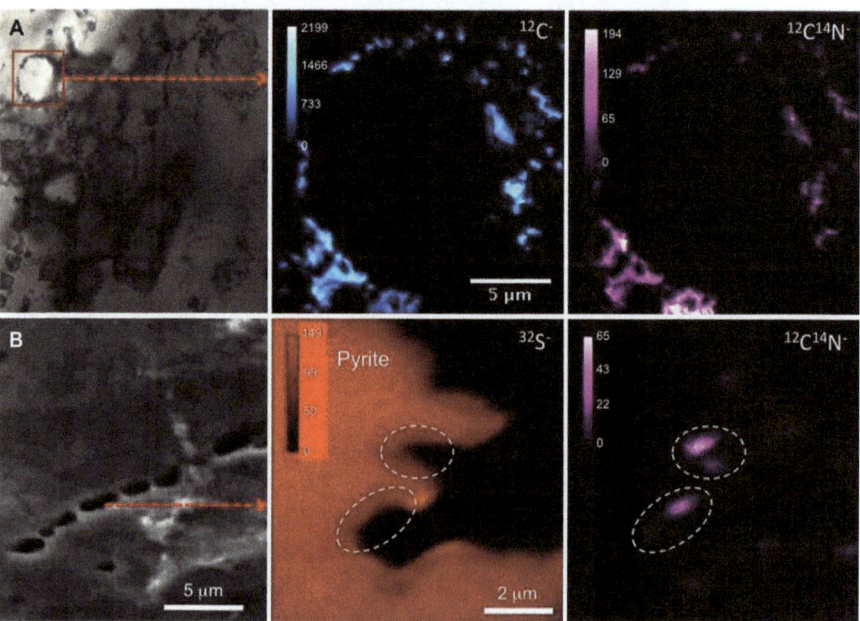

Figure 1.8 Early life on Earth from the ∼3430 Ma Strelley Pool Formation, Western Australia. (A) Thin-section photomicrograph and NanoSIMS ion images showing a spheroidal microfossil with walls composed of carbon and nitrogen. Modified from Wacey *et al.*[50] (B) SEM image and NanoSIMS ion images of sulfur and nitrogen from microbial etch pits within a detrital pyrite grain. Modified from Wacey *et al.*[51]

able to assign a likely metabolism to these microbes (sulfate reduction and/ or sulfur disproportionation) based on NanoSIMS and regular SIMS sulfur isotope data from pyrite intimately associated with the microfossils. This provided the most robust evidence yet for life at around 3500 Ma. Linked to this discovery, Wacey *et al.*[51] showed that detrital pyrite grains in the Strelley Pool Formation contained chains of pits and channels and had laminated carbonaceous coatings. NanoSIMS was used to show enrichments of carbon and nitrogen in the pits and channels (Figure 1.8b) and, along with data from laser Raman and TEM, led to an interpretation that these were trace fossils caused by the attachment of iron- or sulfur-oxidising microbes to the pyrite surface. This work further extended the diversity of microbial metabolisms reported from early Archean rocks.

Wacey *et al.*,[52,53] also working on the Strelley Pool Formation, combined NanoSIMS ion mapping of carbon, nitrogen, and biologically important trace elements (*e.g.* cobalt, iron, nickel, zinc), with carbon isotope analysis to show that there was a biological component to the formation of ambient inclusion trails (AIT). This provided the first geochemical data to support a long-standing biological hypothesis for AIT formation[54] and introduced AIT as a new potential trace fossil in ancient rocks.

In younger Archean rocks, NanoSIMS has been used in the debate over when some key microbial groups evolved. The discovery of a suite of lipid biomarkers including hopane and sterane in 2700 Ma rocks was thought to indicate that cyanobacteria and eukaryotes had evolved by this time in Earth's history,[55,56] and extended previously known evidence for cyanobacteria and eukaryotes by about 700 and 1000 million years respectively. However, a subsequent study using NanoSIMS[57] has cast serious doubt on these claims. The NanoSIMS data showed that carbon indigenous to the rock had carbon isotope values that were consistent with those expected for rocks of that age, but these values were different from the lipid biomarkers previously extracted. This essentially shows that the biomarkers were probably not indigenous to their host rock, *i.e.* they were younger contaminants, and reopens the debate over when cyanobacteria and eukaryotes first evolved. These authors followed up this breakthrough study with a detailed paper outlining protocols for analysing carbon isotopes by NanoSIMS.[58] Other isotopic systems, notably oxygen[59] and sulfur (see section 3.2.3), can now also be analysed using NanoSIMS.

1.3.2.2 Stromatolites

The nanoscale co-occurrence of carbon, nitrogen, and sulfur has been shown to be a reliable biosignature in modern microbialites where the depositional environment is known to be a low-temperature sedimentary setting (see section 3.1.3). This knowledge has been applied to Precambrian material with some success. Most notably, Wacey[60] studied some of the earliest known stromatolites from the Strelley Pool Formation, also interpreting the nanoscale co-occurrence of carbon, nitrogen, and sulfur as evidence for a biological component to their formation. This reinforced previous biological interpretations of these stromatolites obtained from macroscale studies.[20,61] Recently, an additional novel NanoSIMS method has been developed to look at these stromatolites; this is the microscale sulfur isotope analysis of sulfur incorporated into organic material.[62] This study also concluded that the organic matter in the stromatolites was biological and reinforced earlier conclusions[50] that sulfur-processing microbes were an important part of this ancient biosphere.

1.3.2.3   Biominerals

Microbes frequently leave behind vestiges of their activity as mineral products that show no obvious biological morphology, the most useful of which is pyrite. Since pyrite is common in sedimentary rocks throughout the geological column and can record distinctive sulfur isotope fractionations induced by microbes,[63] it acts as an important record of the evolution of Earth's sulfur cycle.[64] NanoSIMS (indeed SIMS in general) has helped to revolutionise this field, providing *in situ* sulfur isotope data from very small objects that is both complementary to traditional bulk analyses and detects additional trends that bulk analyses artificially homogenise.

NanoSIMS has been applied in two main periods of Earth's early history to inform on changing sulfur cycles. In the early Archean, Wacey *et al.*[65] combined NanoSIMS $\delta^{34}S$ analysis of single pyrite grains with large-radius SIMS $\delta^{34}S$ and $\Delta^{33}S$ to show that two coexisting sulfur-based metabolisms (microbial sulfate reduction and microbial sulfur disproportionation) were operational in beach-like environments at approximately 3430 Ma. This finding was able to resolve some of the controversy surrounding previous work that had provided conflicting data regarding the origin of these two metabolisms.[66–68] McLoughlin *et al.*[11] extended such analyses into the early Archean sub-seafloor environment, showing that microbes metabolising sulfur also inhabited fractures in ~3450 Ma pillow lavas. Most recently, in Palaeoproterozoic rocks of the ~1900 Ma Gunflint chert, Wacey *et al.*[69] performed the first *in situ* sulfur isotope analysis of individual pyritised microfossils. These data showed that carbonaceous microfossils were pyritised rapidly in small anoxic pods of sediment porewater containing limited sulfate ion concentrations.

NanoSIMS has even been used in attempt to identify possible habitats for the origin of life itself. One proposed habitat is in the vesicles of pumice, a low-density volcanic rock that would have floated at the air–water interface for considerable amounts of time on the early Earth.[70] Brasier *et al.*[71] tested the oldest known pumice on Earth, from the ~3460 Ma Apex Formation of Western Australia, to see if there was any evidence for biologically relevant elements or minerals. NanoSIMS ion maps showed intimate associations of carbon, nitrogen, phosphorus, and sulfur, plus potential biominerals such as pyrite (Figure 1.9). Complementary analyses detected the remains of potential catalysts such as clay minerals, zeolites, and titanium oxides, together

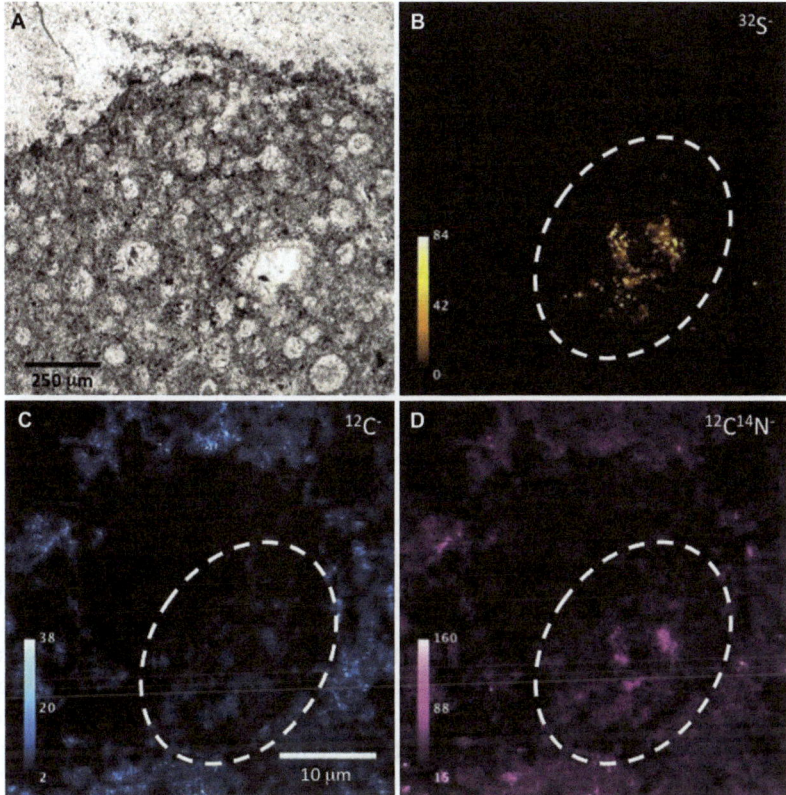

Figure 1.9 Pumice as a habitat for the origin of life. (A) Thin-section photomicrograph of a pumice clast from the 3460 Ma Apex Formation, Western Australia. (B–D) NanoSIMS ion images showing a pumice vesicle permeated by carbon and nitrogen, plus submicron-sized pyrite crystals (dashed circles represent identical areas within each image). Modified from Brasier *et al.*[71]

providing the first geological evidence that the components needed for the earliest stages of life occurred together in pumice on the early Earth.

1.3.2.4 Geochronology

Investigation of the origin and evolution of life on Earth requires a firm understanding of the ages of the rocks in which microfossils and other biosignals are found. As mentioned above, SIMS has become the gold standard for uranium–lead geochronology, using the high sensitivity and specificity of large-geometry (LG) instruments such as SHRIMP and the Cameca IMS1270/80. However, the minerals analysed, such as zircon, commonly exhibit zonation on a scale much smaller than the typical spot size used by LG-SIMS (10–30 μm). There is therefore a need to reduce the size of the analytical area to better understand the often complex history of the host rock. Stern *et al.*[72] demonstrated the possibility of performing lead isotope geochronology by magnetic peak switching on a single detector, with a 3–5 μm spot on zirconolites measuring less than 10 μm. Sano *et al.*[73] extended this technique to the uranium–lead system using a modified multi-collection configuration to allow the simultaneous measurement of ^{204}Pb and ^{206}Pb. Although NanoSIMS is unlikely to offer a real alternative to the LG-SIMS for *in situ* geochronology, a recent study has shown that the distribution of lead in very old high-temperature metamorphic zircon can be highly variable at the micron scale, casting doubt on the reliability of accurate uranium–lead dates.[74] NanoSIMS may offer some potential for advance screening of suitable zircon grains for dating.

1.3.3   Mineralogy and Petrology

1.3.3.1   *Diffusion in Minerals*

Diffusion modelling of minor and trace elements across zoned volcanic minerals can be used to calculate the timescales between the growth of zones and subsequent quenching of the sample during eruption. Plagioclase crystals have long been recognised as an archive of magmatic evolution, but to obtain diffusion timescales in the region of months and days, the spatial resolution of the analytical technique used must have micron or submicron capability, respectively. The sensitivity and lateral resolution of NanoSIMS give it a distinct advantage over other techniques for the investigation of diffusion across compositional interfaces. The submicron probe size interacts with a much smaller volume of sample material than, say, an electron beam at 20 kV or a UV laser attached to an ICP-MS. Furthermore, scanning the beam across the sample surface to produce a linescan (as detailed in section 1.2) allows resolution dictated by beam diameter and pixel size as opposed to resolution limited by motor step size as in most other techniques. Saunders *et al.*[75] measured the relative concentrations of calcium, sodium, silicon, strontium, barium, lithium, titanium, magnesium, and iron across compositional interfaces within zoned plagioclase crystals from the 1980 eruption of Mount St Helens (Washington State, USA) with 300–600 nm resolution. The data suggest the final stages of crystal growth occurred within days to months prior to the material being ejected.

1.3.3.2   *Mapping Invisible Gold in Pyrite*

SIMS has previously been used to analyse gold in solid solution in sulfide minerals such as pyrite (FeS_2) and arsenopyrite (FeAsS) due to the high secondary ion yield of gold.[76] Other surface analytical techniques, such as EPMA, lack the sensitivity to detect very low concentrations of gold, even when the bulk-rock concentration is economically viable. This has led to the term 'invisible gold'. NanoSIMS, with its combination of high sensitivity and

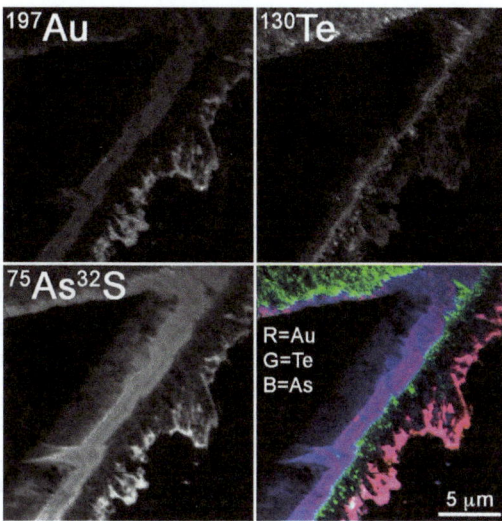

Figure 1.10 Mapping invisible gold in pyrite. NanoSIMS ion images showing the distribution of ^{197}Au, ^{130}Te, and ^{75}As^{32}S in a small pyrite grain. The three-colour overlay (where gold is red, tellurium is green, and arsenic is blue) illustrates the complex relationships between the elements which only becomes evident when imaged at the submicron scale. Modified from Barker *et al.*[77]

high lateral resolution, has proved to be the ideal technique for mapping gold in Carlin-type deposits, where it is often present only as micron-thick rims on sulfide minerals.[77] Furthermore, high sensitivity to other elements common in sulfide deposition, such as tellurium and antimony, means that the intricate relationships between gold and associated elements can be investigated at the submicron scale, revealing changes in the composition of the gold-bearing fluid over time (Figure 1.10).

To further constrain the evolution of the gold-bearing fluid, relative changes in δ^{34}S can also be measured from regions with different trace element compositions. However, care must be taken that the changes in composition do not affect the matrix to the extent that the relative isotopic compositions become inaccurate. As mentioned in the previous section, most natural isotopic variations in terrestrial rocks are beyond the detection limits of the NanoSIMS, yet terrestrial fractionations in sulfur isotopes can be relatively large (>10s of per mil). Many sulfide minerals, including pyrite and arsenopyrite, are semiconductors and can be easily polished, which is conducive to obtaining good isotopic analyses. Barker *et al.*[77] demonstrated that the gold-rich rim of a single grain of pyrite exhibited a difference in δ^{34}S of 15–20‰ compared to the gold-poor core. *In situ* sulfur isotope measurements have become one of the most important contributions of NanoSIMS to the geosciences, as demonstrated in a number of different applications in this chapter.

1.3.4   Cosmochemistry

Cosmochemists were the first to use NanoSIMS for the isotopic measurements of minerals, as the high spatial resolution and imaging capability made it possible to locate individual micron-sized pre-solar dust grains against a background of millions of interplanetary dust particles with solar composition.[78,79] Pre-solar grains exhibit isotopic fractionations of hundreds or thousands of per mil, reflecting the isotopic composition of the star in which they formed. As the isotopic composition of pre-solar material is so different from that of material from within our solar system, the apparent $\sim 1\permil$ limit on precision is insignificant. Tiny grains of diamond or silicon carbide, dissolved out of primitive meteorites and mounted on a flat substrate, can be screened by rastering the primary beam over a relatively large area. Grains that exhibit large isotopic fractionations can then be identified and targeted for more in-depth analysis.[80] Similarly, individual submicron-sized grains from the comet 81P/Wild2 collected by the NASA Stardust mission were analysed to reveal pre-solar oxygen isotope signatures.[81] This 'needle-in-a-haystack' approach has been adopted by the nuclear safeguards community for use with the Cameca IMS1280 ion probes, as it is highly effective at locating enriched uranium particles in environmental dust samples.[4] Large isotopic fractionations have also been found *in situ* within meteorites. Bland *et al.*[82] used NanoSIMS to reveal oxygen isotopic anomalies in micro-calcium–aluminium-rich inclusions (CAIs) and identify pre-solar silicate grains within the carbonaceous chondrite Acfer 94. Busemann *et al.*[83] and Nakamura *et al.*[84] report remnants of the protosolar disk manifest as large hydrogen and nitrogen fractionations in organic material from primitive meteorites.

References

1. A. Benninghoven, F. G. Rüdenauer and H. W. Werner, *Secondary Ion Mass Spectrometry: Basic Concepts, Instrumental Aspects, Applications, and Trends*, Wiley, New York, 1987.
2. T. Ireland and I. S. Williams, Considerations in zircon geochronology by SIMS., *Rev. Min. Geochem.*, 2003, **53**, 215–242.
3. M. Tang, E. Anders, P. Hoppe and E. Zinner, Meteoritic SiC and its stellar sources: implications for galactic chemical evolution., *Nature*, 1989, **339**, 351–354.
4. P. M. L. Hedberg, P. Peres, J. B. Cliff, F. Rabemananjara, S. Littmann, H. Thiele, C. Vincent and N. Albert, Improved particle location and isotopic screening measurements of sub-micron-sized particles by Secondary Ion Mass Spectrometry., *J. Anal. At. Spectrom.*, 2011, **26**, 406–413.
5. R. A. Stern, Introduction to secondary ion mass spectrometry (SIMS) in geology, in *Secondary Ion Mass Spectrometry in the Earth Sciences*, ed. M. Fayek. Mineral. Assoc. Can., Short Course Series, 2009, **41**, 1–18.

6. G. Slodzian, B. Daigne, F. Girard, F. Boust and F. Hillion, Scanning secondary ion analytical microscopy with parallel detection., *Biol. Cell.*, 1992, **74**, 53–60.

7. F. Hillion, B. Daigne, F. Girard and G. Slodzian, A new high performance instrument: the Cameca 'NanoSIMS 50', in *Secondary Ion Mass Spectrometry SIMS IX*, ed. A. Benninghoven, Y. Nihei, N. Shimizu and H. W. Werner. John Wiley and Sons, New York, 1994, pp. 254–257.

8. G. Slodzian, M. Chaintreau, R. Dennebouy and A. Rousse, Precise *in situ* measurements of isotopic abundances with pulse counting of sputtered ions., *Euro. J. App. Phys.*, 1994, **14**, 199–231.

9. G. Slodzian, F. Hillion, F. J. Stadermann and E. Zinner, QSA influences on isotopic ratio measurements., *App. Surf. Sci.*, 2004, **231–232**, 874–877.

10. F. Hillion, M. R. Kilburn, P. Hoppe, S. Messenger and P. K. Weber, The effect of QSA on S, C, O and Si isotopic ratio measurements., *Geochim. Cosmchim. Acta,*, 2008, **72**, A377.

11. N. McLoughlin, E. G. Grosch,, M. R. Kilburn and D. Wacey, Sulfur isotope evidence for a Paleoarchean subseafloor biosphere, Barberton, South Africa., *Geology*, 2012, **40**, 1031–1034.

12. M. A. Chan, B. Beitler, W. T. Parry, J. Ormo and G. Komatsu, A possible terrestrial analogue for haematite concretions on Mars., *Nature*, 2004, **429**, 731–734.

13. K. A. Weber, T. L. Spanbauer,, D. Wacey, M. R. Kilburn, D. B. Loope and R. M. Kettler, Biosignatures link microorganisms to iron mineralization in a paleoaquifer., *Geology*, 2012, **40**, 747–750.

14. P. J. Davies, B. Bubela and J. Ferguson, The formation of ooids., *Sedimentology*, 1978, **25**, 703–730.

15. M. Pacton, D. Ariztegui, D. Wacey, M. R. Kilburn, C. Rollion-Bard, R. Farah and C. Vasconcelos, Going nano: a new step towards understanding the processes governing freshwater ooid formation., *Geology*, 2012, **40**, 547–550.

16. M. R. Walter, R. Buick and J. S. R. Dunlop, Stromatolites 3,400–3,500 Myr old from the North Pole Area, Western Australia., *Nature*, 1980, **246**, 443–445.

17. N. McLoughlin, L. A. Wilson and M. D. Brasier, Growth of synthetic stromatolites and wrinkle structures in the absence of microbes—implications for the early fossil record., *Geobiology*, 2008, **6**, 95–105.

18. D. Wacey, D. Gleeson and M. R. Kilburn, Microbialite taphonomy and biogenicity: new insights from NanoSIMS., *Geobiology*, 2010, **8**, 403–416.

19. T. Bosak, B. Liang, T-D. Wu, S. P. Templer, A. Evans, H. Vali, J.-L. Guerquin-Kern, V. Klepac-Ceraj, M. S. Sim and J. Mui, Cyanobacterial diversity and activity in modern conical microbialites., *Geobiology*, 2012, **10**, 384–401.

20. A. C. Allwood, M. R. Walter, B. S. Kamber, C. P. Marshall and I. W. Burch, Stromatolite reef from the Early Archaean era of Australia., *Nature*, 2006, **441**, 714–718.

21. D. Fike, C. L. Gammon, W. Ziebis and V. J. Orphan, Micron-scale mapping of sulfur cycling across the oxycline of a cyanobacterial mat: a

combined nanoSIMS and CARD-FISH approach., *ISME Journal*, 2008, **2**, 749–759.

22. M. M. M. Kuypers and B. B. Jørgensen, The future of single-cell environmental microbiology., *Environ. Microbiol.*, 2007, **9**, 6–7.

23. V. J. Orphan, C. H. House, K.-U. Hinrichs, K. D. McKeegan and E. F. DeLong, Methane-consuming Archaea revealed by directly coupled isotopic and phylogenetic analysis., *Science*, 2001, **293**, 484–487.

24. R. Levi-Setti and M. LeBeau, Cytogenetic applications of high resolution secondary ion imaging microanalysis: detection and mapping of tracer isotopes in human chromosomes., *Biol. Cell*, 1992, **74**, 51–58.

25. C. Lechene, F. Hillion, G. McMahon, D. Benson, A. M. Kleinfeld, J. P. Kampf, D. Distel, Y. Luyten, J. Bonventre, D. Hentschel, K. M. Park, S. Ito, M. Schwartz, G. Benichou and G. Slodzian, High resolution quantitative imaging of mammalian and bacterial cells using stable isotope mass spectrometry., *J. Biol.*, 2006, **5**, 20.1–20.30.

26. A. E. Dekas and V. J. Orphan, Identification of diazotrophic microorganisms in marine sediment via fluorescence *in situ* hybridization coupled to nanoscale secondary ion mass spectrometry (FISH-NanoSIMS)., *Methods Enzymol.*, 2011, **486**, 281–305.

27. S. Behrens, T. Loesekann, J. Pett-Ridge, P. K. Weber, Wing-On Ng, B. S. Stevenson, I. D. Hutcheon, D. A. Relman and A. M. Spormann, Linking microbial phylogeny to metabolic activity at the single-cell level by using enhanced element labeling-catalyzed reporter deposition fluorescence *in situ* hybridization (EL-FISH) and NanoSIMS., *Appl. Environ. Microbiol.*, 2008, **74**, 3143–3150.

28. N. Musat, H. Halm, B. Winterholler, P. Hoppe, S. Peduzzi, F. Hillion, F. Horreard, R. Amann, B. B. Jørgensen and M. M. M. Kuypers, A single cell view on the ecophysiology of anaerobic phototrophic bacteria., *Proc. Natl. Acad. Sci. U. S. A.*, 2008, **105**, 17861–17866.

29. A. W. Thompson, R. A. Foster, A. Krupke, B. J. Carter, N. Musat, D. Vaulot, M. M. M. Kuypers and J. P. Zehr, Unicellular cyanobacterium symbiotic with a single-celled eukaryotic alga., *Science*, 2012, **337**, 1546–1550.

30. D. Woebken, L. C. Burow, L. Prufert-Bebout, B. M. Bebout, T. M. Hoehler, J. Pett-Ridge, A. M. Spormann, P. K. Weber and S. W. Singer, Identification of a novel cyanobacterial group as active diazotrophs in a coastal microbial mat using NanoSIMS analysis., *ISME Journal*, 2012, **6**, 1427–1439.

31. A. M. Herrmann, K. Ritz, N. Nunan, P. L. Clode, J. Pett-Ridge, M. R. Kilburn, D. V. Murphy, A. G. O'Donnell and E. A. Stockdale, Nanoscale secondary ion mass spectrometry—A new analytical tool in biogeochemistry and soil ecology: A review article., *Soil Biol. Biochem.*, 2007, **39**, 1835–1850.

32. P. L. Clode, M. R. Kilburn, D. L. Jones, E. A. Stockdale, J. B. Cliff, A. M. Herrmann and D. V. Murphy, *In situ* mapping of nutrient uptake in the rhizosphere using nanoscale secondary ion mass spectrometry., *Plant Physiol.*, 2009, **151**, 1751–1757.

33. J. Stolarski, A. Meibom, M. Mazur and R. Przenioslo, A Cretaceous scleractinian coral with a calcitic skeleton., *Science*, 2007, **318**, 92–94.
34. Y. Sano, K. Shirai, N. Takahata, T. Hirata and N. C. Sturchio, Nano-SIMS analysis of Mg, Sr, Ba and U in natural calcium carbonate., *Anal. Sci.*, 2005, **21**, 1091–1097.
35. D. Kunioka, K. Shirai, N. Takahata, Y. Sano, T. Toyofuku and Y. Ujiie, Microdistribution of Mg/Ca, Sr/Ca, and Ba/Ca ratios in *Pulleniatina obliquiloculata* test by using a NanoSIMS: Implication for the vital effect mechanism., *G³*, 2006, 7, Q12P20.
36. A. Meibom, J.-P. Cuif, F. Hillion, B. R. Constantz, A. Juillet-Leclerc, Y. Dauphin, T. Watanabe and R. B Dunbar, Distribution of magnesium in coral skeleton., *Geophys. Res. Lett.*, 2004, **31**, L23306.
37. A. Meibom, S. Mostefaoui,, J.-P. Cuif, Y. Dauphin, F. Houlbreque, R. G. Dunbar and B. Constantz, Biological forcing controls the chemistry of reef-building coral exoskeleton., *Geophys. Res. Lett.*, 2007, **34**, L02601.
38. A. Meibom,, J.-P. Cuif, F. Houlbreque, S. Mostefaoui, Y. Dauphin, M. L. Meibom and R. G. Dunbar, Compositional variations at ultra-structure length scales in coral skeleton., *Geochim. Cosmochim. Acta.*, 2008, **72**, 1555–1569.
39. F. Houlbrèque, A. Meibom, J.-P. Cuif, J. Stolarski, Y. Marrocchi, C. Ferrier-Pagès, I. Domart-Coulon and R. B. Dunbar, Strontium-86 labeling experiments show spatially heterogeneous skeletal formation in the scleractinian coral, *Porites porites. Geophys. Res. Lett.*, 2009, **36**, 1–5.
40. C. Brahmi, I. Domart-Coulon, L. Rougee, D. G. Pyle, J. Stolarski, J. J. Mahoney, R. H. Richmond and G. K. Ostrander, Pulsed 86 Sr-labeling and NanoSIMS imaging to study coral biomineralization at ultra-structural length scales., *Coral Reefs,*, 2012, **31**, 741–752.
41. C. Brahmi, C. Kopp, I. Domart-Coulon, J. Stolarski and A. Meibom, Skeletal growth dynamics linked to trace-element composition in the scleractinian coral Pocillopora damicornis., *Geochim. Cosmochim. Acta*, 2012, **99**, 146–158.
42. A. T. Marshall, P. L. Clode, R. Russel, K. Prince and R. A. Stern, Electron and ion microprobe analysis of calcium distribution and transport in coral tissues., *J. Experiment. Biol.*, 2007, **210**, 2453–2463.
43. S. J. Giovannoni, M. R. Fisk, T. D. Mullins and H. Furnes, Genetic evidence for endolithic microbial life colonizing basaltic glass/seawater interfaces., *Proc. ODP*, 1996, **148**, 207–214.
44. H. Furnes, I. H. Thorseth, O. Tumyr, T. Torsvik and M. R. Fisk, Microbial activity in the alteration of glass from pillow lavas from Hole 896A., *Proc. ODP*, 1996, **148**, 191–206.
45. N. McLoughlin, D. Wacey, C. Kruber, M. R. Kilburn, I. H. Thorseth and R. B. Pedersen, A combined TEM and NanoSIMS study of endolithic microfossils in altered seafloor basalt., *Chem. Geol.*, 2011, **289**, 154–162.
46. D. Z. Oehler, F. Robert, S. Mostefaoui, A. Meibom, M. Selo and D. S. McKay, Chemical mapping of Proterozoic organic matter at sub-micron spatial resolution., *Astrobiology*, 2006, **6**, 838–850.

47. M. D. Brasier, N. McLoughlin, O. R. Green and D. Wacey, A fresh look at the fossil evidence for early Archaean cellular life., *Phil. Trans. R. Soc. B*, 2006, **361**, 887–902.
48. D. Z. Oehler, F. Robert, M. R. Walter, K. Sugitani, A. Allwood, A. Meibom, S. Mostefaoui, M. Selo, A. Thomen and E. K. Gibson, NanoSIMS: insights to biogenicity and syngeneity of Archaean carbonaceous structures., *Precamb. Res.*, 2009, **173**, 70–78.
49. D. Z. Oehler, F. Robert, M. R. Walter, K. Sugitani, A. Meibom, S. Mosterfaoui and E. K. Gibson, Diversity in the Archaean biosphere: new insights from NanoSIMS., *Astrobiology*, 2010, **10**, 413–424.
50. D. Wacey, M. R. Kilburn, M. Saunders, J. Cliff and M. D. Brasier, Microfossils of sulfur metabolizing cells in ~3.4 billion year old rocks of Western Australia., *Nat. Geosci.*, 2011, **4**, 698–702.
51. D. Wacey, M. Saunders, M. D. Brasier and M. R. Kilburn, Earliest microbially mediated pyrite oxidation in ~3.4 billion-year-old sediments., *Earth Plan. Sci. Lett.*, 2011, **301**, 393–402.
52. D. Wacey, M. R. Kilburn, N. McLoughlin, J. Parnell, C. A. Stoakes and M. D. Brasier, Use of NanoSIMS in the search for early life on Earth: ambient inclusion trails in a c.3400 Ma sandstone., *J. Geol. Soc. Lon.*, 2008, **165**, 43–53.
53. D. Wacey, M. R. Kilburn, C. A. Stoakes, H. Aggleton and M. D Brasier, Ambient inclusion trails: their recognition, age range and applicability to early life on Earth, in *Links Between Geological Processes, Microbial Activities and Evolution of Life*, ed. Y. Dilek, H. Furnes and K. Muehlenbachs, Springer, New York, 2008, pp. 113–134.
54. A. H. Knoll and E. S. Barghoorn, Ambient pyrite in Precambrian chert: new evidence and a theory., *Proc. Nat. Acad. Sci. U. S. A.*, 1974, **71**, 2329–2331.
55. J. J. Brocks, G. A. Logan, R. Buick and R. E. Summons, Archean molecular fossils and the early rise of eukaryotes., *Science*, 1999, **285**, 1033–1036.
56. J. J. Brocks, R. Buick, R. E. Summons and G. A. Logan, A reconstruction of Archean biological diversity based on molecular fossils from the 2.78 to 2.45 billion-year old Mount Bruce Supergroup, Hamersley Basin, Western Australia., *Geochim. Cosmochim. Acta*, 2003, **67**, 4321–4335.
57. B. Rasmussen, I. R. Fletcher, J. J. Brocks and M. R. Kilburn, Reassessing the first appearance of eukaryotes and cyanobacteria., *Nature*, 2008, **455**, 1101–1104.
58. I. R. Fletcher, M. R. Kilburn and B. Rasmussen, NanoSIMS μm-scale *in situ* measurement of $^{13}C/^{12}C$ in early Precambrian organic matter, with permil precision., *Int. J. Mass Spectrom.*, 2008, **278**, 59–68.
59. D. Schumann, T. D. Raub, R. E. Kopp, J-L. Guerquin-Kern, T-D. Wu, I. Rouiller, A. V. Smirnov, S. K. Sears, U. Lucken, S. M. Tikoo, R. Hesse, J. L. Kirschvink and H. Vali, Gigantism in unique biogenic magnetite at the Paleocene-Eocene thermal maximum., *Proc. Nat. Acad. Sci. U. S. A.*, 2008, **105**, 17648–17653.

60. D. Wacey, Stromatolites in the ∼3400 Ma Strelley Pool Formation, Western Australia: examining biogenicity from the macro- to the nanoscale., *Astrobiology*, 2010, **10**, 381–395.
61. H. J. Hofmann, K. Grey, A. H. Hickman and R. Thorpe, Origin of 3.45 Ga coniform stromatolites in Warrawoona Group, Western Australia., *Bull. Geol. Soc. Am.*, 1999, **111**, 1256–1262.
62. T. R. R. Bontognali, A. L. Sessions, A. C. Allwood, W. W. Fischer, J. P. Grotzinger, R. E. Summons and J. M. Eiler, Sulfur isotopes of organic matter preserved in 3.45-billion-year-old stromatolites reveal microbial metabolism., *Proc. Nat. Acad. Sci. U. S. A.*, 2012, **109**, 15146–15151.
63. I. R. Kaplan and S. C. Rittenberg, Microbiological fractionation of sulphur isotopes., *J. Gen. Microbiol.*, 1964, **34**, 195–212.
64. D. E. Canfield and R. Raiswell, The evolution of the sulfur cycle., *Am. J. Sci.*, 1999, **299**, 697–723.
65. D. Wacey, N. McLoughlin, M. J. Whitehouse and M. R. Kilburn, Two coexisting sulfur metabolisms in a ca. 3,400 Ma sandstone., *Geology*, 2010, **38**, 1115–1118.
66. P. Philippot, M. van Zuilen, C. Thomazo, J. Farquhar and M. J. Van, Kranendonk, Early Archaean microorganisms preferred elemental sulfur, not sulfate, *Science*, 2007, **317**, 1534–1537.
67. Y. Shen, J. Farquhar, A. Masterson, A. J. Kaufman and R. Buick, Evaluating the role of microbial sulfate reduction in the early Archean using quadruple isotope systematics., *Earth Planet. Sci. Lett.*, 2009, **279**, 383–391.
68. Y. Ueno, S. Ono, D. Rumble and S. Maruyama, Quadruple sulfur isotope analysis of ca. 3.5 Ga Dresser Formation: New evidence for microbial sulfate reduction in the early Archean., *Geochim. Cosmochim. Acta*, 2008, **72**, 5675–5691.
69. D. Wacey, N. McLoughlin, M. R. Kilburn, M. Saunders, J. Cliff, C. Kong, M. E. Barley and M. D. Brasier, Nano-scale analysis reveals differential heterotrophic consumption in the ∼1.9 Ga Gunflint Chert., *Proc. Nat. Acad. Sci. U. S. A.*, 2013, **110**, 8020–8024.
70. M. D. Brasier, R. Matthewman, S. McMahon and D. Wacey, Pumice as a remarkable substrate for the origin of life., *Astrobiol.*, 2011, **11**, 725–735.
71. M. D. Brasier, R. Matthewman, S. McMahon, M. R. Kilburn and D. Wacey, Pumice from the ∼3460 Ma Apex Basalt, Western Australia: a natural laboratory for the early biosphere., *Precamb. Res.*, 2013, **224**, 1–10.
72. R. A. Stern, I. R. Fletcher, B. Rasmussen, N. J. McNaughton and B. J. Griffin, Ion microprobe (NanoSIMS 50) Pb-isotope geochronology at < 5 μm scale., *Int. J. Mass Spectrom.*, 2005, **244**, 125–134.
73. Y. Sano, N. Takahata, Y. Tsutsumi and T. Miyamoto, Ion microprobe U-Pb dating of monazite with about five micrometer spatial resolution., *Geochem. J.*, 2006, **40**, 597–608.
74. M. A. Kusiak, M. J. Whitehouse, S. A. Wilde, A. A. Nemchin and C. Clark, Mobilization of radiogenic Pb in zircon revealed by ion imaging: Implications for early Earth geochronology., *Geology*, 2013, **41**, 291–294.

75. K. Saunders, B. Buse, M. R. Kilburn, S. Kearns and J. Blundy, Nanoscale characterisation of crystal zoning, *Chem. Geol.*, 2014, **364**, 20–32.

76. R. H. Fleming and B. M. Bekken, Isotope ratio and trace element imaging of pyrite grains in gold ores., *Int. J. Mass Spectrom.*, 1995, **143**, 213–224.

77. S. L. L. Barker, K. A. Hickey, J. S. Cline, G. M. Dipple, M. R. Kilburn, J. R. Vaughan and A. A. Longo, Uncloaking invisible gold: use of nano-sims to evaluate gold, trace elements, and sulfur isotopes in pyrite from Carlin-type gold deposits., *Econ. Geol.*, 2009, **104**, 897–904.

78. S. Messenger, L. P. Keller, F. J. Stadermann, R. M. Walker and E. Zinner, Samples of stars beyond the solar system: Silicate grains in inter-planetary dust., *Science*, 2003, **300**, 105–108.

79. A. N. Nguyen and E. Zinner, Discovery of ancient silicate stardust in a meteorite., *Science*, 2004, **303**, 1496–1499.

80. P. Hoppe and A. Besmehn, Evidence for extinct vanadium-49 in presolar silicon carbide grains from supernovae., *Astrophys. J.*, 2002, **576**, L69–L72.

81. K. D. McKeegan, J. Aléon, J. Bradley, D. Brownlee, H. Busemann, A. Butterworth, M. Chaussidon, S. Fallon, C. Floss, J. Gilmour, M. Gounelle, G. Graham, Y. Guan, P. R. Heck, P. Hoppe, I. D. Hutcheon, J. Huth, H. Ishii, M. Ito, S. B. Jacobsen, A. Kearsley, L. A. Leshin, M.-C. Liu, I. Lyon, K. Marhas, B. Marty, G. Matrajt, A. Meibom, S. Messenger, S. Mostefaoui, S. Mukhopadhyay, K. Nakamura-Messenger, L. Nittler, R. Palma, R. O. Pepin, D. A. Papanastassiou, F. Robert, D. Schlutter, C. J. Snead, F. J. Stadermann, R. Stroud, P. Tsou, A. Westphal, E. D. Young, K. Ziegler, L. Zimmermann and E. Zinner, Isotopic compositions of cometary matter returned by Stardust., *Science*, 2006, **314**, 1724–1728.

82. P. A. Bland, F. J. Stadermann, C. Floss, D. Rost, E. P. Vicenzi, A. T. Kearsley and G. K. Benedix, A cornucopia of presolar and early solar system materials at the micrometer size range in primitive chondrite matrix., *Meteorit. Planet. Sci.*, 2007, **42**, 1417–1427.

83. H. Busemann, A. F. Young, C. M. O'D. Alexander, P. Hoppe, S. Mukhopadhyay and L. R. Nittler, Interstellar chemistry recorded in organic matter from primitive meteorites., *Science*, 2006, **312**, 727–730.

84. K. Nakamura-Messenger, S. Messenger, L. P. Keller, S. J. Clemett and M. E. Zolensky, Organic globules in the Tagish Lake meteorite: remnants of the protosolar disk., *Science*, 2006, **314**, 1439–1442.

CHAPTER 2

Clumped Isotope Geochemistry

ALLAN R. CHIVAS* AND FLORIAN W. DUX

GeoQuEST Research Centre, School of Earth & Environmental Sciences, University of Wollongong, NSW 2522, Australia
*Email: toschi@uow.edu.au

2.1 Introduction

The understanding of recent past climates and environments is reliant on secure estimates of past temperatures and rainfall. Although sediments from the ocean basins now provide a fair indication of global changes from millennial through to million-year timescales, difficulties remain in understanding the effect of changing ocean–atmosphere dynamics upon regional terrestrial climates. Palaeoclimate archives from ice-free continents include lake sediments, speleothems, and ancient soils; however, such archives typically offer indirect, non-linear, or ambiguous evidence for past climate change. There remain few, if any, truly robust means of estimating past air temperature on land, backed with the support of both theoretical and empirical evidence. Nevertheless, it is self-evident that the refinement of terrestrial temperature proxies is of critical importance, given that the bulk of human activities and occupation occur on land, and that climate impact on soil and water resources, agriculture, and habitability may change in the future. Reconstruction of past climate from sedimentary archives is the only means through which our understanding of the global climate system over timescales greater than about 100 years can be challenged, and through which the models used to predict future change can be tested. Developing techniques that provide robust estimates of past temperature and hydrological change is therefore a challenge of great importance.

RSC Detection Science Series No. 4
Principles and Practice of Analytical Techniques in Geosciences
Edited by Kliti Grice
© The Royal Society of Chemistry 2015
Published by the Royal Society of Chemistry, www.rsc.org

The study[1-3] of multiply substituted isotopologues, commonly called clumped isotopologues, or clumped isotopes of CO_2 liberated from carbonate minerals provides a new palaeothermometer that will greatly assist in the investigation of various geochemical palaeoclimatic problems.

2.2 Background

The development of oxygen isotope geochemistry fundamentally altered the discipline of palaeoclimatology. Sixty years after the pioneering work[4-6] of the Chicago group following Harold Urey, who established the modern theory and application of stable isotope fractionation to geochemistry, oxygen isotope ratios are now measured routinely in a wide range of materials by a variety of mass spectrometric techniques. Arguably the most valuable and unquestionably the most measured archives of oxygen isotope ratios are carbonate and carbonate-bearing minerals, which form in a wide range of environments by both biological and abiotic processes. The oxygen isotope composition (expressed as $\delta^{18}O$; a measure of $^{18}O/^{16}O$) of carbonate minerals is a function of the $\delta^{18}O$ value of the fluid from which they formed and the temperature at which precipitation occurred. Oxygen isotope ratios are thus commonly applied either as a palaeothermometer for waters from which carbonate minerals precipitated or as a tracer for the source or modification (*e.g.* through evaporation) of that water. Whereas the application of oxygen isotope geochemistry has and continues to be a fundamental tool in palaeoenvironmental studies, its utility for establishing unambiguous absolute palaeotemperatures is forever hindered by the requirement to measure independently, or more commonly, to assume, the $\delta^{18}O$ value of the original fluid from which carbonates formed. For marine calcites deposited in the past few million years, confident assumptions of initial $\delta^{18}O_{fluid}$ values may be attempted, but robust approximations for older marine sediments and terrestrial carbonates are virtually impossible to make.

The definition of the $\delta^{18}O$ value (expressed in per mil, or ‰) for palaeotemperature work is by comparison to the now exhausted PDB carbonate standard (a sample of the belemnite *Belemnitella americana* from the Cretaceous Pee Dee formation of South Carolina).

$$\delta^{18}O_{VPDB}(in\ ‰) = \left[\frac{^{18}O/^{16}O_{sample} - {^{18}O/^{16}O_{VPDB}}}{^{18}O/^{16}O_{VPDB}} \right] \times 1000 \qquad (2.1)$$

If we allow the ratio $^{18}O/^{16}O$ to be R, equation (2.1) can also be expressed as

$$\delta^{18}O_{VPDB}(in\ ‰) = \left[\frac{R_{sample}}{R_{standard}} - 1 \right] \times 1000 \qquad (2.2)$$

In practice, the currently available material[7-9] for interlaboratory comparison is NBS-19, which has a $\delta^{18}O$ VPDB value of -2.2 ‰.

To ensure highly precise data, it is necessary to use two standards, as the measured delta span between two samples commonly varies slightly (up to a few per cent) depending on the mass spectrometer used. Accordingly both NBS-19 and NBS-18 ($\delta^{18}O = -23.0$ ‰ VPDB) are needed. The standard form of mass spectrometry[10] is by electron bombardment (commonly called electron impact (EI), or gas-source mass spectrometry) and uses a dual inlet to permit measurement cycling between the sample CO_2 and a standard CO_2 of known isotopic composition. The CO_2 is generated by reaction[4] of the carbonate mineral with phosphoric acid, originally at 25 °C although more recently commonly at higher temperatures.

A typical $\delta^{18}O$ palaeotemperature equation[5,11] is of the form

$$T(°C) = 16.9 - 4.2(\delta^{18}O_c - \delta^{18}O_w) + 0.13(\delta^{18}O_c - \delta^{18}O_w)^2 \qquad (2.3)$$

and refers to calcitic molluscs. There is a family of virtually parallel curves for, among others, aragonitic molluscs,[12] foraminifers,[13] and ostracods.[14,15] Wanamaker *et al.* report updated data[16] for the calcitic portion of the blue mussel *Mytilus edulis*.

In equation (2.3), $\delta^{18}O_c$ refers to the $\delta^{18}O$ value of the carbonate, and can be measured; $\delta^{18}O_w$ refers to the isotopic composition of the water. Except in modern monitored aquatic environments, or laboratory growth or culturing experiments, $\delta^{18}O_w$ may need to be estimated. The offsets on $\delta^{18}O_c$ for different organisms and compared to inorganic carbonate precipitation are commonly referred to as 'vital effects'.

A promising new technique to avoid the prerequisite of knowing the oxygen isotope composition of source fluids to determine palaeo-temperature from carbonates of virtually any origin has recently been developed[17,18] by John Eiler and colleagues at the California Institute of Technology. Known as the clumped isotope method, this technique is founded on theoretical estimates based on lattice dynamics of the relative abundance of the isotopologues of carbonate ($^{60-67}CO_3$), not just the most abundant which are typically measured on the CO_2 ($^{44-46}CO_2$) produced by reaction of carbonate minerals in phosphoric acid (for both 'conventional' carbon and oxygen isotope analyses). Clumped isotope measurements are typically expressed in per mil variation of the relative abundance of a specific isotopologue (chiefly $^{47}CO_2$) from the theoretically predicted relative abundance based on a random distribution.[19,20] The clumped isotope palaeothermometer has now been calibrated for inorganic calcite, biogenic aragonite,[17] fish otoliths,[21] foraminifers, and coccoliths,[22,23] molluscs and brachiopods,[24-26] deep-sea corals,[27] and inorganic siderite[28] ($FeCO_3$) and should be applicable to many carbonate minerals that were formed in the 0–150 °C temperature range. Moreover, simultaneous determinations of $^{47}CO_2$ and $\delta^{18}O$ for carbonates will constrain the $\delta^{18}O$ of the water from which they precipitated, which is a sensitive tracer of atmospheric and surface hydrological change. Accordingly, clumped isotope thermometry is rapidly becoming established as a tool through which accurate estimates of past temperature can be derived,[1,2,29] providing an unique and powerful tool that can address key questions on global climate history.

2.3 How to Measure 'Clumped' Isotopes

Carbonate 'clumped' isotope geochemistry "is concerned with the state of order of rare isotopes within natural materials. That is, it examines[1] the extent to which heavy isotopes (^{13}C, ^{18}O) bond with or near each other rather than with the sea of light isotopes in which they swim". The proportion of ^{13}C–^{18}O bonds in carbonate minerals (mass 47 in CO_2, *i.e.* ^{13}C–^{18}O–^{16}O), extracted from $CaCO_3$, is sensitive to their growth temperatures, largely independent of bulk isotopic composition. The reaction temperature[30] between H_3PO_4 and $CaCO_3$ which liberates CO_2 is required to be controlled and constant for a given series of samples and standards.

The various isotopologues[1] of all six masses of CO_2, and their relative abundances are given in Table 2.1.

The temperature signal is derived from the deviation of the ratio of $^{47}CO_2$ in a sample from its stochastic or random isotopologue distribution. This is termed Δ_{47} and $\Delta_{47} = 0$‰ for the stochastic distribution and is readily produced in the laboratory (and used as a standard) by heating any sample of natural CO_2 to 1000 °C for 2 hours. Thus, where R_{47} is the sample and R_{47*} is the stochastic CO_2,

$$\Delta_{47} = \left[\left(\frac{R_{47}}{R_{47*}} - 1 \right) - \left(\frac{R_{46}}{R_{46*}} - 1 \right) - \left(\frac{R_{45}}{R_{45*}} - 1 \right) \right] \times 1000 \qquad (2.4)$$

The theoretical basis for this temperature-dependent isotopologue fractionation has been well discussed[19,20,31–33] and reviewed.[3,34] Additional, albeit small, factors that affect Δ_{47} include the speciation of the dissolved inorganic carbon species (*i.e.* $H_2CO_3/HCO_3^-/CO_3^{2-}$), which is pH dependent,

Table 2.1 Stochastic abundances of the isotopologues of carbon dioxide.[a]

Mass	Isotopologue	Relative abundance[b]
44	$^{12}C^{16}O_2$	98.40%
45	$^{13}C^{16}O_2$	1.11%
	$^{12}C^{17}O^{16}O$	748 ppm
46	$^{12}C^{18}O^{16}O$	0.40%
	$^{13}C^{17}O^{16}O$	8.4 ppm
	$^{12}C^{17}O_2$	0.142 ppm
47	$^{13}C^{18}O^{16}O$	44.4 ppm
	$^{12}C^{17}O^{18}O$	1.5 ppm
	$^{13}C^{17}O_2$	1.6 ppb
48	$^{12}C^{18}O_2$	3.96 ppm
	$^{13}C^{17}O^{18}O$	16.8 ppb
49	$^{13}C^{18}O_2$	44.5 ppb

[a]Source: Eiler.[1]
[b]Assuming $^{17}O/^{16}O$ and $^{18}O/^{16}O$ ratios equal to the VSMOW standard and $^{13}C/^{12}C$ ratio equal to the VPDB standard.

and salinity during carbonate precipitation,[33] and have yet to be addressed and incorporated in empirical temperature calculations.

Ultra-high-sensitivity gas-source mass spectrometry is required, measuring the equivalent $\delta^{18}O$ (*i.e.* CO_2 of mass 46/mass 44) and Δ_{47} to $\pm 0.01\permil$. Hydrocarbon and chlorine contaminants (*e.g.* $^{12}C^{35}Cl$), even at the parts per billion level, are fatal to the analysis. Accordingly, CO_2 masses 48 and 49 are also measured simultaneously (requiring a six-collector array for *m/e* of 44, 45, 46, 47, 48, 49), and CO_2 purification by gas chromatography is mandatory.[35] Several commercially available dual-inlet machines have been used[1,17,35–38] and several instrumental corrections[35,38–40] need to be carefully applied. Long analysis times (2 h) are required to gain the necessary precision to declare a carbonate growth temperature of ± 1.5 °C. Standardisation is by calibrated standard gases, and constant repeat analyses with overlapping unknowns. It is further recommended that Δ_{47} be reported in an absolute reference frame[41] ($\Delta_{47\text{-RF}}$) by the use of a second reference standard, by equilibration of CO_2 with H_2O at 25 °C.

The full range of Δ_{47} values in nature varies from 0‰ to 0.8‰, with an even smaller range from 0.8‰ to 0.5‰ for carbonate precipitated in the Earth-surface temperature range of 0–60 °C. Accordingly, a clumped isotope (palaeo)temperature equation (see below) will only be of use if Δ_{47} can be measured to about $\pm 0.01\permil$.

There have been several attempts at such calibration, for inorganic calcite precipitation[17] with the most recent[42] (equation 2.5) over a larger temperature range, producing a curve (Figure 2.1) that is in better agreement with many biogenic carbonates.

$$\Delta_{47} = 0.053 \times 10^6/T^2 + 0.052 \tag{2.5}$$

Figure 2.1 indicates that a variety of carbonate-secreting organisms grown at controlled temperature, or harvested from environments where the water temperature was monitored, display a Δ_{47}–T relationship similar to that of inorganic calcite. However, Δ_{47} data for shallow-water fast-growing corals,[43] not plotted, lie above the carbonate line, and speleothem calcite Δ_{47} values commonly lie below the line. In each case, kinetic isotope effects related to rapid CO_2 degassing or rapid hydration of CO_2, respectively, are suspected.[3] Notwithstanding such difficulties with these two media, the majority of carbonates appear amenable to palaeotemperature investigation and with the advantage that mineralogical variability (calcite, aragonite, or siderite) and vital effects among most $CaCO_3$-secreting organisms are minimal. The following section provides some examples of successful and plausible applications.

2.4 Geological Applications

A number of studies in the area of palaeoceanography have been chosen to successfully highlight key changes in Earth climates even in the older parts

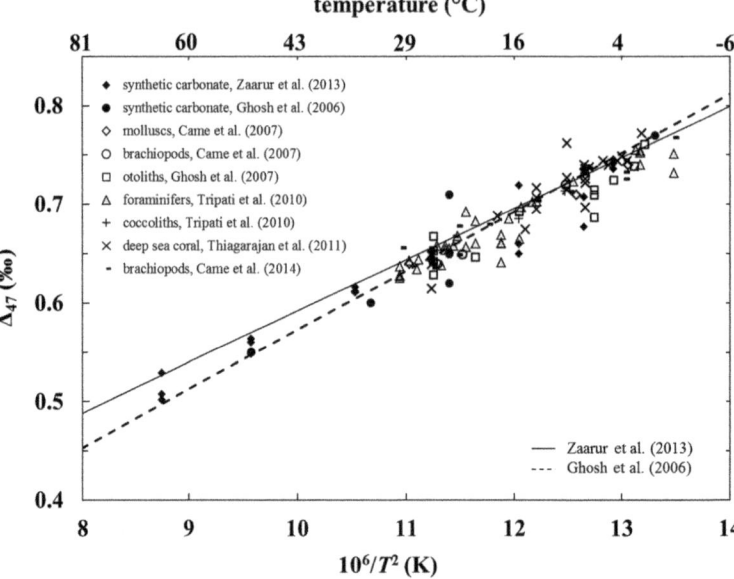

Figure 2.1 Δ_{47}–T calibration line for combined synthetic[17,42] and biogenic carbonates.[21,22,26,27,45] The compilation is from Zaarur *et al.*,[42] and all data are plotted into the Gosh *et al.*[17] reference frame.

of the geological record. Using clumped isotopes, applied to a variety of biota including corals and brachiopods, a short interval of cooling by 5 °C (from 32 to 37 °C) is demonstrated[44] for tropical oceans at the time of the Late Ordovician glaciation, approximately 445 million years ago. Warmer-than-present ocean temperatures are demonstrated[45] in the Early Silurian ($\sim$430 Ma), and exceedingly warm tropical sea-surface temperatures (to 39 °C) are proposed[46] for the end-Permian extinction. Other studies have investigated the North American Cretaceous interior seaway,[47] environmental changes at the Cretaceous–Paleogene (66 Ma) boundary,[48] the warm conditions[49–51] of the Late Cretaceous ($\sim$140 Ma) and Eocene ($\sim$50 Ma), and sea-surface cooling[52] during the last glacial maximum.

In the freshwater domain, clumped isotope data on molluscs[53] from the Canadian Arctic show temperatures during the Pliocene up to 15 °C warmer than today. Aragonitic land snails indicate calcification temperatures at higher than ambient conditions, reflecting snail body temperature[54] at the time of calcification. The carbonate phase of bioapatite is amendable to clumped isotope palaeothermometry, thereby placing vertebrate palaeontology within the realm of investigation. One of the more spectacular results indicates that some dinosaurs were warm-blooded[55,56] and therefore not reptiles.

Speleothems commonly display Δ_{47} disequilibrium and are not yet readily usable in isolation,[57–62] and may need supplementation by an additional independent analytical technique to realise their potential. By contrast, soil carbonates have proven fruitful, with a number of studies[63–69] spanning the

Cenozoic rock record. Several investigations have used carbonate from palaeosols to infer uplift histories[18,70–73] for the Andes and the Colorado Plateau.

Whereas the foregoing studies have attempted to elucidate past Earth-surface temperatures there is scope to look at somewhat deeper geological processes where temperatures up to 100–200 °C are attained. This is the domain of shallow hydrothermal vein formation, mineralised fault planes, diagenesis, low-temperature metamorphism, and ore deposition.[74–81] It also provides an opportunity to examine how higher-temperature carbonates retain or adjust their Δ_{47} values at progressively shallower depths during erosion.[82] The initial studies of rates of cooling and kinetics[83] in such systems show promise and complexity, and further advances can be anticipated. The reliable upper temperature limit for carbonate clumped isotopes will depend on the rate of heating and/or cooling of each rock system. Initial experiments[84] indicate that calcite maintained at around 100 °C for 10^6 to 10^8 years should retain its initial clumped isotope signature.

The original applications[3,85,86] of Δ_{47} were to modern atmospheric CO_2. As a tracer, $\Delta^{47}/\delta^{18}0_{CO_2}$ plots readily distinguish among CO_2 sources from car-exhaust pollution, respiration, and the troposphere. Other atmospheric 'clumped' systems are in development. The minor isotopologues of atmospheric O_2 ($^{18}O^{18}O$ and $^{17}O^{18}O$) are being developed[87] to study its global budget and the technique may be expanded to the study of trapped atmospheric samples in ice cores. The advent of high-resolution gas-source mass spectrometry[88,89] has permitted investigation of the isotopologues of methane and the establishment[90] of a thermometer for methane formation applicable to atmospheric, biological, and geological fields.

References

1. J. M. Eiler, 'Clumped-isotope' geochemistry—The study of naturally-occurring, multiply substituted isotopologues, *Earth Planet. Sci. Lett.*, 2007, **262**, 309–327.
2. J. M. Eiler, Paleoclimate reconstruction using carbonate clumped isotope thermometry, *Quat. Sci. Revs.*, 2011, **30**, 3575–3588.
3. J. M. Eiler, The isotopic anatomies of molecules and minerals, *Annu. Rev. Earth Planet. Sci.*, 2013, **41**, 411–414.
4. J. M. McCrea, On the isotopic chemistry of carbonates and a paleo-temperature scale, *J. Chem. Phys.*, 1950, **18**, 849–857.
5. S. Epstein, R. Buchsbaum, H. Lowenstam and H. C. Urey, Revised carbonate-water isotopic temperature scale, *Bull. Geol. Soc. Am.*, 1953, **64**, 1315–1326.
6. H. Urey, The thermodynamic properties of isotopic substances, *J. Chem. Soc.*, 1947, 562–581.
7. R. Gonfiantini, Standards for stable isotope measurements in natural compounds, *Nature*, 1978, **271**, 534–536.

8. T. B. Coplen, C. Kendall and J. Hopple, Comparison of stable isotope reference samples, *Nature*, 1983, **302**, 236–238.

9. T. B. Coplen, P. De Bièvre, H. R. Krouse, R. D. Vocke Jr., M. Gröning and K. Rozanski, Ratios for light-element isotopes standardized for better interlaboratory comparison, *Eos, Trans. Am. Geophys. Union*, 1996, 77, 255.

10. C. R. McKinney, J. M. McCrea, S. Epstein, H. A. Allen and H. C. Urey, Improvements in mass spectrometers for the measurement of small differences in isotope abundance ratios, *Rev. Sci. Instrum.*, 1950, **21**, 724–730.

11. H. Craig, The measurement of oxygen isotope paleotemperatures, in *Stable Isotopes in Oceanographic Studies and Paleotemperatures*, ed. E. Tongiorgi, CNR-Laboratorio di Geologica Nucleare, Pisa, 1965, pp. 161–182.

12. Y. Horibe and T. Oba, Temperature scales of aragonite-water and calcite-water systems, *Fossils*, 1972, **23/24**, 69–79.

13. J. Erez and B. Luz, Experimental paleotemperature equation for planktonic foraminifera, *Geochim. Cosmochim. Acta*, 1983, **47**, 1025–1031.

14. J. Xia, E. Ito and D. R. Engstrom, Geochemistry of ostracode calcite: part 1. An experimental determination of oxygen isotope fractionation, *Geochim. Cosmochim. Acta*, 1997, **61**, 377–382.

15. A. R. Chivas, P. De Deckker, S. X. Wang and J. A. Cali, Oxygen-isotope systematics of the nektic ostracod *Australocypris robusta,* in *The Ostracoda—Applications in Quaternary Research*, ed. J. A. Holmes and A. R. Chivas, Geophysical Monograph 131, American Geophysical Union, Washington DC, 2002, 301–313.

16. A. D. Wanamaker Jr., K. J. Kreutz, H. W. Borns Jr., D. S. Introne, S. Feindel, S. Funder, P. D. Rawson and B. J. Barber, Experimental determination of salinity, temperature, growth and metabolic effects on shell isotope chemistry of *Mytilus edulis* collected from Maine and Greenland, *Paleoceanography*, 2007, **22**, PA2217.

17. P. Ghosh, J. Adkins, H. P. Affek, B. Balta, W. Guo, E. A. Schauble, D. Schrag and J. M. Eiler, ^{13}C-^{18}O bonds in carbonate materials: A new kind of paleothermometer, *Geochim. Cosmochim. Acta*, 2006, **70**, 1439–1456.

18. P. Ghosh, C. Garzione and J. M. Eiler, Rapid uplift of the Altiplano revealed through ^{13}C-^{18}O bonds in paleosol carbonates, *Science*, 2006, **311**, 511–515.

19. Z. Wang, E. A. Schauble and J. M. Eiler, Equilibrium thermodynamics of multiply substituted isotopologues of molecular gases, *Geochim. Cosmochim. Acta*, 2004, **68**, 4779–4797.

20. E. A. Schauble, P. Ghosh and J. M. Eiler, Preferential formation of ^{13}C-^{18}O bonds in carbonate minerals, estimated using first-principles lattice dynamics, *Geochim. Cosmochim. Acta*, 2006, **70**, 2510–2529.

21. P. Ghosh, J. M. Eiler, S. E. Campana and R. F. Feeney, Calibration of the carbonate 'clumped isotope' paleothermometer for otoliths, *Geochim. Cosmochim. Acta*, 2007, **71**, 2736–2744.

22. A. K. Tripati, R. A. Eagle, N. Thiagarajan, A. C. Gagnon, H. Bauch, P. R. Halloran and J. M. Eiler, ^{13}C–^{18}O isotope signatures and 'clumped isotope' thermometry in foraminifera and coccoliths, *Geochim. Cosmochim. Acta*, 2010, **74**, 5697–5717.

23. A-L. Grauel, T. W. Schmid, B. Hu, C. Bergami, L. Capotondi, L. Zhou and S. M. Bernasconi, Calibration and application of the 'clumped isotope' thermometer to foraminifera for high-resolution climate reconstructions, *Geochim. Cosmochim. Acta*, 2013, **108**, 125–140.

24. R. A. Eagle, J. M. Eiler, A. K. Tripati, J. B. Ries, P. S. Freitas, C. Heibenthal, A. D. Wanamaker Jr., M. Taviani, M. Elliot, S. Marenssi, K. Nakamura, P. Ramirez and K. Roy, The influence of temperature and seawater carbonate saturation state on ^{13}C-^{18}O bond ordering in bivalve mollusks, *Biogeosci. Discuss.*, 2013, **10**, 157–194.

25. G. A. Henkes, B. H. Passey, A. D. Wanamaker Jr., E. L. Grossman, W. G. Ambrose Jr and M. L. Carroll, Carbonate clumped isotope compositions of modern marine mollusk and brachiopod shells, *Geochim. Cosmochim. Acta*, 2012, **106**, 307–325.

26. R. E. Came, U. Brand and H. P. Affek, Clumped isotope signatures in modern brachiopod carbonate, *Chem. Geol.*, 2014, 377, 20–30.

27. N. Thiagarajan, J. Adkins and J. M. Eiler, Carbonate clumped isotope thermometry of deep-sea corals and implications for vital effects, *Geochim. Cosmochim. Acta*, 2011, **75**, 4416–4425.

28. A. Fernandez, J. Tang and B. Rosenheim, Siderite 'clumped' isotope thermometry: A new paleoclimate proxy for humid continental environments, *Geochim. Cosmochim. Acta*, 2014, **126**, 411–421.

29. H. P. Affek, Clumped isotope paleothermometry: principles, applications, and challenges, *The Paleontological Society Papers*, 2012, **18**, 101–114.

30. U. Wacker, J. Fiebig and B. R. Schoene, Clumped isotope analysis of carbonates: comparison of two different acid digestion techniques, *Rapid Commun. Mass Spectrom.*, 2013, **27**, 1631–1642.

31. W. Guo, J. L. Mosenfelder, W. A. Goddard III and J. M. Eiler, Isotopic fractionations associated with phosphoric acid digestion of carbonate minerals: Insights from first-principles theoretical modeling and clumped isotope measurements, *Geochim. Cosmochim. Acta*, 2009, **73**, 7203–7225.

32. X. Cao and Y. Liu, Theoretical estimation of the equilibrium distribution of clumped isotopes in nature, *Geochim. Cosmochim. Acta*, 2012, 77, 292–303.

33. P. S. Hill, A. K. Tripati and E. A. Schauble, Theoretical constraints on the effects of pH, salinity, and temperature on clumped isotope signatures of dissolved inorganic carbon species and precipitating carbonate minerals, *Geochim. Cosmochim. Acta*, 2014, **125**, 610–652.

34. M. A. Webb and T. F. Miller III, Position-specific and clumped stable isotope studies: comparison of the Urey and Path-Integral approaches for carbon dioxide, nitrous oxide, methane, and propane, *J. Phys. Chem.*, 2014, **118**, 467–474.

35. K. W. Huntington, J. M. Eiler, H. P. Affek, W. Guo, M. Bonifacie, L. Y. Yeung, N. Thiagarajan, B. Passey, A. Tripati, M. Daëron and R. Came, Methods and limitations of 'clumped' CO_2 isotope (Δ_{47}) analysis by gas-source isotope ratio mass spectrometry, *J. Mass Spectrom.*, 2009, **44**, 1318–1329.

36. T. W. Schmid and S. M. Bernasconi, An automated method for 'clumped-isotope' measurements on small carbonate samples, *Rapid Commun. Mass Spectrom.*, 2010, **24**, 1955–1963.

37. N. Yoshida, M. Vasilev, P. Ghosh, O. Abe, K. Yamada and M. Morimoto, Precision and long-term stability of clumped-isotope analysis of CO_2 using a small-sector isotope ratio mass spectrometer, *Rapid Commun. Mass Spectrom.*, 2013, **27**, 207–215.

38. B. E. Rosenheim, J. Tang and A. Fernandez, Measurement of multiply substituted isotopologues ('clumped isotopes') of CO_2 using a 5 kV compact isotope ratio mass spectrometer: Performance, reference frame, and carbonate paleothermometry, *Rapid Commun. Mass Spectrom.*, 2013, **27**, 1847–1857.

39. B. He, G. A. Olack and A. S. Colman, Pressure baseline corrections and high-precision CO_2 clumped-isotope (Δ_{47}) measurements in bellows and micro-volume modes, *Rapid Commun. Mass Spectrom.*, 2012, **26**, 2837–2853.

40. S. M. Bernasconi, B. Hu, U. Wacker, J. Fiebig, S. F. M. Breitenbach and T. Rutz, Background effects on Faraday collectors in gas-source mass spectrometry and implications for clumped isotope measurements, *Rapid Commun. Mass Spectrom.*, 2013, **27**, 603–612.

41. K. J. Dennis, H. P. Affek, B. H. Passey, D. P. Schrag and J. M. Eiler, Defining an absolute reference frame for 'clumped' isotope studies of CO_2, *Geochim. Cosmochim. Acta*, 2011, **75**, 7117–7131.

42. S. Zaarur, H. P. Affek and M. T. Brandon, A revised calibration of the clumped isotope thermometer, *Earth Planet. Sci. Lett.*, 2013, **382**, 47–57.

43. C. Saenger, H. P. Affek, T. Felis, N. Thiagarajan, J. M. Lough and M. Holcomb, Carbonate clumped isotope variability in shallow water corals: Temperature dependence and growth-related vital effects, *Geochim. Cosmochim. Acta*, 2012, **99**, 224–242.

44. S. Finnegan, K. Bergman, J. M. Eiler, D. S. Jones, D. A. Fike, I. Eisenman, N. C. Hughes, A. K. Tripati and W. W. Fischer, The magnitude and duration of Late Odovician-Early Silurian glaciation, *Science*, 2011, **331**, 903–906.

45. R. E. Came, J. M. Eiler, J. Veizer, K. Azmy, U. Brand and C. R Weidman, Coupling of surface temperatures and atmospheric CO_2 concentrations during the Palaeozoic era, *Nature*, 2007, **449**, 198–201.

46. U. Brand, R. Posenato, R. Came, H. Affek, L. Angiolini, K. Azmy and E. Farabegoli, The end-Permain mass extinction: A rapid volcanic CO_2 and CH_4-climatic catastrophe, *Chem. Geol.*, 2012, **322–323**, 121–144.

47. K. J. Dennis, J. K. Cochran, N. H. Landman and D. P. Schrag, The climate of the Late Cretaceous: New insights from the application of the carbonate clumped isotope thermometer to Western Interior Seaway macrofossil, *Earth Planet. Sci. Lett.*, 2013, **362**, 51–65.

48. T. S. Tobin, G. P. Wilson, J. M. Eiler and J. H. Hartman, Environmental change across a terrestrial Cretaceous-Paleogene boundary section in eastern Montana, USA, constrained by carbonate clumped isotope paleothermometry., *Geology*, 2014, **42**, 351–354.

49. G. D. Price and B. H. Passey, Dynamic polar climates in a greenhouse world: Evidence from clumped isotope thermometry of Early Cretaceous belemnites, *Geology*, 2013, **41**, 923–926.

50. C. R. Keating-Bitonti, L. C. Ivany, H. P. Affek, P. Douglas and S. Samson, Warm, not super-hot, temperatures in the early Eocene subtropics, *Geology*, 2011, **39**, 771–774.

51. C. M. Frantz, V. A. Petryshyn, P. J. Marenco, A. Tripati, W. M. Berelson and F. A. Corsetti, Dramatic local environmental change during the Early Eocene Climatic Optimum detected using high resolution chemical analyses of Green River Formation stromatolites, *Palaeogeogr. Palaeoclimatol. Palaeoecol.*, 2014, **405**, 1–15.

52. A. K. Tripati, S. Sahany, D. Pittman, R. A. Eagle, J. D. Neelin, J. L. Mitchell and L. Beaufort, Modern and glacial tropical snowlines controlled by sea surface temperature and atmospheric mixing., *Nat. Geosci.*, 2014, 7, 205–209.

53. A. Z. Csank, A. K. Tripati, W. P. Patterson, R. A. Eagle, N. Rybczynski, A. P. Ballantyne and J. M. Eiler, Estimates of Arctic land surface temperatures during the early Pliocene from two novel proxies, *Earth Planet. Sci. Lett.*, 2011, **304**, 291–299.

54. S. Zaarur, G. Olack and H. P. Affek, Paleo-environmental implication of clumped isotopes in land snail shells, *Geochim. Cosmochim. Acta*, 2011, **75**, 6859–6869.

55. R. A. Eagle, E. A. Schauble, A. K. Tripati, T. Tütken, R. C. Hulbert and J. M. Eiler, Body temperatures of modern and extinct vertebrates from ^{13}C-^{18}O bond abundances in bioapatite, *Proc. Natl. Acad. Sci. U. S. A.*, 2010, **107**, 10377–10382.

56. R. A. Eagle, T. Tütken, T. S. Martin, A. K. Tripati, H. C. Fricke, M. Connely, R. L. Cifelli and J. M. Eiler, Dinosaur body temperatures determined from isotopic (^{13}C-^{18}O) ordering in fossil biomaterials, *Science*, 2011, **333**, 443–445.

57. M. Daëron, W. Guo, J. Eiler, D. Genty, D. Blamart, R. Boch, R. Drysdale, R. Maire, K. Wainer and G. Zanchetta, $^{13}C^{18}O$ clumping in speleothems: Observations from natural caves and precipitation experiments, *Geochim. Cosmochim. Acta*, 2011, **75**, 3303–3317.

58. H. P. Affeck, M. Bar-Matthews, A. Ayalon, A. Matthews and J. M. Eiler, Glacial/interglacial temperature variations in Soreq cave speleothems as recorded by 'clumped isotope' thermometry, *Geochim. Cosmochim. Acta*, 2008, **72**, 5351–5360.

59. H. P. Affek, Clumped isotopic equilibrium and the rate of isotope exchange between CO_2 and water, *Am. J. Sci.*, 2013, **313**, 309–325.

60. K. Wainer, D. Genty, D. Blamart, M. Daëron, M. Bar-Matthews, H. Vonhof, Y. Dublyansky, E. Pons-Branchu, L. Thomas, P. van Calsteren, Y. Quinif and N. Caillon, Speleothem record of the last 180 ka in Villars cave (SW France): Investigation of a large ^{18}O shift between MIS6 and MIS5, *Quat. Sci. Revs.*, 2011, **30**, 130–146.

61. T. Kluge and H. P. Affek, Quantifying kinetic fractionation in Bunker Cave speleothems using Δ_{47}, *Quat. Sci. Revs.*, 2012, **49**, 82–94.

62. T. Kluge, H. P. Affek, Y. G. Zhang, Y. Dublyansky, C. Spötl, A. Immenhouser and D. K. Richter, Clumped isotope thermometry of cryogenic cave carbonates, *Geochim. Cosmochim. Acta*, 2013, **61**, 3461–3475.

63. J. Quade, C. Garzione and J. M. Eiler, Paleoelevation reconstruction using pedogenic carbonates, *Rev. Mineral. Geochem.*, 2007, **66**, 53–87.

64. J. Quade, J. M. Eiler, M. Daëron and H. Achyuthan, The clumped isotope geothermometer in soil and paleosol carbonate, *Geochim. Cosmochim. Acta*, 2013, **105**, 92–107.

65. B. H. Passey, N. E. Levin, T. E. Cerling, F. H. Brown and J. M. Eiler, High-temperature environments of human evolution in East Africa based on bond ordering in paleosol carbonates, *Proc. Natl. Acad. Sci. U. S. A.*, 2010, **107**, 11245–11249.

66. K. E. Snell, B. L. Thrasher, J. M. Eiler, P. L. Koch, L. C. Sloan and N. J. Tabor, Hot summers in the Bighorn Basin during the early Paleogene, *Geology*, 2013, **41**, 55–58.

67. N. A. Peters, K. W. Huntington and G. D. Hoke, Hot or not? Impact of seasonally variable soil carbonate formation paleotemperature and O-isotope records from clumped isotope thermometry, *Earth Planet. Sci. Lett.*, 2012, **361**, 208–218.

68. J. H. VanDeVelde, G. J. Bowen, B. H. Passey and B. B. Bowen, Climatic diagenetic signals in the stable isotope geochemistry of dolomitic paleosols spanning the Paleocene-Eocene boundary. *Geochim. Cosmochim. Acta*, 2013, **109**, 254–267.

69. B. G. Hough, M. Fan and B. H. Passey, Calibration of the clumped isotope geothermometer in soil carbonate in Wyoming and Nebraska, USA: Implications for paleoelevation and paleoclimate reconstruction., *Earth Planet. Sci. Lett.*, 2014, **391**, 110–120.

70. K. W. Huntington, B. P. Wernicke and J. M. Eiler, Influence of climate change and uplift on Colorado Plateau paleotemperatures from carbonate clumped isotope thermometry, *Tectonics*, 2010, **29**(TC3005), 1–19.

71. J. Quade, D. O. Breecker, M. Daëron and J. Eiler, The paleoaltimetry of Tibet: an isotopic perspective. *Am. J. Sci.*, 2011, **311**, 77–115.

72. A. Leier, N. McQuarrie, C. Garzione and J. Eiler, Stable isotope evidence for multiple pulses of rapid surface uplift in the Central Andes, Bolivia, *Earth Planet. Sci. Lett.*, 2013, **371–372**, 49–58.
73. C. N. Garzione, D. J. Auerbach, J. J.-S. Smith, J. J. Rosario, B. H. Passey, T. E. Jordan and J. M. Eiler, Clumped isotope evidence for diachronous surface cooling of the Altiplano and pulsed surface uplift of the Central Andes, *Earth Planet. Sci. Lett.*, 2014, **393**, 173–181.
74. J. M. Ferry, B. H. Passey, C. Vasconcelos and J. M. Eiler, Formation of dolomite at 40–80°C in the Latemar carbonate buildup, Dolomites, Italy, from clumped isotope thermometry, *Geology*, 2011, **39**, 571–574.
75. T. F. Bristow, M. Bonifacie, A. Derkowski, J. M. Eiler and J. P. Grotzinger, A hydrothermal origin for isotopically anomalous cap dolostone cements from south China, *Nature*, 2011, **474**, 68–71.
76. K. W. Huntington, D. A. Budd, B. P. Wernicke and J. M. Eiler, Use of clumped-isotope thermometry to constrain the crystallization temperature of diagenetic calcite, *J. Sediment. Res.*, 2011, **81**, 656–669.
77. E. M. Swanson, B. P. Wernicke, J. M. Eiler and S. Losh, Temperatures and fluids on faults based on carbonate clumped-isotope thermometry, *Am. J. Sci.*, 2012, **312**, 1–21.
78. S. J. Loyd, F. A. Corsetti, J. M. Eiler and A. K. Tripati, Determining the diagenetic conditions of concretion formation: assessing temperatures and pore waters using clumped isotopes, *J. Sediment. Res.*, 2012, **82**, 1006–1016.
79. D. A. Budd, E. L. Frost III, K. W. Huntington and P. F. Allwardt, Syndepositional deformation features in high-relief carbonate platforms: long-lived conduits for diagenetic fluids, *J. Sediment. Res.*, 2013, **82**, 12–36.
80. A. Dale, C. M. John, P. S. Mozley, P. C. Smalley and A. H. Muggeridge, Time-capsule concretions: Unlocking burial diagenetic processes in the Mancos Shale using carbonate clumped isotopes, *Earth Planet. Sci. Lett.*, 2014, **394**, 30–37.
81. S. C. Bergman, K. W. Huntington and J. G. Crider, Tracing paleofluid sources using clumped isotope thermometry of diagenetic cements along the Moab Fault, Utah, *Am. J. Sci.*, 2013, **313**, 490–515.
82. K. J. Dennis and D. P. Schrag, Clumped isotope thermometry of carbonatites as an indicator of diagenetic alteration, *Geochim. Cosmochim. Acta*, 2010, **74**, 4110–4122.
83. B. H. Passey and G. A. Henkes, Carbonate clumped isotope bond reordering and geospeedometry, *Earth Planet. Sci. Lett.*, 2012, **351–352**, 223–236.
84. G. A. Henkes, B. H. Passey, E. L. Grossman, B. J. Shenton, A. Pérez-Huerta and T. E. Yancey, Temperature limits for the preservation of primary calcite clumped isotope paleotemperatures, *Geochim. Cosmochim. Acta*, 2014, doi: http://dx.doi.org/10.1016/j.gca.2014.04.040.
85. J. M. Eiler and E. A. Schauble, $^{18}O^{13}C^{16}O$ in Earth's atmosphere, *Geochim. Cosmochim. Acta*, 2004, **68**, 4767–4777.

86. H. P. Affek and J. M. Eiler, Abundance of mass 47 CO_2 in urban air, car exhausts, and human breath, *Geochim. Cosmochim. Acta*, 2006, **70**, 1–12.

87. L. Y. Yeung, E. D. Young and E. A. Schauble, Measurements of $^{18}O^{18}O$ and $^{17}O^{18}O$ in the atmosphere and the role of isotope-exchange reactions, *J. Geophys. Res.*, 2012, **117**, D18306.

88. J. M. Eiler, M. Clog, P. Magyar, A. Piasecki, A. Sessions, D. Stolper, M. Deerberg, H.-J. Schueter and J. Schwieters, A high-resolution gas-source isotope ratio mass spectrometer, *Int. J. Mass Spectrom.*, 2013, **335**, 45–56.

89. J. M. Eiler, B. Bergquist, I. Bourg, P. Cartigny, J. Farquhar, A. Gagnon, W. Guo, I. Halevy, A. Hofmann, T. E. Larson, N. Levin, E. A. Schauble and D. Stolper, Frontiers of stable isotope geoscience. *Chem. Geol.*, 2014, **372**, 119–143.

90. D. A. Stolper, A. L. Sessions, A. A. Ferreira, E. V. Santos Neto, A. Schimmelmann, S. S. Shusta, D. L. Valentine and J. M. Eiler, Combined ^{13}C-D and D-D clumping in methane: Methods and preliminary results, *Geochim. Cosmochim. Acta*, 2014, **126**, 169–191.

CHAPTER 3

Application of Radiogenic Isotopes in Geosciences: Overview and Perspectives

SVETLANA TESSALINA,*[a] FRED JOURDAN,[a] LAURIE NUNES,[a] ALLEN KENNEDY,[a] STEVEN DENYSZYN,[b] IGOR PUCHTEL,[c] MATHIEU TOUBOUL,[c] ROBERT CREASER,[d] MAUD BOYET,[e] ELENA BELOUSOVA[f] AND ANNE TRINQUIER[g]

[a] John de Laeter Centre for Mass Spectrometry, Curtin University, Perth, WA, Australia; [b] School of Earth and Environment, University of Bergen, The University of Western Australia, Australia; [c] Department of Geology, University of Maryland, 237 Regents Drive, College Park, MD 20742, USA; [d] Department of Earth & Atmospheric Sciences, University of Alberta, 126 Earth Sciences Building, Edmonton, Alberta, Canada; [e] Laboratoire Magmas et Volcans, Université Blaise Pascal, CNRS UMR 6524, 5 rue Kessler, 63038, Clermont-Ferrand, France; [f] Macquarie University, North Ryde, NSW, Australia; [g] Thermo Fisher Scientific, Hanna-Kunath-Str. 11, 28199, Bremen, Germany
*Email: Svetlana.Tessalina@curtin.edu.au

3.1 Introduction

Isotope geology is an integral part of the earth sciences, but its development would be impossible without the concepts of nuclear physics and chemical methods. The invention of the first mass spectrometer at the beginning of the 20th century made it possible to measure of the isotopic compositions of

RSC Detection Science Series No. 4
Principles and Practice of Analytical Techniques in Geosciences
Edited by Kliti Grice
© The Royal Society of Chemistry 2015
Published by the Royal Society of Chemistry, www.rsc.org

the chemical elements that make up natural systems. Based on these measurements, the nuclear physics principle of radioactive decay enabled the direct age determination of rocks and minerals, which transformed geology into a quantitative science. Nowadays isotope geology addresses several fundamental questions of earth and planetary sciences, from astrophysics, the Earth's structures and internal dynamics to applied studies of the formation of mineral and oil deposits. The concept of isotopic tracers has allowed scientists to address the problems of environmental studies, erosion, and the transport of materials. This chapter brings together the recent advances in the field of isotope geology, but it is not meant to replace existing texts[1-3] covering basic notions and techniques.

3.2   Principles of Radioactive Decay

Some natural elements are unstable (see Table 3.1) and can transform spontaneously, by liberation of one or more particles and types of radiation. The principles of radioactive decay are described in detail elsewhere[1-4] and are only briefly summarised here.

A 'parent' element P (*e.g.* ^{187}Re) can decay to the 'daughter' element D (here ^{187}Os). During a unit of time, each atom of P has the same probability to decay. The number of disintegrations in time, $-P/dt$ (the minus indicates the decrease) is a product of this probability by the number of atoms in the system. The law of radioactive decay is expressed in the following manner (*Eqn* 3.1). For any particular element, it is always the same fraction of parent element P that decays during a unit of time. This is expressed by equation 3.1 and can be rewritten as equation 3.1a, where λ is a constant of radioactive decay which is expressed in units inverse to the time:

$$\frac{dP}{dt} = -\lambda P \tag{3.1}$$

$$\frac{dP}{P} = -\lambda dt \tag{3.1a}$$

Table 3.1   Examples of radiogenic/radioactive isotopes.

H																	He
Li	*Be*		Re	Radioactive (P)		Os	Radiogenic (D)			B	*C*	N	O	F	Ne
Na	*Mg*									*Al*	Si	P	S	*Cl*	Ar
K	Ca	Sc	Ti	V	*Cr*	*Mn*	Fe	Co	Ni	Cu	Zn	Ga	Ge	As	Se	Br	Kr
Rb	Sr	Y	Zr	Nb	Mo	*Tc*	Ru	Rh	*Pd*	*Ag*	Cd	In	Sn	Sb	Te	I	Xe
Cs	Ba	Lu	Hf	Ta	W	Re	Os	Ir	Pt	Au	Hg	Tl	*Pb*	Bi	Po	At	Rn
Fr	*Ra*	Lr	*Rf*	Db	*Sg*	Bh	Hs	Mt	*Uun*	*Uuu*	*Uub*		*Uuq*				

La	Ce	Pr	Nd	*Pm*	Sm	Eu	Gd	Tb	Dy	Ho	Er	Tm	Yb
Ac	**Th**	Pa	U	*Np*	Pu	*Am*	*Cm*	*Bk*	*Cf*	*Es*	*Fm*	*Md*	*No*

C	Cosmogenic nucleides	U	U-series (all radioactive)	*Es*	Not present on Earth

This equation may be integrated between time 0 and time t to give equation 3.2:

$$P(t) = P(0)e^{-\lambda t} \tag{3.2}$$

Most commonly we use T instead of λ, which is the time period during which element $P(0)$ decays by half, otherwise known as the half-life. The relationship between λ and T can be established using the definition in equation 3.3. Applying equation 3.2 to the period T, an equation for time can be obtained (equation 3.4). This equation means that after 10 units of time, $P(t)$ is decreased by a factor of approximately 1024 (2^{10}).

$$\frac{P(T)}{P(0)} = \frac{1}{2} \tag{3.3}$$

$$T = \frac{\ln(2)}{\lambda} \tag{3.4}$$

The number of daughter element (D) atoms freshly formed is equal to the number of decayed atoms of the parent element, as described in equation 3.5. The total number at time t, $D(t)$, is equal to the initial number of $D(0)$, already present in the system, plus the number of decayed parent isotopes (see equation 3.6). Using equation 3.2, we obtain equation 3.7, which is known as a practical expression of the radioactive decay law. It can be used in practice because, in general, $P(t)$ is the actual value measured using the mass spectrometer, even if $P(0)$ is unknown.

$$dF + dP = 0 \tag{3.5}$$

$$F(t) = F(0) + [P(0) - P(t)] \tag{3.6}$$

$$D(t) = D(0) + P(t)[e^{\lambda t} - 1] \tag{3.7}$$

This radioactive decay law may be used to date the systems containing the isotopes (rocks, minerals, *etc.*) using the following assumption: if we can estimate the variation of quantity of daughter element since the system's formation, we can calculate the age of this system.

Several types of radioactive decay have been recognised (Table 3.2).

3.3 Mass Spectrometry

3.3.1 General Principles

Ion source mass spectrometry using mass-to-charge separation of ionic species in a magnetic field was the outcome of pioneering work by Thomson,[5] Dempster,[6] Aston,[7] Bainbridge and Jordan,[8] and Nier.[9]

The aim of mass spectrometry is to disperse ions of differing masses and focus them onto a collection/detector system. This is achieved by accelerating an ion beam, containing ions of mass m with a velocity v and a charge state q, through a potential difference V and then through a uniform

Table 3.2 Types of radioactive decay.

Type of decay	Decay equation	Example of isotope systematics
Alpha decay (α)	${}_Z^A A \rightarrow {}_{Z-2}^{A-4} B + {}_2^4 He$	${}^{147}Sm \rightarrow {}^{143}Nd + {}_2^4 He$
Beta-minus decay (β^-)	${}_Z^A A \rightarrow {}_{Z-1}^A B + e + \bar{v}, \text{or}$ $n \rightarrow p + \beta + \bar{v}$	${}^{87}Rb \rightarrow {}^{87}Sr + \beta + \bar{v}$
Beta-plus decay (β^+)	${}_Z^A A \rightarrow {}_{Z-1}^A B + e^+ + v$	${}^{187}Re \rightarrow {}^{187}Os + e^+ + v$
Electron capture	${}_Z^A A + e^- \rightarrow {}_{Z-1}^A B + v, \text{or}$ $p + e \rightarrow n + v$	${}^{40}K + e^- \rightarrow {}^{40}Ar + v$
Gamma-decay (γ)	${}_Z^A A \rightarrow {}_Z^A B + \gamma$	${}^3He \rightarrow {}^3He + \gamma$
Spontaneous fission	Fission is a chain reaction caused by neutrons when they have sufficient energy	$U \rightarrow Kr + Xe$

n, neutron; p, proton; e, electron; v, neutrino; $\bar{v}$, antineutrino; β, beta particle (high-speed electron or positron); γ - photon; A, neutron number; Z, proton number.
After Allègre.[2]

magnetic field B. The kinetic energy, KE, acquired by each ion can be expressed as described in equation 3.8. If the velocity vector of the charged particle is perpendicular to the uniform magnetic field vector, the particle will move in a circular path of radius of curvature r, with radial and centripetal forces F given by equations 3.9 and 3.10 respectively.

$$KE = \frac{1}{2}mv^2 = qV \tag{3.8}$$

$$F = qvB \tag{3.9}$$

$$F = \frac{mv^2}{r} \tag{3.10}$$

By equating 3.9 and 3.10, the velocity vector v of a charged particle of mass m can be expressed as equation 3.11. If 3.11 is substituted into 3.8, we obtain equation 3.12. Therefore the radius of curvature of each ion of mass m can be written as described in equation 3.13.

$$qvB = \frac{mv^2}{r} \rightarrow v = \frac{qBr}{m} \tag{3.11}$$

$$\frac{m}{q} = \frac{r^2 B^2}{2V} \tag{3.12}$$

$$r = \frac{1}{B}\sqrt{\left(\frac{m}{q}\right) \times 2V} \tag{3.13}$$

Heavier ion beams will therefore have a larger radius of deflection in relation to lighter ion beams, resulting in a dispersion of the composite beam.[10] However, given the inherent angular spread of the composite ion beam (in the horizontal plane), aberrations in the dispersed beam will produce slightly different radial paths for ions with the same mass, thereby leading to an unfocused ion beam at the detector. Therefore, to minimise these aberrations and maximise dispersion, a magnetic sector field is used such that dispersed ion beams come into focus outside the magnetic field. In addition to this dispersion and focusing of ion beams moving in the horizontal plane, the magnetic sector field has the property that it can focus the ion beam in the z-plane given the inherent fringing property of the magnetic field, thereby vertically deflecting diverging ion beams.[11] This arrangement is referred to as 'stigmatic double focusing' where the focusing action will occur when $\Phi = 90°$; $L' = L'' = 2r$; $\varepsilon_1 = \varepsilon_2 = 26.5°$; $\Omega = 37°$. A schematic of this arrangement is shown in Figure 3.1.

Dispersed ions that emerge from the magnetic sector field are then collected *via* a detection/collection system outside the magnetic field. The collection system will either be multi-collection, whereby different mass ion beams are collected simultaneously, or peak jumping, whereby the ion beam of a single mass is collected at a time. For the latter, the magnetic field strength is varied to change the radii of the path of the ions through the magnet so that ions of different m/q ratios can be directed into a detector slit in succession.

Mass spectrometers operate in a high vacuum system in order to prevent collisions of ions with residual gas molecules during the flight from the ion source to the detector. The vacuum should be such that the mean free path of the ion is orders of magnitude longer than the distance from the ion source to the detector.

In thermal ionisation mass spectrometry (TIMS), a charged ion beam is generated when a filament assembly, containing a suitable sample deposit, is heated under vacuum conditions. The probability that an atom is ionised

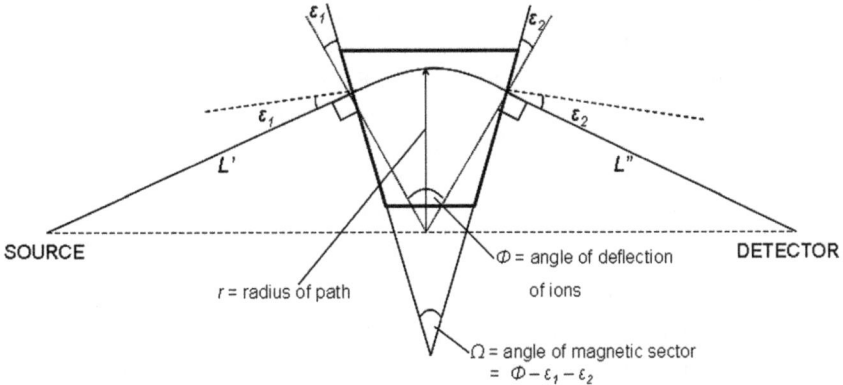

Figure 3.1   General schematic of a mass spectrometer magnetic sector arrangement.
Reproduced with permission from Cross.[11]
Copyright © 1951, AIP Publishing LLC.

and ejected with low energy from the filament surface can be determined using the Saha–Langmuir equation (equation 3.8), which represents the ionisation efficiency of positively charged ions in a thermal ionisation source (equation 3.14).

$$E = \frac{N^+}{N^o} A e^{[(\phi - I)/kT]} \tag{3.14}$$

where N^+ is the number of positively charged ions leaving the filament, N^o is the number of neutral atoms, A is a constant, ϕ is the work function of the filament, I is the ionisation potential of the ionised particles, k is the Boltzmann constant, and T is the temperature. The degree of ionisation (N^+/N^o) is therefore dependent upon the ionisation potential of the element I: if this is low, $(\phi - I)$ is positive and the source is very efficient. For elements with a high ionisation potential, *i.e.* $(\phi - I)$ is negative, then the maximum efficiency is obtained when T is as large as possible.

Significant proportions of present-day mass spectrometers employ thermal ionisation or inductively coupled plasma ionisation techniques.

3.3.2 Thermal Ionisation and Noble Gas Mass Spectrometry

Thermal ionisation is achieved by passing a current through a refractory metal filament on which the sample is deposited.[12] This versatile and robust method has been extensively used in multiple scientific applications. because of the high ionisation potentials of some elements, a number of strategies have been developed to enhance ionisation efficiency, either by addition of colloidal solutions, the use of multiple filament assemblies, or negative ion analysis. Thermal ionisation efficiencies range from 0.001% to 40% of the sample load. Memory effects from previous samples are negligible and isobaric interferences are mainly a function of analyte purity. Once ionised, the sample is accelerated in an electrical potential and the resulting ion beams are deflected in a magnetic field according to the mass-to-charge ratios. To avoid collision with gas molecules, the ion beams travel in a high vacuum. The relative isotopic abundance is measured by the comparison of ion beam intensities. The collectors are either Faraday cups, or, for low-intensity beams, ion counters. Faraday cups convert the incident charges into a current, which in turn is amplified. Low-intensity ion beams generate electric currents in ion counters through emission of secondary electrons, which are subsequently amplified. TIMS analyses introduce a mass-dependent analytical bias (fractionation), which occurs upon evaporation and evolves with the consumption of the sample. The magnitude of fractionation is of the order of 1% by mass unit for light elements (*e.g.* lithium), and 0.1% by mass unit for heavy elements (*e.g.* uranium). The fractionation can be accurately modelled by the Rayleigh diffusion law and can be corrected for by internal normalisation, double spike techniques, or total evaporation. The magnitude of the mass-dependent fractionation in plasma source mass spectrometers is up to 10 times greater.

In its simplest form, a sample is dried on the filament surface of a single filament assembly as a solution of the chloride, nitrate, or oxide of the element concerned. In some cases, an ionisation enhancer is used to improve the ionisation efficiency on the filament.[13-15 and references therein] The sample is then carefully taken to dryness and the assembly is mounted into the mass spectrometer ion source. The filament is then heated under vacuum conditions leading to evaporation/vaporisation of the sample. Generated neutral atoms then interact with the hot surface of the filament material and may be positively ionised.

If the sample being analysed has a high ionisation potential, high temperatures are required in order to achieve sufficient ionisation efficiency. However, this may result in the filament melting and/or the sample evaporating before ionisation can occur. Therefore, two- and three-filament assemblies are used, allowing the heating of each filament in the assembly to be independently controlled. The two-filament assembly consists of an evaporation filament which the sample is mounted on to and evaporated at low temperature, and an ionisation filament which ionises evaporated neutral atoms and is maintained at a higher temperature. In the triple-filament arrangement, the sample is mounted on two side filaments, which, when heated, vaporise the sample; evaporated neutral atoms are then ionised by the central filament.[10,16]

Rhenium is often preferred as the filament material because of its high work function (4.98 eV) and high melting point (3180 K).[10] Other filament materials that are used are tantalum, tungsten, and platinum.

Noble gas mass spectrometers (*i.e.* for ^{40}Ar/^{39}Ar isotope measurement) follow the same general principle as TIMS, with the exception that a gaseous sample consisting of noble gases is purified in an extraction line and introduced directly into the ion source chamber. A tungsten coiled-wire filament continuously heated at high temperature emits electrons that are accelerated by electric potential. Electrons, with an initial energy of 50–100 eV and a small permanent magnet to increase the length of their trajectory through the gas, hit and ionise (single-charge) noble gas atoms. The ions are then accelerated through plates with a variable voltage of several thousand volts, whereas the electrons emitted from the filament are recovered by the electron trap.

3.3.3   Isotopic Dilution

In order to learn the particular elemental contents (*e.g.* osmium) of a sample with known isotopic ratios (measured using the mass spectrometer), the quantity of osmium isotopes must be known. TIMS instruments especially provide excellent measurements of isotopic ratios but signal levels depend poorly on the size of the loads. Isotope dilution is a way to measure contents very precisely using the measurements of isotopic ratios. Addition of a precise quantity of ^{190}Os, which is a stable osmium isotope, allows us to estimate the quantity of ^{188}Os by measuring the ^{190}Os/^{188}Os isotope ratio.

This addition of enriched isotope tracer is referred to as the isotopic dilution technique. In order to know the contents of other radiogenic isotopes we use the same technique, adding a precise amount of enriched stable isotope and measuring the isotopic ratios of the mixture. It is important to notice that the isotopic equilibrium should be achieved between the sample and spike.

The ^{190}Os/^{188}Os ratio of the mixture may be expressed as:

$$\left(\frac{^{190}Os}{^{188}Os}\right)_{mixture} = \frac{(^{190}Os)_{sample} + (^{190}Os)_{spike}}{(^{188}Os)_{sample} + (^{188}Os)_{spike}}$$

$$= \frac{\left(\frac{^{190}Os}{^{188}Os}\right)_{spike} + \left(\frac{^{190}Os}{^{188}Os}\right)_{sample}\left[\frac{^{188}Os_{sample}}{^{188}Os_{spike}}\right]}{1 + \left[\frac{^{190}Os_{sample}}{^{190}Os_{spike}}\right]} \qquad (3.15)$$

After dividing the top and bottom of this equation by ^{188}Os and other minor manipulations, we obtain for the ^{188}Os contents of the rock under study as:

$$^{188}Os_{sample} = {}^{188}Os_{spike}\left(\frac{^{190}Os}{^{188}Os}\right)_{spike}\frac{\left(\frac{^{190}Os}{^{188}Os}\right)_{spike} - \left(\frac{^{190}Os}{^{188}Os}\right)_{mixture}}{\left(\frac{^{190}Os}{^{188}Os}\right)_{mixture} - \left(\frac{^{190}Os}{^{188}Os}\right)_{sample}} \qquad (3.16)$$

Using this equation with known ^{188}Os content of the spike and measuring the isotopic ratios of the mixture, the Os content of the studied sample can be calculated.

This method is applicable for a range of elements with several stable isotopes. Spikes are usually artificially enriched in one or more stable isotopes. The isotopic dilution method therefore has several advantages. First, highly accurate precision is attained after isotopic equilibrium with the sample is achieved, even if the chemistry does not yield a 100% score. Another advantage is the great precision in measuring very low-level trace elements due to the high precision of isotopic ratio measurements using mass spectrometry. The best results are achieved by calibrating each mass spectrometer against the international isotopic standards.

3.3.4   International Isotopic Standards

Each mass spectrometer has its own characteristics and it is quite difficult to get a precisely consistent isotopic ratio among many mass spectrometers from different manufacturers. A reference sample should be used to find the instrumental bias for an individual mass spectrometer. For example, for the isotopic composition of neodymium, the La Jolla neodymium standard prepared by Lugmair and Carlson[17] is mostly used to monitor the instrumental bias. The reference rock material with homogeneous contents and isotopic ratios of elements of interest are also used to estimate the precision of techniques used, accuracy of spike calibration, and chemical preparation.

In what follows, several short and long-lived isotopic systems and their application in earth sciences are described in detail.

3.4   Short-Lived Isotopes

3.4.1   Extinct Radioactivity

An extinct radioactive decay system is one where the radioactive parent isotope may have been present at the time the solar system was formed, 4.56 billion years ago, but has now decayed to undetectable levels. Since the parent isotope is no longer present an absolute age cannot be determined and only a relative chronology can be obtained. A critical assumption of radiometric decay systems is that the parent and daughter elements were initially homogeneous in isotopic composition throughout the system under investigation. Also, the decay system must remain closed to loss or gain of the parent and daughter elements after formation. If these constraints hold, then the abundance of the daughter isotope, which changes rapidly over short time intervals in these systems, can be measured and used to calculate relative ages that have excellent temporal resolution. A large number of these systems have been studied in meteorites and this has produced a fine-scale chronology of early solar system processes that fractionate the parent–daughter ratios of each system before the parent isotope became extinct. Examples of extinct parent isotope–daughter isotope systems studied in meteorites are ^{10}Be–^{10}B, ^{26}Al–^{26}Mg, ^{41}Ca–^{41}K, ^{53}Mn ^{53}Cr, ^{60}Fe–^{60}Ni, ^{107}Pd–^{107}Ag, ^{129}I–^{129}Xe, ^{182}Hf–^{182}W, and ^{146}Sm–^{142}Nd (Table 3.3). These decay systems have half-lives ranging from 0.1 Ma for ^{10}Be to 68 Ma for ^{146}Sm. We describe some of these in more detail below.

3.4.2 ^{26}Al: Isotopic Anomalies in ^{26}Mg

Aluminium possesses a radioactive isotope ^{26}Al, with a short half-life of 0.72 Ma. More than 99% of the ^{26}Al will decay to produce ^{26}Mg in 5 million years, and therefore ^{26}Al rapidly becomes extinct. ^{26}Al was produced by pre-solar stellar nucleosynthesis and incorporated into the solar nebula just prior to condensation of CAI (refractory, high temperature, calcium–aluminium-rich inclusions found in chondritic meteorites), magnesium-rich chondrules, and fine matrix materials. Accretion of these components and dust from the nebula led to the formation of chondritic meteorites, and ultimately, the rocky planets of our solar system. Magnesium and aluminium are both abundant in the solar system; aluminium has one stable isotope, ^{27}Al, and magnesium has three, ^{24}Mg, ^{25}Mg, and ^{26}Mg in the following proportions: 79%, 10%, and 11%. In contrast, ^{26}Al is rare, as the ^{26}Al/^{27}Al ratio at the time of formation of the first solar system objects was measured to be $\sim 5.23 \times 10^{-5}$.[18–21] Hence, an excess of ^{26}Mg produced by radioactive decay of ^{26}Al is extremely difficult to identify in magnesium-bearing material.

Table 3.3 Summary of selected radiogenic systematics with their respective decay type and half-lives.

Isotopic system	Type of decay	Half-life
Short-lived isotope systematics		
^{10}Be–^{10}B	β^- : ^{10}Be $\rightarrow$ ^{10}B $+ \beta^-$	0.1 Ma
^{41}Ca–^{41}K	Electron capture: ^{41}Ca $+ e^- \rightarrow$ ^{41}K $+ \nu$	0.1 Ma
^{26}Al–^{26}Mg	Electron capture: ^{26}Al $+ e^- \rightarrow$ ^{26}Mg $+ \nu$ and β^+: ^{26}Al $\rightarrow$ ^{26}Mg $+ e^+ + \nu$	0.7 Ma
^{60}Fe–^{60}Ni	*series of* β^- : ^{60}Fe$\rightarrow$^{60}Co $+ e + \bar{\nu} \rightarrow$ ^{60}Ni $+ e + \bar{\nu}$	1.5 Ma
^{53}Mn–^{53}Cr	Electron capture: ^{53}Mn $+ e^- \rightarrow$ ^{53}Kr $+ \nu$	3.7 Ma
^{107}Pd–^{107}Ag	*Combined (electron capture + γ-rays):* ^{108}Pd $+ e^- \rightarrow$ ^{108}Ag $+ \nu + \gamma$	6.5 Ma
^{182}Hf–^{182}W	β^- : ^{182}Hf $\rightarrow$ ^{182}W $+ \beta^-$	8.9 Ma
^{129}I–^{129}Xe	β^- : ^{129}I $\rightarrow$ ^{129}Xe $+ \beta^-$	17 Ma
^{146}Sm–^{142}Nd	α : ^{146}Sm $\rightarrow$ ^{142}Nd $+ {}^4_2$He	68 Ma
Long-lived isotope systematics		
^{190}Pt–^{186}Os	α : ^{147}Sm $\rightarrow$ ^{143}Nd $+ {}^4_2$He	469 Ga
^{147}Sm–^{143}Nd	α : ^{147}Sm $\rightarrow$ ^{143}Nd $+ {}^4_2$He	106 Ga
^{87}Rb–^{87}Sr	β^- : ^{87}Rb $\rightarrow$ ^{87}Sr $+ \beta^-$	49 Ga
^{187}Re–^{187}Os	β^+ : ^{187}Re $\rightarrow$ ^{187}Os $+ e^+ + \nu$	42 Ga
^{176}Lu–^{177}Hf		37 Ga
^{40}K–^{40}Ar	Electron capture: ^{40}K $+ e^- \rightarrow$ ^{40}Ar $+ \nu$	12 Ga
^{40}K–^{40}Ca	β^- : ^{40}K $\rightarrow$ ^{40}Ca $+ \beta^-$	1.4 Ga
U–Th–Pb systems (uranium series decay: α^+ spontaneous fusion)		
^{232}Th–^{208}Pb	^{232}Th $\rightarrow$ ^{208}Pb $+ 6{}^4_2He + 4\beta^-$	14 Ga
^{235}U–^{207}Pb	^{235}U $\rightarrow$ ^{207}Pb $+ 7{}^4_2He + 4\beta^-$	704 Ma
^{238}U–^{206}Pb	^{238}U $\rightarrow$ ^{206}Pb $+ 8{}^4_2He + 6\beta^-$	4468 Ma

However, an understanding of the distribution of ^{26}Al is critical to models of planetary formation and solar system evolution. If accretion occurred shortly after formation of ^{26}Al, then the decay of ^{26}Al would produce sufficient heat to produce melting and igneous differentiation of planetary bodies.[22]

The first evidence of ^{26}Mg from decay of ^{26}Al was found in early-condensed CAI, which is aluminium-rich material that is virtually devoid of magnesium phases within CAI.[23,24] The presence of excess of ^{26}Mg shows these materials formed very early in the evolution of the solar system, less than a million years after nucleosynthesis of ^{26}Al and quite possibly within a 20 000 year time window.[18] Although finding evidence for decay of ^{26}Mg is extremely difficult in magnesium-rich materials, recent high-precision magnesium isotope analysis has shown the magnesium-rich phases in some chondrules did form with elevated ^{26}Al/^{27}Al ratios, with values up to $\sim 0.7 \times 10^{-5}$ and an initial ^{26}Mg/^{24}Mg ratio close to the measured initial solar system value,[19] suggesting they formed at a much earlier time than most chondrules. Other studies of magnesium-rich chondrules measure much lower initial ^{26}Al abundances, $\leq 1 \times 10^{-6}$, and this is consistent with the later formation of these chondrules.[19,25] A 1.2–4 Ma age difference between CAIs and chondrule formation is the conclusion of many ^{26}Al studies interpreted under the assumption of an homogeneous distribution of ^{26}Al in the inner solar system.[19-21,26] Villeneuve *et al.*[19] also show there are a number of peaks in the distribution of ^{26}Al–^{26}Mg ages for chondrules from a single parent body, and this suggests that chondrule formation occurred in pulses over a period of several million years. The assumption of a solar nebula with a homogeneous initial magnesium isotope composition is incorrect, as there are numerous examples of low-level (100 ppm) heterogeneity of initial magnesium isotopic composition (*i.e.* variable $(\delta^{26}Mg^*)_0$) in high-precision studies.[22,27] However, this does not prevent the use of the ^{26}Al–^{26}Mg geochronometer, but simply restricts its use to the comparison of materials that are shown to have the same $(\delta^{26}Mg^*)_0$.

Magnesium isotopic compositions are typically expressed as isotopic ratios relative to a reference isotopic composition. The δ^{26}Mg notation gives the relative deviation, in parts per thousand (‰), of the ^{26}Mg/^{24}Mg ratio from the reference isotopic composition, typically an international reference material.

Figure 3.2 shows examples of ^{26}Al–^{26}Mg isochrons. If the data of Figure 3.2 is interpreted in a simple model where time zero equates to the homogenisation of the nebular source with $(^{26}Al/^{27}Al)_i = 5.23 \times 10^{-5}$ and $(\delta^{26}Mg^*)_i = -0.038$‰, the three isochrons show that crystallisation of the 3535–1 CAI, the 3665A CAI, and the Semarkona chondrule occurred at three different points in time during the evolution of the solar system, at ~ 0.05 Ma, 0.29 Ma, and 2.5 Ma, respectively, after time zero.

In summary, the ^{26}Al–^{26}Mg cosmochronometer, which is based on decay of an extinct isotope, allows us to precisely define relative time differences of approximately 10 000 years, 4.567 billion years ago, for materials with identical initial magnesium isotopic compositions, and this allows us to follow and understand the earliest history of the solar system in great detail.

3.4.3 ^{182}Hf–^{182}W Isotope System

3.4.3.1 *Introduction*

The short-lived ^{182}Hf–^{182}W isotopic system is characterised by a short half-life of 8.9 Ma,[28] which is well suited for timescales of the formation and earliest evolution of solar system objects. The main interest in the Hf–W system was initially related to its potential for dating core formation.[29-31] Hafnium and tungsten are both refractory elements and are presumed to occur in chondritic relative abundances in bulk planetary objects. Tungsten is a moderately siderophile element (MSE) and is largely, but not completely, drained into the metal during core segregation.[32,33] In contrast, hafnium is strictly lithophile and is completely retained in silicate portions of planetary bodies. Planetary cores, which contain virtually no hafnium and hence no radiogenic ^{182}W, have frozen the tungsten isotopic composition at the time of their formation. The tungsten isotopic compositions of magmatic iron meteorites[34-38] are close to the initial isotopic composition of chondrites

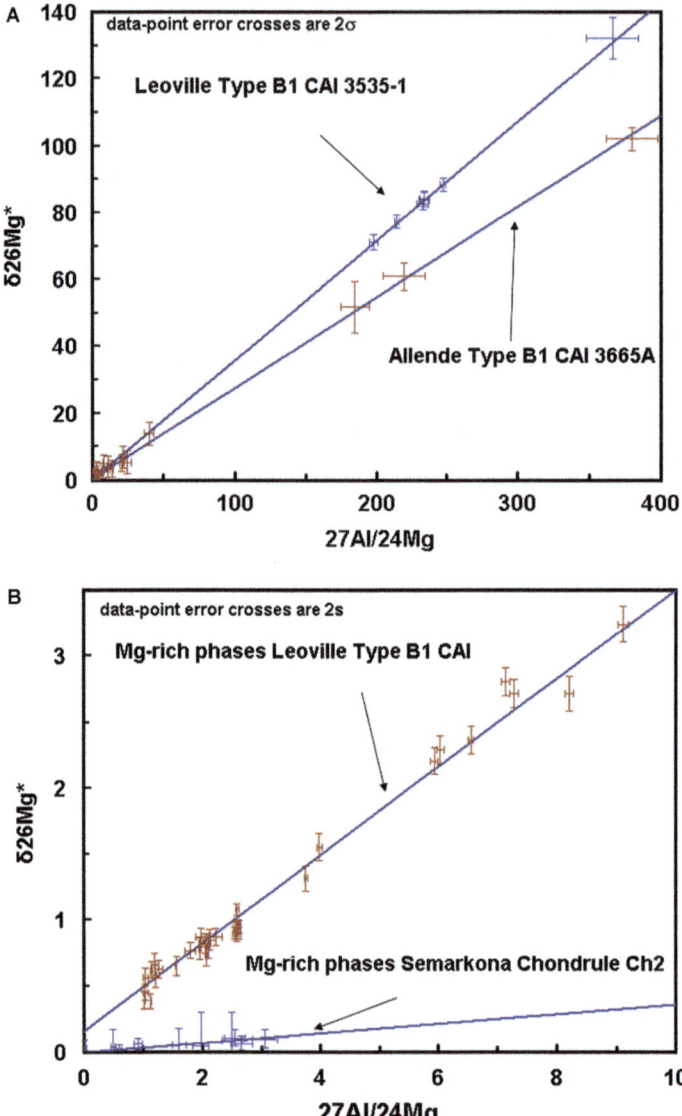

Figure 3.2 The ^{26}Al–^{26}Mg isochrones: (A) ^{26}Al–^{26}Mg isochrons measured in two type B1 CAI, 3535-1 from Leoville CV3 carbonaceous chondrite (Kita *et al.*[21]) and 3665A from the Allende CV3 carbonaceous chondrite (Kennedy *et al.*[27]). The 3535-1 isochron has an initial $(^{26}\text{Al}/^{27}\text{Al})_0$ of $5.002 \pm 0.65 \times 10^{-5}$ and a $(\delta^{26}\text{Mg}^*)_0$ of $0.06 \pm 0.08‰$, while 3665A has $(^{26}\text{Al}/^{27}\text{Al})_0$ of $3.95 \pm 4.2 \times 10^{-5}$ and a $(\delta^{26}\text{Mg}^*)_0$ of $0.44 \pm 0.38‰$. (B) ^{26}Al–^{26}Mg isochrons for magnesium-rich phases from the Leoville CV3 carbonaceous chondrite (Kita *et al.*[21]) and a chondrule (Ch2) from the Semarkona LL3 ordinary chondrite (Villeneuve[19]). The CAI isochron has an initial $(^{26}\text{Al}/^{27}\text{Al})_0$ of $5.002 \pm 0.65 \times 10^{-5}$ and a $(\delta^{26}\text{Mg}^*)_0$ of $0.06 \pm 0.08‰$, while the chondrule isochron has a $(^{26}\text{Al}/^{27}\text{Al})_0$ of $5.071 \pm 1.8 \times 10^{-6}$ and a $(\delta^{26}\text{Mg}^*)_0$ of $-0.003 \pm 0.01‰$.

4568 Ga ago, as determined from hafnium–tungsten internal isochrons of CAI,[39,40] indicating core formation and hence accretion of some differentiated planetesimals less than 1 Ma after solar system formation.

3.4.3.2 Core Formation

Model ages of core formation can also be inferred using the tungsten isotope composition of silicate portions of planetary bodies. Terrestrial rocks[41–43] and Martian meteorites[39,44,45] have radiogenic $^{182}W/^{184}W$ ratios relative to bulk chondrites (Figure 3.3). The higher than chondritic $^{182}W/^{184}W$ ratios of the terrestrial and Martian mantles have been interpreted to reflect core segregation and generation of a highly suprachondritic hafnium/tungsten ratio in the silicate portion of terrestrial planets, while ^{182}Hf was still extant. Combined with current best estimates for the hafnium/tungsten ratio of bulk Martian[46] and terrestrial mantles,[47] these yield hafnium–tungsten model ages

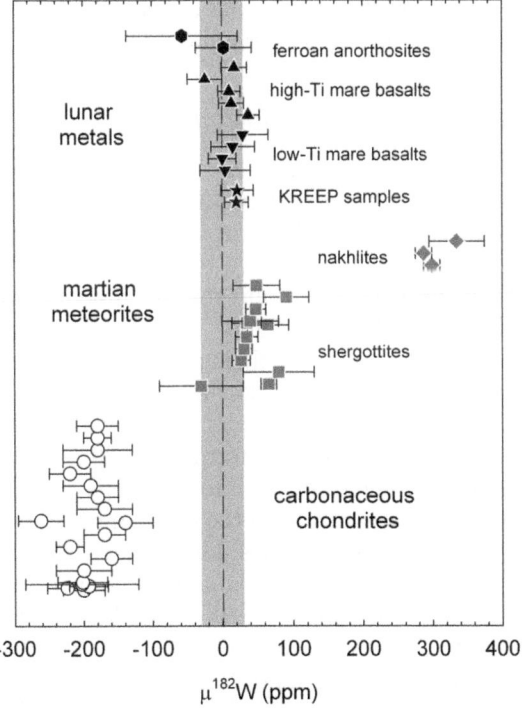

Figure 3.3 $^{182}W/^{184}W$ ratios of carbonaceaous chondrites (open circles), Martian meteorites (grey symbols) and lunar samples (closed symbols), expressed in $\mu^{182}W$ notation, which corresponds to parts per million deviation of the isotpic ratio of the sample from terrestrial standards. Data are from Yin *et al.*,[43] Kleine *et al.*,[39] Foley *et al.*,[45] and Touboul *et al.*[48,49] Grey box indicates the 2δ external reproducibility classically reached during these studies (~ ±30 ppm).

of ~ 2 and ~ 30 Ma for core formation on Mars and Earth, respectively, assuming these processes occur as single instantaneous events. This assumption might be justified for Mars, which likely corresponds to a planetary stranded embryo, but it is difficult to reconcile with the occurrence of giant impacts which characterised subsequent accretion of larger planets such as Earth. The calculated hafnium–tungsten model age of the Earth's core is very sensitive to the timing of these large collisions and to the degree of equilibration of impactor cores with the proto-Earth mantle and range from ~ 30 Ma to more than 100 Ma after solar system formation. Therefore, hafnium–tungsten systematics of the Earth's mantle does not provide a unique time constraint for the formation of the Earth. Nevertheless, the identical $^{182}W/^{184}W$ ratios of the lunar and terrestrial mantles[48,49] suggest that the giant Moon-forming impact and the termination of Earth's core formation occurred after extinction of ^{182}Hf (*i.e.* more than ~ 50 Ma after CAIs). Furthermore, this isotopic similarity requires that the Moon consists predominantly of terrestrial material, consistent with the most recent dynamical simulations of the Moon-forming giant impact.[50,51] Alternatively, tungsten isotopes might have been equilibrated between the proto-lunar disk and Earth's magma ocean, as proposed for oxygen isotopes by Pahlevan and Stevenson.[52]

3.4.3.3   Early Mantle Differentiation

In addition to core formation, the ^{182}Hf–^{182}W system can also be used for exploring early mantle differentiation. Tungsten is one of the most incompatible elements, similar to thorium and uranium. Hafnium is also an incompatible element; however, relative to tungsten, it is preferentially incorporated in minerals such as clinopyroxene, ilmenite, garnet, and magnesium perovskite.[53,54] Crystallisation of these phases (*e.g.* during solidification of a magma ocean) or their partial melting (*e.g.* during crustal extraction) can result in the production of mantle reservoirs with variable hafnium/tungsten ratios. If these reservoirs formed within the lifetime of ^{182}Hf (less than ~ 60 Ma after solar system formation), differences in their *Hf/W* hafnium/tungsten ratio would translate into variations of their $^{182}W/^{184}W$ ratio. ^{182}W heterogeneities are present within the Martian mantle, as revealed by the difference of tungsten isotopic composition between shergottites and nakhlites (Figure 3.3).[39,44,45] This indicates a solidification of a Martian magma ocean during the effective lifetime of ^{182}Hf, most likely ~ 40 Ma after the start of the solar system. In contrast to Mars, there is no $^{182}W/^{184}W$ variation in the lunar mantle,[48,49] indicating magma ocean solidification later than ~ 60 Ma, consistent with the ~ 150 Ma timescale derived from ^{146}Sm–^{142}Nd chronometry[55] and the young age of 4.360 Ga for the ferroan anorthosite 60025.[56]

The possibility of identifying early magmatic processes within the Earth's mantle has led to intensive tungsten isotopic analysis of diverse terrestrial materials. Until recently these studies have reported no resolved ^{182}W isotopic heterogeneities among terrestrial rocks,[57–59] but recent improvements in precision of tungsten isotope measurements[60] have enabled evidence for

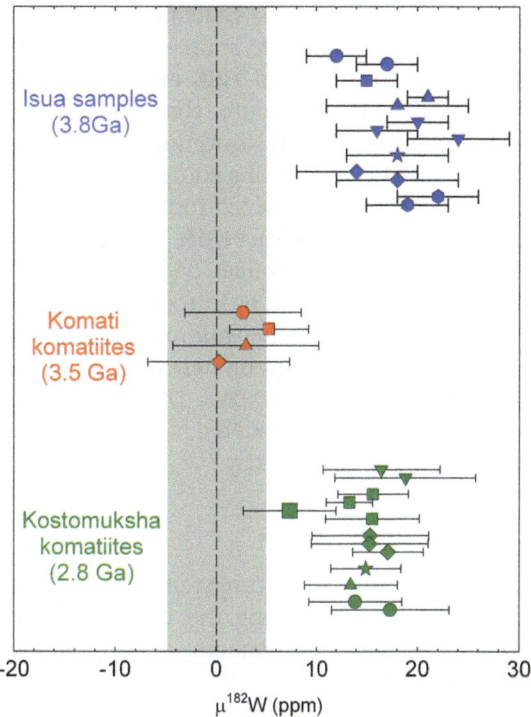

Figure 3.4 High-precision μ^{182}W measurements of Archean terrestrial samples. The blue symbols correspond to the 3.8 Ga Isua samples (Willbold *et al.*[61]) the red symbols to the 3.5 Ga komatiites from Komati and the green symbols to the 2.8 Ga komatiites from Kostomuksha (Touboul *et al.*[63]). Each symbol represents a different sample and identical symbols indicate replicated analysis of the same sample. Note the 2δ external reproducibility (grey box) of the new analytical techniques ($\sim \pm 5$ ppm), which is about ~ 6 times better than earlier measurements (see Figure 3.3).

the existence of subtle, previously undetectable ^{182}W heterogeneities (Figure 3.4). Willbold *et al.*[61] first reported 13 ± 4 ppm 182W enrichments in rocks 3.8 Ga old from the Isua greenstone belt (Greenland). These positive ^{182}W anomalies were interpreted as reflecting the tungsten isotope composition of the Earth's mantle before the addition of extraterrestrial materials *via* late accretion. This process, which corresponds to the addition of $\sim 0.5\%$ of the total mass of the mantle to Earth by continued accretion of materials with bulk chondritic compositions, subsequent to cessation of core formation, is commonly invoked to account for the relatively high absolute and chondritic relative abundances of highly siderophile elements (HSE) in the modern mantle (*e.g.* Walker[62]). Touboul *et al.*[63] also found 15 ± 4 ppm ^{182}W excesses in 2.8 Ga komatiites from the Kostomuksha supracrustal belt (Baltic Shield) but no ^{182}W anomaly in 3.5 Ga komatiites from Komati (South Africa). Compared to highly metamorphosed rocks from Isua, komatiites are better preserved and have well-behaved HSE systematics, making it possible to access HSE abundances of their mantle source. The mantle source of Kostomuksha komatiites has 80% of total HSE content relative to the modern mantle, inconsistent with a mantle source that has been preserved from late accretion. Positive [182]W anomalies in Kostomuksha komatiites instead reflect the presence, in their mantle source, of a component derived from a reservoir with a high hafnium/tungsten ratio which formed less than 60 Ma after solar system formation. This represents either a deep mantle region that underwent metal-silicate equilibration, or a product of large-scale magmatic differentiation of the mantle. The preservation, in rocks dated at 2.8 Ga, of isotopic anomalies produced during the first 60 million years of solar system history implies that Earth was never totally molten early in its history, even subsequent to the Moon-forming giant impact. During this time convection was much more sluggish than so far considered,[64] allowing isotopic signatures to be preserved, perhaps to the present day.[65]

3.4.3.4   Meteorite Dating

The fractionation of hafnium from tungsten is not restricted to planetary differentiation but also occurs among the constitutive minerals of many meteorites and some of their components, making it possible to obtain a precise internal isochron. Owing to its high closure temperatures (*e.g.* between 800 and 900 °C in H chondrites), the hafnium–tungsten system provides time constraints on the earliest high-temperature evolution of planetesimals. Combined with thermal modelling, ages determined using hafnium–tungsten internal isochrons indicate that the H chondrite parent body accreted $\sim$2.5 Ma[66] and the acapulcoite–lodranite parent body $\sim$1.5–2 Ma after solar system formation.[67] Coupled with hafnium–tungsten ages of $\sim$1 Ma for differentiation of the parent bodies of magmatic iron meteorites, these chronological constraints reveal an inverse correlation between accretion age of asteroids and peak temperature in their interiors, suggesting that the different thermal histories of most meteorite parent bodies primarily reflect variations in their initial abundance of heat-producing radioisotopes (*e.g.* [26]Al), which is determined by their accretion time. Nevertheless, impact-related processes are an important heat source for the subsequent evolution of asteroids. For instance, the ages obtained from hafnium–tungsten internal isochrons of winonites[68,69] and from tungsten isotopic composition of eucrite metals[70] postdate solar system formation by $\sim$15 Ma and $\sim$20 Ma respectively and probably reflect shock-related secondary perturbation evidenced by the fact that almost all eucrites are breccias.

3.4.4   [146]Sm–[142]Nd Isotope System

3.4.4.1   Introduction

The [146]Sm–[142]Nd isotope system consists of two highly refractory elements belonging to the rare earth group. The half-life of [146]Sm has been recently revised and the new value of 68 Ma[71] replaces the older value of 103 Ma. Deviations of the ^{142}Nd/^{144}Nd ratio are always very small and the analytical precision was a critical issue in the development of the systematics. Samarium and neodymium have very close partition coefficients and, like most of the short-lived systems, the initial abundance of the parent element was very low. This is why the ^{146}Sm–^{142}Nd systematics have mainly been investigated during the past decade, with the development of the new generation of thermal ionisation mass spectrometers allowing a precision of ± 5 ppm for the ^{142}Nd/^{144}Nd ratio.

3.4.4.2 Early Silicate Differentiation and Mantle-Crust Evolution

This systematics offers a unique opportunity to investigate early silicate differentiation. The samarium/neodymium ratio generally is not changed during condensation processes and since these elements have no affinity for the metal phase, core formation processes would not modify this ratio either. However, because neodymium is slightly more compatible than samarium during magmatic processes (partial melting and fractional crystallisation), they can be used for tracing silicate differentiation and mantle-crust evolution. Measurement of the ^{142}Nd isotope can then provide powerful indications on the Hadean period ($\sim$ the ^{146}Sm lifetime), a geological period still largely unknown because of the scarcity of very old rock records.

^{142}Nd data is now available for a large range of terrestrial samples including rocks collected in the oldest terranes as well as more recent samples collected in a large range of tectonic settings.[72–85] Small deviations of the ^{142}Nd/^{144}Nd ratios, lower than 20 ppm relative to the neodymium terrestrial standard, have been measured in samples collected in two areas: the Itsaq complex in south-west Greenland[73,78–82] and the Nuvvuagittuq Greenstone Belt in northern Quebec, Canada (Figure 3.5).[83–85] The age of the differentiation event is calculated by coupling data obtained on both ^{146}Sm–^{142}Nd and ^{147}Sm–^{143}Nd systematics. Because resolving the age equations requires different assumptions (composition of the Earth at the time of silicate differentiation, single differentiation event) the age of the differentiation event cannot be precisely defined but it must have occurred in the first 150 Ma of the solar system's history. Both excesses and deficits in ^{142}Nd have been measured, suggesting that early-formed chemically differentiated reservoirs would have been formed during the crystallisation of a terrestrial magma ocean following the giant impact event forming the Moon. So far no ^{142}Nd anomalies have been detected in samples younger than 3.4 Ga, suggesting that no early-formed samarium/neodymium chemical heterogeneities have been preserved in the convective mantle during more than 1 Ga.

Boyet and Carlson[86] showed that chondrites, which are assumed to be the building blocks of rocky planets, have an average ^{142}Nd/^{144}Nd ratio $\sim$ 20 ppm lower than terrestrial samples. They explain this isotope signature

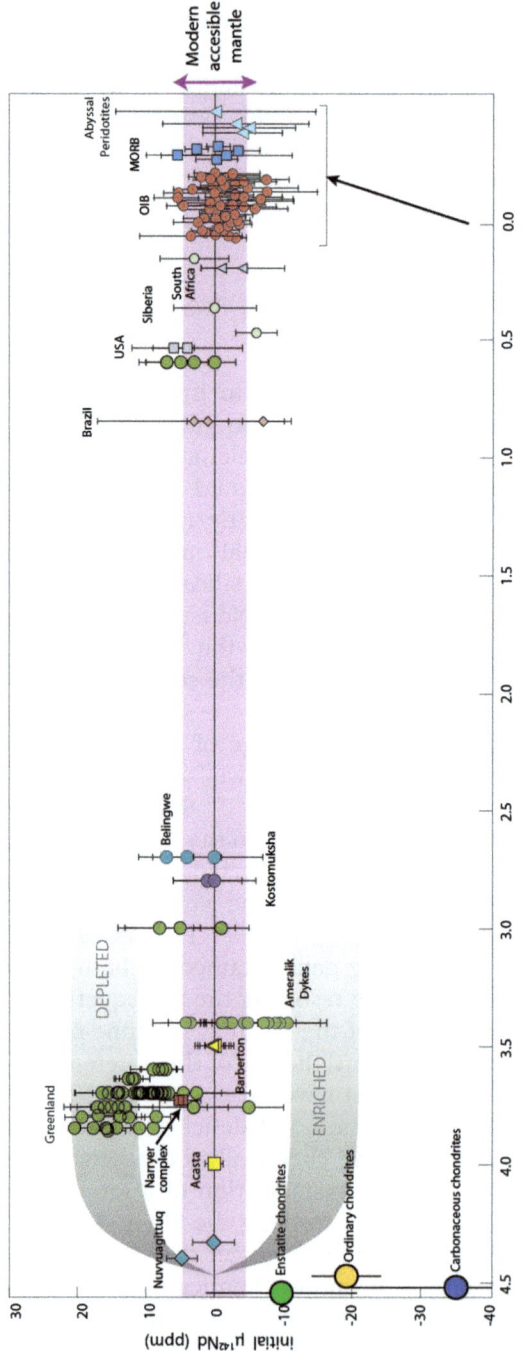

Figure 3.5 Compilation of all published initial $^{142}Nd/^{144}Nd$ ratios for terrestrial samples. Figure modified from Rizo *et al.*,[82] the data sources are cited in the text. Means for chondrites are from Gannoun *et al.*[90] Measured $^{142}Nd/^{144}Nd$ ratios for chondrites are corrected to a common $^{147}Sm/^{144}Nd$ ratio of 0.196. Ages of chondrite formation are not respected in this figure. The grey shaded area in the middle represents the external analytical error of ± 5 ppm. Positive ^{142}Nd anomalies are formed in a mantle source depleted in incompatible elements (high samarium/neodymium ratio) and formed during the ^{146}Sm lifetime. Contrarily, a source enriched in incompatible elements will develop through time deficit in ^{142}Nd.

by a global differentiation of the silicate Earth within 30 Ma of the Earth's formation. The ^{142}Nd signature of terrestrial samples requires that a complementary reservoir has been stored and isolated from the convective mantle since its formation. Alternative scenarios propose that the chondritic model is not appropriate for the bulk silicate Earth.[87] More detailed investigations on chondritic material have revealed that different groups of chondrites have different ^{142}Nd isotope compositions.[88–90] These results confirm that the estimated ^{142}Nd/^{144}Nd ratio of the bulk Earth changes according to the group of meteorites considered. In conclusion, recent studies based on the measurement of stable neodymium and samarium isotope ratios highlight the difficulty of interpreting the ^{142}Nd difference between terrestrial samples and chondrites. The incomplete mixing of material formed in different stellar environments creates significant ^{142}Nd variation and the initial isotope composition of the Earth remains difficult to assess.

3.5   Radioactivity of Long-Lived Chronometers

The development of modern physics would not have been possible without reliable chronometers. In earth sciences, the geological timescale is different from that of humans, requiring appropriate chronometers for us to be able to measure geological history. The development of geochronology was a great breakthrough in being able to quantify geological time. It is now recognised that the Precambrian represents about 90% of Earth's history, although it was limited to a couple of pages in old geological textbooks. The development of geochronology could be the topic of a separate book and so here the discussion is limited to general principles of radiogenic isotopes.

3.5.1   Principles of Isotope Dating

The main principle is as follows: if one can estimate the quantity of radiogenic daughter and radioactive parent isotopes, and knowing the decay constant λ, it is possible to calculate the age of the system since its formation, based on the Curie–Rutherford–Soddy equation introduced above (equation 3.7). This equation may be transformed as following:

$$t = \frac{1}{\lambda} \ln \left\{ \left[\frac{D(t) - D(0)}{P(t)} \right] + 1 \right\} \qquad (3.17)$$

The only problem is knowing the D(0) component of the system, *i.e.* the initial quantity of daughter isotope present in the system before the 'radioactive clock' started and produced the purely radiogenic part of the same isotope. To overcome this difficulty, a concrete example of the rubidium–strontium isotope system is presented in the following section.

It is important to note, however, that the system should have remained closed since its formation, with nothing going in or out. This can be applied to any geological object, *i.e.* rocks, minerals, the Earth, the solar system, *etc.*

3.5.1.1 Rubidium–Strontium System

Rubidium has two natural isotopes: ^{85}Rb and ^{87}Rb (27.8%). The latter is radioactive and produces the ^{87}Sr radiogenic isotope by beta-minus decay: ^{87}Rb $\rightarrow$ ^{87}Sr $+ \beta^-$.

Strontium consists of four stable isotopes in the following proportions: ^{84}Sr(0.56%), ^{87}Sr(7.0%), ^{86}Sr(9.9%), and ^{88}Sr(82.6%). Only ^{87}Sr is radiogenic, produced by radioactive decay of rubidium. According to equation 3.7 the present-day ^{87}Sr concentrations obtained are:

$$^{87}\text{Sr} = {}^{87}\text{Sr}_0 + {}^{87}\text{Rb}(e^{\lambda t} - 1) \tag{3.18}$$

Given that the mass spectrometer allows measurement of the isotopic ratios (see section 3.2), it would be convenient to transform equation 3.8 by dividing each member by the non-radiogenic strontium isotope ^{86}Sr, assuming that $^{86}\text{Sr}_0 = {}^{86}\text{Sr}_t$ (where 0 = initial time and t = present time). Doing this, one obtains:

$$\left(\frac{^{87}\text{Sr}}{^{86}\text{Sr}}\right)_t = \left(\frac{^{87}\text{Sr}}{^{86}\text{Sr}}\right)_0 + \left(\frac{^{87}\text{Rb}}{^{86}\text{Sr}}\right)_t (e^{\lambda t} - 1) \tag{3.19}$$

Note that this equation contains an elemental ratio (*i.e.* rubidium–strontium) which often limits the precision; this is not the case for the Pb–Pb system. For a fixed value of t, (3.19) is an equation of a straight line (with variables ^{87}Sr/^{86}Sr and ^{87}Rb/^{86}Sr) which can be transformed as:

$$Y = Y_0 + X(e^{\lambda t} - 1) \tag{3.20}$$

The decay constant λ for rubidium is known (Table 3.3), thus the $(e^{\lambda t} - 1)$ term, which represents the slope of the isochron, is a constant value because it is similar for all samples of the same age. The initial ratio at the time of system formation, $(^{87}\text{Sr}/^{86}\text{Sr})_0$, may be estimated from this equation.

To obtain the straight line, at least two samples with different ^{87}Rb/^{86}Sr ratios should be analysed. Taking into account the analytical uncertainty, more than two samples should be analysed to obtain better precision, verify the accuracy of the results, and check if the system has stayed closed during the decay. It could be either different minerals from the same rock with different ^{87}Rb/^{86}Sr ratios or samples of different lithology from the same geological formation in order to obtain quite a large spectrum of ^{87}Rb/^{86}Sr ratios. The rubidium–strontium system has been one of the major chronometers used to date the returned Moon samples.[91] In recent years, however, the use of the rubidium–strontium dating technique has been restricted mostly to the study of planetary material and it is now seldom used for terrestrial systems because of its sensitivity to open behaviour during alteration processes.

3.5.2 Uranium–Lead Isotope System

3.5.2.1 Decay Constants

One reason the uranium-lead method is so powerful is that unlike other radioisotopic systems, multiple isotopes of the same element are measured and act on a cross-check on each other, and have decay schemes that are well-constrained. ^{238}U decays to stable daughter ^{206}Pb with a half-life of $4.4683 \pm 0.0024 \times 10^9$ years; ^{235}U decays to stable daughter ^{207}Pb with a half-life of $7.0381 \pm 0.0046 \times 10^8$ years, measured by direct decay counting by Schoene *et al.*[92] The decay constant $(\ln_2/t_{1/2})$ of ^{235}U has been refined empirically by Mattinson[93] to $0.98671 \pm 0.00012 \times 10^9$ years^{-1}, within the uncertainties of Schoene *et al.*[92] but improving their $\sim 0.1\%$ precision. Though imperfect (see *e.g.* Schoene *et al.*[94]), these fundamental parameters for the calculation of ages based on isotopic ratios are better established than other radioisotopic systems. ^{208}Pb results from the decay of ^{232}Th, and ^{204}Pb is not radiogenic, and so has a roughly constant abundance on Earth (except for extraterrestrial sources), presently about 1.4%.

3.5.2.2 Analytical Techniques

Different analytical techniques are used in uranium–lead geochronology. The most accurate and precise of these is isotope dilution thermal ionisation mass spectrometry (ID-TIMS). First applied to zircon ($ZrSiO_4$) geochronology in the 1950s,[95–97] ID-TIMS refers to the measurement on a thermal ionisation mass spectrometer of a sample that has been spiked with an isotopic tracer. This method is insensitive to mass spectrometric sensitivities and abundance, and because of the tracer it is not reliant for calibration on external standards, which must themselves be well characterised isotopically. Precision is therefore limited primarily by the instrument, and can be better than 0.1%. When applied to zircon, and using appropriate laboratory techniques, ID-TIMS uranium–lead dating of zircon can reliably produce ages virtually throughout geological time at $\sim \pm 1$ Ma or better, making it uniquely powerful among radioisotopic geochronological methods. Compston *et al.*[98] applied secondary ionisation mass spectrometry (SIMS) to geochronology with the development of the sensitive high-mass resolution ion microprobe (SHRIMP), which uses a beam of oxygen ions to ablate a small pit ~ 20 μm in diameter and a few microns deep, the contents of which are then analysed by a sector mass spectrometer. Precision is limited to about 1%, due to the small sample being measured and the reliance upon external calibration standards. Laser-ablation inductively coupled mass spectrometry (LA-ICPMS) has been more recently employed as an *in situ* method for uranium–lead geochronology. This has similar limitations to the SHRIMP though poorer sensitivity hampers precision; however, it is a fast and relatively inexpensive method that is has seen an increase in use recently. The *in situ* methods are most useful when there is a need for rapid analysis of many grains (*e.g.* detrital mineral analysis), when the crystal must be preserved for other geochemical analyses, or when cost is a factor. TIMS produces more precise ages, but completely destroys the sample, requires clean laboratory work and laborious chemical procedures, and is therefore more expensive.

An important advance has been the realisation, illustrated by Mundil *et al.*,[99] that ages obtained from multi-crystal zircon fractions (where several mineral grains are dissolved together and analysed in bulk) may not be accurate. Multi-crystal fractions have been used in order to improve the strength of the radiogenic lead signal and therefore improve analytical precision. However, it is apparent that analysing a population of grains that may include both older, xenocrystic or inherited grains and apparently younger grains that have suffered varying degrees of lead loss will yield an average that may be concordant, but inaccurate. Therefore, single-crystal analyses have become the standard in obtaining ages that may be assessed realistically and with statistical rigour, though even single grains can include inherited cores, or may themselves produce an age that represents an average of a protracted growth history, *e.g.* in a magma chamber (residence time).

3.5.2.3   Dating Applications of the Uranium–Lead Technique

In recent decades, advances in technology and in our understanding of mineralogical and petrological processes have led to the uranium–lead system becoming the benchmark in geochronology, at present the method capable of obtaining the most precise and accurate age information on most minerals. Common accessory phases that are used include titanate, monazite, apatite, xenotime, perovskite, and others, though these minerals contain initial lead which must be corrected for.

Minerals such as zircon ($ZrSiO_4$) and baddeleyite (ZrO_2) incorporate uranium and, to a lesser extent, thorium into their crystal structure during growth (typically 10–1000 ppm uranium), and exclude most other elements, including lead. This means that virtually all lead in these minerals is radiogenic, and corrections for initial lead are negligible. For this reason these minerals, of which zircon is more common, are preferred for use in uranium–lead geochronological analysis.

Closure temperatures for the uranium–lead isotopic system are typically high (~ 900 °C for zircon,[100] and at least that for baddeleyite; see *e.g.* Davidson and Van Breemen[101]) and it is important to bear this in mind when comparing age results with those obtained from other isotopic systems. The uranium–lead age is generally the high-temperature constraint on the formation of a geological unit, and can also reflect hydrothermal events, detrital or inherited histories, or various stages in the development of magma bodies, as reflected in the growth of mineral phases.

Generally speaking, the uranium–lead dating method can be applied to any mineral that contains uranium. For example, CAIs in meteorites have been used to date the oldest solids formed in our solar system,[102] and authigenic xenotime has been used to date sedimentary rock formation.[103] Hydrothermal events, such as those related to ore-forming processes, have been dated using hydrothermal zircon or rutile.[104] Ore deposits have been dated using columbite,[105] or monazite[106] grown *in situ* during the formation of these deposits. Uranium–lead dating has been applied to Quaternary materials,[107,108] and to the oldest continental Earth materials known, detrital zircons from the Jack Hills of Western Australia.[109]

The high precision possible with this method enables the resolution of magmatic events that were not previously distinguishable from one another, and when used in parallel with lower-temperature geochronometers such as the ^{40}Ar–^{39}Ar method allow the creation of thermal histories of magmatic systems.[110] It allows the establishment of connections between magmatic events and biotic crises such as mass extinctions,[111] and of a geological timescale based upon absolute ages determined with high precision.[112]

As long as uranium-bearing minerals can be extracted from a rock, whether it is baddeleyite from a basalt, zircon from a granite, or perovskite from the mantle, the uranium–lead method can be used to obtain a reliable age for the material. Care must be taken in interpreting the data in order to ensure that a system closed to lead and uranium is maintained, or at least that open-system behaviour is understood.

3.5.3 Potassium–Argon and $^{40}Ar/^{39}Ar$ Dating Techniques

3.5.3.1 Potassium–Argon Technique

The potassium–argon technique is based on the natural radioactive decay of ^{40}K to $^{40}Ar^*$ and $^{40}Ca^*$ with a combined half-life of ~ 1.25 Ga. Because potassium is in solid form and argon is a gas, potassium and argon must be measured using different machines on different aliquots. Potassium is mostly measured using either flame photometry or isotopic dilution and argon is measured using a noble gas spectrometer. This method, however, encounters immediate limitations as (1) it requires relatively large aliquots to homogenise the sample composition, (2) there is no way of determining if the sample has remained closes to external perturbations (*e.g.* alteration, recrystallisation) or (3) has incorporated initial ^{40}Ar and ^{36}Ar with a composition different from the assumed $^{40}Ar/^{36}Ar$ atmospheric ratio (~ 299).[113] This technique has been progressively replaced by the ^{40}Ar–^{39}Ar approach over the years, except in a few cases such as the dating of clay where the ^{40}Ar–^{39}Ar technique has technical limitations and the potassium–argon technique is still better suited.

3.5.3.2 ^{40}Ar–^{39}Ar Technique

The methodology behind the ^{40}Ar–^{39}Ar technique represents a drastic improvement over the potassium–argon method. In essence, the idea is to transform ^{39}K into ^{39}Ar by bombarding the sample with neutrons in the core of a nuclear reactor. As a result, the measurement of the ^{40}Ar/^{39}Ar ratio makes it possible to calculate an age by comparison with a standard of a known age. For the interested reader, the technique is described in greater detail in the book by McDougall and Harrison.[114] The direct advantage of this approach is the ability to measure an age using a single grain of a given sample, thus circumventing sample heterogeneity. More importantly, step-heating *via* laser or furnace allows obtaining a detailed distribution of the ^{40}Ar/^{39}Ar ratio in the sample and testing if the sample has been perturbed after its formation, or if a robust age can be extracted from the set of analyses. When the ^{40}Ar/^{39}Ar ratio is homogenous throughout most of the age spectrum, it yields a so-called plateau. A plateau age is a succession of steps including at least 70% of the total ^{39}Ar gas released and giving the same ^{40}Ar/^{39}Ar ratio, and hence the same age. Such a plateau can only be achieved through stringent sample preparation where samples must be devoid of alteration products and must be homogenous. Otherwise, structured profiles usually indicate processes such as thermally activated diffusion loss of ^{40}Ar*, alteration, excess ^{40}Ar*, mixture of phases with different time–temperature histories, and ^{39}Ar or ^{37}Ar recoil processes.[114] The step-heating approach, combined with stringent statistical χ^2-tests (MSWD and *P*-values where $P \geq 0.05$) to define a plateau (*e.g.* Baksi[115]) represents a robust and objective mean of testing the validity of an age. The problem of inherited or excess argon (*i.e.* parentless ^{40}Ar present in the system at the time of formation) can be in most case easily addressed using the inverse isochron approach, which, contrary to the plateau age spectrum approach, does not assume the composition of the trapped ^{40}Ar* but measured it directly *via* the ^{40}Ar/^{36}Ar ratio. Hence, for rocks containing abundant amount of trapped ^{40}Ar, the inverse isochron approach is usually preferred. Note that the same statistical tests employed for the plateau method are applicable to the inverse isochron approach.

3.5.3.3   Decay Constants

Until recently, ^{40}Ar–^{39}Ar (and potassium–argon) ages were calculated using the decay constants (^{40}Ar* and ^{40}Ca*) recommended by Steiger and Jäger.[116] Over the years, argon geochronologists noted that the ^{40}Ar–^{39}Ar ages seem to be systematically younger by about 1% compared to the ^{238}U–^{206}Pb ages (*e.g.*[117,118]). As such, a recent study by Renne *et al.*[119] using ^{40}Ar–^{39}Ar and ^{238}U–^{206}Pb data from rocks that have recorded the same event, and direct activity measurement of the ^{40}K total decay constant, led to the calculation of a new set of ^{40}K decay constants. Due to the joint use of both the ^{238}U decay constant and the measured ^{40}K activity, the set of decay constants proposed by Renne *et al.*[119] make it possible to calculate ^{40}Ar–^{39}Ar ages that, for potassium-rich minerals, are equally (or in some cases more) precise than uranium–lead ages measured with TIMS, at least for the Phanerozoic period. For example, Renne *et al.*[119] showed that the age of the Permo–Triassic boundary measured using sanidine ^{40}Ar–^{39}Ar step-heating analyses is 252.3 ± 0.4 Ma by comparison with zircon ^{238}U–^{206}Pb ages of 253.4 ± 0.4 Ma (both ages include all source of errors and are reported at 2σ).

3.5.3.4 Dating Applications of the ^{40}Ar–^{39}Ar Technique

In essence, ^{40}Ar–^{39}Ar dating can be applied to all minerals and rocks that contain a measurable amount of potassium oxide (*e.g.* sanidine, micas), from few thousand years old (*e.g.* Renne[120]) to as old as the solar system itself (4.56 Ga; cf. review by Bogard[121]), but in theory there is no real older limit. Even minerals with very low abundance of potassium oxide, such as plagioclase or pyroxene, can be analysed and yield accurate age data, albeit generally less precise than for potassium-rich minerals. Broadly speaking, potassium oxide concentration generally correlates with analytical precision. Because most crustal minerals contain potassium, ^{40}Ar–^{39}Ar dating is applicable to a large range of geological processes. A good example of the range of possible applications is given by various papers in the book by Jourdan *et al.*[122] In the following sections we provide a few examples of the most common applications.

Planetary Sciences. ^{40}Ar–^{39}Ar dating can be applied to the study of planets, moons, and asteroids. In particular, it is relevant to the study of the bombardment, volcanic, and time–temperature histories of any planetary body.[121,123,124] Due to its sensitivity to moderate thermal perturbations, the ^{40}Ar–^{39}Ar technique is particularly well suited to the dating of major hyper-velocity impact events. For example a well-defined age of 4507 ± 20 Ma has been recorded in plagioclase from a eucrite meteorite thought to come from the asteroid 4 Vesta.[125] High-precision dating of lunar (impact) melt breccia has shown that most ^{40}Ar–^{39}Ar ages cluster around 3.8–4.0 Ga,[126] possibly suggesting an increase in the bombardment of the Moon at that time.[127] Impacts have also been recorded in spherules formed during rapid quenching of silicate melt during impact. ^{40}Ar–^{39}Ar analyses of hundreds of spherules recovered from three distinct Apollo landing sites have yielded ages ranging from 4.4 Ga to the present, indicating the constant bombardment of the Moon since its formation.[128-130] Furthermore, these studies have shown an abundance of spherules with ages older than 3–2.5 Ga and a sharp spike of spherules with ages younger than about 400 Ma.

If the system has not been perturbed by impact events, then the ^{40}Ar–^{39}Ar technique can be applied to date extraterrestrial basaltic eruptions (see *e.g.* Cohen *et al.*[131]) and the cooling rate of asteroids.[132] For example, Jourdan[124] reinterpreted the ^{40}Ar–^{39}Ar data obtained on lunar basalts from the Mare Crisium by Cohen *et al.*[131] to illustrate a sequence of four eruptions with well-defined ages between about 3.22 and 3.33 Ga.

Volcanic Eruptions. The entire range of volcanic products can be dated with the ^{40}Ar–^{39}Ar technique, from basalts to rhyolitic tuffs. Young rocks between a few thousand years old and up to a few tens of millions of years can be dated on groundmass with excellent accuracy and good precision,[133,134] and the technique has been applied to problems such as the study of an active volcanic island like Hawaii.[135] For rocks older than ~30 Ma, it is preferable to isolate datable minerals due to the increasingly undesirable effect of pervasive alteration, and where alteration products are hard if not impossible to remove from groundmass.[136,137] For basalts older than 30 Ma, the target mineral by default is plagioclase. Because plagioclase contains only around 0.05% of potassium oxide, it generally provides ages that are not the of utmost precision. On the other hand, plagioclase data can be very accurate, as the presence of sericite alteration can be easily identified on these phases using a step-heating approach.[137] Plagioclase ^{40}Ar–^{39}Ar dating is widely used in the investigation of the duration timing of large igneous provinces (*e.g.* [138–141]). ^{40}Ar–^{39}Ar dating of basalts has largely contributed to the idea that large igneous provinces have been responsible for mass extinctions throughout Earth's history, due to their identical timing.[142]

For more evolved rocks, minerals such as hornblende, biotite, or sanidine yield very precise and robust ages (Figure 3.6). For example, the dating of sanidine has been recently used to obtain an age of 73.9 ± 0.6 ka (2σ) for the infamous Toba supereruption.[143] Such precision can be very important when studying processes such as the risks associated with a given volcanic region. ^{40}Ar–^{39}Ar dating of volcanic products, in particular pyroclastic tuffs rich in sanidine, is important to calibrate the geological timescale and crucial in pinpointing the age of archaeological artefacts by dating

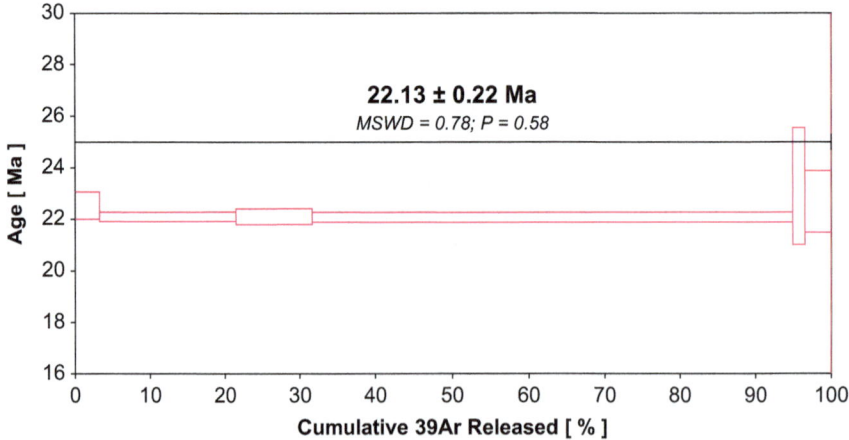

Figure 3.6   Example of ^{40}Ar–^{39}Ar age spectra where the individual steps are plotted as a function of the ^{39}Ar released. The ^{40}Ar–^{39}Ar plateau age has been obtained on sanidine from a syenite volcanic rock (Africa). Unpublished data from the archive of F. Jourdan. Note, the uncertainty is reported at 2σ and includes all sources of errors. The mean square weighted deviation (MSWD) and P values indicated that this age is statistically valid.

bracketing volcanic layers (*e.g.* the age of the *Australopithecus* Lucy at 3.18 ± 0.01 Ma.[144]

Thermochronology Applied to Tectonic Processes. One of the major applications of the ^{40}Ar–^{39}Ar technique relies on the various closure temperatures of different minerals. The closure temperature is an approximate temperature below which a given mineral will start to retain most of its radiogenic ^{40}Ar*. ^{40}Ar–^{39}Ar thermochronology studies can be applied to a range of ubiquist minerals with distinct closure temperatures ranging from up to ~600 °C (hornblende) down to ~150 °C (feldspar).[114] Each mineral will give the age at which it reached its closure temperature. In other words, the higher the closure temperature, the older the apparent age, except in the case where all the apparent ages are indistinguishable, indicating extremely fast cooling. This behaviour is usually observed when a rock cools down to ambient temperature after having undergone deformation or slow cooling. This property is abundantly exploited during the study of orogeny (mountain formation), when rocks are usually strongly deformed through metamorphism process and follow complex time–temperature histories. Such histories can be recovered by analysing a suite of minerals with different closure temperatures, and/or by inverting results from age spectra that display clear diffusion profiles (*i.e.* progressive ^{40}Ar* loss). Major orogens have been studied with this approach such as the relatively recent Himalaya and Alps chains,[145–147] or much older orogens such as the Paleoproterozoic Mt Wood Inlier in Australia[148] or the East African Orogen in Mozambique.[149]

Mineral Deposits. ^{40}Ar–^{39}Ar is a powerful tool for dating mineral deposits. In essence, several key potassium-rich minerals tend to form during the hydrothermal process associated with many ore deposits (*e.g.* copper, gold, molybdenum). For example, intrusive-related mineral and metal deposits are associated with porphyry bodies and epithermal processes. Hydrothermal alteration, consisting of one or several temporally distinct pulses, can be dated using the ^{40}Ar–^{39}Ar technique using minerals such as sericite (alteration of plagioclase),[137] adularia,[150,151] and alunite, all of which have been successfully used in several ^{40}Ar–^{39}Ar studies related to the timing of ore deposits. For example, gold mineralisation in the Liba goldfield (China) has been dated at 216.4 ± 0.7 Ma using hydrothermal mica minerals.[152] Although the minerals mentioned above tend to yield well-defined plateau ages, the real difficulty resides in the interpretation of those ages. Indeed, one needs to keep in mind that what is really dated is the formation of these minerals during hydrothermal activity. If the hydrothermal activity is brief, on the order of a few hundred thousand years, then the age of the confidently dated mineral gives the age of the ore formation. On the other hand, if the hydrothermal activity is episodic and lasts over a few million years, then the ^{40}Ar–^{39}Ar age can provide a range of possible ages within this timeframe. This range includes the age of the end of the hydrothermal activity, which can be a few million years younger than the age of the ore formation itself. Such a phenomenon has been observed, for example, for the Xihuashan tungsten deposit in China.[153]

3.5.4 Lutetium–Hafnium Isotope System

3.5.4.1 Introduction

The behaviour of the whole-rock samarium–neodymium isotopic system in magmatic rocks closely parallels that of the whole-rock lutetium–hafnium system,[154,155] and the latter is strongly controlled by the mineral zircon. Once it crystallises from a magma, zircon is stable up to high metamorphic grades, and because of its very low lutetium/hafnium ratio, can preserve the $^{176}Hf/^{177}Hf$ ratio of the host magma at the time of crystallisation. Thus the link between the age and the isotopic composition of the magma is more likely to be preserved than in whole-rock isotopic systems. High values of $^{176}Hf/^{177}Hf$ indicate a 'juvenile' or mantle-derived origin for the magma, while low values imply the reworking of older crustal material. Recent studies of oxygen and hafnium isotopes in single zircon grains[156] have shown that a high $^{176}Hf/^{177}Hf$ ratio is linked with mantle-like oxygen-isotope values. These studies confirm the utility of hafnium isotope data in distinguishing between juvenile and recycled components in crustal rocks. However, more detailed work is needed to understand the oxygen–hafnium systematics of crustal recycling processes, and to use these data to better constrain the range of hafnium isotope compositions that represents juvenile sources.

3.5.4.2 Interrelation with Other Isotopic Systems

Zircon is a common accessory mineral in many igneous rocks, and has become one of the most important tools in the earth sciences; thousands of grains are analysed yearly in laboratories worldwide, mainly for dating. The isotopic systematics of oxygen and the uranium–thorium–lead and lutetium–hafnium systems in zircon provide a range of information on the age and sources of melts and have been extensively used for studying processes of crustal and mantle evolution.[157–159] *In situ* microanalysis of the uranium–lead system allows us to measure the age of an individual zircon grain. The isotopic composition of hafnium and oxygen can tell us whether a magmatic rock was generated in the mantle, or by remelting of older crust.

3.5.4.3 Application for Crustal–Mantle Evolution

The lutetium–hafnium system in zircon therefore is a powerful tool for studying processes of crustal and mantle evolution,[155,160–165] and new technology makes if feasible to do this on a global scale. To understand the growth of a crustal block, we need to determine the sources of magmatic material over its history: whether it was juvenile (derived from the convecting mantle or from recently formed crust) or recycled (remelting of older crust), or possible a mix of those two. Early models of crustal evolution made extensive use of the samarium–neodymium isotopic system; age data (typically from uranium–lead dating of zircons) were combined with neodymium isotope analysis of the host rocks to define the source material. This approach is powerful, but it has several drawbacks: it is time consuming, so data for any area are necessarily concentrated on relatively few selected samples, and the whole-rock samarium–neodymium system is prone to disturbance by metamorphism or weathering.

A number of recent studies have provided extensive datasets on a global scale; one of these, by Belousova,[166] used 23 000 analyses. This study took a new approach to the modelling of hafnium isotope data, and came to two major conclusions, each of which broke with conventional thinking:

- At least 70% of all exposed continental crust was originally generated in Archean time (>2.5 Ga; Figures 3.7 and 3.8) and has been reworked to different degrees during younger tectonic events.

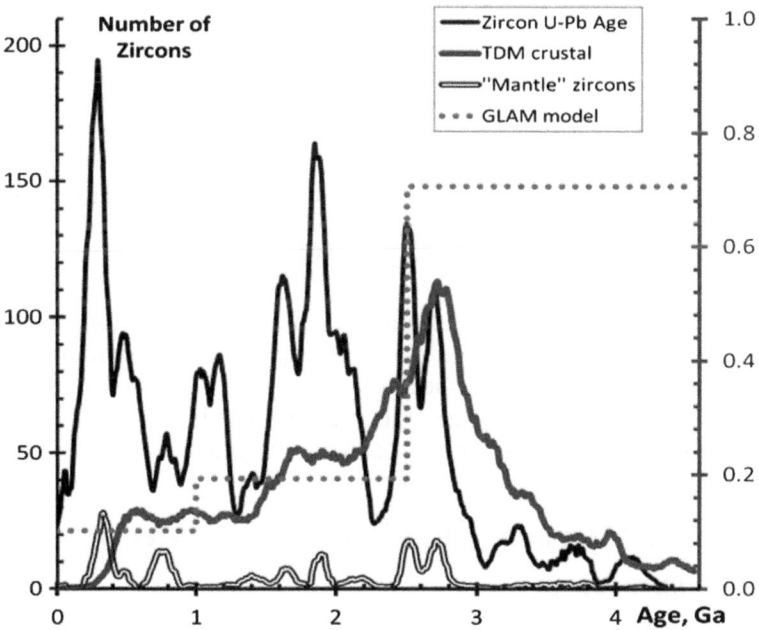

Figure 3.7 Relative probability curves (left scale) of uranium-lead zircon ages), T_{DM}^{C} model ages and number of zircons with juvenile hafnium isotope compositions (defined as grains with $\varepsilon_{Hf} > 0.75 \times \varepsilon_{Hf}^{DM}$). Proportions of the continental lithosphere formed during three major time intervals derived from GLAM mapping (Belousova *et al.*[166]) are shown by the dashed line (right axis). A compilation of zircon uranium-lead and hafnium isotope data illustrates the important difference between 'age peaks' and episodes of 'crustal generation'.

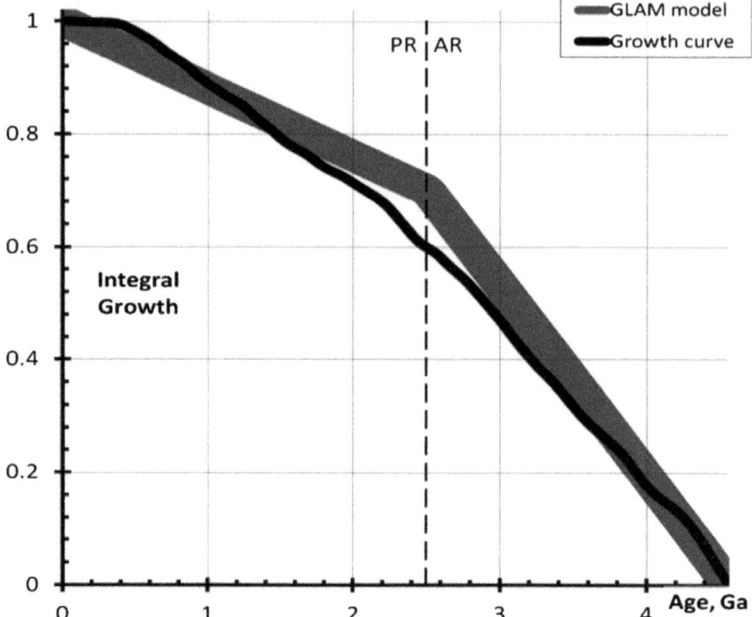

Figure 3.8 The integrated crustal growth curve (black line) using combined uranium–lead and hafnium isotope data, compared with the GLAM model (Belousova *et al.*[166]).

- The prominent peaks in the age distribution of zircon-bearing rocks do not reflect periods of net crustal growth, but simply record magmatic activity that largely reworked pre-existing crust.

This study treated the zircon database on a worldwide scale. It was followed by one[167] that used the paleogeographic distribution of different hafnium isotope signatures in specific time intervals, to identify and analyse two fundamentally different tectonic systems that have organised Earth's crustal evolution for at least the last 500 million years.

3.5.5 ^{190}Pt–^{186}Os and ^{187}Re–^{187}Os Isotope Systems

3.5.5.1 Introduction

In the past decade, as a result of tremendous analytical advances, the rhenium–osmium and platinum–osmium isotopic systems have seen an explosion of applications in geochemistry and cosmochemistry. ^{187}Re decays to ^{187}Os with a half-life of 41.6 Ga, whereas ^{190}Pt decays to ^{186}Os with a half-life of 469 Ga (Table 3.3). Both systems are unique in that rhenium, platinum, and osmium belong to the HSE, *i.e.* they strongly partition into metal or sulfide liquid relative to silicate melt. As a result, these elements were nearly quantitatively extracted from silicate parts of planetary bodies into the core during planetary differentiation. Additionally, during mantle melting, osmium is compatible with the melting residue, whereas platinum is weakly incompatible and rhenium moderately incompatible during low to moderate degrees of partial melting, such as those involved in generation of basalts. These elements, therefore, can provide unique insights into certain processes, to which the more traditional, lithophile element-based isotopic systems, are not sensitive. These processes include planetary accretion and differentiation, crust–mantle and core differentiation, and core–mantle exchange. Excellent reviews of both isotope systems are available.[168–172] Below, we provide a brief overview of the most common applications.

3.5.5.2   Rhenium–Osmium and Platinum-Osmium Chronometry

The uniqueness of the ^{190}Pt–^{186}Os and ^{187}Re–^{187}Os isotope systems as chronometers stems from their applications to materials that cannot be dated using other isotopic systems. For example, they are the only systems able to date directly metal phases, as has been largely illustrated by the dating of iron meteorites, which, in turn, was the first application of this chronometer. A number of common applications include the following.

Dating Volcanic Sequences Consisting of Mafic-Ultramafic Lavas, Such as Komatiites and Picrites. During differentiation of komatiite and picrite lava flows, osmium is usually compatible with the fractionating mineral assemblage, while both platinum and rhenium are incompatible. This contrasting behaviour results in large variations in platinum/osmium and rhenium/osmium ratios between different portions of individual lava flows, which, in turn, permits generation of isochrons. The latter can be used to obtain precise chronological information, assess closed-system behaviour of HSEs during post-magmatic processes, and precisely determine the initial osmium isotopic compositions and time-integrated rhenium/osmium and platinum/osmium ratios in the sources of the lavas. The correlations between magnesium oxide and HSE, combined with the osmium isotopic data, can also be used to precisely calculate the absolute HSE contents of the sources of komatiite lavas.[173–175]

Dating Ore Genesis. The rhenium–osmium and platinum-osmium isotopic systems can be directly used to study ore-forming minerals, rather than host silicates. One of the first applications of the rhenium–osmium isotopic system was to date molybdenites, which have very high rhenium and very low osmium concentrations, so that nearly all osmium in the sample is radiogenic. Recent developments in this technique allow accurate and precise chronological information to be obtained.[176] Another important ore mineral is chromite, which, unlike molybdenite, has a very low rhenium/osmium ratio. As a result, this mineral, which is highly resistant to alteration, freezes the initial osmium isotopic composition of the magma it crystallises from, thus providing both chronological and petrogenetic information.[177–179] Finally, sulfides are also widely used to obtain both chronological and petrogenetic information.[172,180–182]

Dating Diamonds. Diamonds often contain sulfide inclusions with high (up to ppm level) osmium concentrations. This allows the rhenium–osmium system to be applied to obtain both chronological and petrogenetic information on single crystals. A recent study of sulfide inclusions in diamonds from the Panda kimberlite in Canada[183] made it possible to trace the source of the diamonds to the fluids rising from the subducting slab and argues for the onset of plate tectonics as early as 3.52 Ga.

Dating Organic-Rich Sediments. One of the peculiar features of the HSE is their affinity for organic-rich matter within sedimentary sequences, such as black shales, coal, and oil-hosting sands. Provided the rhenium–osmium system in such organic-rich materials has remained closed since the time of deposition, a rhenium–osmium study may provide valuable chronological information, as well as help obtain the osmium isotopic composition of the source of organic-rich matter.[184,185]

Dating Oil. The ^{187}Re–^{187}Os isotope system has been successfully applied to petroleum systems in recent years, both as a process tracer using ^{187}Os/^{188}Os and as a geochronometer using rhenium–osmium decay relationships. The first detailed report of rhenium and osmium elemental and isotopic systematics in natural crude oils showed that a very high proportion of both elements resides in the asphaltene fraction, and that the ^{187}Os/^{188}Os ratio of modern crude oils is highly variable, and broadly correlates with source rock age.[186] These observations allow for application of osmium isotopes as a tracer of source rock identity[187] and of evaluation of fluid interaction following reservoir charging.[188] Importantly, the concentration of osmium in the asphaltene fraction makes the osmium isotope system an ideal tracer for heavy biodegraded petroleum from which traditional tracing information, such as biomarkers, may have been lost. Further experimental work studied the rhenium–osmium elemental and isotopic relationships between source rock and oil by hydrous pyrolysis.[189] Several studies have shown that ^{187}Re–^{187}Os isotope system is applicable to petroleum systems as a geochronometer. Bitumen from the Polaris carbonate-hosted lead–zinc deposit recorded a rhenium–osmium age of ~374 Ma.[190] Bitumen here is the biodegraded product of migrated oil, interpreted to have charged the carbonate reservoir together with aqueous fluids that produced lead–zinc mineralisation dated at ~366 Ma by the rubidium-strontium method. Mineralisation at Polaris also yields a ~367 Ma palaeomagnetic age, such that three methods independently yield a Late Devonian age fluid flow at Polaris. Subsequently, a rhenium–osmium age of ~110 Ma was determined for the giant Western Canadian Oil Sands.[191] The UK Atlantic Margin yielded an age of ~68 Ma for oil generation.[187] This rhenium–osmium age compares well with other independent methods of oil generation dating such as ^{40}Ar–^{39}Ar and basin thermal modelling. Although application of the rhenium–osmium isotope system to petroleum systems is still in its infancy, the potential is large for both process tracing and geochronology.

3.5.5.3 Rhenium–Osmium and Platinum-Osmium Isotope Systems as Tracers of Planetary Processes

Chronometry of Mantle Differentiation Processes. Unlike traditional isotopic systems involving lithophile trace elements, which concentrate in the melt to nearly the same degree and are present in low abundances in mantle residues, osmium is concentrated in mantle residues, whereas rhenium (and platinum) are preferentially incorporated in the melt. This feature of the rhenium–osmium isotopic system has proved uniquely suited to study the history of melt extraction from the mantle.[192] These studies provide evidence for strong correlation between the age of continental crust and the timing of melt depletion and formation of subcontinental, residual mantle. In addition, melts formed during remelting of the latter are important sources of HSE mineralisation.

Crust–Mantle and Core–Mantle Interaction. Oceanic crust, consisting mainly of basalt and sediment, has high rhenium/osmium ratio. Upon returning back into the mantle at the subduction zones, it transforms into eclogite and forms isolated domains that can be stored deep in the mantle for up to 1.8 Ga.[193] Given enough time, these domains will develop highly radiogenic $^{187}Os/^{188}Os$. The presence of aged oceanic crust in the sources of plume magmas, such as ocean island basalts (OIB) and komatiites, can be detected using the rhenium–osmium isotopic system in combination with lithophile trace element and isotopic studies.[194–196] Due to a lower incompatibility of platinum during mantle melting and a much greater half-life of ^{190}Pt, the deviation of $^{186}Os/^{188}Os$ in the aged oceanic crust from the mantle reference will be much smaller compared to $^{187}Os/^{188}Os$. However, some plume-derived lavas show coupled radiogenic initial $^{186}Os/^{188}Os$ and $^{187}Os/^{188}Os$ ratios.[197] In order to explain this correlation, several hypotheses have been put forward, such as core–mantle interaction[198] or long-term preservation of early mantle differentiation products.[63] Establishing the nature of osmium isotopic anomalies in the mantle is crucial for our understanding of the Earth's geodynamics, its thermal history, and the timing of planetary differentiation.

HSE in Extraterrestrial Materials: Timing of Accretion in the Solar System. Core formation must have nearly quantitatively removed HSE from the silicate portions into cores of planetary bodies, leaving highly-fractionated HSE patterns. However, HSE abundances in the terrestrial mantle are two to three orders of magnitude higher than could be expected from low-pressure metal-silicate partitioning experiments,[199] and the osmium isotopic composition is roughly chondritic,[200] implying long-term chondritic platinum/osmium and rhenium/osmium ratios in the terrestrial mantle. In order to explain this controversy, a late accretion hypothesis was put forward, whereby 0.5–1% of chondritic materials were added to the mantle after the last major equilibration between the core and mantle.[201]

Unlike the dynamic Earth, the surface of the Moon still holds evidence of late accretion in the form of impact craters and impact melt breccias. Studies of osmium isotopic composition and absolute and relative HSE abundances in the impact melts allows reconstruction of the HSE composition of the impactors,[202–204] which has far-reaching implications for understanding the early history of the solar system and the origin of life.

3.6 Environmental Isotopic Tracers

The ability to trace the sources of dust and pollution deposited in a particular region allows an understanding of the atmospheric transport paths that are involved. Often the elemental concentrations and mineralogy of deposited aerosols have been used for this purpose. However, these systems can be altered by environmental processes, *e.g.* changes in accumulation in a region will impact on measured elemental concentrations. Therefore the use of isotopic systematics to characterise dust and pollution has emerged as an important method. Grousset and Biscaye[205] reviewed the use of strontium, neodymium, and lead isotopic systematics to trace sources of natural aerosols and anthropogenic pollution. Lead isotopes have been shown to be particularly successful in the tracing of atmospheric impurities preserved in glacial snow and ice.

3.6.1 Lead Isotope Systems

The concept of using lead isotopes as an environmental tracer is based on lead isotopic variations that exist in the Earth's crust as a result of the radioactive decay of uranium and thorium. These variations are dependent on the geological history and age of crustal regions. For lead ores, the isotopic composition has remained the same since mineral formation when the ore was geochemically separated from surrounding source rocks containing uranium and thorium. There exist variations in isotopic composition of lead ore bodies as a result of the mixing of primordial lead and radiogenic lead in the surrounding source region, before the separation of the mineral from the surrounding system. The subsequent isotopic composition of an ore body and its geological age is then dependent on the time of this isolation.[4]

The relationship between the isotopic composition of terrestrial lead and its geological age can be illustrated using a radiogenic growth curve (Figure 3.9). The ages shown in the figure (in years before present, BP) indicate the isotopic composition of the ore body at the time of isolation from

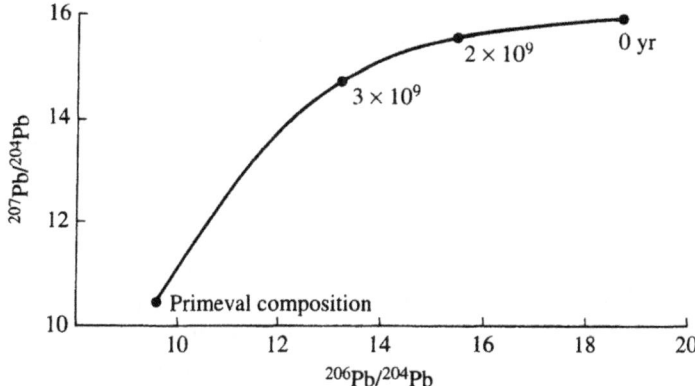

Figure 3.9 Common lead growth curve; ages are given in years before present (BP) Adapted from Figure 19.2 in Faure,[4] Copyright © 1977, 1986, by John Wiley & Sons, Inc.

the source region.[4] Older lead ore bodies, for which the geochemical separation occurred billions of years ago, have an 'unradiogenic' isotopic composition with $^{206}Pb/^{207}Pb$ ratios as low as ~1. Younger lead ore bodies, for which geochemical separation occurred in more recent geological history, and crustal rocks, are characterised by a more 'radiogenic' isotopic composition, with $^{206}Pb/^{207}Pb$ ratios of ~1.20 and reaching as high as ~1.30.

Regional variations in lead isotopic composition that exist between different lead ore bodies, and between lead ore bodies and crustal material, allow different sources of lead to be distinguished. Assuming that the mineralogy and isotopic composition of atmospheric lead reflect the geographical location of its source area and is not substantially modified during transport,[206,207] the isotopic composition of lead measured in environmental samples, such as glacial snow and ice, can be used to trace the source region of the dust and therefore infer atmospheric transport patterns. These techniques have been successfully used to trace the sources of anthropogenic and natural impurities preserved in glacial snow and ice in regions such as Antarctica,[208–210] Greenland,[211] the Himalayas,[212] Bolivia,[213] and Europe,[214] and also the pollution of the oceans.[215]

3.7 Progress in Analytical Techniques

3.7.1 TIMS

Innovations in automation[216,217] and the release of commercially available TIMS instruments have brought about improvements in electronics, magnet and detector geometry, and ion optics. The advent of multiple collector geometries (multi-collection) together with the advent of low-noise high-stability electronics enabled simultaneous ion beam detection, thereby mitigating the effects of ion beam intensity fluctuations. As a result, analytical precision improved from the 0.1–0.01% range in the 1960s and 1970s (automation), over 0.1–0.01‰ in the 1980s and 1990s (magnet geometry, multi-collection), and into the 0.001‰ range since the early 2000s (ion optics, software-controlled movable Faraday cups). The advent of detectors with a uniform response, software-controlled amplifier–detector associations, and correction of the electronic drift of the amplifier resistors has pushed long-term reproducibility to the statistical counting limits.

Means to meet the analytical demands of the mass spectrometry community for lower levels of detection, without sacrificing precision and accuracy, include dynamic ranges spanning several orders of magnitude, ion-counting multi-collection systems, and minimisation of peak tailing effects. The ions per atom used is increasing steadily over time, currently reaching percent levels for most elements in MC-ICPMS instruments. Precision will ultimately be limited by the number of atoms present in the amount of sample processed for analysis and the relevant concentration of the element under investigation. Reducing the procedural blank accordingly is not a minor problem in this context.

3.7.2   New Generation of Noble Gas Mass Spectrometers

Perhaps the most important recent technical development of the ^{40}Ar–^{39}Ar technique is the availability of a new generation of noble gas mass spectrometers. Unlike the older instruments (*e.g.* MAP 215–50, VG5400) which require peak-hopping measurement of the five argon isotopes (^{40}Ar to ^{36}Ar) on a single multiplier, the new machines allow multi-collection in real time, by using a combination of Faraday cups and ion-counting multipliers. Whereas multi-collection using only multipliers is not yet beneficial to the ^{40}Ar–^{39}Ar technique because of the uncertainties associated with the multiplier calibrations, multi-collection using five Faraday cups or four Faraday cups and one ion-counting detector have shown that great improvements can be made to the analytical precision. In some case, an improvement in precision of one order of magnitude compared to the old machines has been measured, similar to the evolution of TIMS instruments. However, the most striking improvement to date concerns young volcanic rocks with large amount of trapped ^{40}Ar, where a precision improvement of ~20-fold has been measured. Phillips[218] compared an age of 300 ± 60 ka obtained using a VG5400 on a low-potassium tholeiitic volcanic rock (a type of rock particularly notorious for giving low-precision ^{40}Ar–^{39}Ar ages), with an age of 300 ± 2 ka when the same rock is ^{40}Ar–^{39}Ar dated using an ARGUS VI machine. At the time of writing, no data obtained with this new generation of machine have yet been published, particularly those using a combination of Faraday cups and multipliers, but it is obvious that these instruments will revolutionise the field of ^{40}Ar–^{39}Ar dating and when combined with the revision of the ^{40}K decay constants,[119] will provide data with a precision difficult to rival, particularly for the Phanerozoic timescale.

References

1. A. P. Dickin, *Radiogenic Isotope Geology*, Cambridge University Press, New York, 2nd edn, 2005.
2. C. J. Allègre, *Isotope Geology*, Cambridge University Press, New York, 2008.
3. G. Faure and T. M. Mensing, *Isotopes: Principles and Applications*, 2004.
4. G. Faure, in *Principles of Isotope Geology*, John Wiley & Sons, New York, 2nd edn, 1986, pp. 309–340.
5. J. J. Thomson, *Proc. R. Soc.*, 1913, **A 89**, 1.
6. A. J. Dempster, *Phys. Rev.*, 1918, **11**, 316.
7. F. W. Aston, *Philos. Mag. Series 6*, 1919, **38**, 707.
8. K. T. Bainbridge and E. B. Jordan, *Phys. Rev.*, 1936, **50**, 282.
9. A. O. Nier, *Rev. Sci. Instrum.*, 1940, **11**, 212.
10. J. R. de Laeter, *Applications of Inorganic Mass Spectrometry,* John Wiley & Sons, New York, 2001, pp. 3–18.
11. W. G. Cross, *Rev. Sci. Instrum.*, 1951, **22**, 717.
12. W. A. Chupka and M. G. Inghram, *J. Chem. Phys.*, 1953, **21**, 1313.
13. A. E. Cameron, D. H. Smith and R. L. Walker, *Anal. Chem.*, 1969, **41**, 525.
14. G. F. Kessinger and J. E. Delmore, *Int. J. Mass Spectrom.*, 2002, **213**, 63.
15. T. Miyazaki, T. Shibata, M. Yoshikawa, T. Sakamoto, K. Iijima and Y. Tatsumi, in *Frontier Research on Earth Evolution*, vol. **2**, Institute for Frontier Research on Earth Evolution, Yokosuka, Japan, 2007, p. 1.
16. H. E. Duckworth, R. C. Barber and V. S. Venkatasubramanian, in *Mass Spectroscopy*, Cambridge University Press, Cambridge, 2nd edn, 1986, p. 42.
17. G. W. Lugmair and R. W. Carlson, *Proc. 9th Lunar and Planet. Sci. Conf.*, Houston, TX, 1978, pp. 689–704.
18. B. Jacobsen, Q.-Z. Yin, F. Moynier, Y. Amelin, A. N. Krot, K. Nagashima, I. D. Hutcheon and H. Palme, *Earth Planet. Sci. Lett.*, 2008, **272**, 353.
19. J. Villeneuve, M. Chaussidon and G. Libourel, *Science*, 2009, **325**, 985.
20. K. K. Larsen, A. Trinquier, C. Paton, M. Schiller, D. Wielandt, M. A. Ivanova, J. N. Connelly, A. Nordlund, A. N. Krot and M. Bizzarro, *Astrophys. J., Lett.*, 2011, **735**, L37.
21. N. T. Kita, T. Ushikubo, K. B. Knight, A. Ruslan, A. Mendybaev, A. M. Davis, F. M. Richter and J. H. Fournelle, *Geochim. Cosmochim. Acta*, 2012, **86**, 37.
22. G. J. Wasserburg, J. Wimpenny and Q.-Z. Yin, *Meteorit. Planet. Sci.*, 2012, **47**, 1980.
23. C. M. Gray and W. Compston, *Nature*, 1974, **251**, 495.
24. T. Lee, D. A. Papanastassiou and G. J. Wasserburg, *Astrophys. J., Lett.*, 1977, **211**, L107.
25. E. Kurahashi, N. T. Kita, H. Nagahara and Y. Morishita, *Geochim. Cosmochim. Acta*, 2008, 72, 3865.
26. A. N. Krot, Y. Amelin, P. Bland, F. J. Ciesla, J. Connelly, A. M. Davis, G. R. Huss, I. D. Hutcheon, K. Makide, K. Nagashima, L. E. Nyquist,

S. S. Russell, E. R. D. Scott, K. Thrane, H. Yurimoto and Q.-Z. Yin, *Geochim. Cosmochim. Acta*, 2009, **73**, 4963.

27. A. K. Kennedy, J. R. Beckett and I. D. Hutcheon, *Geochim. Cosmochim. Acta*, 1997, **61**, 1541.
28. C. Vockenhuber, F. Oberli, M. Bichler, I. Ahmad, G. Quitté, M. Meier, A. N. Halliday, D. C. Lee, W. Kutschera, P. Steier, R. J. Gehrke and R. G. Helmer, *Phys. Rev. Lett.*, 2004, **93**, 172501.
29. D. C. Lee and A. N. Halliday, *Nature*, 1995, **378**, 771.
30. C. L. Harper and S. B. Jacobsen, *Geochim. Cosmochim. Acta*, 1996, **60**, 1131.
31. M. F. Horan, M. I. Smoliar and R. J. Walker, *Geochim. Cosmochim. Acta*, 1998, **62**, 545.
32. M. J. Walter and Y. Thibault, *Science*, 1995, **270**, 1186.
33. M. J. Walter, H. E. Newsom, W. Ertel and A. Holzheid, in *Origin of the Earth and Moon*, ed. R. M. Canup and K. Righter, Lunar and Planetary Institute, Houston, TX, 2000.
34. T. Kleine, K. Mezger, H. Palme, E. Scherer and C. Münker, *Geochim. Cosmochim. Acta*, 2005, **69**, 5805.
35. A. Q. Markowski, G., A. N. Halliday and T. Kleine, *Earth Planet. Sci. Lett.*, 2006, **242**, 1.
36. A. Schérsten, T. Elliott, C. Hawkesworth, S. S. Russell and J. Masarik, *Earth Planet. Sci. Lett.*, 2006, **241**, 530.
37. L. P. Qin, N. Dauphas, M. Wadhwa, J. Masarik and P. E. Janney, *Earth Planet. Sci. Lett.*, 2008, **273**, 94.
38. T. S. Kruijer, P. Sprung, T. Kleine, I. Leya, C. Burkhardt and R. Wieler, *Geochim. Cosmochim. Acta*, 2012, **99**, 287.
39. T. Kleine, K. Mezger, C. Münker, H. Palme and A. Bischoff, *Geochim. Cosmochim. Acta*, 2004, **68**, 2935.
40. C. Burkhardt, T. Kleine, H. Palme, B. Bourdon, J. Zipfel, J. Friedrich and D. Ebel, *Geochim. Cosmochim. Acta*, 2008, 72, 6177.
41. T. Kleine, C. Münker, K. Mezger and H. Palme, *Nature*, 2002, **418**, 952.
42. R. Schoenburg, B. S. Kamber, K. D. Collerson and O. Eugster, *Geochim. Cosmochim. Acta*, 2002, **66**, 3151.
43. Q.-Z. Yin, S. B. Jacobsen, K. Yamashita, J. Blichert-Toft, P. Télouk and F. Albarède, *Nature*, 2002, **418**, 949.
44. Lee, A. N. Halliday, G. A. Snyder and L. A. Taylor, *Science*, 1997, **278**, 1098.
45. C. N. Foley, M. Wadhwa, L. E. Borg, P. E. Janney, R. Hines and T. L. Grove, *Geochim. Cosmochim. Acta*, 2005, **69**, 4557.
46. N. Dauphas and A. Pourmand, *Nature*, 2011, **473**, 489.
47. R. J. Arevalo and W. F. McDonough, *Earth Planet. Sci. Lett.*, 2008, **272**, 656.
48. M. Touboul, T. Kleine, B. Bourdon, H. Palme and R. Wieler, *Nature*, 2007, **450**, 1206.
49. M. Touboul, T. Kleine, B. Bourdon, H. Palme and R. Wieler, *Icarus*, 2009, **199**, 245.

50. R. M. Canup, *Science*, 2012, **338**, 1052.
51. M. Ćuk and S. T. Stewart, *Science*, 2012, **338**, 1047.
52. K. Pahlevan and D. J. Stevenson, *Earth Planet. Sci. Lett.*, 2007, **262**, 438–449.
53. K. Righter and C. K. Shearer, *Geochim. Cosmochim. Acta*, 2003, **67**, 2497.
54. A. Corgne, C. Liebske, B. J. Wood, D. C. Rubie and D. J. Frost, *Geochim. Cosmochim. Acta*, 2005, **69**, 485.
55. A. D. Brandon, T. J. Lapen, V. Debaille, B. L. Beard, K. Rankenburg and C. Neal, *Geochim. Cosmochim. Acta*, 2009, **73**, 6421.
56. L. E. Borg, J. N. Connelly, M. Boyet and R. W. Carlson, *Nature*, 2011, **477**, 70.
57. A. Scherstén, T. Elliott, C. Hawkesworth and M. Norman, *Nature*, 2004, **427**, 234.
58. F. Moynier, Q.-Z. Yin, K. Irisawa, M. Boyet, B. Jacobsen and M. T. Rosing, *Proc. Natl Acad. Sci. U. S. A.*, 2010, **107**, 10810–10814.
59. T. Iizuka, S. Nakai, Y. V. Sahoo, A. Takamasa, T. Hirata and S. Maruyama, *Earth Planet. Sci. Lett.*, 2010, **291**, 189.
60. M. Touboul and R. J. Walker, *Int. J. Mass Spectrom.*, 2012, **309**, 109.
61. M. Willbold, T. Elliott and S. Moorbath, *Nature*, 2011, **477**, 195.
62. R. J. Walker, *Chem. Erde Geochem*, 2009, **69**, 101.
63. M. Touboul, I. S. Puchtel and R. J. Walker, *Science*, 2012, **335**, 1065.
64. V. S. Solomatov and C. C. Reece, *J. Geophys. Res.: Solid Earth*, 2008, **113**, B07408.
65. M. G. Jackson, R. W. Carlson, M. D. Kurz, P. M. Kempton, D. Francis and J. Bluszajn, *Nature*, 2008, **466**, 853.
66. T. Kleine, M. Touboul, J. A. Van Orman, B. Bourdon, C. Maden, K. Mezger and A. Halliday, *Earth Planet. Sci. Lett.*, 2008, **270**, 106.
67. M. Touboul, T. Kleine, B. Bourdon, J. A. Van Orman, C. Maden and J. Zipfel, *Earth Planet. Sci. Lett.*, 2009, **284**, 168.
68. T. Schulz, C. Münker, K. Mezger and H. Palme, *Geochim. Cosmochim. Acta*, 2010, **74**, 1706.
69. T. Schulz, D. Upadhyay, C. Münker and K. Mezger, *Geochim. Cosmochim. Acta*, 2012, **85**, 200.
70. T. Kleine, K. Mezger, H. Palme, E. Scherer and C. Münker, *Earth Planet. Sci. Lett.*, 2005, **231**, 41.
71. N. Kinoshita, M. Paul, Y. Kashiv, P. Collon, C. M. Deibel, B. DiGiovine, J. P. Greene, D. J. Henderson, C. L. Jiang, S. T. Marley, T. Nakanishi, R. C. Pardo, K. E. Rehm, D. Robertson, R. Scott, C. Schmitt, 3. D. Tang, R. Vondrasek and A. Yokoyama, *Science*, 2012, **335**, 1614.
72. M. Boyet and R. W. Carlson, *Earth Planet. Sci. Lett.*, 2006, **250**, 254–268.
73. G. Caro, B. Bourdon, J. L. Birck and S. Moorbath, *Geochim. Cosmochim. Acta*, 2006, **70**, 164.
74. R. Andreasen, M. Sharma, K. V. Subbarao and S. G. Viladkar, *Earth Planet. Sci. Lett.*, 2008, **266**, 14.
75. D. T. Murphy, A. D. Brandon, V. Debaille, R. Burgess and C. J. Ballentine, *Geochim. Cosmochim. Acta*, 2010, **74**, 738.

76. A. Cipriani, E. Bonatti and R. W. Carlson, *Geochem. Geophys. Geosyst.*, 2011, **12**, 1.
77. M. G. Jackson and R. W. Carlson, *Geochem. Geophys. Geosyst.*, 2012, **13**.
78. M. Boyet, J. Blichert-Toft, M. Rosing, M. Storey, P. Télouk and F. Albarède, *Earth Planet. Sci. Lett.*, 2003, **214**, 427.
79. G. Caro, B. Bourdon, J.-L. Birck and S. Moorbath, *Nature*, 2003, **423**, 428.
80. C. L. Harper and S. B. Jacobsen, *Nature*, 1992, **360**, 728.
81. H. Rizo, M. Boyet, J. Blichert-Toft and M. Rosing, *Earth Planet. Sci. Lett.*, 2011, **312**, 267.
82. H. Rizo, M. Boyet, J. Blichert-Toft, J. O'Neil, M. Rosing and J.-L. Paquette, *Nature*, 2012, **491**, 96.
83. J. O'Neil, R. W. Carlson, D. Francis and R. K. Stevenson, *Science*, 2008, **321**, 1828.
84. J. O'Neil, R. W. Carlson, J. Paquette and D. Francis, *Precambrian Res.*, 2012, **220–221**, 23.
85. A. S. G. Roth, B. Bourdon, S. J. Mojzsis, M. Touboul, P. Sprung, M. Guitreau and J. Blichert-Toft, *Earth Planet. Sci. Lett.*, 2013, **361**, 50.
86. M. Boyet and R. W. Carlson, *Science*, 2005, **309**, 576.
87. G. Caro, B. Bourdon, A. N. Halliday and G. Quitté, *Nature*, 2008, **452**, 336.
88. R. Andreasen and M. Sharma, *Science*, 2006, **314**, 806.
89. R. W. Carlson, M. Boyet and M. F. Horan, *Science*, 2007, **316**, 1175.
90. A. Gannoun, M. Boyet, H. Rizo and A. El Goresy, *PNAS*, 2011, **108**, 7693.
91. D. A. Papanastassiou and G. J. Wasserburg, *Earth and Planetary Science Letters*, 1972, **13**, 368–374.
92. A. H. Jaffey, K. F. Flynn, L. E. Glendenin, W. C. Bentley and A. M. Essling, *Phys. Rev.*, 1971, **4**, 1889.
93. J. M. Mattinson, *Chem. Geol.*, 2010, **275**, 186.
94. B. Schoene, J. L. Crowley, D. J. Condon, M. D. Schmitz and S. A. Bowring, *Geochim. Cosmochim. Acta*, 2006, **70**, 426.
95. G. R. Tilton, C. Patterson, H. Brown, M. G. Inghram, R. Hayden, D. Hess and E. Larsen Jr, *Geol. Soc. Am. Bull.*, 1955, **66**, 1131.
96. G. R. Tilton, G. W. Wetherill, G. L. Davis and C. A. Hopson, *Geol. Soc. Am. Bull.*, 1958, **69**, 1469.
97. G. W. Wetherill, *Trans., Am. Geophys. Union*, 1956, **37**, 320.
98. W. Compston, I. S. Williams and C. Meyer, *J. Geophys. Res.*, **89**(S02), B525–B534.
99. R. Mundil, I. Metcalfe, K. R. Ludwig, P. R. Renne, F. Oberli and R. S. Nicoll, *Earth Planet. Sci. Lett.*, 2001, **187**, 131.
100. Lee, I. S. Williams and D. J. Ellis, *Nature*, 1997, **390**, 159.
101. A. Davidson and O. Breemen, *Contrib. Mineral. Petrol.*, 1988, **100**, 291.
102. Y. Amelin, A. Kaltenbach, T. Iizuka, C. H. Stirling, T. R. Ireland, M. Petaev and S. B. Jacobsen, *Earth Planet. Sci. Lett.*, 2010, **300**, 343.
103. I. R. Fletcher, B. Rasmussen and N. J. McNaughton, *Aust. J. Earth Sci.*, 2000, **48**, 845.

104. E. S. Schandl, D. W. Davis and T. E. Krogh, *Geology*, 1990, **18**, 505.
105. R. L. Romer and J. E. Wright, *Geochim. Cosmochim. Acta*, 1992, **56**, 2137.
106. R. R. Parrish, *Can. J. Earth Sci.*, 1990, **11**, 1431.
107. D. A. Richards, S. H. Bottrell, R. A. Cliff, K. Ströhle and P. J. Rowe, *Geochim. Cosmochim. Acta*, 1998, **62**, 3683.
108. J. Woodhead and R. Pickering, *Chem. Geol.*, 2012, **322**, 290.
109. S. A. Wilde, J. W. Valley, W. H. Peck and C. M. Graham, *Nature*, 2001, **409**, 175.
110. S. W. Denyszyn, R. Mundil, S. J. Brownlee and P. R. Renne, *Can. J. Earth Sci.*, 2011, **48**, 557.
111. S. L. Kamo, G. K. Czamanske, Y. Amelin, V. A. Fedorenko, D. W. Davis and V. R. Trofimov, *Earth Planet. Sci. Lett.*, 2003, **214**, 75.
112. R. Mundil, J. Pálfy, P. R. Renne and P. Brack, in *The Triassic Timescale*, Special Publication, Geological Society, London, 2010, **334**, 41.
113. J.-Y. Lee, K. Marti, J. P. Severinghaus, K. Kawamura, H.-S. Yoo, J. B. Lee and J. S. Kim, *Geochim. Cosmochim. Acta*, 2006, **70**, 4507.
114. I. McDougall and T. M. Harrison, *Geochronology and Thermochronology by the $^{40}Ar/39Ar$ method*, Oxford University Press, Oxford, 1999.
115. A. K. Baksi, in *The Origin of Melting Anomalies, Plates, Plumes and Planetary Processes*, ed. G. R. Foulger and D. M. Jurdy, Geological Society of America Special Paper, 2007, vol. **430**, pp. 285–304.
116. R. H. Steiger and E. Jäger, *Earth Planet. Sci. Lett.*, 1977, **36**, 359.
117. K. Min, R. Mundil, P. R. Renne and K. R. Ludwig, *Geochim. Cosmochim. Acta*, 2000, **64**, 73.
118. F. Jourdan, A. Marzoli, H. Bertrand, S. Cirille, L. Tanner, D. J. Kontak, G. McHone, P. R. Renne and G. Bellieni, *Lithos*, 2009, **110**, 167.
119. P. R. Renne, R. Mundil, G. Balco, K. Min and K. R. Ludwig, *Geochim. Cosmochim. Acta*, 2010, **74**, 5349.
120. P. R. Renne, W. D. Sharp, A. L. Deino, G. Orsi and L. Civetta, *Science*, 1997, **277**, 1279.
121. D. D. Bogard, *Meteoritics*, 1995, **30**, 244.
122. F. Jourdan, D. F. Mark and C. Verati, *Advances in $^{40}Ar/^{39}Ar$ Dating from Acheology to Planetary Sciences*, Special Publication, Geological Society, London, 2013.
123. G. Turner, J. C. Huneke, F. A. Podosek and W. G. J., *Earth Planet. Sci. Lett.*, 1971, **12**, 19.
124. F. Jourdan, *Aust. J. Earth Sci.*, 2012, **59**, 199.
125. T. Kennedy, F. Jourdan, A. W. R. Bevan and M. A. M. Gee, Frew, *Geochim. Cosmochim. Acta*, 2013, **115**, 162–182.
126. M. D. Norman, R. A. Duncan and J. J. Huard, *Geochim. Cosmochim. Acta*, 2006, **70**, 6032.
127. E. K. Jessberger, J. C. Hueneke, F. A. Podosek and G. Wasserburg, *Proc. 5th Lunar Sci. Conf.*, Pergamon, New York, 1974, pp. 1419–1449.
128. T. S. Culler, T. A. Becker, M. R. A. and P. R. Renne, *Science*, 2000, **287**, 1785.

129. N. E. B. Zellner, J. W. Delano, T. D. Swindle, F. Barra, E. Olsen and D. C. B. Whittet, *Geochim. Cosmochim. Acta*, 2009, **73**, 4590.

130. S. Hui, M. Norman and F. Jourdan, *Proc. 7th Australian Space Sci. Conf.*, Sydney, 2010.

131. B. A. Cohen, G. A. Snyder, C. M. Hall, L. A. Taylor and M. A. Nazarov, *Meteorit. Planet. Sci.*, 2001, **36**, 1345.

132. M. Trieloff, E. K. Jessberger, I. Herrwerth, J. Hopp, C. Fleni, M. Ghells, M. Bourot-Denise and P. Pellas, *Nature*, 2003, **422**, 502.

133. S. M. Aciego, F. Jourdan, D. J. DePaolo, M. Kennedy, P. R. Renne and K. W. W. Sims, *Geochem. Geophys. Geosyst.*, 2010, **74**, 1620.

134. A. Hicks, J. Barclay, D. F. Mark and S. Loughlin, *Geology*, 2012, **40**, 723.

135. W. D. Sharp and P. R. Renne, *Geochem. Geophys. Geosyst.*, 2005, **6**.

136. C. Hofmann, G. Feraud and V. Courtillot, *Earth Planet. Sci. Lett.*, 2000, **180**, 13.

137. C. Verati and F. Jourdan, in $^{40}Ar/^{39}Ar$ *Dating: From Geochronology to Thermochronology, from Archaeology to Planetary Sciences*, ed. F. Jourdan, D. Mark and C. Verati, Special Publication 378, Geological Society, London, 2014.

138. A. Marzoli, P. R. Renne, E. M. Piccirillo, M. Ernesto, G. Bellieni and A. De Min, *Science*, 1999, **284**, 616.

139. M. K. Reichow, A. D. Saunders, R. V. White, M. S. Pringle, A. I. Al'Mukhamedov, A. I. Medvedev and N. P. Kirda, *Science*, 2002, **296**, 1846.

140. F. Jourdan, G. Féraud, H. Bertrand, M. K. Watkeys and P. R. Renne, *Geochem. Geophys. Geosyst.*, 2008, **9**.

141. C. Cucciniello, L. Melluso, F. Jourdan, J. J. Mahoney, T. Meisel and V. Morra, *Geol. Mag.*, 2013, **150**, 1.

142. V. E. Courtillot and P. R. Renne, *C. R. Geosci*, 2003, **335**, 113–140.

143. M. Storey, R. G. Roberts and M. Saidin, *Proc. Natl. Acad. Sci. U. S. A.*, 2012, **109**, 18684.

144. R. C. Walter, *Geology*, 1994, **22**, 6.

145. B. Carrapa, J. Wijbrans and G. Bertotti, *Geology*, 2003, **31**, 601.

146. G. Sanchez, Y. Rolland, J. Schneider, M. Corsini, E. Oliot, P. Goncalves, C. Verati, J. M. Lardeaux and D. Marquer, *Lithos*, 2011, **125**, 521.

147. C. Wobus, M. Pringle, K. Whipple and K. Hodges, *Earth Planet. Sci. Lett.*, 2008, **269**, 1.

148. C. J. Forbes, D. Giles, F. Jourdan, K. Sato, S. Omori and M. Bunch, *Precambrian Res.*, 2012, **200–203**, 209.

149. K. Ueda, J. Jacobs, R. J. Thomas, J. Kosler, M. S. A. Horstwood, J.-A. Wartho, F. Jourdan, B. Emmel and R. Matola, *J. Geol.*, 2012, **120**, 507.

150. O. Oliveros, D. Tristá-Aguilera, G. Féraud, D. Morata, L. Aguirre, S. Kojima and F. Ferraris, *Miner. Deposita*, 2008, **43**, 61.

151. I. Márton, R. Moritz and R. Spikings, *Tectonophysics*, 2010, **483**, 240.

152. Q. Zeng, T. C. McCuaig, C. J. R. Hart, F. Jourdan, J. Muhling and L. Bagas, *Miner. Deposita*, 2012, **47**, 799.

153. R.-Z. Hu, W.-F. Wei, 3.-W. Bi, J.-T. Peng, Y.-Q. Qi, L.-Y. Wu and Y.-W. Chen, *Lithos*, 2012, **150**, 111.

154. J. D. Vervoot and J. Blichert-Toft, *Geochim. Cosmochim. Acta*, 1999, **63**, 553.

155. J. D. Vervoot, D. Jeff, P. J. Patchett, J. Blichert-Toft and F. Albarède, *Earth Planet. Sci. Lett.*, 1999, **168**, 79.

156. A. L. S. Kemp, C. J. Hawkesworth, B. Paterson and P. Kinny, *Nature*, 2006, **439**, 580.

157. C. J. Hawkesworth and A. L. S. Kemp, *Chem. Geol.*, 2006, **226**, 144.

158. E. Belousova, W. L. Griffin and S. Y. O'Reilly, *J. Petrol.*, 2006, **47**, 329.

159. A. L. S. Kemp, C. J. Hawkesworth, G. L. Foster, B. A. Paterson, J. D. Woodhead, J. M. Hergt, C. M. Gray and M. J. Whitehouse, *Science*, 2007, **315**, 980.

160. K. C. Condie, E. Beyer, E. Belousova, W. L. Griffin and S. Y. O'Reilly, *Precambrian Res.*, 2005, **139**, 42.

161. W. L. Griffin, N. J. Pearson, E. A. Belousova, S. R. Jackson, E. van Achterbergh, S. Y. O'Reilly and S. R. Shee, *Geochim. Cosmochim. Acta*, 2000, **64**, 133.

162. W. L. Griffin, E. Belousova, S. G. Walters and S. Y. O'Reilly, *Aust. J. Earth Sci.*, 2006, **53**, 125.

163. P. J. Patchett, O. Kouvo, C. E. Hedge and M. Tatsumoto, *Contrib. Mineral. Petrol.*, 1981, **78**, 279.

164. P. E. Smith, M. Tatsumoto and R. Farquhar, *Contrib. Mineral. Petrol.*, 1987, **97**, 93.

165. E. Scherer, C. Muenker and K. Mezger, *Science*, 2001, **293**, 683.

166. E. Belousova, Y. A. Kostitsyn, W. L. Griffin, G. C. Begg, S. Y. O'Reilly and N. J. Pearson, *Lithos*, 2010, **119**, 457.

167. W. J. Collins, E. A. Belousova, A. L. S. Kemp and J. B. Murphy, *Nat. Geosci.*, 2011, **4**, 333.

168. S. B. Shirey and R. J. Walker, *Annu. Rev. Earth Planet. Sci.*, 1998, **26**, 423–500.

169. L. Reisberg and T. Meisel, *Geostand. Newsl.*, 2002, **26**, 249–267.

170. R. W. Carlson, *Lithos*, 2005, **82**, 249–272.

171. R. W. Carlson, S. B. Shirey and M. Schonbachler, *Elements*, 2008, **4**, 239–245.

172. R. J. Walker, J. W. Morgan, E. S. Beary, M. I. Smoliar, G. K. Czamanske and M. F. Horan, *Geochim. Cosmochim. Acta*, 1997, **61**, 4799–4807.

173. I. S. Puchtel, R. J. Walker, C. R. Anhaeusser and G. Gruau, *Chem. Geol.*, 2009, **262**, 355–369.

174. I. S. Puchtel, R. J. Walker, A. D. Brandon and E. G. Nisbet, *Geochim. Cosmochim. Acta*, 2009, **73**, 6367–6389.

175. B. D. Connolly, I. S. Puchtel, R. J. Walker, R. J. Arevalo, P. M. Piccoli, G. R. Byerly, C. Robin-Popieul and N. T. Arndt, *Earth Planet. Sci. Lett.*, 2011, **311**, 253–263.

176. H. J. Stein, R. J. Markey, J. W. Morgan, J. L. Hannah and A. Schersten, *Terr. Nova*, 2001, **13**, 479–486.

177. T. E. McCandless, J. Ruiz, B. I. Adair and C. Freydier, *Geochim. Cosmochim. Acta*, 1999, **63**, 911–923.
178. R. J. Walker, H. M. Prichard, A. Ishivatari and M. Pimentel, *Geochim. Cosmochim. Acta*, 2002, **66**, 329–345.
179. I. S. Puchtel, G. E. Brügmann, A. W. Hofmann, V. S. Kulikov and V. V. Kulikova, *Contrib. Mineral. Petrol.*, 2001, **140**, 588–599.
180. R. J. Walker, J. W. Morgan, A. J. Naldrett, C. Li and J. D. Fassett, *Earth Planet. Sci. Lett.*, 1991, **105**, 416–429.
181. R. J. Walker, J. W. Morgan, E. J. Hanski and V. F. Smolkin, *Ontario Geol. Surv. Spec. Publ.*, 1994, **5**, 343–355.
182. J. Kirk, J. Ruiz, J. Chesley, J. Walshe and G. England, *Science*, 2002, **297**, 1856–1858.
183. K. J. Westerlund, S. B. Shirey, S. H. Richardson, R. W. Carlson, J. J. Gurney and J. W. Harris, *Contrib. Mineral. Petrol.*, 2006, **152**, 275–294.
184. A. S. Cohen, A. L. Coe, J. M. Bartlett and C. J. Hawkesworth, *Earth Planet. Sci. Lett.*, 1999, **167**, 159–173.
185. D. Selby and R. A. Creaser, *Geology*, 2005, **33**, 545.
186. D. Selby, R. A. Creaser and M. G. Fowler, *Geochim. Cosmochim. Acta*, 2007, **71**, 378.
187. A. J. Finlay, D. Selby and M. J. Osborne, *Earth Planet. Sci. Lett.*, 2012, **313–314**, 95.
188. A. J. Finlay, D. Selby, M. J. Osborne and D. Finucane, *Geology*, 2010, **38**, 979.
189. D. Rooney, D. Selby, M. D. Lewan, P. G. Lillis and J.-P. Houzay, *Geochim. Cosmochim. Acta*, 2012, **77**, 275.
190. D. Selby, R. A. Creaser, K. Dewing and M. Fowler, *Earth Planet. Sci. Lett.*, 2005, **235**, 1.
191. D. Selby and R. A. Creaser, *Science*, 2005, **308**, 1293.
192. R. J. Walker, R. W. Carlson, S. B. Shirey and F. R. Boyd, *Geochim. Cosmochim. Acta*, 1989, **53**, 1583–1595.
193. E. H. Hauri and S. R. Hart, *Chem. Geol.*, 1997, **139**, 185–205.
194. J. C. Lassiter and E. H. Hauri, *Earth Planet. Sci. Lett.*, 1998, **164**, 483–496.
195. J. C. Lassiter, E. H. Hauri, P. W. Reiners and M. O. Garcia, *Earth Planet. Sci. Lett.*, 2000, **178**, 269–284.
196. R. J. Walker, L. M. Echeverria, S. B. Shirey and M. F. Horan, *Contrib. Mineral. Petrol.*, 1991, **107**, 150–162.
197. A. D. Brandon, R. J. Walker, J. W. Morgan, M. D. Norman and H. M. Prichard, *Science*, 1998, **280**, 1570–1573.
198. I. S. Puchtel, A. D. Brandon, M. Humayun and R. J. Walker, *Earth Planet. Sci. Lett.*, 2005, **237**, 118–134.
199. J. W. Morgan, G. A. Wanderless, R. K. Petrie and A. J. Irving, *Tectonophysics*, 1981, **75**, 47–67.
200. J. W. Morgan, *Nature*, 1985, **317**, 703–705.
201. C.-L. Chou, D. M. Shaw and J. H. Crocket, *J. Geophys. Res.*, 1983, **88**, A507-A518.

202. M. D. Norman, V. C. Bennett and G. Ryder, *Earth Planet. Sci. Lett.*, 2002, **202**, 217–228.
203. I. S. Puchtel, R. J. Walker, O. B. James and D. A. Kring, *Geochim. Cosmochim. Acta*, 2008, **72**, 3022–3042.
204. M. Fischer-Gödde and H. Becker, *Geochim. Cosmochim. Acta*, 2012, 77, 135–156.
205. F. E. Grousset and P. E. Biscaye, *Chem. Geol.*, 2005, **222**, 149.
206. H. Maring, D. M. Settle, P. Buat-Ménard, F. Dulac and C. C. Patterson, *Nature*, 1987, **300**, 154.
207. W. T. Sturges and L. A. Barrie, *Nature*, 1987, **329**, 144.
208. P. Vallelonga, K. Van de Velde, J. P. Candelone, V. I. Morgan, C. F. Boutron and K. J. R. Rosman, *Earth Planet. Sci. Lett.*, 2002, **204**, 291.
209. P. Vallelonga, P. Gabrielli, K. J. R. Rosman, C. Barbante and C. F. Boutron, *Geophys. Res. Lett.*, 2005, **32**, L01706.01701.
210. L. J. Burn-Nunes, P. Vallelonga, R. D. Loss, G. R. Burton, A. Moy, M. Curran, S. Hong, A. M. Smith, R. Edwards, V. I. Morgan and K. J. R. Rosman, *Geochim. Cosmochim. Acta*, 2011, **75**, 1.
211. G. R. Burton, J.-P. Candelone, K. J. R. Rosman, L. J. Burn, C. F. Boutron and S. Hong, *Earth Planet. Sci. Lett.*, 2007, **259**, 557.
212. K. Lee, S. D. Hur, S. Hou, L. J. Burn-Nunes, S. Hong, C. Barbante, C. F. Boutron and K. J. R. Rosman, *Sci. Total Environ.*, 2011, **412–413**, 194.
213. K. J. R. Rosman, S. Hong, G. R. Burton, L. J. Burn, C. F. Boutron, C. P. Ferrari, L. G. Thompson, L. Maurice-Bourgoin and B. Francou, *J. Phys. France IV*, 2003, **107**, 1157.
214. K. J. R. Rosman, C. V. Ly, K. P. Van de Velde and C. F. Boutron, *Earth Planet. Sci. Lett.*, 2000, **176**, 413.
215. R. M. Sherrell, E. A. Boyle and B. Hamelin, *J. Geophys. Res.*, 1992, **97**, 11257–11268.
216. W. Compston and P. A. Arriens, *Can. J. Earth Sci.*, 1968, **5**, 561.
217. G. J. Wasserburg, D. A. Papanastassiou, E. V. Nenow and C. A. Bauman, *Rev. Sci. Instrum.*, 1969, **40**, 288.
218. D. Phillips, in *34th IGC Conference*, Brisbane, 2012.

CHAPTER 4

Advances in Fluorescence Spectroscopy for Petroleum Geosciences

KEYU LIU,*[a,b] NEIL SHERWOOD[c] AND MENGJUN ZHAO[b]

[a] CSIRO Earth Science and Resource Engineering, P.O. Box 1130, Bentley, WA 6112, Australia; [b] Research Institute of Petroleum Exploration and Development, PetroChina, P.O. Box 910, Beijing 100083, China; [c] CSIRO Earth Science and Resource Engineering, P.O. Box 136, North Ryde, NSW 1670, Australia
*Email: Keyu.Liu@csiro.au

4.1   Introduction

Fluorescence was first described by Sir George Stokes over 160 years ago.[1] It is a phenomenon in which a material absorbs light of one colour (wavelength) and emits it at a different colour (wavelength). Excitation occurs when an incoming photon causes an electron to move from a stable ground state to a higher energy, unstable 'excited' state. When the excited electron returns to the ground state it emits a photon (fluorescence). Some energy is lost to heat in the process, such that the fluorescence emitted has less energy (longer wavelength) than the original photon.

The field of fluorescence spectroscopy has continued to grow steadily since its discovery, especially in the last few decades with developments in sophisticated instrumentation, in knowledge of fundamental aspects and in applications in analytical, physical and organic chemistry, molecular

RSC Detection Science Series No. 4
Principles and Practice of Analytical Techniques in Geosciences
Edited by Kliti Grice
© The Royal Society of Chemistry 2015
Published by the Royal Society of Chemistry, www.rsc.org

sciences, biology, biomedicine, and medical research. In the last decade there have been over a dozen books on fluorescence analyses published by Springer alone.[2]

Fluorescence has been used in detecting the presence of oil in rocks for petroleum exploration for over 100 years, as summarised in Riecker.[3] However, it was not used quantitatively in petroleum exploration until the first dedicated fluorimeter for well site 'logging' was developed in the 1950s.[4] Texaco Oil Corporation further refined the fluorimeter technology of Zierfuss and Coumou[4] and developed their patented Quantitative Fluorescence Technology (QFT) and QFT-II.[5,6] More recently Schlumberger developed an innovative downhole fluid analysis (DFA) optical tool.[7] Sophisticated fluorescence spectrophotometers have also been used to detect traces of oils *via* remote sensing (*e.g.* airborne laser fluorescence[8]) and to characterise oils,[9–11] oil inclusions,[12–14] and water[15,16] in laboratory-based studies. As summarised in Table 4.1, fluorescence technologies are now widely used in every stage of petroleum exploration and development, from regional exploration to appraisal, production, and post-drilling environmental monitoring.

This chapter aims to provide an overview of some recent developments in fluorescence techniques in the field of petroleum exploration and production (E&P). It is structured in three parts: (1) a brief review of the fundamentals of fluorescence; (2) evolution of fluorescence spectroscopic instrumentation and techniques in petroleum E&P; and (3) two novel developments in fluorescence technologies for petroleum system analysis, including fluorescence alteration of multiple macerals (FAMM) for source rock maturity evaluation, and quantitative grain fluorescence (QGF) and total scanning fluorescence (TSF) for detecting oil migration and accumulation and characterisation.

Table 4.1 Summary of fluorescence applications in petroleum E&P including surface geochemical exploration, petroleum system analysis, and petroleum production.

Petroleum E&P	Applications	References
Surface geochemical exploration	Airborne laser fluorescence (ALF) Seaboard fluorescence (TSF) Surface geochemical exploration (TSF)	8,9,27
Petroleum system evaluation	Source rock maturity (FAMM) Oil–source correlation (TSF) Migration pathway detection (QFT, DFA, GOI, QGF, QGF-E, TSF) Petroleum accumulation and preservation (QFT, DFA, QGF, QGF-E, TSF)	6,7,11,13,14,57,60,88,91
Petroleum production	Petroleum property determination (TSF; FLIM) Producing oil and water (TSF; DFA) Environmental monitoring (TSF)	7,9,11,15,16,23,28,43

4.2   Basis for Fluorescence Spectroscopic Methods

Fluorescence spectroscopy is one of the most sensitive, rapid, versatile, and inexpensive screening techniques used in petroleum E&P and environmental monitoring. It is much more sensitive than chromatography (GC) and even GC with mass spectrometry (GC-MS) in detecting trace amounts of hydrocarbons in sediments or fluids. Most of the modern spectrophotometers can detect picomolar $(10^{-12}$ M) fluorescent compounds in solution.[17,18] Laboratory experiments indicate that they are able to detect sub-parts per million (ppm) to parts per billion (ppb) levels of oil in solvents and water.[11,13,19]

The fluorescing components in petroleum are primarily aromatic and polar components (resin and asphaltene).[7,9,20] When excited with UV light these compounds emit fluorescence. The fluorescence signature can be used to determine the types and amounts of compounds in petroleum,[11,20–22] and fluorescence parameters can thus be used to detect oils and determine their types and abundances.

Petroleum can be detected both in undiluted (neat) and diluted forms and as surface coatings and inclusions.[13,20,23] A variety of methods have been developed to measure fluorescence from these different sources.

4.2.1   Fluorescence and Fluorescence Spectrophotometry

Fluorescence spectrophotometry has developed rapidly over the last few decades concomitantly with the invention and improvement of sophisticated spectrophotometers.[2,24,25] For petroleum exploration and development various fluorescence methods have been used in the laboratory as a characterisation technique to evaluate the attributes of oil,[9,11] and in the field to detect the presence of hydrocarbons in air,[6,26] seawater,[27–29] and unconsolidated sediments.

Fluorescence spectroscopy has been also extensively used to characterise petroleum inclusions trapped in minerals in petroleum reservoirs.[14] However, one key problem with current spectrophotometers, when analysing solid samples such as reservoir grains, is the high level of stray light due to scattering. The stray light renders the instruments unable to record adequate emission spectra from solid samples directly. By using a customised external bandwidth filter, however, Liu *et al.*[30,31] successfully obtained fluorescence emission spectra on bulk reservoir grains directly using a fluorescence spectrophotometer without using thin sections under UV microscopes.

4.2.2   Advances in Instrumentation and Software

With the development of fluorometer technology,[25] modern fluorescence spectrophotometers use wide ranges of excitation and emission wavelengths, from shortwave UV to visible light and into the IR range (220–1100 nm). Most commercially available spectrophotometers use a long-life xenon flash lamp light source and have high sensitivity and high signal-to-noise ratios.

Modern xenon lamp technology allows rapid capture of a data point every few milliseconds and scans at a few tens of thousands of nanometres per minute without peak shifts.[17] Most currently available spectrophotometer instruments are equipped with a microplate reader, offer automated full wavelength scanning of up to 384 samples at one time, and have accessories available for analysing fluids and solid samples.

Most spectrophotometers have proprietary software for data acquisition and post processing including 2D and 3D visualisation. The data can easily be exported to other software such as neural network packages for chemometric analysis[32] and statistical packages for further batch processing and analysis such as principal component analysis (PCA).

4.2.3 Advances in Fluorescence Technologies

Various hydrocarbon detecting techniques have been developed based on the modern spectrophotometers (Table 4.2). The initial fluorescence analysis used in petroleum exploration is UV fluorescence detection, with photography, commonly used at well sites ('UV box') for imaging core and cuttings.

The development and evolution of fluorescence spectrophotometry instrumentation are summarised by Rhys-Williams.[17] Emission spectroscopy is the most commonly used method for rapidly detecting oil in petroleum systems although excitation (adsorption) spectroscopy is also used. These two techniques are mainly applied for rapid characterisation of oil and oil extracts semiquantitatively.[10,11] They are also applied to characterise petroleum inclusions,[33,34] and coatings of reservoir grains (*e.g.* QGF and QGF-Extract/QGF-E)[13] and source rocks (*e.g.* FAMM).[35,56,58–60] 3D excitation–emission matrix (EEM), or synchronous scanning, has been used to obtain spectrograms for determining physical attributes and thermal maturity of oils.[9,27,36] Synchronous fluorescence scanning was also developed to further characterise oil[32,37] and oil inclusions.[32,38] The instrument evolution of fluorescence microscopy, including epifluorescence, wide-field, scanning, confocal, one-photon excitation, and multiphoton excitation spectroscopy, is

Table 4.2 Summary of various fluorescence spectroscopic technologies in petroleum exploration and production and their evolution.

Method	Application	References
Well site UV box	Well site oil trace identification	3,4,17
UV light imaging	Oil saturation detection in core	3,4,17
UV emission spectroscopy	Oil identification and properties	5,13
Excitation–emission matrix	Oil identification and properties	9,36
Total synchronous fluorescence scanning	Oil identification and properties	27,37
UV emission micro-spectroscopy	Oil inclusion identification and properties	12,21
Confocal laser scanning microscopy/spectroscopy	Oil identification and properties PVT modelling	39–41
Fluorescence lifetime imaging	Oil properties	21,43

well illustrated by Masters.[24] Laser confocal microscopy (LSM) enables imaging of petroleum inclusions and volumetric constructions of petroleum inclusions and associated gas bubbles which are essential for pressure–volume–temperature (PVT) analysis.[39–41] Fluorescence lifetime imaging systems have recently been developed to better differentiate hydrocarbon compounds and their attributes.[42–44] The development of multiphoton laser confocal microscopy allows imaging of fluorescence decay of petroleum inclusions using fluorescence lifetime imaging (FLIM) techniques.[42–44]

4.3 Overview of the Fluorescence Alteration of Multiple Macerals (FAMM) Technique for Thermal Maturity Analyses

4.3.1 Introduction

The change of fluorescence emission from organic matter (OM) in rocks over the duration of constant UV or blue light excitation has been studied since the early 1970s.[45,46] Subsequently, additional studies have described characteristics of 'fluorescence alteration' of macerals and relationships with thermal maturity/coal rank.[47–52] In addition, studies on the maceral vitrinite have shown that the intensity of its fluorescence emission is directly related to hydrogen content[53] and coal rank (see Taylor *et al.*[54] and references therein). The fluorescence alteration of multiple macerals (FAMM™) technique was introduced in the early 1990s in order to quantify these observations and to develop an analytical tool to solve some common problems in thermal maturity assessment.[35,53,55–61]

The major challenge in accurately measuring thermal maturity is to differentiate effects of maturity from those inherited from the progenitor plant material (defined by 'OM/maceral type'). Vitrinite reflectance (VR) is one of the most reliable indicators of thermal maturity of coals and other petroleum source rocks, largely because a specific OM type is analysed. Vitrinite varies in chemical composition at a given thermal maturity level; however, such that the reflectance may be anomalously low ('VR suppression') or high ('VR enhancement'). Vitrinites having suppressed reflectances are perhydrous and yield anomalously high fluorescence intensities compared to 'normal' (orthohydrous) vitrinite;[59,60] vitrinites having enhanced reflectance are subhydrous and yield anomalously low fluorescence intensities.[35,62,63] The FAMM method was developed primarily to assess and correct for suppressed VR values measured from source rocks within basins of the Northwest Shelf of Australia.

4.3.2 Background

Primary fluorescence of macerals originates from chemical structures of the precursor plant material[51,54] and accounts for the fluorescence of vitrinite

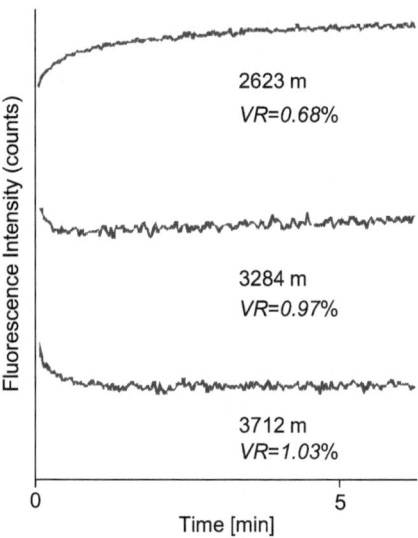

Figure 4.1   Fluorescence alteration curves for vitrinites from the Dampier-1 well, Carnarvon Basin, Australia, showing variation from positive alteration for the sample from 2623 m to negative for the sample from 3712 m (modified after Wilkins *et al.*[57])

precursors in peats and associated sediments. Primary fluorescence intensity of vitrinite decreases with increases in thermal maturity, to a minimum at about 0.4% VR. With further increases in maturity, this fluorescence is extinguished, and secondary fluorescence, which is due to generation of incipient bitumens/mobile components within the vitrinite,[51,54,64] becomes prevalent. Secondary fluorescence intensity increases with thermal maturity to a maximum at about 1% VR. The subsequent reduction is probably caused by thermal cracking of the molecules that cause the fluorescence.

Fluorescence intensities of macerals vary over time upon excitation, and the nature of this 'fluorescence alteration' systematically changes according to thermal maturity level (Figure 4.1). The change in fluorescence intensity with excitation is due to photochemical oxidation reactions;[65] the chemical basis of fluorescence alteration has been further discussed by Pradier *et al.*[66,67]

The FAMM technique is based on measuring the changes in fluorescence intensity from various macerals over time and calibrating against reflectance of orthohydrous vitrinite.

4.3.3   Analytical Method

FAMM analyses are carried out on polished blocks of whole rock samples prepared using the standard methods employed for VR analyses. A laser microprobe with a 488 nm excitation source is used to collect fluorescence data over 400 s (700 s in early studies) from measurement areas ∼2 μm in diameter. The emission is measured at a wavelength of 620–630 nm, which in general, corresponds to the peak intensity from macerals over the calibrated thermal maturity range of the FAMM technique. Where possible all maceral groups are measured and samples are analysed 'in air' rather than in oil immersion as is the case for standard VR analyses.

Two variables are extracted from the fluorescence measurements: 'final intensity' at 400 s (F_{400}) which is mainly controlled by inherited chemistry from the progenitor plant material, and the fluorescence alteration ratio $(F_{400}/$initial intensity; $F_{400}/F_o)$, which is mainly controlled by thermal maturity for a given OM type. These two variables are plotted on fluorescence alteration diagrams which are calibrated according to VR. The calibration is based on measurements from a suite of orthohydrous vitrinites in the range of 0.4–1.2% reflectance, which define a near-vertical line on the diagram (Figure 4.2). Fluorescence data from hydrogen-rich macerals, such as liptinite and perhydrous vitrinite, plot to the right of this 'normal' vitrinite line and hydrogen-poor macerals, such as subhydrous vitrinite and most inertinites, plot to the left. Vitrinites of the same thermal maturity, but having

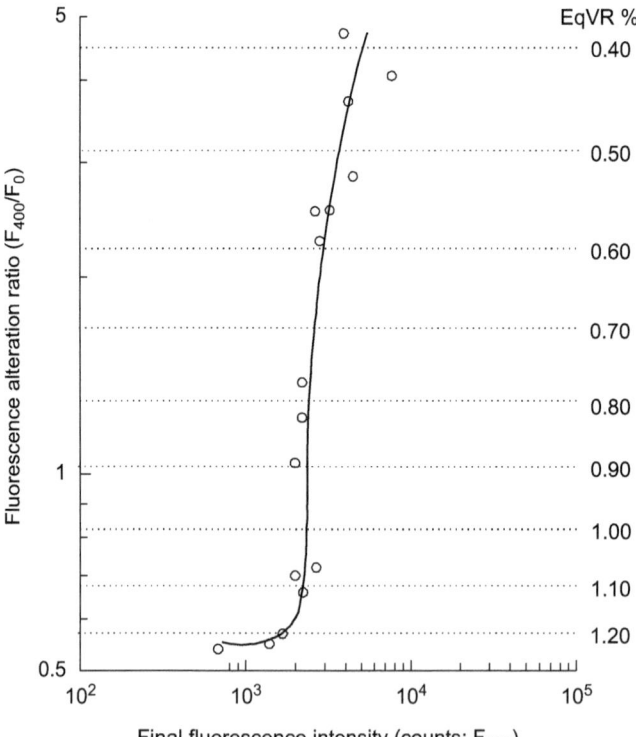

Figure 4.2 FAMM calibration curve for 'normal' (orthohydrous) vitrinite mainly based on a suite of Permo-Triassic coals from eastern Australia. Each point represents an average value for 10 telovitrinite grains in the coal samples.

different compositions, in particular hydrogen contents, have similar fluorescence alteration ratios. This is designated on the FAMM diagram by horizontal isoequivalent VR (EqVR) lines. EqVR therefore is representative of VR for orthohydrous vitrinite and it can be determined on the basis of the average plotted position of vitrinites relative to the iso-EqVR lines, or on the position of a polynomial curve drawn through data points from a suite of macerals.[56,59,60]

4.3.3.1 Data Quality

Because F_{400} for vitrinite is directly related to hydrogen content (Figure 4.3 and Sherwood *et al.*[68]) and to the amount of VR suppression/enhancement, the difference between EqVR and measured VR can be estimated on the basis of plotted positions of vitrinites on FAMM diagrams with respect to iso-correction curves (Figure 4.4 and Wilkins *et al.*[56,59]). These curves are based on best-fit lines through vitrinite points where corresponding VR and FAMM analyses have been done, including samples of various ages.[59,62,69] As shown in Figure 4.5, these estimations of VR suppression/enhancement can be used for cross-checking consistency of FAMM EqVR with measured VR. If the EqVR values are within 0.1% absolute of the measured VR values corrected for suppression/enhancement, a high level of confidence applies to both sets of determinations. The main causes for disagreements between these data sets are:

- different populations of vitrinite measured with the two techniques— this problem can be exacerbated if different operators are involved but can be solved by carrying out parallel VR measurements on the same grains as those measured with the FAMM technique

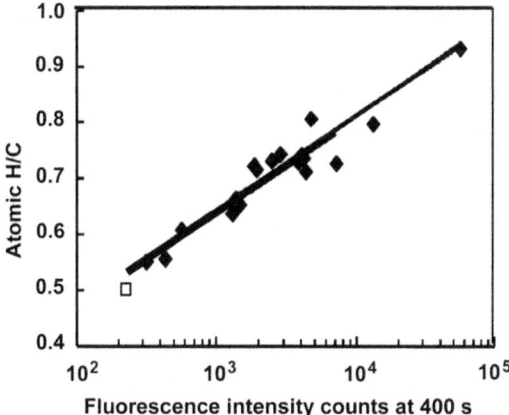

Figure 4.3 Plot of atomic H/C *versus* fluorescence intensity at 400 s, for numerous vitrinite concentrates and one inertinite concentrate (square symbol) from various coals from northern China (after Sherwood *et al.*[68]).

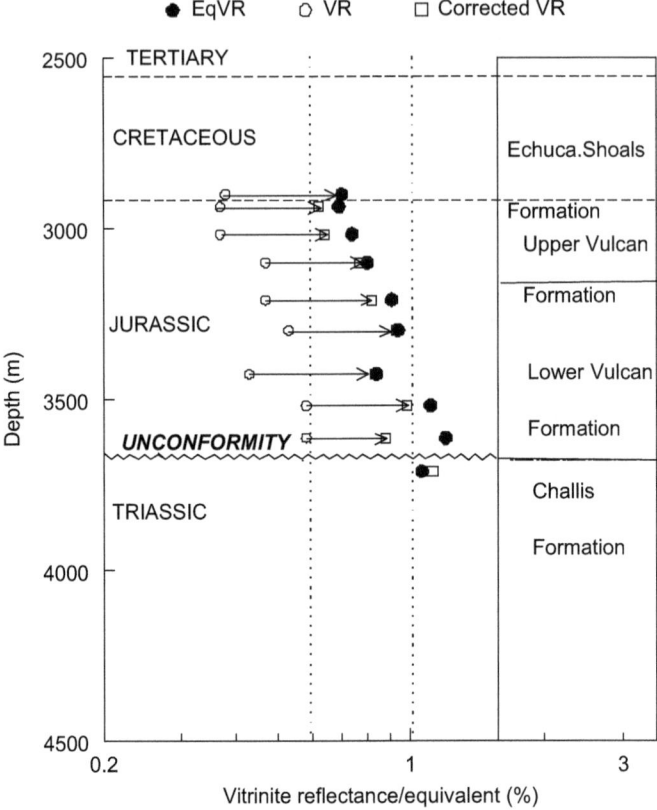

Figure 4.4 FAMM diagram showing plotted vitrinites from a sample from the MITI
Nishi-Kubiki well, Japan (cross represents average). A perhydrous composition is indicated and the estimated VR suppression correction is
about 0.3%. Considering the estimated suppression, the 0.95% EqVR is
consistent with the measured VR of 0.6% (after Ujiié *et al.*[63]).

- inaccuracies in VR data, most likely due to poor polish or misidentification of indigenous vitrinite
- inaccuracies in the EqVR data, most likely due to post-maturation oxidation of the sample, thermal maturities near the limits of calibration,
  or misidentification of indigenous OM.

4.3.4   Some Case Studies

Numerous examples of applying FAMM analyses for thermal maturity
evaluation of petroleum source rocks have been published. Most of these
studies have emphasised the potentially profound effects that VR suppression may have on thermal maturity evaluation and thermal history
modelling. They have also discussed application of FAMM analyses to solve
these problems and assist in identification of indigenous vitrinite.

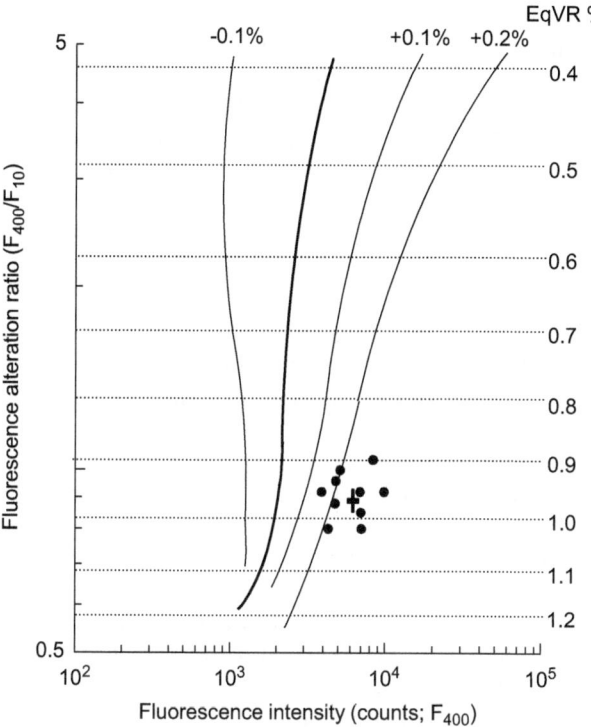

Figure 4.5 Plot of depth against FAMM-derived equivalent vitrinite reflectance (VR), measured VR, and VR corrected for suppression, for samples from Maple 1, Bonaparte Basin, Australia. The arrows indicate the amount of estimated correction on the basis of plotted positions of vitrinite on FAMM diagrams with respect to iso-correction curves. With the exception of the second deepest sample, the corrected VR values are in acceptable agreement with the EqVR values. The suppression correction for the lowermost Vulcan Formation sample may be underestimated due to oxidation effects (after Wilkins *et al.*[55]).

4.3.4.1 Identification of VR Suppression, Cavings, and Reworking

Wilkins and co-authors[60] used FAMM analyses to address problems of VR suppression using a suite of cuttings samples from Dampier-1, a petroleum exploration well in the Carnarvon Basin of the Northwest Shelf of Australia. Vitrinite in these samples plots in the perhydrous field of FAMM diagrams and the EqVR values are up to ∼0.4% higher than the measured VR values. Samples from two additional wells in the basin (Jupiter-1, Barrow-1) and one in the Bonaparte Basin (Flamingo-1) were studied by Wilkins *et al.*,[57] again finding VR suppression of up to 0.4%. The suppression identified in both studies is associated with deltaic/marginal marine and marine sequences.

Thermal maturity and petroleum generation modelling of some other offshore Carnarvon Basin wells were discussed by Wilkins *et al.*,[61] who found that VR suppression is widespread for the Jurassic and Cretaceous sequences and that significant offsets in the VR profiles with depth are due to differences in lithology/depositional environment rather than to differences in thermal history.

Ellacott *et al.*[70] applied FAMM analyses for troubleshooting VR data from the Gippsland and Otway Basins. The fluorescence studies were used to help delineate cavings from indigenous vitrinite, as well as to address problems associated with VR suppression.

The FAMM technique was further applied to a variety of samples from south-east Asian petroleum exploration wells by Wilkins *et al.*[35] It was mainly used to identify complexities in thermal maturity evaluation due to VR suppression and the presence of reworked OM, cavings, and mud additives.

Thermal maturity assessment employing combined VR and FAMM analyses was carried out by Ujiié *et al.*[63] for samples from two petroleum exploration wells in the Niigata Basin and two in the Akita Basin, both located in northern Honshu Island, Japan. This work showed widespread VR suppression for many of the Neogene rocks studied; the FAMM-derived EqVR values are up to 0.35% higher than the measured VR. This is consistent with the high H/C ratios, volatile matter contents, and fluidities during carbonisation, reported for Japanese Tertiary coals by Van Krevelen.[71] Because of variations in vitrinite composition for the sequences studied, it was emphasised that the amount of VR suppression could not be assumed for a given formation. They also discussed problems that can arise in using FAMM analyses for rocks having maturities less than the lower limit of the calibration range (*i.e.* <0.4% EqVR).

In 2009, Petersen and co-authors[72] identified problems with VR data for Cenozoic sections intersected by a suite of petroleum exploration wells in the north-eastern Vietnamese part of the Malay Basin. The problems are mainly associated with VR suppression and identification of cavings and bitumens. Evaluating the VR data along with associated FAMM data indicated VR suppression of up to $\sim 0.15\%$ for an alginite-bearing mudstone and allowed determinations of more reliable thermal maturity gradients than previously calculated.

Grosjean and co-authors[73] used VR and FAMM data to reassess thermal maturity of the basal Kockatea Shale in the offshore part of the northern Perth Basin (Australia). They found that VR data for samples studied from the Hovea Member indicate immaturity to marginal maturity for oil generation but the FAMM data shows that the VR underestimates maturity by up to 0.3% and consequently the sequence studied is 'early mature'.

4.3.4.2   *Implications for Heat Flow Modelling*

The implications of VR suppression for modelling heat flow for the Malay Basin was discussed by Waples,[74] who showed that the modelling based on

FAMM data justified the use of a more constant heat flow than that based on the suppressed VR values.

Okui and co-authors[75] found that FAMM data for samples from various south-east Asian exploration wells, together with temperature measurements, allowed the use of a high constant heat flow for modelling, which is more consistent with the geological setting than the heat flow implied by the suppressed VR data.

Okui and co-authors[76] used FAMM data to constrain basin models for the petroleum system of the Khmer Trough (Cambodia) because the VR values are lower than predicted from corresponding temperature data. The models based on VR led to interpretation of low palaeo-heat flow followed by a rapid rise from the Pleistocene, which could not be geologically justified. The FAMM analyses indicate VR suppression of up to 0.3%, consistent with a more constant heat flow.

4.3.4.3 Extending FAMM Analyses to Cenozoic Rocks

Rock-Eval, VR, and FAMM analyses were compared by Teerman *et al.*[62] and used to derive a FAMM calibration for a suite of Tertiary coals from Indonesia. Except for a minor deviation to the 'subhydrous' field in the 0.9–1.2% EqVR range, the calibration curve for 'normal' vitrinites is similar to the original FAMM calibration based largely on Australian Permo-Triassic coals.[57,60] They also found that the coals heated by igneous intrusions are characterised by subhydrous vitrinite having enhanced VR.

4.3.4.4 Comparison with Other Thermal Maturity Techniques

Cooper and co-authors[77] used a combination of apatite fission track thermochronology (AFTT), VR, and FAMM analyses to assess thermal maturity of a suite of samples from the Otway Ranges, Victoria, Australia. For samples from the south-east and south-west areas, the results from the three types of analyses are in general agreement but, due to subhydrous vitrinite compositions, FAMM EqVR values are significantly lower than the VR values for samples from the north. Where AFTT data are available for these samples they are in agreement with the FAMM data.

Comparisons between various thermal maturity indicators, including VR, spore coloration index, Rock-Eval T_{max}, and FAMM were discussed by Wilkins,[55] using data from strata intersected by three petroleum exploration wells in the North-West Shelf of Australia (Bowers-1, Carnarvon Basin; Flamingo-1, Bonaparte Basin; Kalyptea-1, Browse Basin). Because of dissimilarities between activation energies for the organic reactions on which the various methods are based, the author emphasised that discrepancies between thermal maturity values from the different techniques are not surprising. In addition, complexities due to the presence of cavings, mud additives, and oxidised OM, as well as identification of indigenous OM, contribute to discrepancies.

On the basis of analyses of a suite of coals from the Sydney Basin (Australia), Wilkins *et al.*[58] gave a detailed account of comparisons of two different petrographic methods for thermal maturity/rank evaluation: FAMM and paired reflectance-fluorescence analyses, in part similar to the methods of Quick,[78,79] Newman,[80,81] and Newman *et al.*[82] They found that both methods can only be applied to OM that has not been affected by post-sampling oxidation and that the methods generally yield similar results although significant differences exist for some samples.

4.3.4.5   Rank Evaluation of Coals

The FAMM technique has also been applied for rank evaluation of coals. For example, VR, FAMM, and Rock-Eval results from a suite of Carboniferous, Late Cretaceous, and Tertiary coals and source rocks from North America indicate that most of the vitrinites studied have 'normal' (orthohydrous) compositions.[83] The study also showed that the original FAMM calibration[57,60] is suitable for the North American samples. In addition the potential for errors in FAMM results was identified for samples having maturities near the calibration limits of the technique.

A similar study on Carboniferous coals from the Netherlands by Veld *et al.*[69] showed that the 'normal' telovitrinites analysed are more perhydrous than those used in the original FAMM studies, such that a revised calibration curve, applicable for the studied coals, was proposed. In addition, FAMM-derived EqVR results were applied for thermal maturity evaluation of samples from the Hengevelde-1 well, where the VR is commonly anomalously low.

Using combined VR and FAMM data, Kalkreuth *et al.*[84] found that a suite of Permian coals studied from the Paraná Basin (Brazil) contain orthohydrous and perhydrous vitrinites such that VR values are suppressed by up to 0.2%. The suppression is not definitively related to liptinite contents nor with mineral contents. In the Candiota Coalfield the VR suppression was previously suspected because of inconsistencies between VR and calorific values.

4.3.5   Limitations and Benefits

The main limitation of the FAMM method is that it is only calibrated for the EqVR range of 0.4–1.2% for Carboniferous–Tertiary rocks, although meaningful results have been obtained for Devonian sections that contain higher plant-derived OM.[85]

The FAMM technique is only applicable to rocks that have EqVR values greater than 0.4% because the method is largely based on secondary fluorescence of vitrinite. At EqVR values of about 0.4% the primary fluorescence disappears and secondary fluorescence develops. If low maturity is not recognised, anomalously high EqVR values may be produced and major inconsistencies will arise between the indicated FAMM-derived EqVR and VR corrected according to iso-correction curves on FAMM diagrams. This problem can be addressed by carrying out parallel VR measurements for samples suspected to have low thermal maturities. For rocks having EqVR values greater than 1.2%, fluorescence intensity of vitrinite abruptly decreases.

Maceral fluorescence is sensitive to oxidation, the main consequence being a decrease in fluorescence intensity. Rocks that are likely to have been substantially oxidised after maturation such as outcrop samples, porous sandstones, and iron-stained samples, should therefore be avoided for FAMM analyses. Because the mineral matrices of claystones effectively protect OM from oxidation,[56] shaly lithologies sampled from the subsurface are the most appropriate for FAMM analyses. Because of the risk of oxidation after sample preparation, polished blocks need to be stored under vacuum and analysed shortly after preparation. If oxidation of the prepared samples occurs, a fresh surface can be achieved by repolishing.

As with all methods based on OM analyses for thermal maturity evaluation, confidence in FAMM data may be lower for samples complicated by an abundance of reworked OM, cavings, and mud additives, or samples that contain only traces of very finely disseminated OM.

The FAMM technique affords the benefits of other petrological methods such as VR, in that measurements are made on specific OM types and identification of complicating effects of contamination, reworking, bitumens, and cavings is possible through optical examination. In addition, minimal sample preparation is required and analyses are non-destructive. As with the VR method, FAMM analyses are possible on a wide variety of sample types, including cuttings and OM-poor rocks. An additional advantage of the FAMM method is that, if higher levels of plant-derived inertinite and liptinite are present, the use of 'multimaceral curves' on fluorescence alteration diagrams[35,56,58–60] may allow estimations of EqVR to be made for vitrinite-poor or vitrinite-free samples.

The main benefit of the FAMM method, however, is that it involves two parameters, one that is mainly controlled by OM type and one mainly controlled by thermal maturity. This enables solutions to many of the limitations of other techniques, such as delineating complex OM assemblages and, most importantly, VR suppression/enhancement.

4.4 Overview of Quantitative Grain Fluorescence and Total Scanning Fluorescence

4.4.1 Introduction

Petroleum or oil inclusions trapped in petroleum reservoirs or along oil migration pathways by mineral diagenesis can preserve important information on the pressure and temperature conditions and compositions of the fluids at the time of trapping.[86,87] They provide crucial information for petroleum system analysis[88] with regard to hydrocarbon source, charge timing, and preservation potential.[14,40,89–93] Fluid inclusion data have been widely used to reconstruct palaeo-temperature and pressure histories of petroleum reservoirs,[40,41] and to determine hydrocarbon migration and accumulation[95,96] and charge timing,[96] as well as fluid evolution within a reservoir.[90,97]

Conventionally, petroleum fluid inclusions are primarily analysed with microscopic,[87,98] spectroscopic,[33,34,90] and molecular geochemical[99–102] methods. Oil inclusion abundances are normally estimated microscopically by grain counting.[95] The 'grains containing oil inclusions' (GOI™) method[94,98] has proved to be particularly useful in identifying palaeo-oil columns and has been widely applied.[90,91,94] Oil inclusion abundances can also be proxied using the n-alkane yields from extracts after crushing, *e.g.* the 'molecular composition of inclusions' (MCI) technique using GC-MS.[99] The MCI data can also provide detailed information on source and thermal maturity of inclusion oils using biomarkers.[102,103]

Conventional microscopic and spectroscopic methods generally involve measurement of a limited population of oil inclusions identified under the microscope under long-wavelength UV ($\sim$365 nm) in a thin section. Such an approach is labour intensive and requires expert knowledge to identify individual oil inclusions before a measurement can be performed. Although MCI analysis yields detailed information, it is also laborious and costly and is not suitable for routine analyses of large numbers of samples. A widespread screening technique termed fluid inclusion stratigraphy (FIS)[104] detects hydrocarbons and inorganic compounds by bulk analysis of volatiles liberated from cleaned core chips or cuttings that are crushed using pneumatic rams. Crushing small quantities of a large number of samples can create geochemical logs that unravel more complex fluid transition zones than suggested by resistivity logs; an example of this application has been illustrated for the Magnus oil field in the North Sea (UK).[104] The quantitative grain fluorescence (QGF)[13,105] and TSF techniques for inclusions[37,38] provide alternative rapid, cost-effective solutions for studying oil inclusions using fluorescence spectrophotometry.

4.4.2   Background

Hydrocarbons present in pore spaces of reservoir rocks are investigated routinely by solvent extraction. Core extracts are enriched in polar components compared with drill stem test (DST) recovered oils,[106–108] and sequential extracts using increasingly polar solvents contain more polar compounds.[109,110] This observation suggests that polar compounds in crude oil have an affinity for mineral surfaces[108,111] and that their abundance may be used as a proxy for hydrocarbons in reservoir pore spaces.

UV fluorescence spectroscopy can not only detect the presence of oil but can also be used to deduce bulk chemical and physical properties.[22,112] To study thermal maturation, degradation, and physical attributes, fluorescence spectra can be measured on solvent-diluted (*e.g.* hexane, heptane, or dichloromethane) whole oil, on oil fractions,[9,20,113,114] and on undiluted whole oils using microscopic spectrometry[10,11] and laser fluorescence spectroscopy.[7,43]

Oils and individual hydrocarbon compounds fluoresce differently and have distinct spectra when they are measured as undiluted films rather than in solvents.[7] Liu and Eadington[13] developed techniques to detect traces of oil and hydrocarbon compounds trapped as inclusions in reservoir grains or adsorbed on grains by directly illuminating the grains (QGF) or solvents after extraction (QGF-E). The characteristic spectra of known oils were used to recognise various fluorescence spectra emitted from inclusion oils and hydrocarbons adsorbed on reservoir grains. The TSF technique was developed[37,38] to fingerprint fluid inclusion oils by using fluorescence spectroscopy for analysis after crushing the rock samples.[100]

4.4.3   Analytical Methods

The QGF, QGF-E, and TSF techniques for analyses of oil inclusions involve standardised procedures for sample preparation, analysis, data processing, and interpretation.[13,37,38,105]

4.4.3.1   Sample Preparation

For QGF and QGF-E analysis, core and cuttings samples are disaggregated into individual grains by crushing and gentle agitation. The reservoir grains are then sieved to select the sand fraction (63 μm–1 mm) to obtain quartz and feldspar grains using electromagnetic separation or other physical methods. The concentrated grains are then cleaned using dichloromethane (DCM), diluted hydrogen peroxide, and hydrochloric acid in an ultrasound bath to remove any drilling additives and soluble hydrocarbons.[13] The cleaned grains are extracted a second time with DCM to recover hydrocarbon compounds adhering to the grain surface. The extract is preserved for QGF-E analysis, whereas the cleaned quartz grains are dried for QGF analysis.

For TSF analysis, the QGF cleaned samples are further cleaned using hot aqua regia and concentrated hydrogen peroxide to further remove any hydrocarbons or mineral impurities on grain surfaces.[37,38] The dried grains are extracted with DCM for analysis using a fluorescence spectrophotometer. Once the grains are deemed to be thoroughly cleaned, they are measured to obtain QGF$^+$ spectra. The thoroughly cleaned grains are then crushed in DCM to extract the fluid inclusion oil for TSF analysis.

4.4.3.2   Instrument Setup and Measurement Procedure

An Agilent Varian Cary-Eclipse fluorescence spectrophotometer is used to analyse reservoir grains (QGF) and surface extracts (QGF-E), thoroughly cleaned grains (QGF$^+$), and inclusion extracts (TSF). The instrument is fitted with an external bandwidth filter to minimize stray light and a customised microplate sampling stage which allows multiple aliquots to be analysed automatically during QGF and QGF$^+$ analyses. The QGF-E and TSF extracts are analysed in a UV-transparent quartz cuvette cell.

An excitation wavelength of 254 nm is employed for QGF and QGF$^+$ analysis. This wavelength has been selected to generate fluorescence in a wide range of compounds and to produce maximum intensity. An excitation wavelength of 260 nm is used for QGF-E analysis. The emission fluorescence spectra are recorded at 300–600 nm.

For TSF analysis, spectral measurements are made using a 3.5 mL quartz cuvette cell. Consecutive excitation from 220 nm to 340 nm at 2 nm intervals is used, while the corresponding emission spectra are recorded between 250 and 540 nm synchronously to produce 3D fluorescence spectrograms.

4.4.3.3   Data Processing and Interpretation

The acquired fluorescence spectra are processed to obtained some key spectral parameters and produce plots for geological interpretation. For the QGF (QGF$^+$) spectra, the parameters derived include maximum intensity, QGF (QGF$^+$) index (ratio between the average intensity for 375–475 nm and that at 300 nm), spectral peak (λ_{max}) and 'full width at half maximum' (FWHM or $\Delta\lambda$).[13] Only two parameters are used for interpreting QGF-E spectra: the maximum fluorescence intensity normalised to weight (grains) and volume (solvent), and λ_{max}. For TSF spectral analysis, the key parameters used are normalised maximum intensity, excitation and emission wavelength corresponding to the spectral peak (Ex/Em), and a ratio of the emission intensities at 360 nm to that at 320 nm corresponding to the excitation wavelength of 270 nm (TSF R$_1$).[27,37]

The QGF (QGF$^+$) index and QGF-E intensity parameters for samples within reservoir intervals are normally plotted with depth against geophysical logs (*e.g.* gamma-ray log). Other QGF and QGF-E parameters such as $\Delta\lambda$ and λ_{max} may also be plotted with depth or as cross plots to determine the (palaeo-) oil–water contact (OWC) or gas–oil contact (GOC). QGF and QGF-E spectra are also presented alongside the QGF index and QGF-E intensity-depth profiles at appropriate depths to help differentiate water zone, current residual zone, and palaeo-oil zone samples.[13,105]

Interpretation of QGF and QGF-E data is primarily based on spectral characteristics and depth profiles of QGF index values and QGF-E intensities. A local baseline for the reservoir investigated is ideally established using samples from the water zone. In the case of an unknown water zone or a lack of samples, an empirical database is used as a reference. Liu and Eadington[13] found that in current producing oil wells in the North-West Shelf of Australia the QGF index values are generally greater than 4 photometer counts (pc), whereas the QGF-E intensities are greater than 40. However, it should be pointed out that there are no uniform global cut-off thresholds for either QGF index or QGF-E intensity data as both are affected by a number of geological variables that may be unique for each reservoir investigated.[105]

4.4.4 Some Case Studies

In a series of publications and conference presentations, Liu and co-authors applied the QGF (QGF$^+$), QGF-E, and TSF techniques to investigate hydrocarbon migration, accumulation, and preservation, as well as oil families for a number of basins in Australasia, south-east Asia, and China.[13,32,37,38,102,115–120] This section presents some case studies of the use of those techniques to investigate various petroleum system elements and processes.[88]

4.4.4.1 *Detecting Residual and Palaeo-Oil*

By using the QGF and QGF-E techniques Liu and co-authors[13,105] found widespread residual and palaeo-oil zones in currently producing basins in Australia and Papua New Guinea. Residual oil zones are characterised by elevated QGF-E intensities in excess of 40 pc with upwards-increasing depth profiles that are commonly below current OWC or free-water level (FWL). Palaeo-oil zones are featured with upwards-increasing QGF index-depth profiles from a baseline defined in the water leg, which is normally less than 4. Palaeo-oil zones have been found in current water zones below OWC, current gas zones above GOC, and in sections intersected by dry wells. The presence of both residual oil and palaeo-oil zones in wells indicates re-adjustment of reservoirs after the initial charge due to seal breaching, late gas flushing, or reservoir dynamic readjustment. Figure 4.6 shows the presence of a palaeo-oil zone below the current OWC and residual oil in the seal unit for the Gidgealpa-17 well in the Cooper Basin (Central Australia), based on QGF and QGF-E analyses. The initial oil accumulation was leaked through the seal due to poor caprock integrity.

4.4.4.2 *Current Oil and Palaeo-Oil Correlation*

TSF spectral signatures of oils and inclusion oil extracts provide fingerprints to understand similarities and differences between current day reservoir oil and palaeo-oil (inclusion oil). In the hydrocarbon charge history study of the Lunnan Oilfield (Tarim Basin, western China), Gong *et al.*[121] used the QGF$^+$ and TSF techniques to document the contrasting differences and similarities between currently reservoired oil and fluid inclusion oil in the Ordovician and Triassic reservoir intervals (Figure 4.7). In the Triassic reservoir interval, the TSF signature of the current reservoir ($R_1 = 2.5$) is similar to that of the petroleum inclusions ($R_1 = 2.9$), indicating that the reservoir has not experienced any alteration since the initial charge. In contrast, in the Ordovician reservoir interval, the TSF spectral signature of the current oil ($R_1 = 1.5$) is different from that of the inclusion oil ($R_1 = 3.4$), suggesting that the reservoir was initially charged with a less mature oil with a subsequent late charge of high mature oil or condensate. The findings from the TSF spectral signatures are consistent with those from the MCI-derived biomarkers.

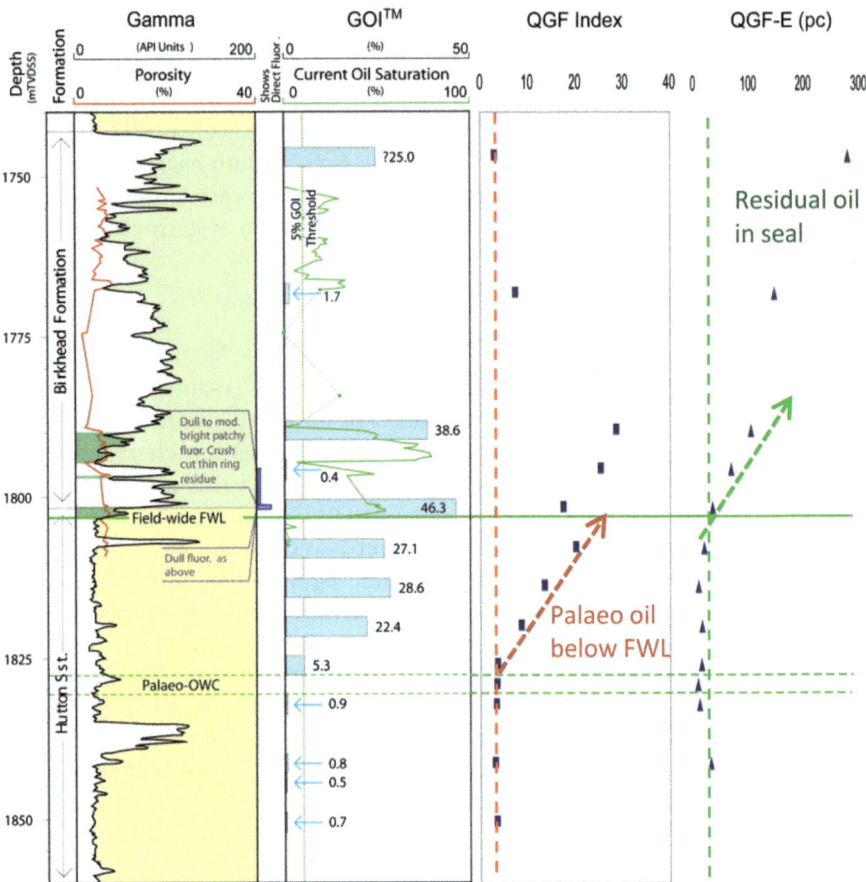

Figure 4.6 Depth profiles of QGF-E intensity and QGF Index from Gidgealpa-17, Cooper Basin, central Australia showing (1) the presence of a palaeo-oil zone below the current OWC (FWL) as delineated by the QGF Index profile, and (2) the presence of residual oil in the seal (leaked) as shown by the elevated QGF-E intensities. Both QGF Index and QGF-E intensity depth profiles display distinct upwards-increasing trends.

4.4.4.3   Inclusion Oil Properties

QGF$^+$ and TSF have been useful in differentiating oils with different American Petroleum Institute (API) gravities and maturities.[38,118] Figure 4.8 summarises three categories of correlations between the QGF$^+$ spectral signatures, TSF spectrograms, and n-alkane profiles for 43 samples.[38] In general, condensates have n-alkane maxima at n-C_{12} to n-C_{14}, corresponding to QGF$^+$ spectral peaks below 420 nm and TSF R_1 values less than 2.0. Most of the normal to light oils have a unimodal n-alkane distribution with maxima at about n-C_{16} to n-C_{23} and corresponding to QGF$^+$ fluorescence spectral peaks at 420–450 nm and TSF R_1 values of 2.0–3.0. Medium to heavy

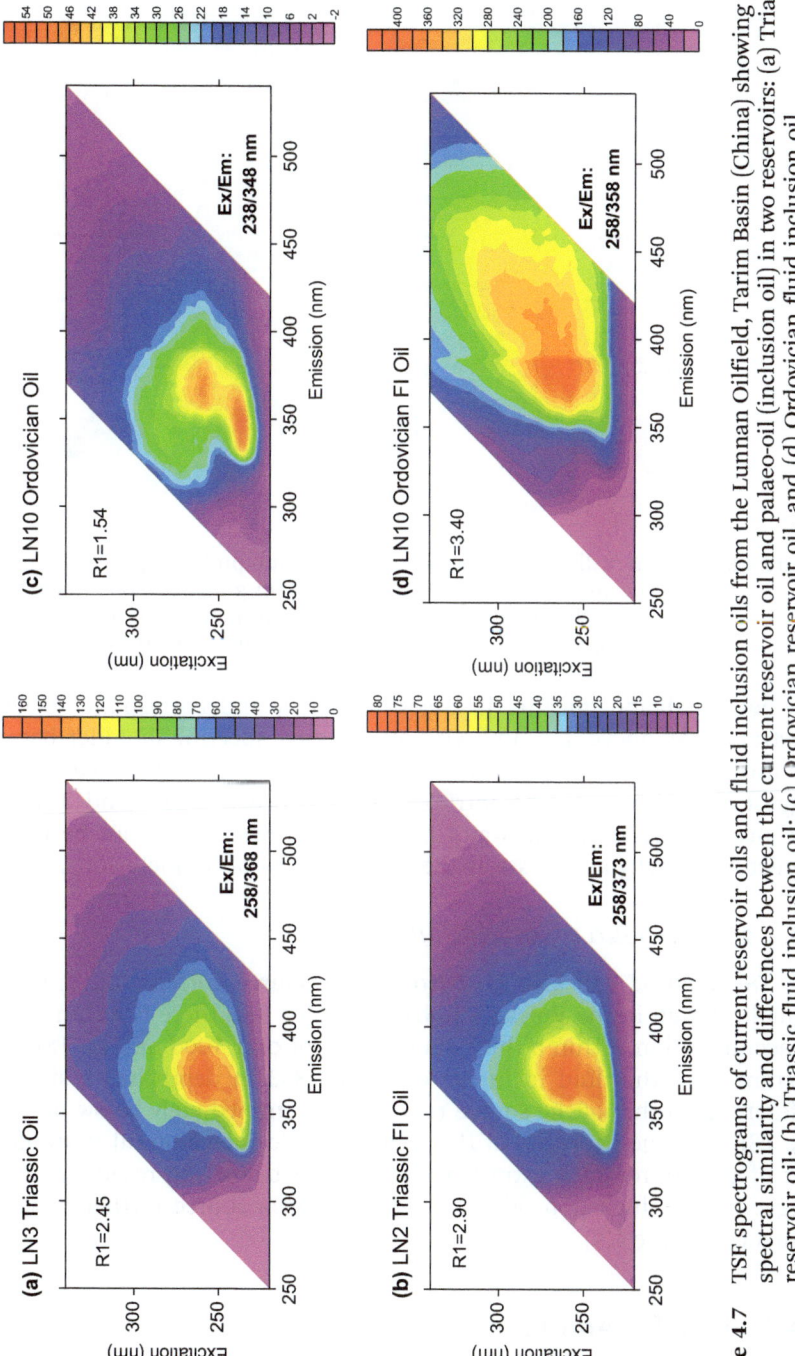

Figure 4.7 TSF spectrograms of current reservoir oils and fluid inclusion oils from the Lunnan Oilfield, Tarim Basin (China) showing TSF spectral similarity and differences between the current reservoir oil and palaeo-oil (inclusion oil) in two reservoirs: (a) Triassic reservoir oil; (b) Triassic fluid inclusion oil; (c) Ordovician reservoir oil, and (d) Ordovician fluid inclusion oil.

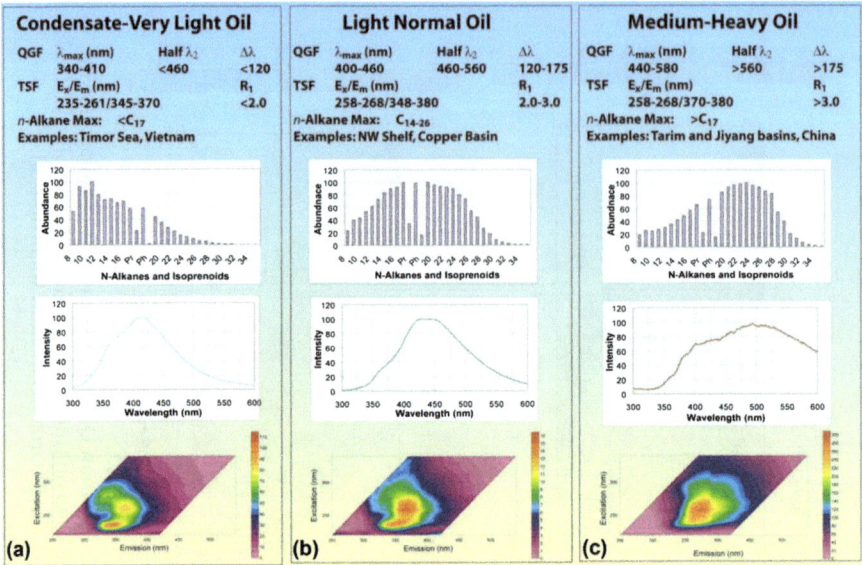

Figure 4.8 Summary diagrams of n-alkane distribution profiles, QGF⁺ spectra, and TSF spectrograms for three categories (API gravity) of oils (inclusions), indicating the fluorescence techniques can be used to differentiate oil (inclusion) types, in a way similar to the n-alkane distribution.

oils have n-alkane maxima in excess of n-C$_{24}$ and correspond to QGF$^+$ spectral peaks above 450 nm, with an elevated spectral shoulder between 450 nm and 600 nm and TSF R$_1$ values greater than 3.0. These correlations suggest that there may be intrinsic links between QGF$^+$ and TSF fluorescence spectral signatures and n-alkane profiles, which can be used to predict API gravity and possibly thermal maturity of oil inclusions.

4.4.4.4 Oil–Source Correlation

The TSF method can also be used for oil–source correlation.[122] For the Paqualin-1 well in the Vulcan Sub-basin, Timor Sea, there is a thin current oil accumulation within a thick source rock of over 1000 m. The TSF spectrogram from the fluid inclusion oil from 2853 m depth has a signature different from that of the source rock extracts immediately below or above. The TSF R$_1$ value (R$_1$ = 2.0) of the reservoir fluid inclusion oil is similar to that of the source rock extract about 1000 m below the reservoir unit,[122] suggesting the inclusion oil migrated oil from a source further down the sequence (Figure 4.9)

4.4.5 Benefits and Limitations

As demonstrated by Liu and co-authors,[13,37,38,105] QGF and TSF analyses offer rapid and versatile screening techniques that are able to address issues

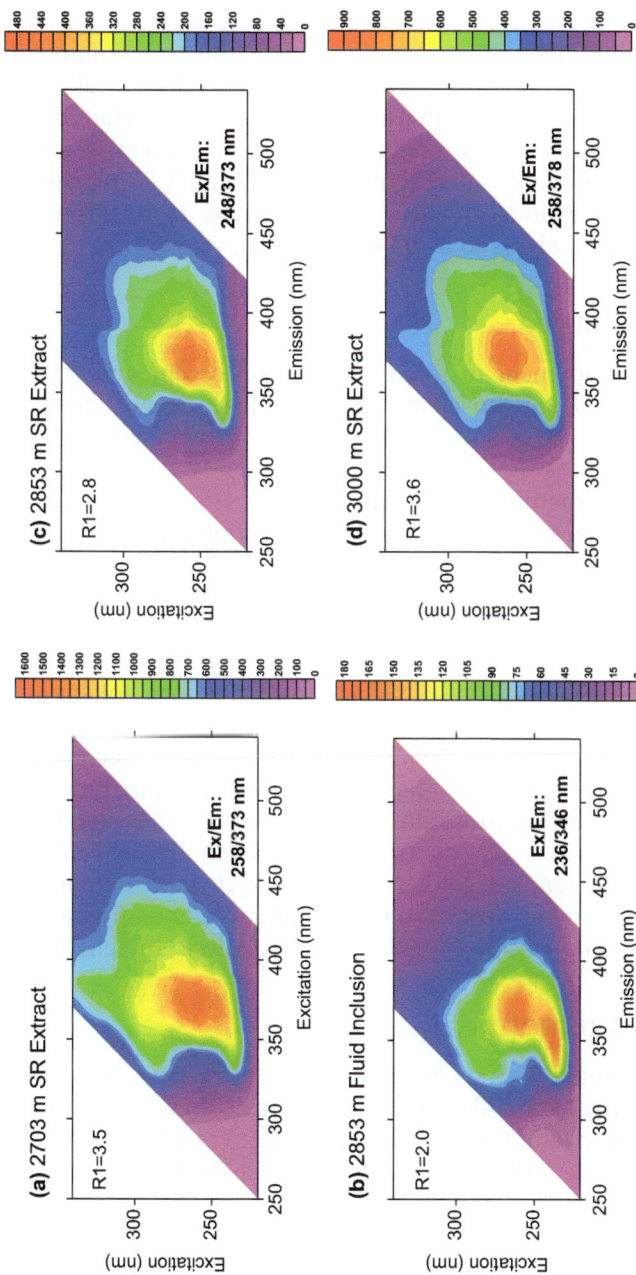

Figure 4.9 TSF spectrograms of source rock extracts and fluid inclusions from Paqualin-1, Vulcan Sub-basin, Timor Sea showing the reservoir oil trapped in fluid inclusions are migrated from mature source rock, not from the adjacent ones.

relating to petroleum system analysis including hydrocarbon migration, accumulation and preservation, reservoir evolution, and oil–oil and oil–source correlation. They provide an alternative to the conventional petrographic and molecular geochemical and isotopic methods

The main limitation to the application of these techniques is the requirement for large numbers of analyses to establish local baselines for interpretation. Other factors that may affect the interpretation of the fluorescence data are potential complications from mineral fluorescence. It should also be borne in mind that the QGF-E and TSF intensities may be affected by fluorescence quenching. Care needs to be taken in sample preparation and analyses to avoid such pitfalls.

4.5 Summary

The recent development in fluorescence spectroscopy and related techniques offers petroleum geoscientists and engineers some versatile tools in characterising the key elements in petroleum system analysis, including

- source rock maturity (*e.g.* FAMM)
- oil migration, accumulation, and preservation (ALF, SLF, QGF and TSF)
- oil properties (TSF, STFS)
- specification of petroleum species (TSF and FLIM).

Combined with advanced analytical software packages, fluorescence spectroscopic methods provide crucial information for a holistic approach to petroleum system analysis.

Further development of fluorescence spectroscopic techniques lies in developing more robust field-based and downhole instrumentation, and compound specific hydrocarbon sensors based on TSF and FLIM for detecting oils in oceans and on land. As demonstrated by Mullins and co-workers,[7] integration with Raman, UV-vis, and IR spectroscopic techniques to determine mineral species allows analyses of non-fluorescing hydrocarbons to be more useful in reservoir fluid analysis and characterisation.

Acknowledgements

The authors would like to acknowledge constructive comments on this overview from CSIRO colleagues Nigel Russell, Dave Whitford, Se Gong, Richard Kempton, and Julien Bourdet, and from Lily Gui from the Research Institute of Petroleum Exploration and Development, PetroChina.

References

1. G. G. Stokes, *Philos. Trans. R. Soc. London*, 1852, **142**, 463–562.
2. *http://link.springer.com/bookseries/4243*.
3. R. E. Riecker, *AAPG Bull.*, 1962, **46**, 60–75.

4. H. Zierfuss and D. Coumou, *AAPG Bull.*, 1956, **40**, 2724–2734.
5. M. V. Reyes, *SPE Form. Eval.*, 1994, **9**, 300–305.
6. K. K. Spilker, P. L. DeLaune and H. L. McKinzie, *Google Patents*, 1997.
7. O. C. Mullins, *The Physics of Reservoir Fluids: Discovery Through Downhole Fluid Analysis*, Schlumberger, Houston, TX, 2008.
8. J. F. Fantasia, *Google Patents, US Patent 3899213*, 1975.
9. T. Barwise and S. Hay, in *Hydrocarbon Migration And Its Near-Surface Expression*, ed. D. Schumacher and M. A. Abrams, AAPG Memoir 66, AAPG, Tulsa, OK, 1996, pp. 363–371.
10. H. Hagemann and A. Hollerbach, *Org. Geochem.*, 1986, **10**, 473–480.
11. A. G. Ryder, in *Reviews in Fluorescence 2005*, ed. C. D. Geddes and J. R. Lakowicz, Springer, New York, 2005, pp. 169–198.
12. L. D. Stasiuk and L. R. Snowdon, *App. Geochem.*, 1997, **12**, 229–241.
13. K. Liu and P. Eadington, *Org. Geochem.*, 2005, **36**, 1023–1036.
14. N. J. Blamey and A. G. Ryder, in *Reviews in Fluorescence 2007*, ed. C. D. Geddes and J. R. Lakowicz, Springer, New York, 2009, pp. 299–334.
15. P. G. Coble, *Marine Chem.*, 1996, **51**, 325–346.
16. P. G. Coble, S. A. Green, N. V. Blough and R. B. Gagosian, *Nature*, 1990, **348**, 432–434.
17. A. T. Rhys-Williams, *An Introduction to Fluorescence Spectroscopy*, PerkinElmer, Inc., United Kingdom, 2000.
18. B. R. Masters, *The Development of Fluorescence Microscopy*, eLS, J. Wiley & Sons, 2010, DOI: 10.1002/9780470015902.a0022093, pp. 1–9.
19. C. Y. Ralston, X. Wu and O. C. Mullins, *Appl. Spectrosc.*, 1996, **50**, 1563–1568.
20. A. Ryder, T. Glynn, M. Feely and A. Barwise, *Spectrochim. Acta, Part A*, 2002, **58**, 1025–1037.
21. A. G. Ryder, *Appl. Spectrosc.*, 2002, **56**, 107–116.
22. I. B. Berlman, *Handbook of Fluorescence Spectra of Aromatic Molecules*, Academic Press, New York, 2nd edn, 1971.
23. L. Stasiuk, T. Gentzis and P. Rahimi, *Fuel*, 2000, **79**, 769–775.
24. B. R. Masters, *Encyclopedia of Life Science*, 2010, 1–9.
25. J. R. Lakowicz, in *Principles of Fluorescence Spectroscopy*, Springer, New York, 1999, pp. 25–61.
26. S. R. Rogers, T. Webster, W. Livingstone and N. J. O'Driscoll, *Estuaries and Coasts*, 2012, **35**, 959–975.
27. J. M. Brooks, L. Barnard, B. Carey, G. Denoux and M. Kennicutt, *Proc. Offshore Technol. Conf.*, Houston, TX, 1983.
28. J. M. Brooks, M. C. Kennicutt and B. Carey, *Oil Gas J.*, 1986, **84**, 66–72.
29. R. Barbini, F. Colao, R. Fantoni, A. Palucci and S. Ribezzo, *Int. J. Remote Sensing*, 1999, **20**, 2405–2421.
30. K. Liu, J. Kurusingal, P. J. Eadington and D. Coghlan, *International PCT Patent Application, No PCT/AU02/00060*, 2002.
31. K. Liu, J. Kurusingal, P. J. Eadington and D. Coghlan, *US Patent Application No 10/470,136*, 2006.

32. K. Liu, S. M. Li, X. Q. Pang, S. Fenton, D. Mills and S. Gong, in *24th International Org. Geochem.*, Bremen, Germany, 2009, p. 497.

33. L. Stasiuk and L. Snowdon, *Appl. Geochem.*, 1997, **12**, 229–241.

34. T. Tsui, *AAPG Bull.*, 1990, **74**, 781.

35. R. Wilkins, N. Sherwood, M. Faiz, S. Teerman and C. Buckingham, *1997, Proc. Petroleum Systems of SE Asia and Australasia Conf.*, May 1997.

36. B. Li, P. W. Ryan, M. Shanahan, K. J. Leister and A. G. Ryder, *Appl. Spectrosc.*, 2011, **65**, 1240–1249.

37. K. Liu, S. Fenton, T. Bastow, B. Van Aarssen and P. Eadington, *APPEA J.*, 2005, **45**, 1–17.

38. K. Liu, C. George, S. Li, X. Pang, S. Fenton, H. Volk and M. Ahmed, *22nd International Meeting on Org. Geochem.*, Seville, Spain, 2005.

39. R. Thiéry, J. Pironon, F. Walgenwitz and F. Montel, *Mar. Petrol. Geol.*, 2002, **19**, 847–859.

40. H.-Y. Tseng and R. J. Pottorf, *Mar. Petrol. Geol.*, 2002, **19**, 797–809.

41. H.-Y. Tseng and R. J. Pottorf, *J. Geochem. Explor.*, 2003, **78**, 433–436.

42. N. J. Blamey, A. G. Ryder, M. Feely and P. Owens, *23rd International Meeting on Org. Geochem. (IMOG)*, Torquay, UK, 9–14 Sept., 2007.

43. A. G. Ryder, *Appl. Spectrosc.*, 2004, **58**, 613–623.

44. A. G. Ryder, M. A. Przyjalgowski, M. Feely, B. Szczupak and T. J. Glynn, *Appl. Spectrosc.*, 2004, **58**, 1106–1115.

45. P. van Gijzel, in *Sporopollenin*, ed. J. Brooks *et al.*, Academic Press, London, 1971, pp. 659–685.

46. P. van Gijzel, *Petrographie organique et potential petrolier*, ed. B. Alpern, CNRS, Paris, 1975, pp. 67–91.

47. M. Teichmüller and K. Ottenjann, *Erdöl Kohle*, 1977, **30**, 387–398.

48. K. Ottenjann, *Org. Geochem.*, 1988, **12**, 309–321.

49. J. Quick, A. Davis and R. L., *47th Ironmaking Conf.*, 1988, pp. 331–337.

50. J. C. Quick, *3rd Coal Research Conference*, Wellington, New Zealand, 1988.

51. M. Teichmüller, *Fluoreszenzmikroskopische Änderungen von Liptiniten und Vitriniten mit zunehmendem Inkohlungsgrad und ihre Beziehungen zu Bitumenbildung und Verkokungsverhalten*, Geologisches Landesamt Nordrhein-Westfalen, 1982.

52. M. Teichmuller and K. Ottenjann, *AAPG Bull,*, 1977, **30**, 387–398.

53. C. F. K. Diessel, *Coal-Bearing Depositional Systems*. Springer Verlag, Berlin, 1992.

54. G. Taylor, M. Teichmüller, A. Davis, C. Diessel, R. Littke and P. Robert, *Organic Petrology*, Gebrüder Borntraeger, Berlin, 1998.

55. R. Wilkins, *AGSO J. Aust. Geol. Geophys.*, 1999, **17**, 67–76.

56. R. Wilkins, C. Buckingham, N. Sherwood, N. Russell, M. Faiz and J. Kurusingal, *APPEA J.*, 1998, **38**, 421–437.

57. R. Wilkins, J. Wilmshurst, G. Hladky, M. Ellacott and C. Buckingham, *APPEA J.*, 1992, **32**, 300–300.

58. R. W. Wilkins, C. F. Diessel and C. P. Buckingham, *Int. J. Coal Geol.*, 2002, **52**, 45–62.

59. R. W. Wilkins, J. R. Wilmshurst, G. Hladky, M. V. Ellacott and C. P. Buckingham, *Org. Geochem.*, 1995, **22**, 191–209.
60. R. W. Wilkins, J. R. Wilmshurst, N. J. Russell, G. Hladky, M. V. Ellacott and C. Buckingham, *Org. Geochem.*, 1992, **18**, 629–640.
61. R. W. T. Wilkins, N. J. Russell and M. V. Ellacott, in *The Sedimentary Basins of Western Australia, Proc. West Australian Basins Symp.*, Perth, WA, 1994, pp. 415–432.
62. S. Teerman, M. Ellacott, J. Wilmshurst and R. Wilkins, in *Organic Geochemistry; Developments and Applications to Energy, Climate, Environment and Human History*, Asociacion Iberica de Geoquimica Organica Ambiental, Donostia-San Sebastian, 1995, pp. 466–468.
63. Y. Ujiié, N. Sherwood, M. Faiz and R. W. Wilkins, *AAPG Bull.*, 2004, **88**, 1335–1356.
64. R. Lin and A. Davis, *Org. Geochem.*, 1988, **12**, 363–374.
65. A. Davis, R. Rathbone, R. Lin and J. Quick, *Org. Geochem.*, 1990, **16**, 897–906.
66. B. Pradier, P. Bertrand, L. Martinez and F. Laggoun-Defarge, *Org. Geochem.*, 1991, **17**, 511–524.
67. B. Pradier, C. Largeau, S. Derenne, L. Martinez, P. Bertrand and Y. Pouet, *Org. Geochem.*, 1990, **16**, 451–460.
68. N. Sherwood, J. Kurusingal and N. Zhong, in *16th Annual Meeting of the Society for Organic Petrology*, Snowbird, UT, 1999.
69. H. Veld, R. Wilkins, X. Xianming and C. Buckingham, *Org. Geochem.*, 1997, **26**, 247–255.
70. M. Ellacott, N. T. Russell and R. Wilkins, *APPEA J.*, 1994, **34**, 216–216.
71. D. W. van Krevelen, *Coal: Typology, Physics, Chemistry, Constitution*, Elsevier, Amsterdam, 1993.
72. H. Petersen, N. Sherwood, A. Mathiesen, M. B. W. Fyhn, N. Dau, N. Russell, J. Bojesen-Koefoed and L. Nielsen, *Mar. Petrol. Geol.*, 2009, **26**, 319–332.
73. E. Grosjean, C. J. Boreham, A. Jones, J. Kennard, D. Mantle and D. Jorgensen, *PESA News Resources WA Supplement*, 2012, pp. 21–25.
74. D. Waples, M. Ramily and W. Leslie, *Geol. Soc. Malaysia Bull.*, 1995, **37**, 269–284.
75. A. Okui, in *Proc. Petroleum Systems of SE Asia and Australasia Conf.*, May 1997, ed. J. V. C. Howes and R. M. Noble. (eds), Indonesian Petroleum Association, 1997, pp. 913–917.
76. A. Okui, A. Imayoshi and K. Tsuji, in *Proc. Petroleum Systems of SE Asia and Australasia Conf.*, May 1997, ed. J. V. C. Howes and R. M. Noble. (eds), Indonesian Petroleum Association, 1997, pp. 365–379.
77. G. T. Cooper, P. B. O'Sullivan, N. Sherwood and K. C. Hill, *Petrol. Explor. Soc. Austr. J.*, 1998, **26**, 159–168.
78. J. C. Quick, in *Vitrinite Reflectance as Maturity Parameter*, ACS Symposium Series 570, 1994, pp. 64–75.
79. J. C. Quick in *9th Annual Meeting. The Society for Organic Petrology*, 1992, pp. 53–55.

80. J. Newman, *APPEA J.*, 1997, **27**, 524–535.
81. J. Newman, *7th New Zealand Coal Conference*, 1997.
82. J. Newman, K. Eckersley, D. Francis and N. Moore, *2000 New Zealand Petroleum Conference Proceedings*, 2000.
83. H. Lo, R. Wilkins, M. Ellacott and C. Buckingham, *Int. J. Coal Geol.*, 1997, **33**, 61–71.
84. W. Kalkreuth, N. Sherwood, G. Cioccari, Z. Corrêa da Silva, M. Silva, N. Zhong and L. Zufa, *Int. J Coal Geol.*, 2004, **57**, 167–185.
85. H. I. M. Struckmeyer, C. J. Boreham, I. Deighton, S. Fenton, R. H. Kempton, J. R. Laurie, M. Faiz, R. Helby, R. S. Nicoll, R. Purcell, N. Russell, G. J. Ryan and N. Sherwood, *New Datasets for the Arafura Basin* (CD). Record 2006/006. Geoscience Australia, Canberra, 2006.
86. R. Burruss, K. Cercone and P. Harris, *Geology*, 1983, **11**, 567–570.
87. J. Parnell, *Geofluids*, 2010, **10**, 73–82.
88. L. Magoon and W. Dow, *AAPG Memoir 60*, 1994, 3–24.
89. R. J. Bodnar, *Mineral. Mag.*, 1990, **54**, 295–304.
90. J. Bourdet, P. Eadington, H. Volk, S. George, J. Pironon and R. Kempton, *Mar. Petrol. Geol.*, 2012.
91. M. Lisk, M. Brincat, P. Eadington and G. O'Brien in *The sedimentary basins of Western Australia 2: Proceedings of the West Australian Basins Symposium*, 1998.
92. J. Parnell, P. Carey and B. Monson, *Chem. Geol.*, 1996, **129**, 217–226.
93. R. Swarbrick, M. Osborne, D. Grunberger, G. Yardley, G. Macleod, A. Aplin, S. Larter, I. Knight and H. Auld, *Mar. Petrol. Geol.*, 2000, **17**, 993–1010.
94. M. Lisk, G. O'Brien and P. Eadington, *AAPG Bull.*, 2002, **86**, 1531–1542.
95. N. H. Oxtoby, A. W. Mitchell and J. G. Gluyas, in *The Geochemistry of Reservoirs*, ed. J. M. Cubitt and W. A. England, Special Publication 86, Geological Society, London, 1995, pp. 141–157.
96. R. K. Mclimans, *Appl. Geochem.*, 1987, **2**, 585–603.
97. S. Burley, J. t. Mullis and A. Matter, *Mar. Petrol. Geol.*, 1989, **6**, 98–120.
98. P. J. Eadington, M. Lisk and F. W. Krieger, *Google Patents*, 1996.
99. S. George, P. Greenwood, G. Logan, R. Quezada, L. Pang, M. Lisk, F. Krieger and P. Eadington, *APPEA J*, 1997, **37**, 490–503.
100. S. George, H. Volk, T. Ruble, M. Lisk, M. Ahmed, K. Liu, R. Quezada, A. Dutkiewicz, M. Brincat and S. Smart, in *21st Int. Meeting on Org. Geochem.*, Krakow, Poland, 2001.
101. S. C. George, M. Lisk and P. J. Eadington, *Mar. Petrol. Geol.*, 2004, **21**, 1107–1128.
102. S. C. George, H. Volk and M. Ahmed, *J. Petrol. Sci. Eng.*, 2007, **57**, 119–138.
103. S. C. George, T. E. Ruble, A. Dutkiewicz and P. J. Eadington, *Appl. Geochem.*, 2001, **16**, 451–473.
104. S. Barclay, R. Worden, J. Parnell, D. Hall and S. Sterner, *AAPG Bull.*, 2000, **84**, 489–504.
105. K. Liu, P. Eadington, H. Middleton, S. Fenton and T. Cable, *J. Petrol. Sci. Eng.*, 2007, **57**, 139–151.

106. I. Horstad, S. Larter, H. Dypvik, P. Aagaard, A. Bjørnvik, P. Johansen and S. Eriksen, *Org. Geochem.*, 1990, **16**, 497–510.
107. D. A. Karlsen and S. R. Larter, *Org. Geochem.*, 1991, **17**, 603–617.
108. L. S. Pang, S. C. George and R. A. Quezada, *Org. Geochem.*, 1998, **29**, 1149–1161.
109. L. Schwark, D. Stoddart, C. Keuser, B. Spitthoff and D. Leythaeuser, *Org. Geochem.*, 1997, **26**, 19–31.
110. A. Wilhelms, I. Horstad and D. Karlsen, *Org. Geochem.*, 1996, **24**, 1157–1172.
111. T. T. Schowalter and P. D. Hess, *AAPG Bull.*, 1982, **66**, 1302–1327.
112. E. J. Billo, *Excel® for Chemists: A Comprehensive Guide*, Wiley, New York, 2nd edn, 2002, pp. 339–348.
113. T. D. Downare and O. C. Mullins, *Appl. Spectrosc.*, 1995, **49**, 754–764.
114. A. G. Ryder, *J. Fluoresc.*, 2004, **14**, 99–104.
115. K. Liu, S. George and P. J. Eadington, in *21st Meeting on Organic Geochemistry*, Krakow, Poland, 2003, vol. Part II, pp. 177–178.
116. K. Liu, X. Pang and S. Fenton, in *AAPG International Conference*, Perth, Australia, 2006, pp. 5–8.
117. K. Liu, X. Pang, Z. Jiang and J. Zhang, in *AAPG Annual Convention*, Houston, TX, 2006.
118. K. Liu, X. Xiao, D. Mills, S. C. George, H. Volk and S. Gong, *J. Geochem. Explor.*, 2009, **101**, 62.
119. K. Liu and P. Eadington, *J. Geochem. Explor.*, 2003, **78–79**, 389–394.
120. K. Liu, P. Eadington and D. Coghlan, *J. Petrol. Sci. Eng.*, 2003, **39**, 275–285.
121. S. Gong, S. C. George, H. Volk, K. Liu and P. A. Peng, *Org. Geochem.*, 2007, **38**, 1341–1355.
122. K. Liu, X. Pang, S. Fenton, F. Wang and S. Li, in *23rd International Meeting on Organic Geochemistry*, Hallsanney, UK, 2007.

CHAPTER 5

Time-of-Flight Secondary Ion Mass Spectrometry (TOF-SIMS): Principles and Practice in the Biogeosciences

VOLKER THIEL*[a] AND PETER SJÖVALL*[b]

[a] University of Göttingen, Geoscience Center, Geobiology Group, Goldschmidtstraße 3, 37077 Göttingen, Germany; [b] SP Technical Research Institute of Sweden, Chemistry, Materials and Surfaces, P.O. Box 857, SE-501 15 Borås, Sweden
*Email: vthiel@gwdg.de; peter.sjovall@sp.se

5.1 Introduction

Time-of-flight secondary ion mass spectrometry (TOF-SIMS) is a technique to analyse the chemical composition of surfaces. In TOF-SIMS, the sample surface is bombarded *in vacuo* using a beam of high (keV) energy primary ions.[1] At every primary ion pulse, secondary ions are released from the sample surface and are recorded in a TOF mass analyser to produce a mass spectrum. The collection of mass spectra at a (sub-)μm raster pattern resolution enables the operator to probe very small structures of interest as well as obtaining images ('ion maps') of the distribution of user-selected secondary ions emitted from the sample surface.

Using a pulsed primary ion beam in the so-called static SIMS mode,[2] only a small fraction of the molecular surface layer is consumed for analysis, so

RSC Detection Science Series No. 4
Principles and Practice of Analytical Techniques in Geosciences
Edited by Kliti Grice
© The Royal Society of Chemistry 2015
Published by the Royal Society of Chemistry, www.rsc.org

the surface is not significantly damaged by the incoming primary ions (and thus remains 'static'). Static TOF-SIMS, at a typical analysis depth of a few nanometres, therefore allows the detection of quite large organic species, including ions in the molecular or near-molecular mass range of typical biomarker classes such as sterols, pigments, or even intact polar lipids. In dynamic SIMS, on the other hand, at high primary ion dose densities (typically $>10^{13}$ ions cm^{-2}) the sample surface is continuously eroded and only elemental information is obtained (*e.g.* NanoSIMS, see Chapter 1 in this volume).

Initially, TOF-SIMS was used mainly in materials science for chemical surface analysis in application areas like polymer science and semi-conductor characterisation.[3,4] Since the beginning of the 21st century, new instrument developments have adopted cluster ion sources (mostly Au_3^+, Bi_3^+, C_{60}^+) instead of the traditional monoatomic Ar^+, Ga^+, and Xe^+ sources as primary ion projectiles (section 5.2). The new sources have increased the yields of secondary ions by about two orders of magnitude, dramatically improving gain of higher-mass organic ions ($> m/z \sim 200$) and opened up attractive perspectives for the analysis of biological materials.[5–8] Since then, research activities employing TOF-SIMS in the biomedical field have been rapidly expanding, and interesting TOF-SIMS studies on mammalian tissues have been published, such as the identification and mapping of cholesterol and glycero(phospho)lipids in rodent and human tissue sections and single cancer cells.[9–17] For a recent overview and further references on biomedical TOF-SIMS studies, see Fletcher and Vickerman.[18]

The capability to combine spatial and molecular information at a micro-scope resolution makes TOF-SIMS, in principle, an attractive option in the biogeosciences (here used to include the fields of organic geochemistry, geobiology, geomicrobiology, microbial ecology, and biogeochemistry). Organic compounds used as biomarkers in these fields are commonly recruited from the compound class of lipids, and are conventionally identified and quantified using coupled gas chromatography–mass spectrometry (GC-MS) or coupled liquid chromatography–mass spectrometry (LC-MS). These well-established and widely used techniques allow for the mass spectrometric analysis of single compounds after chromatographic separation of an extract fraction. A principal shortcoming of GC-MS and LC-MS is, however, that they require relatively large sample amounts and homogenization of the sample prior to extraction and analysis. This prevents the analysis of very small sample amounts as well as small morphological structures in primarily inhomogeneous natural samples. Likewise, it is very difficult to specify the spatial origin of individual analytes within environmental materials, and to directly couple extract-derived biomarker data to histological or petrological information, which are typically obtained at the microscopic level. In cases where these issues are critical, TOF-SIMS may represent an interesting analytical option.

One important strength of TOF-SIMS is the ability to study inorganic and organic ions simultaneously, which can be very useful for the study of biogeochemical composite samples such as mineral-associated organo-clasts, fluid inclusions, or biomineralisates. On the other hand the technique may be somewhat blind to particular analytes because of as yet poorly constrained matrix effects, and TOF-SIMS is, presently at least, not very well suited for the analysis of compounds at low surface concentrations, due to a limited secondary ion yield under static conditions.[19] To overcome these analytical drawbacks, recent analytical developments have been employing giant cluster ion sources (Ar_n^+ with n up to 3000)[20–22] and external accumulation of secondary ions, including MS-MS capabilities.[23–25] Below, we discuss in more detail the principles and developments of the technique (section 5.2), as well as issues of the evaluation of TOF-SIMS data in practice (section 5.3).

Compared to the large number of published TOF-SIMS experiments on clinical targets, *i.e.* mammalian tissues or cells, investigations that are explicitly aimed at topics in the biogeosciences are relatively sparse. Nonetheless these studies cover quite a broad spectrum of relevant sample types such as microbial mats, sedimentary rocks, petroleum, and oil-bearing fluid inclusions. In section 5.4, we give an overview on some of these applications to illustrate the practical use of static TOF-SIMS in the biogeosciences.

This chapter aims to provide readers who already have some background in the established MS techniques with an idea about the principles, analytical window, and selectivity of TOF-SIMS for organic molecular analyses. Here we use the term 'biomarker' in a geochemical' sense, *i.e.* as a tracer substance for organic matter sources, turnover, and biochemical transformation,[26] not in the clinical sense, *i.e.* as a disease marker. This chapter does not deal with elemental/isotopic analyses using dynamic SIMS; for this second branch of SIMS, there are excellent reviews, partly with a (bio-)geochemical focus;[27–30] see also Chapter 1 in this volume.

5.2 Analysis Principle and Instrumentation

5.2.1 Analysis Principle

5.2.1.1 Overview

In a TOF-SIMS experiment, chemical information about the sample surface is provided by secondary ions that are emitted following the bombardment of the surface by high-energy primary ions (Figure 5.1a). The primary ion collision results in the emission of a variety of particles, including atoms, molecular fragments, intact molecules, and molecular complexes. Most of these particles are neutral but a small fraction comes off as ions. The emitted (secondary) ions, which can be either positive or negative, are analysed in a TOF analyser and, by scanning the primary ion beam over the sample surface, separate mass spectra are acquired from each pixel within the analysed area, providing laterally resolved mass spectrometric data of the sample surface at submicron lateral resolution. The acquired data can be displayed

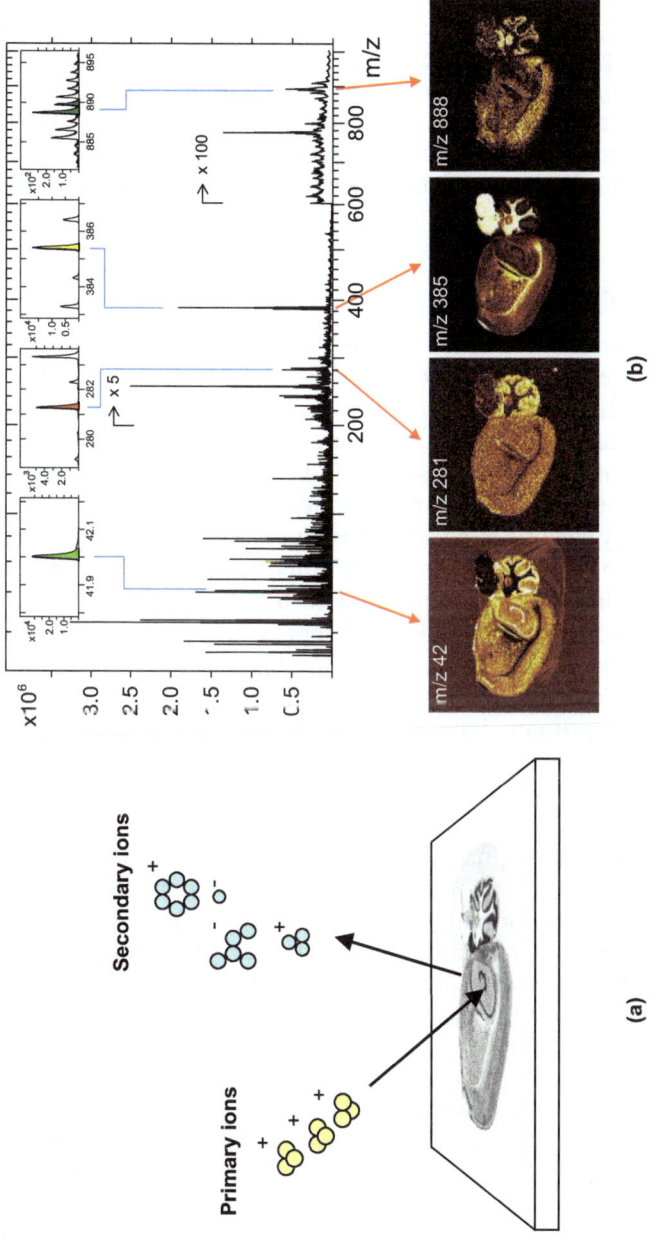

Figure 5.1 Schematic illustration of the TOF-SIMS analysis principle. (a) Secondary ions emitted during the collision of high-energy primary ions with the sample surface are analysed to produce (b) a mass spectrum of the outermost molecular layers of the sample surface. Scanning of the primary ion beam across the analysis area and acquisition of mass spectra from each point make it possible to produce chemically specific ion images, where the signal intensity of selected secondary ions are displayed. The example shows results from a large area analysis of a mouse brain tissue section (12×12 mm^2), including ion images of CNO$^-$ (m/z 281, protein fragment), oleic acid (m/z 281, lipid fragment and/or free fatty acid), cholesterol (m/z 385), and sulfatides (m/z 888, sulfate-containing galactosphingolipid abundant in neural tissue).

in a variety of ways, *e.g.* total area mass spectra and ion images that show the spatial signal intensity distribution of selected secondary ions on the sample surface. The principle is illustrated in Figure 5.1b where the different images of a mouse brain section are generated from the signal intensities of negative ions representing proteins, oleic acid, cholesterol, and sulfatides, respectively. These images were obtained by scanning over a large area of the sample (12×12 mm^2), but measurements are typically made over smaller areas, from 50×50 μm^2 to 500×500 μm^2. The acquired data can also be displayed as mass spectra from selected regions of interest, providing detailed chemical information about specific structures within the analysis area.

5.2.1.2 Molecular SIMS

The chemical information contained in the TOF-SIMS mass spectrum depends critically on the types of secondary ions that are emitted in the SIMS process: that is, if characteristic secondary ions, specific to the different compounds present on the sample surface, are emitted at sufficient yields. Atomic ions and small-fragment ions provide little information about the molecular species, but larger polyatomic fragment ions and molecular ions can be used to identify specific molecules present on the sample surface. For analysis of organic materials, it is therefore important to understand the conditions under which TOF-SIMS spectra with high yields of molecularly specific ions can be obtained. A critical concept in this respect is that of 'static SIMS conditions', which traditionally have been considered essential for obtaining molecular information in SIMS. Another important consideration is the type of primary ions used, since this has a major effect on the yield of molecularly specific secondary ions. In a few cases, the type of primary ions can also relieve or even remove the static SIMS requirement.[31]

The static SIMS concept originates from the fact that the SIMS process is generally considered to be destructive from a molecular point of view.[1,3,32] The incident energy of the primary ions is typically 10–25 keV, which is sufficient to break thousands of chemical bonds, and since this energy is deposited locally at the point of impact of the incident ion, the collision will cause severe fragmentation and a cascade of colliding particles, consisting of atoms and fragments of varying size, near the surface of the sample. A fraction of these energetic particles will escape from the surface to the vacuum, and in the SIMS experiment it is possible to detect those that are ionised in this process. However, another large fraction will remain on the surface and form a molecularly damaged region around the point of impact (around 5 nm in diameter), from which no further molecular information can be obtained (Figure 5.2a). The emission of useful secondary ions thus requires that the majority of primary ions bombard undamaged areas, *i.e.* areas where no previous primary ion collision has occurred. This requirement is fulfilled by completing the analysis before a substantial fraction of

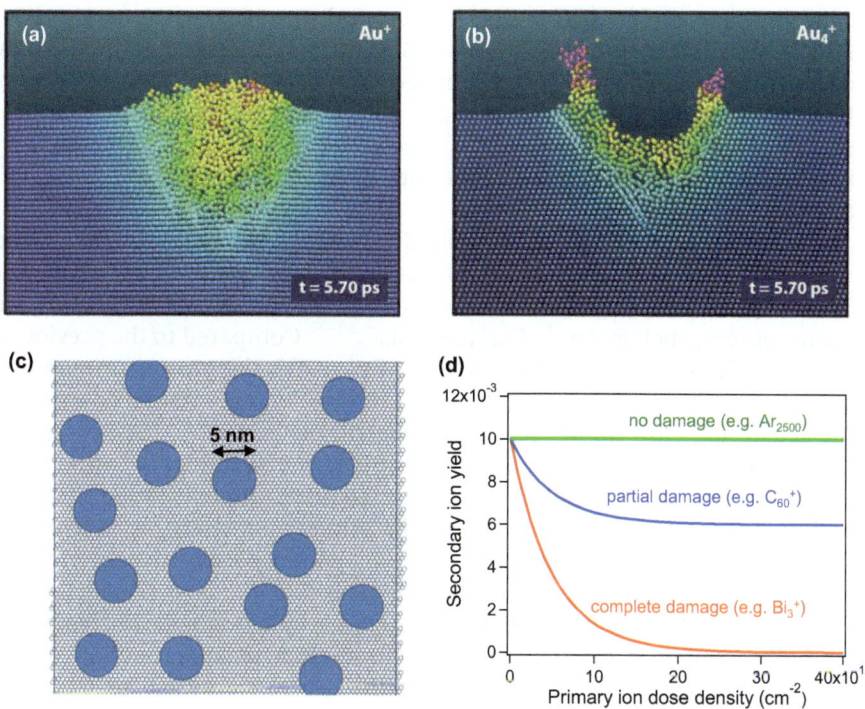

Figure 5.2 Molecular damage and static SIMS. (a, b) Results from molecular dynamics calculations of ion-surface collisions, comparing the effect of atomic (a, Au^+) and cluster (b, Au_4^+) projectiles on a gold sample substrate.[131] The colour code represents temperature, where green, yellow, and red correspond to $1\times$, $1.5\times$, and $2\times$ the melting temperature of gold, respectively. (c) Illustration of the static SIMS conditions showing the damaged area at a primary ion dose density (PIDD) of 1×10^{12} cm^2 assuming a damage cross-section of 20 nm^2 (each primary ion produces a 5 nm diameter molecularly damaged circle on the sample surface). (d) Expected secondary ion signal as a function of PIDD at a 20 nm^2 damage cross-section assuming complete damage (red curve, no molecular signal from damaged area), partial damage (blue curve, reduced molecular signal from damaged area) and no damage (green curve).

the sample surface has been damaged, which happens at an accumulated primary ion dose density (PIDD) in the range of 10^{12}–10^{13} ions cm^{-2}. This accumulated dose density is called the *static limit* and marks the transition from static SIMS (PIDD $< 10^{12}$–10^{13} ions cm^{-2}), for which the analysis is done on a static unaltered surface, to dynamic SIMS (PIDD $> 10^{13}$ ions cm^{-2}), where the sample surface is continuously changing by the primary ion bombardment during the analysis. The situation is illustrated in Figure 5.2c, schematically showing a top view of a sample surface where each large circle represent a damaged area of 20 nm^2 (typical value determined from experiments[7,33,34] produced by one primary ion collision and the concentration of damaged regions correspond to a PIDD of 10^{12} ions cm^{-2}. Furthermore, the red curve in Figure 5.2d shows the expected exponential decay in the signal intensity of a molecularly specific ion as a function of PIDD, assuming that each primary ion collision produces a 20 nm^2 molecularly damaged region.[7,33,35,36] According to this curve, the secondary ion yield has decreased to 82% of its initial value at a PIDD of 10^{12} ions cm^{-2} and to 14% at 10^{13} ions cm^{-2}.

The most critical factor in the SIMS process is the type of primary ion used in the analysis, and an important development in the field of organic TOF-SIMS was the introduction around 10 years ago of cluster primary ion beam sources, such as Au$_n^+$, C$_{60}^+$, and Bi$_n^+$.[35–37] Compared to the previously used atomic primary ion sources (Ga$^+$, Cs$^+$, Ar$^+$), the cluster primary ion beams were shown to increase the yields of large organic fragment ions and molecular ions by up to several orders of magnitude.[7,33] As shown in detail using molecular dynamics computer calculations,[38,39] the main reason for this difference is connected to the penetration of the primary ion and thus how deep into the substrate the primary ion energy is distributed. Light atomic primary ions, such as Ga$^+$, penetrate more deeply into the sample substrate and therefore induce molecular damage relatively deep into the sample, below the outermost molecular layers. Furthermore, since the energy is distributed relatively deep, a smaller fraction of the primary ion energy is transferred to particles that are able to leave the surface and the number of particles actually leaving the surface, and thus the yield of secondary ions, will therefore be relatively low (Figure 5.2a). For heavy cluster primary ions, like Au$_n^+$, Bi$_n^+$ and C$_{60}^+$, the penetration is lower and the incident energy is distributed much closer to the outermost molecular layer, which results in a larger number of particles leaving the surface and a shallower region of molecular damage after the collision (Figure 5.2b).

Compared to Bi$_3^+$, C$_{60}^+$ induces less molecular damage on the sample surface and is even capable of emitting secondary ions originating from undamaged molecular structures in a subsequent primary ion collision.[39] In principle, C$_{60}^+$ is thus capable of providing molecular information in the spectra even at primary ion dose densities above the static limit. The small amount of damage produced by the C$_{60}^+$ primary ions can be rationalised by the immediate dissociation of the C$_{60}$ cluster upon impact and the efficient energy transfer of the primary ion energy that is now distributed between the 60 individual carbon atoms colliding with the surface. This effect is enhanced further in the recently developed giant argon cluster ion sources, for which cluster sizes in the range of 1000–3000 argon atoms have been found to cause the emission of organic molecules and ions essentially without remaining molecular damage.[20–22,40,41] The situation is schematically illustrated in Figure 5.2d, where the curves show the secondary ion yield as a function of PIDD, and the blue and green curves represent typical cases using the C$_{60}^+$ and giant argon cluster ion source for sputtering, respectively. For C$_{60}^+$, the limited damage results in an initial decay in the signal intensity, which, however, stabilises at a non-zero steady-state level.[31,42] The amount of damage, represented by the amount of exponential decay relative to the steady-state level, has been found to vary considerably between different materials/analytes, from almost no damage to complete damage. For giant argon clusters, however, the signal intensity has been found to show essentially no decay from its initial value, indicating very little molecular damage after the collision. In addition, this effect is seen in all cases studied so far, including those where no molecular information has been obtained using C_{60}^{+}. The possibility of continuing the molecular SIMS analysis beyond the static limit and into the dynamic SIMS regime has opened up the possibility for organic depth profiling and 3D analysis;[43,44] see further in section 5.2.2.5.

As mentioned above, only a small fraction of the emitted particles (a few per cent or less) in the SIMS process comes off as ions, while the large majority is neutral. This effect has two important consequences. First, the ionisation probability can be very sensitive to differences in the chemical environment of the analyte, which means that the signal intensity can vary significantly in different samples although the surface concentration is the same. This so-called matrix effect (see section 5.3.1.4) must be carefully considered in all types of evaluation of TOF-SIMS data, including intensity variations in ion images and in comparing spectral intensities between different samples. Secondly, the low ionisation probability provides an opportunity to significantly improve the secondary ion yield if there is a way to increase the ionisation. Different strategies have been developed with this purpose, including laser post-ionisation of the emitted neutrals, also called laser-SNMS (secondary neutral mass spectrometry), where a laser beam parallel to the surface plane is applied to ionise the neutrals after they have been emitted from the sample surface. Various ionisation mechanisms using different laser characteristics (single- and multi-photon processes, resonant and non-resonant, ns to fs pulses) have been applied and it has been demonstrated that the sensitivity can be significantly increased as compared to SIMS.[45–47] Due to the increased ionisation, laser post-ionisation has been shown to reduce the matrix effect, thereby improving the capability for quantification. A disadvantage, however, is that laser post-ionisation also gives rise to extensive fragmentation of the emitted neutrals and the molecular spectral information is therefore considerably reduced. Another strategy to increase the ionisation is to coat the sample surface with a thin layer of matrix, similar to what is done in matrix-assisted laser desorption/ ionisation (MALDI) imaging. Here, the matrix is expected to increase the ionisation probability by providing additional charged particles that can stimulate the ionisation of neutral analytes during the SIMS process. Although the matrix deposition adds complexity to the sample preparation and data evaluation (non-reproducible matrix deposition, matrix crystallisation, *etc.*), matrix-enhanced SIMS (ME-SIMS) has been proved to increase the secondary ion yield considerably and new developments are likely to improve the situation further.[48]

5.2.1.3 *Sensitivity and Lateral Resolution in Static SIMS*

TOF-SIMS is a highly sensitive method with respect to chemical character-isation and molecular detection in very small sample amounts. This also means that very low concentrations can be detected if the analyte is localised to small structures on the sample surface. However, it is not a particularly powerful method for detection of very low concentrations if the analyte is homogeneously distributed in the sample.[10] Furthermore, the detection sensitivity is very different for different analytes and it also varies with the secondary ion used to monitor the analyte. The critical factor with regard to sensitivity is the so-called *secondary ion yield*, which is defined as the number of detected secondary ions, of the specific ion species that is being moni-tored, per incident primary ion.[1,33] Since the number of primary ions that can be applied per unit area is limited by the static SIMS limit, the maximum signal intensity that can be obtained is determined roughly by the secondary ion yield (for a detailed description, see[7,33,49]). The reason for the large variation in sensitivity in TOF-SIMS is that the secondary ion yields of mo-lecularly specific ions can vary by several orders of magnitude for different analytes. In addition, different fragment ions of the same analyte show largely varying secondary ion yields, which frequently leads to the need to compromise between sensitivity (monitoring a less-specific fragment ion with a higher yield) and specificity (monitoring a more-specific ion species with a lower yield). The suitability of using a less-specific fragment ion for monitoring a certain analyte depends on whether the sample contains other molecules that produce the same (or interfering) fragment ion. In that case the detected signal will represent all compounds in the sample producing the monitored ion species.

The lateral resolution that can be obtained in a TOF-SIMS image is limited not only by the focus diameter of the primary ion beam, but equally by the detection sensitivity of the monitored ion species. Even with a very small focus diameter of the primary ion beam, the number of detected ions in one pixel of the corresponding size may not be sufficiently high to produce sig-nificant variations in the signal from one pixel to the next. As an example, the secondary ion yield for the phosphocholine head group ion of palmitoyl-oleoyl-phosphatidylcholine (POPC) in a pure POPC lipid bilayer was deter-mined to be roughly 0.015,[49] which is a relatively high value for a molecularly specific ion. At the static limit of 10^{12} primary ions cm^{-2}, the number of detected secondary ions from a pure POPC structure can then be expected to be around 150 μm^{-2}. In most cases this will be a significant signal, sug-gesting that submicron resolution can be obtained. However, if the con-centration of POPC is only 1%, the expected signal intensity is only 1.5 secondary ions μm^{-2}, which clearly is not a significant signal. Con-sequently, the lateral resolution for detecting changes in this concentration regime is considerably larger than 1 μm^{2}. Since the instrument background in TOF-SIMS is essentially zero, a signal intensity of 4 counts per pixel can, in principle, be considered significant. However, this is only the case if there is no interference in the mass spectra from ions originating from other compounds, and such interferences ('chemical noise') are often present in complex environmental samples, thus requiring higher counts to produce a significant signal. The situation is considerably improved if the analyte is localised (at high concentrations) to a small structure, in which case the data can be extracted from this structure alone, with a much reduced chemical noise (compared to the total chemical noise from the entire analysis area) and a considerably improved overall sensitivity.

The discussion above is based on static conditions as a requirement, which means that only the uppermost 10 nm of the sample surface is probed and available for analysis. If dynamic SIMS conditions can be tolerated, as when C_{60}^+ or giant argon cluster ions are used for sputtering, a thicker portion of the sample surface is available for analysis and considerable improvements in sensitivity can be expected.[18]

5.2.2 Instrumentation and Data Acquisition

5.2.2.1 Instrumentation Overview

Figure 5.3 shows a schematic drawing and a photograph of a TOF-SIMS instrument manufactured by ION-TOF GmbH (Münster, Germany). Together

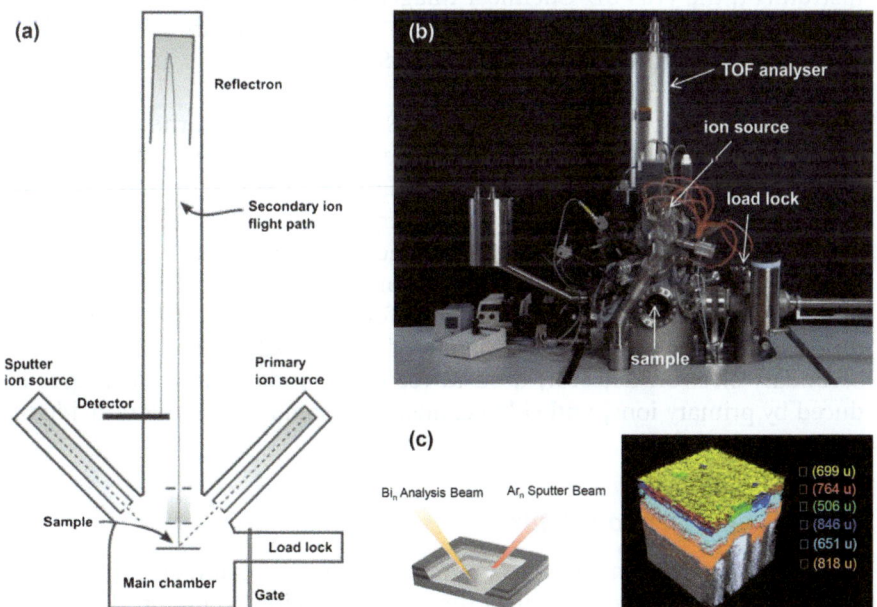

Figure 5.3 (a) Schematic outline and (b) photograph of a TOF-SIMS IV instrument from ION-TOF GmbH. (c) 3D molecular imaging of OLED structure in a dual-beam experiment using giant argon cluster ions for sputtering and Bi_3^+ ions for analysis.[61] The different colours represent signal intensities of (near-)molecular ions, with their peak masses indicated, from the different organic components of the sample (image kindly provided by Dr E. Niehuis).

with similar instruments manufactured by Physical Electronics, Inc. (Chanhassen, MN, USA), this type of instrument has dominated the market for commercial TOF-SIMS instruments during the last 15–20 years, over which period the technique has grown considerably with respect to its use in various academic and industrial research applications. Both these instruments are based on a measurement principle where the primary ions bombard the sample surface in short pulses, enabling the measurement of the flight time of the secondary ions as they travel through the TOF analyser.[1,3,37] For each primary ion pulse, the flight time of each individual secondary ion is determined and all secondary ions thus contribute to the recorded mass spectrum. This leads to a very efficient use of the primary ions, keeping the molecular damage they cause to a minimum, which is important for obtaining sufficient data at PIDDs below the static limit.

During the last few years, instruments with a different measurement principle have been developed. The main difference compared to the earlier type is that the primary ion beam is continuous, *i.e.* not pulsed, and that the secondary ions instead are shaped into short pulses after their emission from the sample surface, before entering the TOF analyser.[23,50] This measurement principle has several advantages over the 'traditional' TOF-SIMS instruments, in particular for 3D analysis of organic samples. A commercial instrument, the J105 3D Chemical Imager is now available from Ionoptika Ltd (Chandlers Ford, Hampshire, UK). These instruments and the differences compared to the 'traditional' TOF-SIMS instruments are further discussed in section 5.2.2.5 below.

The main components of a TOF-SIMS instrument are the primary ion source and the TOF analyser, which are both housed in a high-vacuum chamber that also contains a sample manipulator stage for high-precision positioning of the sample and a low-energy electron gun, used for charge compensation of insulating samples. The instrument may also include a second ion source, which can be used for analysis with an alternative primary ion projectile or, more commonly, for removing material by ion sputtering. Other common components of a TOF-SIMS instrument are a secondary electron detector, used to record secondary electron images induced by primary ions, and video cameras for sample navigation inside the analysis chamber.

5.2.2.2 Primary Ion Source

The purpose of the ion source is to produce a focused beam of short primary ion pulses that can be accurately directed to specific points on the sample surface. This is done in an ion column, where the initially generated ions are shaped into a beam using a variety of devices including electrostatic lenses, deflection plates, apertures, and pulsed electrical choppers. The mechanism by which the ions are generated is different depending on the type of ion source. In a liquid metal ion gun (LMIG), a strong electrostatic field induces the emission of ions from a sharp tip coated with a thin layer of liquid metal.[35,37] The tip and a small adjacent container, housing the metal of which the ions are formed, are heated above the melting point of the metal, resulting in a continuous wetting of the tip by liquid metal. A major advantage of the LMIG is the capability to focus the ion beam to very small diameters with a maintained high ion current, which is due to the very small size of the volume from which the ions are emitted. The most common ion species used in LMIGs for TOF-SIMS analysis are Ga^+, In^+, Au_n^+ ($n = 1$, 3) and Bi_n^+ ($n = 1$, 3, 5 or 7, as well as Bi_3^{++}), of which Bi_n^+ is the most powerful and most commonly used primary ion for static TOF-SIMS analysis of organic materials. The Bi_3^+ beam can be routinely focused to a diameter of 100–200 nm and the state-of-the-art instruments provide a specified focus diameter of 80 nm (TOFSIMS 5, ION-TOF GmbH). A disadvantage of the LMIG, however, is the lack of flexibility with respect to ion species, since only ions from metals that can form a liquid coating on a tip can be used. A more flexible ion formation mechanism is electron ionisation (EI, formerly known as electron impact), for which any atom or molecule that can be ionised by electron bombardment in the gas phase can be used. EI is thus used to generate ions from simple gases, such as Ar^+ and O_2^+, but it is also used for more advanced ion sources, such as the C_{60}^+ and the giant argon cluster ion sources. In the C_{60}^+ source, gas-phase C_{60} molecules are produced by heating of solid C_{60} material and C_{60}^+ ions are generated by electron bombardment.[36] In the giant argon cluster ion source, large argon clusters (Ar_n, $n = 1500$–3500) are formed in a supersonic expansion through a small nozzle aperture.[20] The high pressure required for this type of expansion makes it necessary to use differentially pumped vacuum stages separated by skimmers between the nozzle and the electron impact ionisation region at the top of the ion column. Due to the larger ionisation volume, the EI ion sources cannot be focused to such small beam diameters as the LMIGs, without a severe sacrifice in ion current.[36,51] Finally, in the Cs^+ source a special material is used that emits Cs^+ ions when heated. This source has focusing capability comparable to the EI sources. The loss in ion current connected to beam focusing in the EI and Cs sources is particularly serious when pulsed primary ion beams are used, since the pulsing itself removes a large portion of the generated ion current. However, when continuous beams are used, such as in the NanoSIMS (Chapter 1 of this volume) and the J105 3D Chemical Imager (see 5.2.2.5), focusing in the 100 nm range can often be realised while still keeping the ion current at an acceptable level.

While the focus diameter of the beam is important for the lateral resolution of the analysis, the time duration of the pulses as they collide with the surface limits the mass resolution, since this time duration adds an uncertainty to the determination of the flight time through the TOF analyser. In order to produce short pulses at a maintained high ion current, the pulses are compressed in time by using electrical pulses that temporarily slow down the front ions of the pulse and/or temporarily accelerate the later ions in the pulse so that they all arrive at the 'same' time at the sample surface. This process is called 'bunching' and can typically compress 25 ns pulses to time durations at the sample surface of less than 1 ns, corresponding to a mass resolution ($m/\Delta m$) of around 7000.

A disadvantage of using bunching is that the beam diameter is significantly expanded from the minimum focus diameter, mainly due to space charge effects caused by the high density of ions in the compressed ion pulses. In order to optimise the lateral resolution, bunching must therefore be turned off, which, in turn, reduces the mass resolution of the measurement. TOF-SIMS analysis is therefore typically made with the instrument optimised either for high mass resolution (bunched mode, typical values $m/\Delta m \sim 7000$, $\Delta L \sim 3$ µm) or for high spatial resolution (unbunched mode, typical values $m/\Delta m \sim 300$, $\Delta L \sim 150$ nm); see Figure 5.4 for a comparison between TOF-SIMS data obtained in the two different modes.

5.2.2.3   TOF Analyser

The secondary ions formed in the primary ion collision process on the sample surface are accelerated and focused into the TOF analyser by electrostatic potentials. Inside the TOF analyser, the secondary ions travel at a mainly constant kinetic energy in the field-free region of the flight tube until they reach the detector.[1,37] Their flight times are determined by the familiar expression for the kinetic energy:

$$mv^2/2 = ms^2/2t^2 = zeU$$

where m is the ion mass, v is the ion velocity, s is the length of the flight tube, t is the flight time, z is the charge state of the ion, e is the elementary charge, and U is the electrostatic potential inside the flight tube relative to the sample surface. eU, *i.e.* the kinetic energy of the secondary ions inside the flight tube, is constant and the same for all ions, meaning that the flight time of each ion is proportional to the square root of m/z. Therefore, ions with different m/z values will reach the detector at different times, assuming that they were all generated on the sample surface at the same time and at the same electrical potential.

While the TOF analyser of the ION-TOF instruments is of the reflectron type, in which the ions, after having travelled the full length of the flight tube, are (electrostatically) reflected back to the detector at the entrance part of the analyser (Figure 5.3a), the Physical Electronics instrument has a TRIFT analyser, where the ions instead are (electrostatically) deflected three times by 90° to reach the detector. In addition to reducing the length of the analyser, the electrostatic reflection/deflection stages have an added energy-focusing function to improve the mass resolution, *i.e.* they reduce the spread in flight times caused by slight differences in kinetic energy. Both analysers have similar performance with regard to mass resolution.

Ion images of non-flat sample surfaces normally show significant intensity variations caused by topography effects.[52,53] Typically, 'peaks' in the sample topography provide high secondary ion yields while 'valleys' and sharp 'slopes' display lower yields, although the surface concentration of the

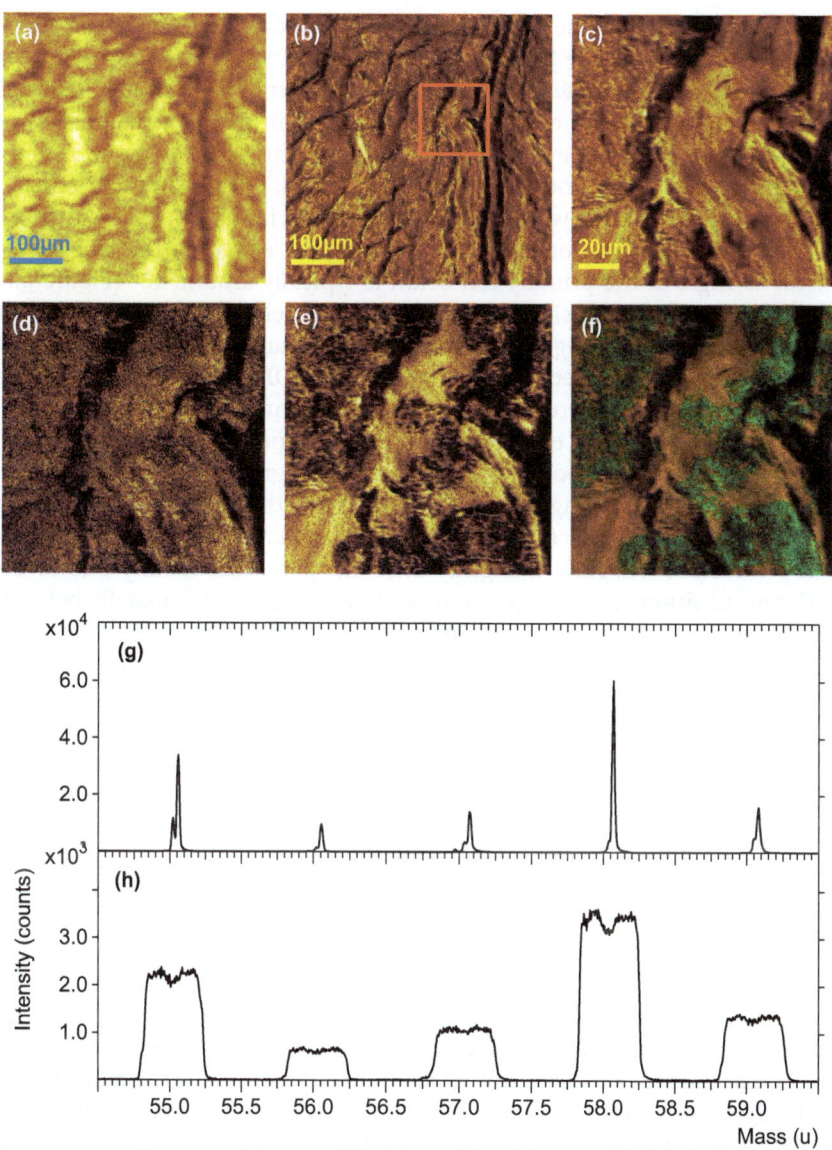

Figure 5.4 Comparison of TOF-SIMS data obtained at high mass resolution (bunched mode) and high spatial resolution (unbunched mode). (a) Total ion image of a mouse muscle tissue section acquired with 25 keV Bi_3^+ primary ions in bunched mode. (b) Total ion image of same area as in (a) acquired at high spatial resolution (25 keV Bi_3^+). (c–f) Zoomed-in ion images of area indicated by a yellow square in (b), where (c) is the total ion image, (d) is a protein fragment ion image (CH_4N^+ and C_4H_8N), (e) is an ion image of phosphatidylcholine fragment ions ($C_3H_8N^+$, $C_5H_{12}N^+$, and $C_5H_{15}NPO_4^+$) and (f) is an overlay image of (d) and (e) with the protein signal in green and the phosphatidylcholine signal in red. (g) Part of mass spectrum acquired in bunched mode to obtain ion image in (a). (h) Corresponding mass spectrum acquired at high spatial resolution (to obtain ion image in (b)).

analyte is constant. One reason for the topography effects is the limited acceptance angle of the TOF analyser, *i.e.* the efficiency by which the secondary ions can be extracted into the TOF analyser is considerably reduced if ion emission angles are too high relative to the direction of the analyser entrance. Since the angular distribution of the emitted secondary ions is typically relatively broad around the surface normal,[54] the extraction efficiency will be reduced in areas with a surface normal deviating from the direction of the analyser. Another source of the topographic effects is the fact that the incident angle of the primary ion beam is typically around 45° (Figure 5.3a), which will give rise to shadowing effects and yield differences due to variations in the incident angle of the individual primary ions on a rough surface.

Insulating samples become charged during TOF-SIMS analysis due to the transport of charged particles to and from the sample, but they can still be analysed if a low-energy electron gun is used to compensate for the charging between each primary ion pulse. A small but significant problem appears, however, when analysing insulating samples with strong topography. As a result of the topography and the insulating properties, different parts of the sample surface will have a slightly different potential, mainly due to their different positions in the extraction field between ground (usually below the insulating sample) and the entrance of the TOF analyser. This leads to a variation in the kinetic energy of the secondary ions in the TOF analyser, which induces a broadening of the peaks in the mass spectrum and thus a significantly reduced mass resolution. Depending on the specific topography of the sample, the effect can give rise to severe peak broadening and even double peaks, where both peaks represent the same ion but emitted from different parts of the surface.

5.2.2.4   *Sputtering and Depth Profiling*

Since the primary ion source (*e.g.* a LMIG) is normally optimised for surface imaging analysis, TOF-SIMS instruments are often equipped with a second ion source, whose main purpose is to remove material from the sample surface by ion sputtering (Figure 5.3a). Although ion sputtering causes molecular damage on the resulting surface and takes away the possibility for static SIMS measurements (see section 5.2.1.2), removal of material from the surface opens up possibilities for a number of additional measurements:

- **depth profiles** in which the signal intensities of selected ions are measured as a function of sputter time, reflecting the concentration of the corresponding chemical specie *versus* sample depth.[20,22,31,42,55]
- **3D imaging analysis** in which 2D ion images are acquired at different sputtering times, resulting in 3D representations of the chemical concentration in the surface region.[31,44,56,57]
- **removal of surface contamination** by sputtering of the sample surface for a brief period before carrying out the surface TOF-SIMS analysis on a 'clean' sample surface.[58]

- **access to hidden structures inside the sample,** which can be obtained by sputtering away material that covers the structure of interest on the initially prepared sample, before starting the surface TOF-SIMS analysis.[59]

With a separate ion source for sputtering, depth profiles and 3D imaging analysis is carried out in a dual-beam measurement, where alternating sequences of analysis and sputtering are executed on the sample. The acquired data will then contain complete 2D mass spectrometric data at different sputtering times, reflecting different sample depths. The results can be represented in various forms, such as depth profiles of specific structures in the 2D image, 2D images at different sample depths, 2D cross-section images, or rendered 3D images showing the 3D distribution of specific ions/chemical species.

An advantage of the dual-beam experiment is that the analysis and sputtering processes can be optimised separately, with respect to both the selection of ion projectile and the measurement parameters used. This is particularly valuable in the analysis of organic materials, where molecular information is often requested. It is then possible, for example, to use an LMIG optimised for high-resolution imaging analysis (such as Bi_3^+) together with a C_{60}^+ or a giant argon cluster ion source for sputtering, which are both capable of retaining molecular information in the mass spectra while sputtering through the sample (see section 5.2.1.2). This arrangement has proved capable of molecular detection of buried organic layers at high depth resolution[22,60] and also 3D imaging of organic materials. Also, it is common in the dual-beam mode to use a significantly larger sputter area $(2-5\times)$ compared to the analysis area in order to avoid crater edge effects. Figure 5.3c shows a rendered 3D image from a TOF-SIMS analysis of an organic light-emitting diode (OLED) device using giant argon cluster ions for sputtering and Bi_3^+ ions for analysis, where the different colours represent the signal from molecular ions of the different organic constituents.[61]

A disadvantage of the dual-beam approach used for depth profiling and 3D analysis, however, is that the material sputtered away during the sputter sequences is not used for secondary ion detection and is in that respect wasted, which may have a negative effect on sensitivity and spatial resolution. This drawback is not present in the new generation TOF-SIMS instruments, where all the secondary ions emitted during sputtering are analysed.[23,50]

5.2.2.5 New Instrumentation

Stimulated by the proven capability of the new cluster ion sources (C_{60}^+, in particular) to provide molecular information even beyond the static limit, the development of instruments with a new measurement principle based on the use of a *continuous* primary ion beam was initiated. Two different instruments based on this principle were developed in parallel by the

Winograd group[50] and the Vickerman group,[23] respectively, and the latter instrument is now manufactured and marketed as the J105 3D Chemical Imager by Ionoptika, Ltd.

Since the beam of primary ions bombarding the sample surface is continuous and the TOF analyser requires short pulses of secondary ions for analysis, the emitted secondary ions are rearranged into short pulses after leaving the sample surface. In contrast to the traditional TOF-SIMS approach, where the secondary ion pulses rely on the pulsing of the primary ions, this new approach results in a decoupling of the mass analysis of the secondary ions and the secondary ion formation process during the primary ion collision on the sample surface. This decoupling has several advantages. In particular, the mass resolution does not need to be compromised by the lateral resolution, as is does with the bunching of the primary ion beam in the traditional approach (see section 5.2.2.2), thus allowing for the acquisition of data at high mass resolution and high lateral resolution. A second advantage is that the degradation in mass resolution caused by strong topography on insulating samples (see section 5.2.2.3) does not occur when the pulsing of the secondary ions is decoupled from the interaction on the sample surface. A third important advantage of the new approach is that 3D analysis measurements can be carried out more efficiently than in a dual-beam measurement, with respect to both analysis time and detection efficiency. The shorter analysis time is due to the intermittent analysis cycles in the dual-beam measurement that are not needed in the new approach. The improved detection efficiency is caused by the fact that all of the sputtered secondary ions contribute to the acquired data, in contrast to the dual-beam approach, where no data is acquired during the sputtering cycles (see section 5.2.2.4). The main disadvantage of the single-beam approach is the need to compromise between optimum sputter and analysis conditions, which can be optimised separately in the dual-beam approach.

In the Winograd instrument,[50] a commercial electrospray ionisation/matrix-assisted laser desorption (ESI/MALDI) mass spectrometer, the QStar instrument from AB Sciex (Framingham, MA, USA), is modified to incorporate a C_{60}^{+} ion source that bombards the sample surface at the MALDI/ESI sample position of the mass spectrometer. The secondary ions are analysed in an identical manner to the ions in a MALDI-MS or ESI-MS measurement, which includes two quadrupole stages, a collision cell and a TOF analyser. In addition to normal TOF analysis (complete transmission through the quadrupole stages), the instrument can thus be used also for MS-MS analysis of the secondary ions. Pulsing of the secondary ions is accomplished at the entrance of the TOF analyser, where portions of the continuous ion beam is pushed in an orthogonal direction into the TOF analyser by short electrical pulses. In the Ionoptika J105 instrument (Figure 5.5), pulsing is accomplished by a buncher, where portions of the continuous secondary ion beam are compressed into short pulses by applying electrical pulses, similar to the bunching of the primary ion beam in the traditional approach.[23] This instrument is also capable of MS-MS

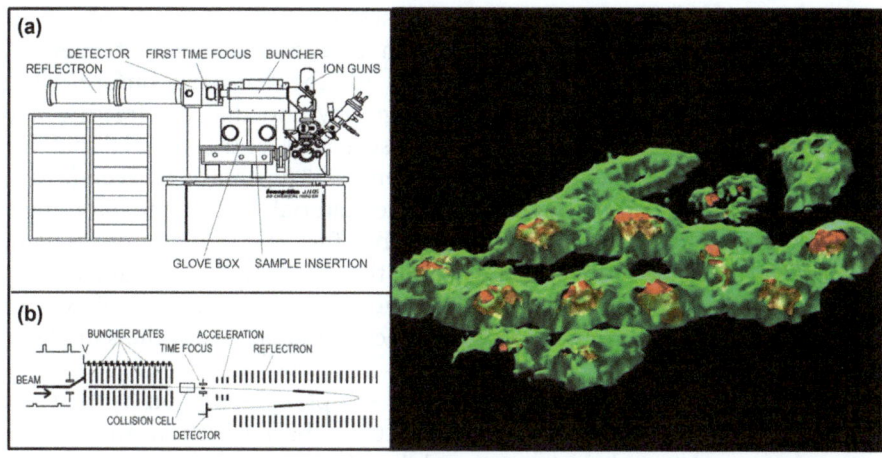

Figure 5.5   (a) Outline of the Ionoptika J105 3D Chemical Imager instrument; reprinted from Fletcher *et al.* (2008), American Chemical Society. (b) Detail showing the principal design of the buncher and TOF analyser. (c) 3D ion image obtained using the Ionoptika J105 instrument of HeLa cells attached on a substrate surface with the signal from phospholipids in green and from adenine in red (visualized area 250×250 μm^2). Reprinted from Fletcher *et al.*,[57] John Wiley and Sons.

analysis of the secondary ions. Figure 5.5c shows an example of 3D imaging using the Ionoptika J105 instrument in the analysis of cells, with the signal from phosphocholine (abundant in cell membranes) displayed in green and that from the nucleic acid adenine (abundant in the cell nucleus) in red.[62]

5.2.3   Sample Preparation

Sample preparation is a critical component of the TOF-SIMS analysis. In general, the analysis does not require any chemical preparation of the sample and since the purpose of the analysis is to probe the chemical characteristics of the native sample surface, treatments involving any chemical modification of the sample should be carefully avoided or minimised. Furthermore, since TOF-SIMS only probes the outermost molecular layers of the sample surface, an important consideration of the sample preparation is to expose the sample structures to be analysed on the sample surface. For the same reason, it is extremely important to minimise contamination of the sample surface before analysis. If possible, it is recommended that the surface to be analysed is prepared immediately before analysis, in order to avoid airborne contamination. Another important consideration is that flat sample surfaces are preferred, owing to the added complexity of the analysis and evaluation of topographically rough surfaces (as discussed in section 5.2.2.3).

Methods of producing freshly exposed surfaces include (1) breaking of the (rod-shaped) sample to produce a fracture surface for analysis,

(2) grinding,[63,64] and (3) cutting using a scalpel or a microtome. The main advantages of (1) is that contamination and relocalisation of chemical components can be essentially excluded for the fractured surface, while disadvantages are the often rough topography and that the fracture may occur in preferential (mechanically weak) structures of the material that may not be representative for the sample in general. The choice between (2) and (3) depends on the properties of the sample material: cutting/microtoming is mainly used for polymers and grinding for mineral and/or metallic materials. Both of these methods have the advantage of producing flat, representative surfaces of the sample; disadvantages are the risks of external contamination and/or smearing of chemical components of the sample on the surface.

Another important consideration for sample preparation is that the analysis is carried out in a vacuum ($< \sim 10^{-6}$ mbar), which means that the samples must be stable in vacuum and that they must not produce excessive outgassing that may compromise the required pressure. The vacuum requirement is particularly relevant for biological samples, such as microbial mats or cell or tissue samples, which need to be either dried or frozen during analysis. A number of preparation protocols have been developed for different types of biological samples. For cells and artificial lipid membrane samples, rapid (plunge) freezing in liquid propane or ethane is often used in order to prevent the formation of large ice crystals, which may otherwise damage the membrane structures.[65] Freeze-fractured cells are produced by placing the cells between two substrate surfaces in a sandwich geometry during freezing and then cracking open the frozen structure by separating the two substrates (preferably inside the TOF-SIMS vacuum chamber) to form fracture surfaces through the ice and hopefully also through some of the cells.[13,62,66] The freeze-fractured cells can be analysed either in the frozen-hydrated state or after freeze drying, although comparisons between these two indicate advantages for analysis in the frozen-hydrated state.[62] Freeze drying of substrate-attached cells and artificial lipid membranes (without freeze fracture) has also been shown to produce well-preserved structures for high-resolution lipid imaging.[34,49,67–69] An important consideration here is the need to rinse away the salt buffer in the sample before plunge freezing; otherwise the dried sample will be coated with a salt layer that makes analysis of the membrane lipids impossible. Artificial lipid membrane samples can be rinsed in deionised water, but cell samples are instead rinsed in a volatile salt buffer such as ammonium formate, which evaporates during the drying process, in order to prevent osmotic rupture of the cells.

Biological tissues and microbial mats[70] are normally prepared for TOF-SIMS analysis by (1) freezing the entire tissue or microbial mat sample, (2) cryosectioning to produce thin slices (5–30 μm), which are placed on a substrate surface, and (3) freeze drying. This protocol has proved to provide high-quality samples for high-resolution lipid imaging in various tissue samples.[16,37,71]

5.3 TOF-SIMS in Practice

5.3.1 Basics of Data Acquisition and Evaluation

5.3.1.1 Sample Navigation and Data Acquisition Procedure

In this section, some of the practical considerations and procedures involved in TOF-SIMS analysis are described, and the discussion applies specifically to the TOF-SIMS IV instrument manufactured by ION-TOF GmbH, which is the instrument model used in our laboratory. Other instruments may have slightly different design solutions and operating procedures, but the main features of the practical operation are similar in all 'traditional' instruments.

The size of the sample surface is typically around 10×10 mm^2, but can be as large as around 7×10 cm^2, although the sample height is normally limited to 5–6 mm. Before analysis, the sample is mounted on a sample holder, which is then inserted into a load lock. After evacuation of the load lock to $\sim <10^{-5}$ mbar, a gate valve to the main analysis chamber is opened and the sample is transferred into the analysis chamber using a transfer arm and placed on a sample manipulator table. The sample holder typically allows for simultaneous mounting of at least 6–8 samples (depending on size) and the time from which the sample is inserted into the load lock until it is ready for analysis is normally around 10 min, although samples requiring extensive outgassing (porous materials) may need longer times for evacuation in the load lock. Controlled sample cooling and heating between -130 °C and 300 °C is possible using a special sample holder with heater elements and cooling arrangements, which can be brought into thermal contact with liquid nitrogen dewars both in the load lock and in the main analysis chamber. The variable-temperature sample holder can, however, only fit one sample sized around 12×12 mm^2 or smaller.

Inside the analysis chamber (vacuum, 10^{-9}–10^{-6} mbar), the sample is moved in the *xyz* planes, rotated and tilted by controlling the sample manipulator table from the computer. Navigation of the sample to suitable areas for analysis is made by the use of two video cameras (macro and micro view), which provide live images of the sample surface in the analysis position. It is also possible to import photographs or other sample images to the acquisition program and, after alignment with the sample position on the sample holder, to use this image for navigation to areas of interest for analysis. Before starting the data acquisition, a number of analysis parameters need to be selected and specified. Assuming that the primary ion to be used is selected and the ion source is properly aligned, the type of analysis must be decided, *i.e.* positive or negative ion detection, bunched (high mass resolution), burst alignment (high lateral resolution), or another analysis mode. The mass range is decided by selecting a value for the cycle time, *i.e.* the time between consecutive primary ion pulses. If an analysis area of 500×500 μm^2 or less is selected, scanning of the analysis area is done by scanning the primary ion beam within the ion source (fast acquisition). Larger analysis areas require movement of the sample holder and patching of multiple data sets into the final data file (slower acquisition). An important consideration in setting up the measurement is the selection of number of pixels in the measurements, *i.e.* the number of data points within the analysis area, because the pixel size should be matched with the diameter of the primary ion beam. Ideally, the pixel size should be equal to or slightly smaller than the diameter of the primary ion beam; a much smaller pixel size is useless since the image resolution will anyway be limited by the primary ion diameter, and a larger pixel size results in an uneven PIDD where the centre of the pixel will be quickly 'burned' (molecularly damaged) while the outer parts will be fresh but unused in the analysis. Finally, the analysis time needs to be considered in order to determine whether the analysis is done under static conditions or not. The PIDD (see section 5.2.1.2) is given by $PIDD = It/eA$, where I is the pulsed primary ion current, t is the acquisition time, e is the elementary charge, and A is the analysis area. For example, assuming a primary ion current of 0.04 pA and an analysis area of 100×100 μm^2, a PIDD of 10^{12} ions cm^{-2} is reached after approximately 400 s. During analysis, it is possible to continuously inspect updated spectra and/or ion images as they are acquired. All data acquired during the measurement is saved in a raw data file, including separate mass spectra from each pixel and from each separate scan of the analysis area. After finished acquisition, the data can be displayed and evaluated in various forms, such as the extraction of ion images of selected peaks, mass spectra from selected regions of interest (ROI), profiles (*i.e.* signal intensity *versus* acquisition time) of selected peaks from selected ROI, as well as line profiles, rendered 3D representation, and multivariate analysis.

5.3.1.2   ToF-SIMS Spectrum

TOF-SIMS primary ion bombardment rarely forms doubly or multiply charged secondary ions, and the *m/z* value of spectral peaks can thus be taken as the exact mass of the respective ions. Both positive and negative ions may be formed, and are usually recorded during two subsequent measurements of the same area. In practice, the actual analysis of, say, a 500×500 μm^2 area usually takes between 3 and 30 min, depending on the pixel size and PIDD selected, but this does not include time needed for adjustment of the instrument, and the search for promising target structures using the built-in instrument optics (see 5.3.1.1).

A principal difference with respect to LC-MS or GC-MS is that TOF-SIMS inherently lacks upstream chromatographic separation. Whereas the former techniques ideally produce spectra representing single compounds after separation from a complex mixture, TOF-SIMS merges all organic and inorganic spectral information obtained from the surface area analysed. An MS-MS option that would make it possible to sort out potential target ions and subject them to secondary fragmentation has recently be introduced (see section 5.2.2.5) but is as yet not available for most groups running the existing traditional instruments. Whereas the possibility to simultaneously detect multiple analytes may in principle be considered advantageous for the study of intricate environmental samples, the difficulty of narrowing down the analytical window implies that TOF-SIMS spectra often represent an overlapping melange of quite complex, inadvertent mixtures. Owing to the high mass resolution, reduced chemical damage, and an increased secondary ion yield obtained with cluster ion sources, the TOF-SIMS spectra obtained from environmental materials usually exhibit a dense forest of peaks. Working out the relevant information from this wealth of data is the essential challenge for the researcher, and may become very difficult if the measurement was aimed at heterogenous environmental samples without a clear objective that fits into the capabilities of the TOF-SIMS design employed.

Figure 5.6 shows a comparison of a TOF-SIMS spectrum of a commercially available standard of ergosterol, a fungal cell membrane sterol, and a 500×500 μm^2 total area spectrum obtained from a microscopic cryosection of a fungus, where the ergosterol is present in the multicompound mixture of biomass.[72] This example also illustrates the property of TOF-SIMS to produce from a single compound diverse ions in the (near-)molecular weight range (*e.g.* molecular, protonated, deprotonated, dehydrated, Na- or

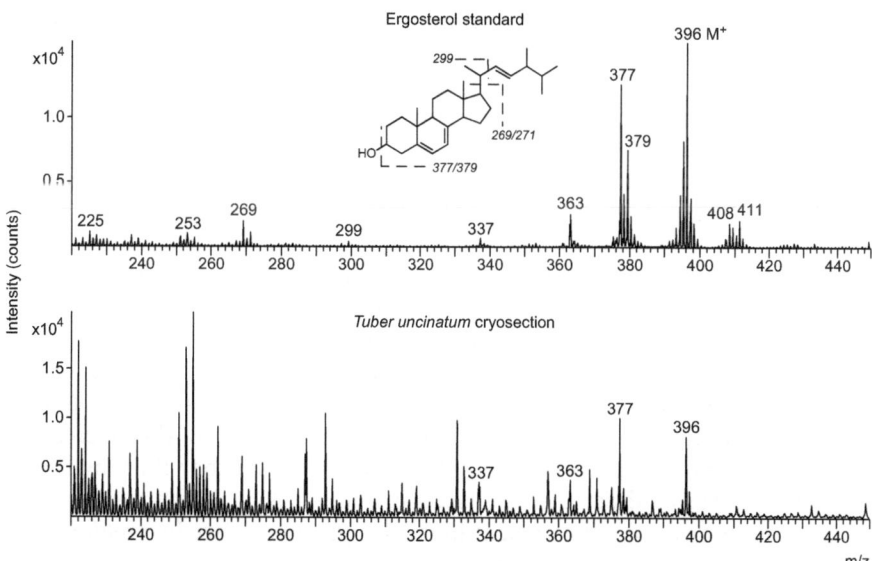

Figure 5.6 Top: Partial TOF-SIMS spectrum (m/z 220–450) of (22*E*)-ergosta-5,7,22-trien-3β-ol (ergosterol) standard (25 keV $Bi_3{}^+$, bunched mode, positive ions). Bottom: Partial TOF-SIMS spectrum (m/z 220–450) obtained from the microscopic 300×300 μm^2 cryosection of a fungus (*Tuber uncinatum*). The ergosterol-derived ions are masked by peaks originating from the complex mixture of co-occurring fungal biomass compounds, but yet still distinguishable when assessed in conjunction with the reference spectrum.

Adapted and reprinted from Leefmann *et al.*,[72] John Wiley and Sons.

K-cationised), and fragment ions. Although such fragmentation tends to further complicate the spectra it can also be employed to the analyst's advantage, because the identifications can be corroborated by checking the presence of more than one diagnostic peak for a particular target compound. Knowledge on the fragmentation behaviour of the compounds of interest through measuring reference spectra of standards is often crucial to extract any useful signals out of the complex TOF-SIMS spectra obtained from environmental samples.[72,73]

In cases where the TOF-SIMS spectra are too complex for a straightforward interpretation, multivariate statistical analysis ('chemometrics') has proved useful to organize and simplify data evaluation, *e.g.*[11,74–78] This approach may be particularly powerful for discriminating systematic spectral differences between different biological samples that are similar in composition and has, for instance, been successfully used to differentiate the cell surfaces of individual yeast[79] and bacterial strains,[78] discriminate the botanical provenances of amber[80] (see section 5.4.3), or identify peaks characteristic of lignin or polysaccharides in solid wood cross-sections[81]. However, TOF-SIMS spectra have to be inspected with great care with respect to peak separation, identification, and topographic effects (see section 5.2.2.3), before they can be used for chemometrics. To ensure that the discrimination of TOF-SIMS data is in fact based on authentic differences in the chemical composition of the sample surface, it is also crucial to carefully exclude sample-specific contamination from the peak set used for multivariate statistics.[74] It has been rightly pointed out that otherwise the consequence can easily be 'rubbish in, rubbish out',[18] and thus, multivariate analysis can usefully support but not replace the researcher's interpretation of the spectra to understand the actual chemical information.

A full TOF-SIMS spectrum of a biogeochemical sample starts at m/z 1 and typically covers the mass range of *e.g.* amino acids, purine bases, carbohydrate monomers, most geological hydrocarbons, pigments (*e.g.* chlorophylls), and intact polar lipids up to $\sim m/z$ 2000. Unlike lipids, proteins and large peptides are not easily sputtered from the sample surface and ionized, and thus cannot be analysed as intact (sequenceable) moieties by TOF-SIMS. In the high-mass region, the mass window is limited by the decrease in desorption and ionisation probability of larger ions rather than by the design of the MS analyser. As discussed in section 5.2.1.3, the more-specific high-mass ions need to be present in fairly high abundance in a region of the surface analysed to provide useful results. If this is not the case, these analytes may still show up as a peak in the TOF-SIMS spectrum (representing the summed signal of the total area), but the signal will be too weak to produce meaningful imaging data. To improve this situation and particularly the yield of large secondary ions, intensive attempts are currently being focused on the development of new primary ion systems, such as giant argon clusters and new instrument designs employing continuous primary ion beams (see section 5.2.2.5).

5.3.1.3   Making Use of Small Ions

Because of their higher probability of formation, ions in the lower mass range ($< \sim m/z$ 100) are usually orders of magnitude more abundant in a TOF-SIMS spectrum than high-mass ions. The small ions largely originate from collisions between primary ions and large surface molecules, or from the fragmentation of large molecules or fragments in the collision cascade. Small, volatile organic molecules are less likely to contribute to the low-mass region of the spectrum, since they are likely to evaporate into the high vacuum of the analysis chamber before analysis. In addition to organic fragments, the low-mass peaks also include monoatomic and cluster element ions, as well as simple inorganic species. In principle, these small ions cannot be regarded as very specific, and it must also be considered that they may originate not only from the sample but also from surface contaminants. With appropriate caution these signals can nevertheless be used as a valuable source of information about the presence of generic organic matter, major organic compound classes, or minerals. For instance, the presence of CH_4N^+ and CNO^- is often considered as indicative of amino acids (proteins), and $C_3H_5O^+$ may be used to track generic carbohydrate-rich matter. Hydrocarbon (C_xH_y) fragment ions with a low C/H ratio, such as $C_4H_9^+$ or $C_5H_9^+$, indicate alkyl chains and are amply produced by compounds such as aliphatic hydrocarbons and fatty acyl lipids. In turn, cyclics and particularly aromatic compounds are characterised by low-mass fragments with a C/H ratio at or above 1, such as $C_7H_7^+$ or C_3H^-. Table 5.1 provides a list of fragment ions that are typically observed in geochemical and/or biological samples, together with an interpretation of their origin.

Figure 5.7 shows a small region in the low-mass range (m/z 57–60) of an environmental TOF-SIMS spectrum, obtained from a 500×500 μm^2 area of a microscopic cryosection from a phototrophic microbial mat. In practice, this mass region (*i.e.* m/z 60) roughly represents the upper limit below which reliable sum formula assignments for most ions can be made without too much presupposition, provided that a reasonably flat sample surface and proper instrument status and mass calibration[82] enable good mass resolution ($m/\Delta m > 3000$) and thus, sharp peaks. Even in the low-mass region, however, peaks may be only partially resolved or completely merged and thus indistinguishable, an effect that is amplified by peak broadening due to topographical sample roughness (see section 5.2.2.3; $[C_3H_7O]^+$ and $[CH_5N_3]^+$ in Figure 5.7). Figure 5.7 also illustrates the co-occurrence of inorganic and organic ions in a TOF-SIMS spectrum. The inorganics can usually be resolved from organic ions because these peaks appear somewhat below the nominal mass, due to the mass defect of the respective nucleoids. Organic peaks, in contrast, tend to be shifted right of the integer, owing to the higher mass of the hydrogen atom. Consequently, highly saturated aliphatic hydrocarbon fragments would appear at the very right of the peak group gathered around each nominal mass (*e.g.* $[C_4H_9]^+$ in Figure 5.7).

Table 5.1   A selection of high-abundance low-mass ions commonly observed in TOF-SIMS analyses of biogeochemical samples.

Ion	Exact mass	Possible source
Na^+	22.99	Salt, minerals, biomass, omnipresent
Mg^+	23.99	Carbonates and other minerals, clastics
Al^+	26.98	Silicates, clastics, very specific for detrital minerals
$C_2H_3^+$	27.02	Hydrocarbons, unspecific
Si^+	27.98	Silicates, clastics, wafer, glass slide
CHO^+	29.00	Carbohydrates, unspecific
$C_2H_5^+$	29.04	Hydrocarbons, unspecific
CH_4N^+	30.03	Generic organic nitrogen, amino acids, proteins
CH_3O^+	31.02	Carbohydrates, unspecific
$C_3H_2^+$	38.02	Cyclics, aromatics, unspecific
K^+	38.96	Salt, minerals, omnipresent, often enriched in biomass
$C_3H_3^+$	39.02	Cyclics, aromatics, unknown specificity
Ca^+	39.96	Minerals, biomass, carbonates
$C_3H_5^+$	41.04	Hydrocarbons, cyclics
$SiCH_3^+$	43.00	PDMS
$C_2H_3O^+$	43.02	Carbohydrates
$C_3H_7^+$	43.05	Aliphatic hydrocarbons, alkyl lipids
$SiOH^+$	44.98	Silicates, clastics, wafer (specimen holder)
$C_2H_5O^+$	45.03	Carbohydrates
$CH_3O_2^+$	47.01	Carbohydrates
$C_4H_3^+$	51.02	Hydrocarbons, cyclics
$C_4H_5^+$	53.04	Hydrocarbons, cyclics
Mn^+	54.94	Manganese minerals, e.g. rhodochrosite
$C_4H_7^+$	55.05	Hydrocarbons, lipids (-alkyl)
Fe^+	56.93	Iron minerals, e.g. pyrite
$CaOH^+$	56.96	$CaCO_3$
$C_3H_5O^+$	57.03	Carbohydrates
$C_4H_9^+$	57.07	Aliphatic hydrocarbons, alkyl lipids
$C_3H_6O^+$	58.04	Carbohydrates
$SiOCH_3^+$	59.00	PDMS
$C_2H_3O_2^+$	59.01	Carbohydrates; carboxylate ion (fatty acids)
$C_3H_7O^+$	59.05	Carbohydrates
PO_3^-	78.96	Phosphates, phospholipids, biomass
PO_4^-	94.95	Phosphates, phospholipids, biomass
SO_4^-	95.95	Sulfates
$Ca_2O_2H^+$	112.91	$CaCO_3$, very specific
Amino acids (proteins)		
CH_4N^+	30.03	Amino acids, proteins, generic organic nitrogen
$C_3H_7^+$	43.05	Leucine, isoleucine
$C_2H_6N^+$	44.05	Alanine
CHS^+	44.98	Cysteine
$C_3H_6N^+$	56.05	Lysine
$CH_5N_3^+$	59.05	Arginine
$C_2H_5S^+$	61.01	Methionine, generic organic sulfur
$C_4H_6N^+$	68.05	Proline

Table 5.1 *(Continued)*

Ion	Exact mass	Possible source
$C_4H_5O^+$	69.03	Threonine
$C_3H_4NO^+$	70.03	Asparagine
$C_4H_8N^+$	70.07	Proline, arginine
$C_3H_3O_2^+$	71.01	Serine
$C_4H_{10}N^+$	72.08	Valine
$C_2H_7N_3^+$	73.06	Arginine
$C_3H_8NO^+$	74.06	Threonine
$C_4H_5N_2^+$	81.05	Histidine
$C_5H_7O^+$	83.05	Valine
$C_4H_6NO^+$	84.04	Glutamine, glutamic acid
$C_5H_{10}N^+$	84.08	Lysine
$C_5H_{12}N^+$	86.10	Leucine, isoleucine (note: also phosphatidylcholine fragment)
$C_3H_7N_2O^+$	87.06	Asparagine
$C_3H_6NO_2^+$	88.04	Asparagine, aspartic acid
$C_7H_7^+$	91.05	Phenylalanine
$C_4H_4NO_2^+$	98.02	Asparagine
$C_4H_{10}N_3^+$	100.09	Arginine
$C_4H_{11}N_3^+$	101.10	Arginine
$C_4H_8NO_2^+$	102.06	Glutamic acid
$C_4H_{10}NS^+$	104.05	Methionine
$C_7H_7O^+$	107.05	Tyrosine
$C_5H_8N_3^+$	110.07	Histidine, arginine
$C_8H_{10}N^+$	120.08	Phenylalanine
$C_5H_{11}N_4^+$	127.10	Arginine
$C_9H_8N^+$	130.07	Tryptophan
$C_9H_7O^+$	131.05	Phenylalanine
$C_9H_8O^+$	132.06	Phenylalanine
$C_8H_{10}NO^+$	136.08	Tyrosine
$C_{10}H_{11}N_2^+$	159.09	Tryptophan
$C_{11}H_8NO^+$	170.06	Tryptophan

PDMS, polydimethylsiloxane (a contaminant).

Above $\sim m/z$ 60, the number of possible sum formulas for a given mass increases substantially and the researcher will often have to start from an educated guess, considering empirical findings and the chemical, biological, and geological relevance or likelihood of the ion species in question. In less well constrained sample types, TOF-SIMS identifications have to be substantiated by comparison with reference spectra and by using complementary analytical techniques.

5.3.1.4 Matrix Effect

The detection of a substance in environmental materials is hampered by matrix effects that alter the secondary ion yield depending on the immediate chemical environment of the analyte studied. The term 'matrix effect' is commonly used to describe variations in the signal intensity in different samples or sample regions that are due to the chemical environment and not

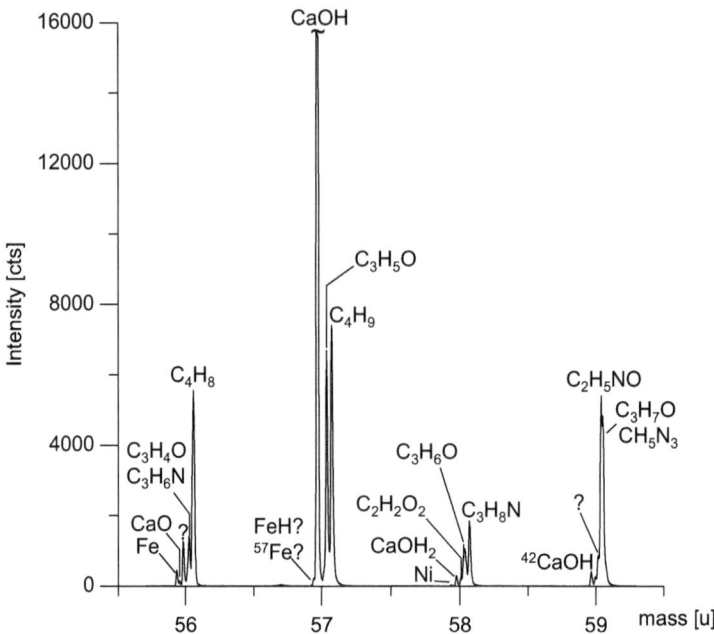

Figure 5.7 Partial TOF-SIMS spectrum (m/z 57–60; 25 keV $Bi_3{}^+$, bunched mode, positive ions) obtained from a 500×500 μm^2 area on a microscopic cryosection prepared from a phototrophic microbial mat (Äspö Hard Rock Laboratory, Sweden). Note the co-occurrence of inorganic and organic ions. See text for further discussion.

to the actual surface concentration of the analyte. The matrix effect is not often discussed in molecular TOF-SIMS studies, but it may evoke orders-of-magnitude differences between the sensitivity to different species, and is ultimately the reason why it is difficult to use TOF-SIMS for quantification, or to prove the absence of a compound in a sample. With respect to the formation of protonated or deprotonated molecular ions, the matrix effect most likely arises from the competition of components for protons, *i.e.* charge exchange processes within the matrix. Some compounds may, so to speak, 'rob' charges from others and turn them into non-detectable neutrals. Phosphatidylcholine-type lipids, for instance, appear to act as a suppressor of ionisation[83,84] in biological tissues. Likewise, salt present on a sample surface will influence the ion formation from organic substances and may severely suppress ionization of the molecules.[85] Salt also gives rise to the formation of sodium or potassium adduct ions, which tend to complicate the spectra if they occur in addition to the protonated molecular ions or fragments.

Several strategies have been suggested to overcome or handle the matrix effect, as a precondition for any use of TOF-SIMS as a reliable quantification method. Laser post-ionisation of the sputtered neutral particles is an effective approach to increase ionisation, thereby reducing the matrix effect and improving the capability for quantification (see section 5.2.1.2., laser-SNMS). However, a critical issue with respect to molecular analyses of organic compounds is that laser-SNMS gives rise to relatively small and thus less-specific ions. Rinsing the surfaces of the samples in 0.15 M ammonium formate[16,86] or analysis of samples in the frozen-hydrated state[87] can reduce the problem of salt-induced secondary ion suppression. Likewise, calibration studies may be performed in order to elucidate the different matrix effects invoked and how they affect the signal strength of the different compounds.[88] However, although this approach will be feasible in a system of known composition containing few compounds, it will be difficult to apply in complex environmental samples. For a detailed discussion of the matrix effect and strategies how to handle it, see Fletcher and Vickerman.[18]

5.3.2 Analysing Traditional Biomarkers with TOF-SIMS

5.3.2.1 Lipids

Many types of glycerolipids, glycerophospholipids, glyceroglycolipids, and sphingolipids have been successfully analysed in biological samples using TOF-SIMS. For a recent, comprehensive list of these compounds, including sum formulas and (near-)molecular ions, see Passarelli and Winograd.[71] In the TOF-SIMS spectra, individual glycerol-based lipids commonly appear in the mass range m/z 500–1000, forming peak clusters that reflect the natural ^{13}C abundance and tailgroup variations with respect to double bonds and carbon chain lengths. With the single-analyser system of the traditional TOF-SIMS instruments, the identification of high-mass lipids in environmental samples is not always straightforward. To a varying degree, it has to rely on circumstantial evidence and probabilistic reasoning, more so if the analytes are not presupposed or have not been precharacterised by preliminary analyses. The following example may illustrate the route along which preliminary assignments can be made in practice, and with a reasonable reliability. Figure 5.8 shows a positive ion mode spectrum obtained from a microscopic section of phototrophic microbial mat (same sample as in Figure 5.7). A prominent peak cluster in the m/z 750 region is suggested to represent sodium adducts of galactolipids, specifically monogalactosyldiacylglycerols (MGDG). This suggestion is based on a number of observations and considerations:

- The masses of the main peaks at m/z 749.53 and 751.55 show only small deviations of 24.7 and 21.9 ppm to the $[M + Na]^+$ ions of MGDG containing $C_{16:1}/C_{16:1}$ and $C_{16:0}/C_{16:1}$ fatty acyl side chains respectively $[C_{41}H_{74}O_{10}Na]^+$, $[C_{41}H_{76}O_{10}Na]^+$).
- The putative MGDG consist of three major peaks separated by 2 u, corresponding to different degrees of fatty acid tailgroup unsaturation, as expected for naturally occurring membrane lipid patterns.

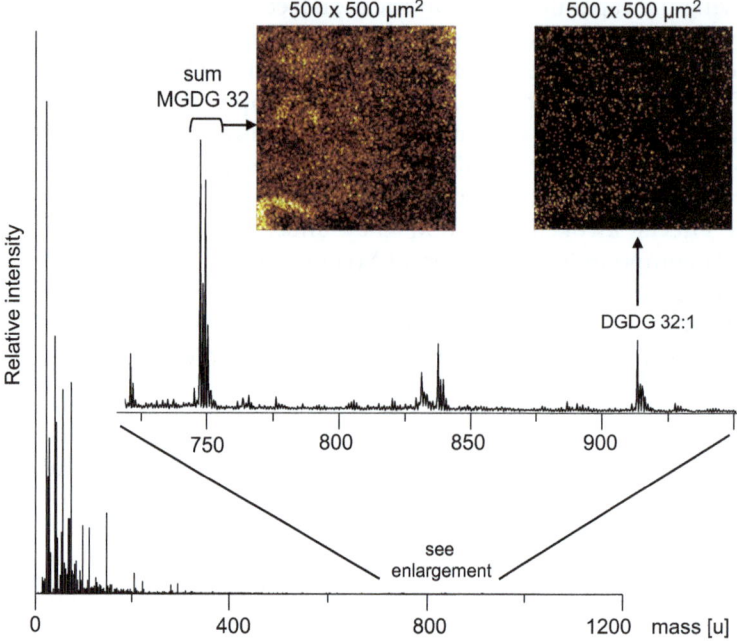

Figure 5.8 Partial TOF-SIMS spectrum (25 keV Bi_3^+, bunched mode, positive ions) obtained from a 500×500 μm^2 area on a microscopic cryosection prepared from a phototrophic microbial mat (same as in Figure 5.7). Note the predominance of low-mass ions. The insert shows a close-up of the m/z 720–950 region, with putative near-molecular ions of MGDG and DGDG lipids and two ion images (normalized for total ion). The images have been prepared from the major individual DGDG (32:1, right image) and the summed MGDG (left image). The former reveals insufficient image contrast due to low secondary ion yield, whereas the latter allows for the recognition of areas with a localized, higher intensity from these lipids (*e.g.* in the lower left corner). See text for further discussion.

- The signals in between the major peaks of the main cluster correspond to ^{13}C-containing molecules. As calculated using the instrument software, the natural isotopic abundance is explained by 100%, *i.e.* these peaks have exactly the size (area) expected for ^{13}C-bearing MGDG molecules.
- A second, much smaller peak cluster at m/z 777.54 and 779.55 is offset by 28 u from the main peak cluster and can be plausibly explained by MGDG homologues containing C_{16} and C_{18} fatty acyl tailgroup moieties, respectively.
- A third series of peaks maximizing at m/z 911.59 and 913.60 is offset by 162 u and can be explained as being derived from the sodium adducts of corresponding digalactosyldiacylglycerols (DGDG). Such co-occurrence of MGDG with DGDG can be anticipated in natural samples.

- Abundant putative sodium-cationized ions derived from the galactose headgroup of MGDG are found in the positive ion mode spectrum at m/z 203.04 $[C_6H_{12}O_6Na]^+$ and m/z 185.04 $[C_6H_{12}O_6Na-H_2O]^+$.
- In the ion images, all areas showing a strong (summed) signal of the suggested MGDG colocalize with the corresponding headgroup ions.
- In the negative ion mode, all areas rich in MGDG also show ample signal from the deprotonated molecular ions $[M-H]^-$ of the corresponding tailgroup fatty acids, with $C_{16:1}[C_{16}H_{29}O_2]^-$ and $C_{16:0}$ $[C_{16}H_{31}O_2]^-$ being most abundant.
- Consistent with the presence of sodium adducts, the Na^+ ion is very abundant, comprising the base peak of the positive ion mode spectrum.
- MGDG and DGDG are abundant components of plasmids and may thus be expected in photosynthetic organisms thriving in the microbial mat analysed.

Provided that a sound assignment can be made based on such multiple circumstantial evidence and, ideally, corroborated by complementary standard and extract analyses, the high-mass peaks of lipids provide a reasonable specificity across samples. The downside of analysing such large lipid ions with static TOF-SIMS is that the secondary ion yields of these heavier species often tend to be low (see section 5.2.1.3). Even if the specific (near-)molecular ions of polar lipids are clearly recognized in the total area spectrum, the intensities in individual pixels may be too low to allow for satisfactory imaging results, at least if the compounds were not concentrated

Table 5.2 A selection of abundant headgroup-derived positive ions commonly observed in TOF-SIMS analyses of glycerophospholipids.

Ion	Exact mass [u]	Tentative assignment
Intact phospholipids (headgroups)		
$C_5H_{12}N^+$	86.10	Phosphatidylcholine, unspecific
$C_2H_5NPO_3^-$	122.00	Phosphatidylethanolamine
$C_2H_7NPO_3^+$	124.02	Phosphatidylethanolamine
$C_2H_7NPO_4^-$	140.01	Phosphatidylethanolamine
$C_2H_9NPO_4^+$	142.03	Phosphatidylethanolamine, phosphonosphingolipids
$C_3H_8PO_6^-$	171.01	Phosphatidylglycerol
$C_5H_{15}NPO_4^+$	184.07	Phosphatidylcholine, sphingomyelin
$C_3H_9PO_6Na^+$	195.00	Phosphatidylglycerol
$C_5H_{14}NPO_4Na^+$	206.06	Phosphatidylcholine, sphingomyelin
$C_8H_{19}NPO_4$	224.11	Phosphatidylcholine
$C_6H_{10}PO_8^-$	241.01	Phosphatidylinositol
$C_6H_{12}PO_9^-$	259.02	Phosphatidylinositol
$C_9H_{16}PO_9^-$	299.05	Phosphatidylinositol

After Ostrowski *et al.*[84]

on a small region within the analysis area. Sometimes, merging of (near-) molecular peaks of related lipids may be performed to sufficiently increase signal strength; the ion images shown in Figure 5.8 illustrate this analytical situation for the microbial mat example. Alternatively, many studies reporting imaging results of lipids reduce their detection to headgroup or fatty acid tailgroup fragments rather than the molecular or near-molecular ions.[5,12,89–91] In the low-mass range the smallest, though unspecific fragments relating to glycerophospholipids are $[PO_3]^-$ at 78.96 u and $[H_2PO_4]^-$ at 96.97 u. More specific polar headgroup fragments that are commonly used for the detection and imaging of these lipids are given in Table 5.2. For published TOF-SIMS spectra of a number of glycerol-based lipids and an interpretation of their fragmentations, see Heim *et al.*[73]

Sterols are of particular interest not only in the biomedical field but also as molecular tracers in the biogeosciences. The most widespread functional cell membrane sterol of animals, cholesterol, has been identified and imaged quite successfully in tissue sections of mammals, as it readily ionises to form strong $[M-OH]^+$ at 369.35 u and $[M-H]^+$ at *m/z* 385.4 u (Figure 5.1b).[16,92] It seems that cholesterol also gives rise to some kind of matrix effect as it has been identified as a chemical environment that promotes the protonation of co-occurring molecules.[93,94] It seems however, that other sterols may show quite different ionisation schemes. Ergosterol (Figure 5.6) reveals a prominent $[M-H_3O]^+$ together with a strong molecular ion $M^{+\cdot}$ (396.35 u) whose occurrence may be explained by the easy removal of an electron from the $\Delta^{5,7}$ double bond system.[72] Like many glycerol-based lipids, sterols can be regarded as rather 'TOF-SIMS friendly' compounds, that can be more easily sputtered and ionised than other molecules. For standard spectra of a number of sterols and other terpenoids, see Leefmann *et al.*[72] and Jetter and Sodhi.[95]

5.3.2.2   Hydrocarbons

Acyclic, aliphatic hydrocarbons have been considered difficult to ionise using TOF-SIMS and thus, to produce rather weak spectra. The reason for this is that the C–C bond itself represents the highest occupied molecular orbital from which electrons are removed during ionisation.[96] Consequently, when measured as pure standards, saturated acyclic hydrocarbons such as *n*-alkanes produce a strong envelope of smaller $[C_xH_y]^+$ fragments, but appear to be somewhat resilient to the formation of single, clearly assignable (near-)molecular ions. This situation seems to be somewhat relaxed when using cluster ion sources, yet still several (near-)molecular ion species are produced, including a weak $M^{+\cdot}$ along with much stronger $[M-H]^+$ and $[M-2H]^+$ ions whose relative intensity appears to be controlled to some extent by the chemical environment and the primary ions used (see spectra published in[95,97,98]). Likewise, in analyses of environmental samples it seems to be virtually impossible with TOF-SIMS to distinguish between isomeric alkanes having the same sum formula. For example, the C_{20}-alkane isomers *n*-icosane, methylnonadecanes, and phytane (3,7,11,15-tetra-methylhexadecane) may all be abundant in biogeochemical samples, but as yet cannot be separated with TOF-SIMS and thus have to be treated as a pooled entity.[99] Aliphatics are preferentially analysed in positive ion mode. For published standard spectra of linear and branched alkanes, see[95–98,100].

Cyclic, aliphatic hydrocarbons appear to provide more distinctive molecular ions than acyclic compounds.[96] Studies employing monoatomic[101] and cluster ion sources[59,99] have shown that TOF-SIMS can be used to identify cyclic terpanes such as hopanes and steranes. In published spectra obtained with Bi_3^+ primary ions $[M-H]^+$ is the most predominant (near)-molecular species by far, and is accompanied by many of the 'classical' fragment ions known from GC-MS (217.20 u, $[C_{16}H_{25}]^+$ for steranes; 191.18 u, $[C_{14}H_{23}]^+$ for hopanes). Both the (near-)molecular species and fragments have been used to detect steranes and hopanes in crude oils,[99] single fluid inclusions,[59] or hydrocarbon spheres.[102] Again, particular care must be taken for a robust identification of these biomarkers with a single-analyser system, as the above-mentioned fragment ions may not be exclusive, and polar and aromatic compounds might contribute to the same *m/z* as the steranes and hopanes in some oils.[99] For TOF-SIMS spectra of geochemically relevant cyclic terpenoid hydrocarbons, see[72,99,101].

Aromatic hydrocarbons are somewhat different from aliphatics because an electron can be more easily removed from the π-system of the aromatic ring.[96] The resulting odd electron is delocalized and remains relatively stable. Consequently, most polyaromatic compounds exhibit the parent ion $M^{+\cdot}$ in relatively high abundance. In addition, aromatics produce positive fingerprint ions with C/H ratios at or above unity, such as $[C_7H_7]^+$ (tropylium ion), $[C_9H_7]^+$, $[C_{10}H_8]^+$, $[C_{11}H_9]^+$, and $[C_{12}H_{10}]^+$, with the former two being usually very abundant. Together with the molecular species, these fragments can therefore be used to analyse polycyclic aromatic hydrocarbons (PAHs)[46,97,103,104] or other, more complex aromatic species, such as melanin, where $[C_xH_y]$ fragments with very high C/H ratios stick out particularly in the negative ion mode[105] (and see section 5.4.2). For spectral data of some aromatic hydrocarbons, see Toporski and Steele.[97] For a detailed discussion of fundamental phenomena controlling the secondary ion emission from aromatics, see also Delcorte *et al.*[106]

5.4 Examples of TOF-SIMS in the Biogeosciences

Although steadily growing, the number of studies that have used TOF-SIMS to tackle questions in the biogeosciences is as yet relatively small. Our compilation provided in Table 5.3 includes an appreciable proportion of these approaches, although it makes no claim for completeness. In the following sections we introduce some of these studies, with a focus on representative sample categories and analytes to illustrate the capabilities of TOF-SIMS in different areas of the biogeosciences.

Table 5.3 Studies reporting the TOF-SIMS analysis of organic molecules in sample types of interest in the biogeosciences. Note that biomedical oriented TOF-SIMS studies on mammal tissues, cancer cells *etc.* are not considered here.

Analytical target	Key compounds detected	Sample type/age	References
Biomarker standards			
Reference spectra of biomarkers	Hopanoid hydrocarbons	Standards	101
Reference spectra of biomarkers	PAH, *N*-heterocyclics, *n*-alkanes, isoprenoid hydrocarbons	Standards	97
Reference spectra of biomarkers	Diverse acyclic glycero(phospho)-lipids, archaeol	Standards	73
Reference spectra of biomarkers	Diverse triterpenoids (steroids, hopanoids) carotene, tocopherol, chlorophyll	Standards	72
Sediments and rocks			
Organics preserved in fossil biofilms and microfossils	C_xH_y, $C_xH_yN_z$, $C_xH_yO_z$ fragments	Lacustrine sedimentary rock, Oligocene	107
Pigments in animal fossils[a]	Melanin fragments	Marine sedimentary rock, Eocene, melanin standards	105
Pigments in animal fossils	Haemoglobin-derived porphyrin	Lacustrine sedimentary rock, Eocene, haem standards	108
Hydrocarbons in fossilized fungal hyphae	C_xH_y fragments (alkanes, aromatics)	Hydrothermal minerals, Ordovician impact crater	132
Organics in fracture fillings of granitic rock, fossil biofilms	C_xH_y, $C_xH_yN_z$, $C_xH_yO_z$ fragments	Calcite veins in granite, Pleistocene (?)	63
Organics in hot spring deposits	Unknown ions, putative hopanoid fragments	Siliceous sinters, Recent to Pleistocene	109
Organics on crack surface in basalt	Unknown ions, putative hopanoid fragments	Pyroxene grain crack surface, subrecent ($\sim$200 yrs)	110
Organics on fracture surface in the deep subsurface	C_xH_y and CN-fragments	Biofilm in active fracture, 2.8 km deep, Recent	111
Organics on aerosol particles (TOF-SIMS/Laser-SNMS)	PAH	Aerosol particles, (pre-sorted, 0.5–3.5 µm), Recent	46
Organics in ancient biofilm	PAH	Kerogen-rich layer in chert, Precambrian (3.3 Ga)	112
Organics in extraterrestrial matter	PAH	Meteorite rock (Martian)	104

Table 5.3   (*Continued*)

Analytical target	Key compounds detected	Sample type/age	References
Organics in extraterrestrial matter	PAH	Comet dust particles	103
Petroleum			
Hydrocarbon patterns in oil-bearing fluid inclusions	C_xH_y fragments	Fluid inclusions in limestone and calcite fracture fillings, Eocene	113
Hydrocarbon biomarkers in crude oil	Steranes and hopanes	Crude oils, Mesozoic	99
Hydrocarbon biomarkers in oil-bearing fluid inclusions[a]	Steranes and hopanes	Fluid inclusions in hydrothermal fluorite, Ordovician	59
Hydrocarbon biomarkers in oil-bearing fluid inclusions	Acyclic and cyclic hydrocarbons, aromatics	Sandstone, Mid-Proterozoic (Roper Group, Australia)	114
Hydrocarbons in microspheres from drillhole fluid	C_xH_y fragments, PAH, putative hopane fragment	Solid 1–50 μm microspheres, expelled from deep (>2 km) Precambrian rocks	102
Biological materials, environmental			
Lipids of methanotrophic archaea	Isopranyl glycerol di- and tetraethers	Extracts, microbial mat cryosections, methane-related carbonate	70,115
Lipids and pigments of diatoms[a]	Fatty acids, acylglycerols, carotenoids, chlorophyll	Single diatom cells in phototrophic microbial mat	116
Plant waxes	n-Alkanes, long-chain fatty acids/acetates	Leaf surfaces, standards	117
Plant waxes	Triterpenoids (glutinol, friedelin), n-alkanes	Leaf surfaces, standards	95
Plant secondary metabolites	Flavonoids, quercetin and kaempferol glycosides	Sections from plant seed (pea)	118
Wood tissue biopolymers	Lignin and polysaccharide fragments, fatty acids, triolein, sterols	Wood sections (spruce), extracts	119
Wood tissue biopolymers	Lignin fragments, guaiacyl and syringyl moieties	Wood sections (*Acer*)	120
Wood tissue biopolymers	Lignin and cellulose fragments	Wood sections (*Pinus*)	81
Wood tissue biopolymers	Cellulase (enzyme) activity on wood	Wood powder (*Picea, Populus*)	121
Cuticular hydrocarbons of insects	Alkanes	Insect cuticles, extracts, standards	100

Table 5.3 (*Continued*)

Analytical target	Key compounds detected	Sample type/age	References
Higher-plant resins[a]	Succinate, plant diterpenoids	Amber, Palaeogene	80
Biological materials, controlled experiments			
EPS characteristics of benthic diatoms	Diverse unknown marker ions, chemometrics based	EPS isolated from cultured benthic diatoms	122
Lipid domain changes in protozoans during mating[a]	Phospholipids (headgroups)	Cultured ciliate (*Tetrahymena*), single-cells	89,123
Discrimination of bacterial strains	Diverse marker ions, chemometrics based	Bacterial isolates, colonies	78,124,125
Discrimination of yeast strains	Diverse marker ions, chemometrics based	Cultured yeast colonies	79
Antibiotics in swarming bacterial communities	Cyclic lipopeptide (surfactin)	Cultured *Bacillus* colonies	126
Antibiotics in soil bacteria	Actinorhodin, tripyrroles (prodiginine and prodigiosin derivatives)	Soil bacteria isolates (*Streptomyces*)	127
Antibiotics in plant root associated bacterial biofilms	Cyclic lipopeptides (surfactins, iturins, and fengycins)	*Bacillus* biofilm on plant root (tomato), lab culture	128
Bacterial and eukaryotic cell residues after detachment	Diverse amino acid fragments (Table 5.1), chemometrics based	Proteins on metal (Ti) surface	129

EPS, extracellular polymeric substances; PAH, polycyclic aromatic hydrocarbons.
[a]See sections 5.4.1–5.4.5.

5.4.1 Liquid Hydrocarbons in Ancient Rocks: Steranes and Hopanes in Oil-Bearing Fluid Inclusions

TOF-SIMS has been employed on ancient rock samples with the aim of detecting biomarkers in individual oil-bearing fluid inclusions selected from petrographic thin sections.[59,114] In addition to their utility for oilfield correlation and exploration purposes, oil-bearing fluid inclusions may provide a contamination-free source of ancient biomarkers, as the compounds have been isolated from the environment since the formation of the inclusion.[130] However, analysis of classical biomarkers in single oil-bearing fluid inclusions, which is often necessary owing to the presence of different generations of inclusions, had not previously been possible because of the small size of most inclusions. A TOF-SIMS dual-beam experiment (see section 5.2.2.4) was employed on an Ordovician rock sample to remove the rock matrix of the target inclusion using a 10 keV C_{60}^+ sputter beam, and to monitor the progress towards the opening of the inclusion using an alternating 25 keV Bi_3^+ analysis beam.[59] After removing a thin organic dirt layer on the thin section

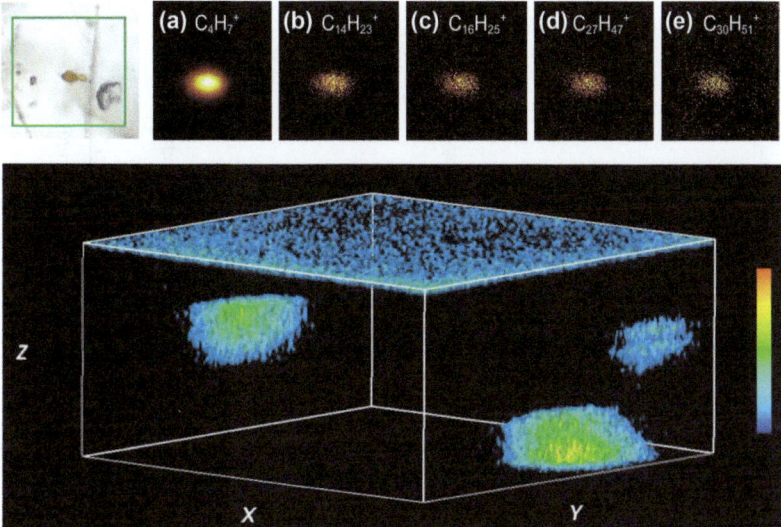

Figure 5.9 Top: Optical micrograph of a target oil-bearing fluid inclusion in fluorite (CaF_2), with the green frame indicating the 100×100 μm² area of analysis (left). The corresponding TOF-SIMS images from the area of analysis show secondary ions from (b) $C_{14}H_{23}^{+}$ (*m/z* 191.19, hopane fragment); (c) $C_{16}H_{25}^{+}$ (*m/z* 217.20, sterane fragment); (d) $C_{27}H_{47}^{'}$ (*m/z* 371.36, sterane deprotonated molecular ion); (e) $C_{30}H_{51}^{+}$ (*m/z* 411.40, hopane deprotonated molecular ion). Note that all ions localize to the area of the inclusion. Bottom: 3D representation of the hydrocarbon-derived $C_4H_7^{+}$ (*m/z* 55.06) signal strength obtained during high-mass resolution depth profiling of a rock section containing several oil bearing fluid inclusions. The z-direction of the cube is the depth axis; hence the top blue side of the cube is the surface (100×100 μm²) of the thin section whereas the blue–green–yellow spheres are oil-bearing fluid inclusions. Red indicates the strongest signal intensity while blue is the weakest. Adapted and reprinted from Siljeström *et al.*,[59] John Wiley and Sons.

surface, the sputter beam reached the target inclusion after approximately 1 h of sputtering at a depth of about 4 μm. Upon opening of the inclusion, secondary ions from the filling of the fluid inclusion suddenly appeared in the spectra recorded, among them diagnostic fragments and deprotonated molecular ions of hopanes and steranes. Imaging showed the exact colocation of the biomarker ions with the site of the fluid inclusion (Figure 5.9, top). The raw data file from a dual-beam TOF-SIMS measurement over a $100 \times 100 \times 10$ μm³ rock volume containing multiple inclusions was used for retrieving molecular data from individual inclusions in 3D (Figure 5.9, bottom).

5.4.2 Pigments in Ancient Rocks: Melanin in Fossils

A recent palaeobiological study employing TOF-SIMS, among other techniques, aimed at analysing fossil micron-sized spherical bodies that commonly comprise the carbonized traces of fossil feathers, hairs, and eyes.

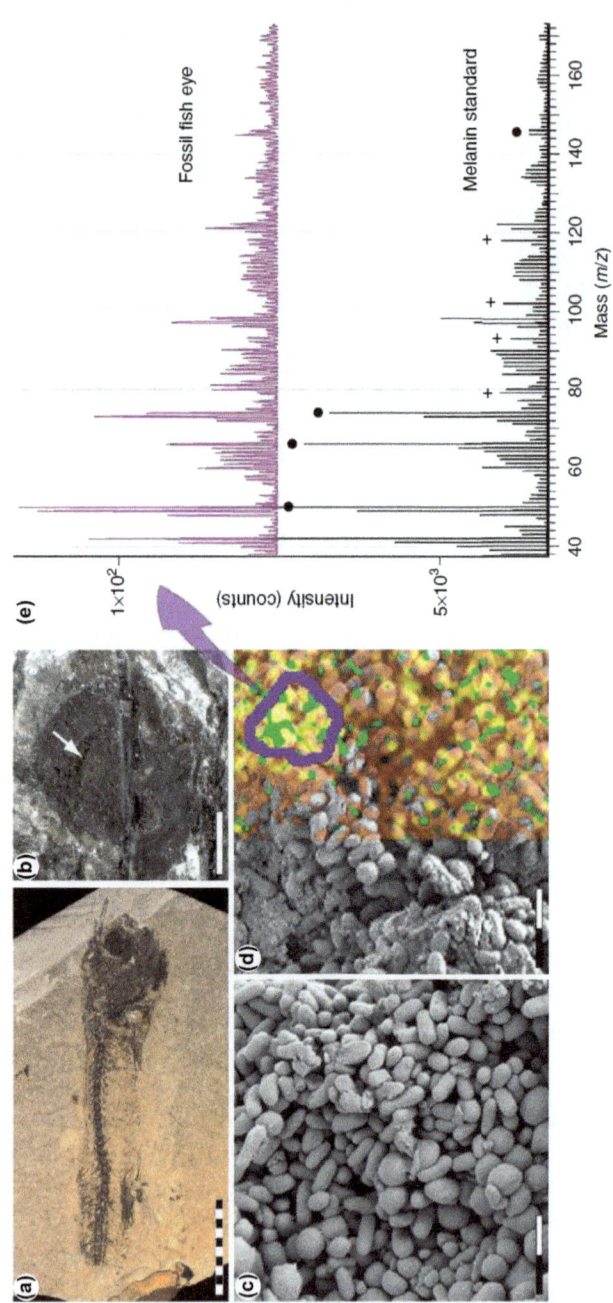

These minute structures are typically 0.5–2 μm in the long axis direction and thus in the typical size range of bacterial cells. Commonly the microbodies were regarded as the remains of biofilms of keratinophilic bacteria, but recently they have been reinterpreted as melanosomes, *i.e.* colour-bearing organelles. Proving the endogeneity of these structures is of great interest in the fields of palaeontology and evolutionary biology given, for example, the possibility of reconstructing integumentary colours and plumage colour patterns. Using a fossil fish eye from the early Eocene of Denmark[105] resolved this fundamental difference in interpretation by combining TOF-SIMS with IR microspectroscopy and electron microscopy. In conjunction with data from natural and synthetic standard compounds, modern retinal melanosomes, and other reference materials, it could be shown that the organic content of melanosome-like microbodies in the ancient fish eye was dominated by the pigment melanin. Simultaneous identification and ion mapping in the ancient sample were performed at high spatial resolution with TOF-SIMS and the ion maps were correlated with scanning electron microscope (SEM) images. This revealed an intimate association of the melanin-derived signal with melanosome-like microbodies (Figure 5.10). It was also evident that other biomolecules, such as proteins and lipids, were essentially absent in the ancient organelles, suggesting that the process of

---

Figure 5.10   Negative ion TOF-SIMS spectra (25 keV Bi_3^+) of melanin and microbodies from a fossil fish eye (Eocene). (a) Optical photograph of the specimen. Scale bar: 10 mm. (b) Close-up of target (brownish matter) located in the orbit; position of the area analysed by scanning electron microscopy (SEM) and TOF-SIMS (Bi_3^+ primary ions) indicated by an arrow. Scale bar: 1 mm. (c) Detail of the eye (SEM image) showing closely spaced, elongate and oblate melanosome-like structures preserved as solid bodies. Scale bar: 2 μm. (d) A semi-transparent ion image showing the distribution of melanin-derived ions (identified from melanin standard) superimposed onto a SEM image of tightly packed melanosome-like bodies. The added signal intensity from C_3N^- (m/z 50), C_3NO^- (m/z 66) and C_5N^- (m/z 74) is shown in orange–yellow, whereas areas high in signal from peaks at $C_{12}H^-$ (m/z 145) and $C_{11}N^-$ (m/z 146) are shown in green. The purple line demarcates the area from which the mass spectrum presented in (e) ('fossil fish eye') was collected. Scale bar: 2 μm. (e) Negative TOF-SIMS spectra representing the fossil fish eye and a natural melanin standard (from *Sepia officinalis*). The fossil fish eye spectrum was reconstructed from a ~ 4 μm^2 area in a measurement at 50×50 μm^2 acquired at high spatial resolution (unbunched mode, mass resolution $m/\Delta m \sim 300$) in order to allow for analysis of TOF-SIMS data specifically from within the selected area marked in (d), whereas the melanin standard spectrum was recorded at high mass resolution (bunched mode, $m/\Delta m \sim 5000$). The close agreement between the two spectra (with regard to both their peak positions and relative intensity distributions) provides compelling evidence for a high melanin content in the eye. Filled circles indicate peaks used to produce the ion image in (d), whereas crosses indicate peaks from inorganic ions that are not part of the melanin structure.
Reprinted from Lindgren *et al.*,[105] Nature Publishing Group.

fossilization purified the melanin, whereas more labile organic molecules were degraded and disappeared. It was suggested that the remarkable stability of the indol-based structural network of melanin may relate to its biological role as an energy transducer, designed to accommodate both optical and chemical energy and dissipate it as heat, thereby greatly reducing the rate of degradation.[105]

5.4.3   Ancient Resins: Succinate and Diterpenes in Amber

Amber consists of polymerized plant resins combining natural products such as terpenoids, carboxylic acids, and associated alcohols. The remarkable persistence of amber in sediments as old as Triassic and its potential to entomb and preserve organisms with great fidelity has made it a fascinating target for palaeobiologists for centuries. A study utilizing TOF-SIMS for the chemical characterisation of amber and discrimination of its botanical provenances was recently introduced by Sodhi *et al.*,[80] who analysed microtomed surfaces of Palaeogene Baltic and Bitterfeld ambers, two of Europe's major amber deposits. Compared to the Baltic amber (Middle Eocene), the Bitterfeld amber (Latest Oligocene) is contained in ~20 Myr younger marine sediments, suggesting a geologically distinct deposit. However, the recognized similarity between insect and spider assemblages from both deposits argues that Bitterfeld amber represents secondary redeposition of Baltic amber, and that the ambers themselves are in fact coeval to each other. Authentic standards of monomethyl succinate and diterpene resin acids were analysed to guide interpretation of the results, and TOF-SIMS spectral data were processed using principal component analysis (PCA). Interestingly, the positive ion mode mass spectra from the three ambers revealed highly complex, yet very similar patterns from which no individual characters could be extracted using explanatory PCA. In contrast, TOF-SIMS spectra obtained in negative polarity mode were much more interpretable in terms both of differentiating the ambers and of identifying major peaks on the basis of the standard spectra. The results highlighted differences that are likely due to age, botanical provenance, and post-depositional history. Importantly, the abundance of succinate was consistently higher in Baltic amber relative to Bitterfeld amber, indicating that they are distinct deposits and not regional variants of each other.

5.4.4   Biomarkers in Environmental Microorganisms: Diatom Cells in a Microbial Mat

TOF-SIMS offers the possibility of directly coupling biomarker information to microscopic, histological, or petrological sample structures at the microscopic level. Such coupling may be very useful for differentiating biomarker sources in typical structures of geobiological interest, *e.g.* finely zoned mineral precipitates, sediments, and microbial mats. A recent study reported the use of TOF-SIMS to differentiate diatom cells as a component of an environmental phototrophic microbial mat by identification of their characteristic lipid and pigment content.[116] TOF-SIMS was applied to a microscopic cryosection of the microbial mat, and results were compared with those of complementary analysis using GC-MS of the mat extract and optical microscopy of the TOF-SIMS analysis area on the cryosection. TOF-SIMS spectra and ion images revealed individual biomarkers, including fatty acids, mono-, di-, and triacylglycerols, phosphatidylglycerol, carotenoids, and chlorophylls to be colocalized with cells of a particular diatom species (*Planotidium lanceolatum*) (Figure 5.11). Comparison of extract data and single-cell mass spectra made it possible to specify the lipid contribution of these algae to the microbial mat system. Notwithstanding many difficulties still accompanying the TOF-SIMS analysis of poorly constrained environmental materials (*e.g.* the matrix effect, see section 5.3.1.4), the results underpin the potential of TOF-SIMS for a clear-cut assignment of organic compounds to distinctive morphological structures and specific microorganisms within complex biogeochemical samples.

5.4.5 Biochemical Process Markers: Phospholipids in Mating Ciliate Cells

Studying not only the content and localisation, but also the dynamics of particular biomarkers in microorganisms can certainly be considered as an area of great potential interest in the biogeosciences, but tools for such investigations are sparse. With an elaborate experimental design, TOF-SIMS has been used to directly observe the continuous modification of the cell membrane composition during the mating (conjugation) event of *Tetrahymena thermophila*.[89,123] Ciliates such as the ubiquitous genus *Tetrahymena* comprise a group of unicellular heterotrophic eukaryotes that have frequently been studied in the biogeosciences, because they play an important role as heterotrophic decomposers at the base of aquatic food chains, and may considerably contribute to sedimentary organic matter. Likewise, these organisms are of particular interest in the biomedical field, where the relatively large (50–100 μm) *Tetrahymena* cells are used as model organisms for the study of cell membrane processes. When mating, conjugating *Tetrahymena* cells form hundreds of 100–200 nm membrane pores at the site of fusion, allowing cells to exchange gametic nuclei. Pore formation is associated with a curved localised membrane structure, which energetically favours certain lipids, excluding others. However, it has been unclear whether the lipids actively modify membrane structure in anticipation of the mating event, or *vice versa*, the structural changes involved in pore formation drive the chemistry of the lipid bilayer. Kurczy and co-workers[89] used freeze-fractured, frozen-hydrated (see section 5.2.3) *Tetrahymena* cells in different mating stages to determine with TOF-SIMS the spatial distribution of the phosphocholine headgroup ion (m/z 184), indicative of the lamellar lipids phosphatidylcholine and sphingomyelin, *versus* the headgroup ion of the

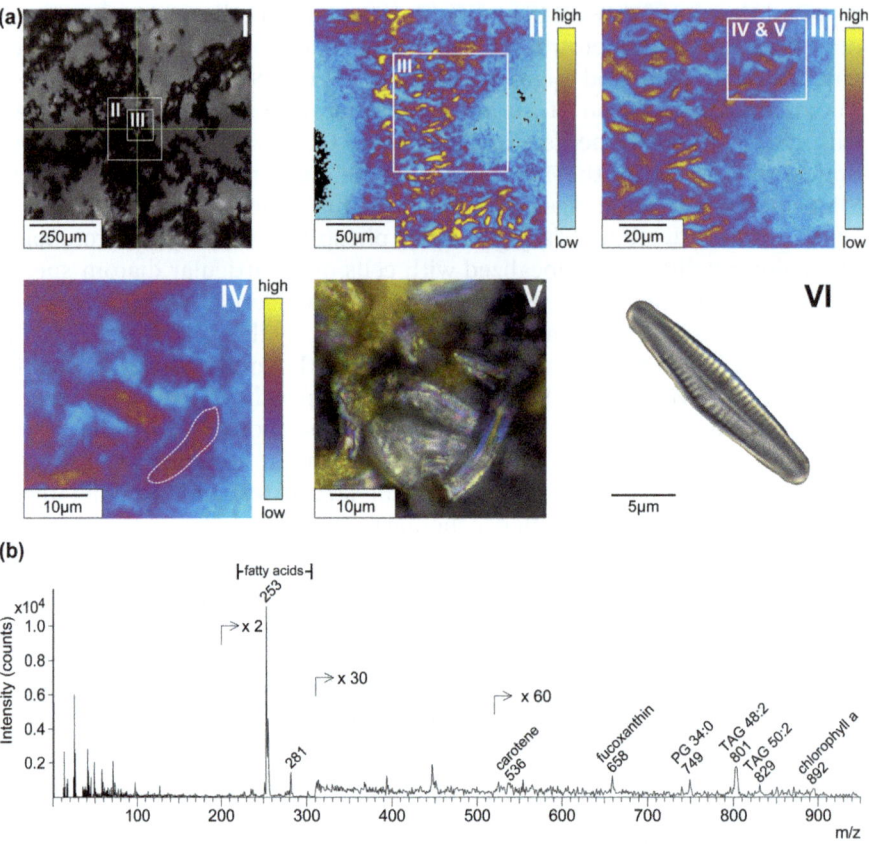

Figure 5.11   Single-cell biomarker imaging. (a) Microscopic and ion images of microbial mat cryosection analysed at high spatial resolution: (I) TOF-SIMS built-in video camera image with areas analysed marked by white frames, (II) ion image of 200×200 μm area showing distribution of summed C_{16} and C_{18} FAs (negative ion mode, m/z 251–258 $+ m/z$ 275–284), (III) ion image of 100×100 μm area showing distribution of summed C_{16} and C_{18} FAs, (IV) enlarged detail of (III) showing enrichment in C_{16} and C_{18} FAs in diatom cell (white, dashed line marks area from which the spectrum in (b) was reconstructed), (V) fluorescence microscopic image from the area corresponding to IV after TOF-SIMS measurements showing accumulation of diatom cells, (VI) single cell of predominant diatom, *Planothidium lanceolatum*, in the phototrophic microbial mat. (b) Partial negative burst alignment mode TOF-SIMS spectrum (m/z 0–950) of single diatom cell in Äspö phototrophic microbial mat cryosection corresponding to the area marked in (a) IV. PG, phosphatidylglycerol; TAG, triacylglycerol). Reprinted from Leefmann *et al.*,[116] Elsevier.

non-lamellar, high-curvature lipid 2-aminoethylyphosphonolipid (2-AEP; m/z 126). It was found that the signal of the curved 2-AEP considerably increased in the cell junction region at the expense of the lamellar phosphatidylcholine (Figure 5.12). Ion images taken at different stages of the ∼3 h mating process indicated that lipid domain formation follows rather than

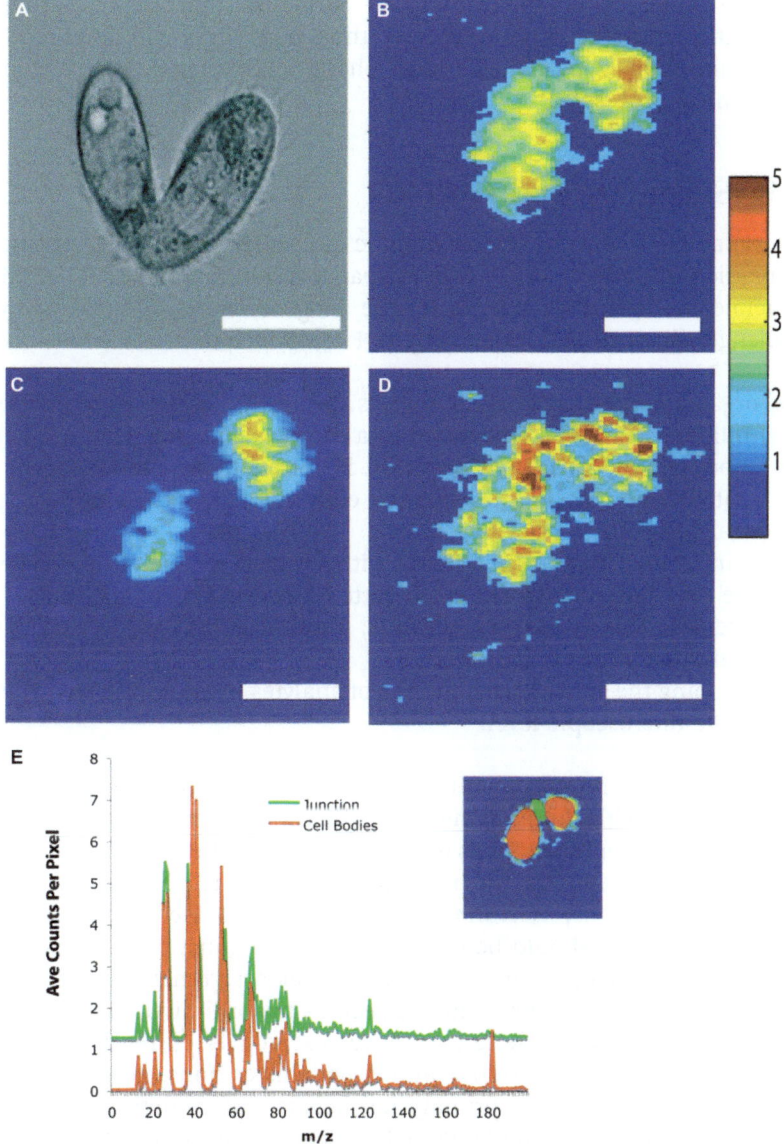

Figure 5.12   Using TOF-SIMS to observe lipid domains in mating *Tetrahymena* cells. (A) Differential interference contrast microscopy image of a mating cell pair (scale bar: 25 μm). (B) Ion image of *m/z* 69 $[C_5H_9]^+$, representing generic acyl lipids. (C) Ion image of *m/z* 184, representing the phosphocholine headgroup of the lamellar lipids phosphatidylcholine and sphingomyeline. (D) Ion image of *m/z* 126 (intensity multiplied by 3), representing the headgroup of the curved lipid 2-aminoethylyphosphonolipid (2-AEP). (E) Region of interest analysis of the same cells. The red mass spectrum (high in lamellar lipids, *m/z* 184) is from the cell bodies, and the green mass spectrum is from the junction. The inset highlights the respective regions on the ion image, red for the cell bodies and green for the junction.
Reprinted from Kurczy *et al.*,[89] National Academy of Sciences.

precedes changes in the membrane structure that occur during the conjugation process. TOF-SIMS was thus used in a 'lifetime mode' to quasi-continuously record chemical changes associated with a complex biochemical process.

5.5   Summary and Outlook

The capability to directly couple organic molecular data to microscopic and inorganic information at a very high spatial resolution makes TOF-SIMS, in principle, an attractive option in the biogeosciences. Typical analytical challenges that can be tackled with TOF-SIMS include

- organic analysis of small targets, typically several micrometres to millimetres in size, embedded in a complex chemical environment.
- spot analysis of isolated particles, if not enough material can be separated from the bulk sample for conventional, extract-based organic analyses.
- comparative fingerprint analysis with a wide-scale analytical window for the differentiation of chemical patterns between samples.
- microanalysis of extracts or fluids, if too little material is available for traditional organic analyses.
- mapping the spatial distributions of analytes in a complex environment at the microscopic level.

The potential and successful use of TOF-SIMS for the microanalysis of biomarkers notwithstanding, the technique cannot, however, be regarded as an analytical 'silver bullet' or as a full substitute for the established methods used in biogeochemistry, such as GC-MS or LC-MS. Due to its inherent analytical limitations, TOF-SIMS must generally be used thoughtfully and data interpretation should be corroborated in conjunction with other techniques. In particular, the uncertainties in peak assignment and the limited secondary ion yield can presently be seen as major limitations of the traditional static TOF-SIMS design. Intense research is currently directed at overcoming these limitations, and new cluster ions and instruments equipped with continuous primary ion beams and tandem analysers (MS-MS) are recently being implemented. Indeed it currently seems as if new primary ions and the decoupling of the ionisation event from mass analysis will make it possible to abandon the static limit, at least to some extent. This will allow for dynamic TOF-SIMS analysis of organic molecules in 3D, achieving both high lateral resolution and high mass resolution at the same time. Given the rapid technical developments and the increasing range of applications in the field, it can be anticipated that the next decade will see a more widespread use and a continuously growing input of TOF-SIMS in the biogeosciences.

Acknowledgements

We are grateful to Rana Sodhi (University of Toronto) for his thoughtful review of the original manuscript. We also thank Per Borchardt, Christine Heim, Jukka Lausmaa, Tim Leefmann, and Sandra Siljeström for many helpful discussions and analytical assistance. This work received financial support from the German Research Foundation (FOR 571, Courant Centre Geobiology), and the Swedish Governmental Agency for Innovation Systems (VINNOVA).

References

1. A. Benninghoven, *Angew. Chem., Int. Ed. Engl.*, 1994, **33**, 1023–1043.
2. A. Benninghoven, *Phys. Status Solidi B*, 1969, **34**, K169-K171.
3. A. M. Belu, D. J. Graham and D. G. Castner, *Biomaterials*, 2003, **24**, 3635–3653.
4. J. Vickerman and D. Briggs, *ToF-SIMS, Surface Analysis by Mass Spectrometry*, IM Publications, Chichester, UK, 2001.
5. A. Brunelle and O. Laprévote, *Anal. Bioanal. Chem.*, 2009, **393**, 31–35.
6. J. S. Fletcher, N. P. Lockyer and J. C. Vickerman, *Surf. Interface Anal.*, 2006, **38**, 1393–1400.
7. D. Touboul, F. Kollmer, E. Niehuis, A. Brunelle and O. Laprevote, *J. Am. Soc. Mass Spectrom.*, 2005, **16**, 1608–1618.
8. J. Xu, S. Ostrowski, C. Szakal, A. G. Ewing and N. Winograd, *Appl. Surf. Sci.*, 2004, **231**, 159–163.
9. C. A. Barnes, J. Brison, M. Robinson, D. J. Graham, D. G. Castner and B. D. Ratner, *Anal. Chem.*, 2012, **84**, 893–900.
10. F. Benabdellah, A. Seyer, L. Quinton, D. Touboul, A. Brunelle and O. Laprévote, *Anal. Bioanal. Chem.*, 2010, **396**, 151–162.
11. M. Brulet, A. Seyer, A. Edelman, A. Brunelle, J. Fritsch, M. Ollero and O. Laprévote, *J. Lipid Res.*, 2010, **51**, 3034–3045.
12. K. Börner, P. Malmberg, J.-E. Månsson and H. Nygren, *Int. J. Mass Spectrom.*, 2007, **260**, 128–136.
13. I. Lanekoff, M. E. Kurczy, R. Hill, J. S. Fletcher, J. C. Vickerman, N. Winograd, P. Sjövall and A. G. Ewing, *Anal. Chem.*, 2010, **82**, 6652–6659.
14. H. Nygren, P. Malmberg, C. Kriegeskotte and H. F. Arlinghaus, *FEBS Letters*, 2004, **566**, 291–293.
15. P. Sjövall, B. Johansson, D. Belazi, P. Stenvinkel, B. Lindholm, J. Lausmaa and M. Schalling, *Appl. Surf. Sci.*, 2008, **255**, 1177–1180.
16. P. Sjövall, J. Lausmaa and B. Johansson, *Anal. Chem.*, 2004, **76**, 4271–4278.
17. D. Touboul, S. Roy, D. P. Germain, P. Chaminade, A. Brunelle and O. Laprévote, *Int. J. Mass Spectrom.*, 2007, **260**, 158–165.

18. J. S. Fletcher and J. C. Vickerman, *Anal. Chem.*, 2013, **85**, 610–639.
19. J. C. Vickerman, *Analyst*, 2011, **136**, 2199–2217.
20. S. Ninomiya, K. Ichiki, H. Yamada, Y. Nakata, T. Seki, T. Aoki and J. Matsuo, *Rapid Commun. Mass Spectrom.*, 2009, **23**, 1601–1606.
21. S. Rabbani, A. M. Barber, J. S. Fletcher, N. P. Lockyer and J. C. Vickerman, *Anal. Chem.*, 2011, **83**, 3793–3800.
22. A. G. Shard, R. Havelund, M. P. Seah, S. J. Spencer, I. S. Gilmore, N. Winograd, D. Mao, T. Miyayama, E. Niehuis, D. Rading and R. Moellers, *Anal. Chem.*, 2012, **84**, 7865–7873.
23. J. S. Fletcher, S. Rabbani, A. Henderson, P. Blenkinsopp, S. P. Thompson, N. P. Lockyer and J. C. Vickerman, *Anal. Chem.*, 2008, **80**, 9058–9064.
24. P. D. Piehowski, A. J. Carado, M. E. Kurczy, S. G. Ostrowski, M. L. Heien, N. Winograd and A. G. Ewing, *Anal. Chem.*, 2008, **80**, 8662–8667.
25. D. F. Smith, E. W. Robinson, A. V. Tolmachev, R. M. A. Heeren and L. Paŝa-Tolić, *Anal. Chem.*, 2011, **83**, 9552–9556.
26. G. Eglinton, P. M. Scott, T. Belsky, A. L. Burlingame and M. Calvin, *Science*, 1964, **145**, 263–264.
27. S. G. Boxer, M. L. Kraft and P. K. Weber, *Annu. Rev. Biophys.*, 2009, **38**, 53–74.
28. J. S. Fletcher, *Analyst*, 2009, **134**, 2204–2215.
29. V. J. Orphan and C. H. House, *Geobiology*, 2009, 7, 360–372.
30. M. Wagner, *Annu. Rev. Microbiol.*, 2009, **63**, 411–429.
31. N. Winograd and B. J. Garrison, *Annu. Rev. Phys. Chem.*, 2010, **61**, 305–322.
32. J. C. Vickerman, *Surf. Sci.*, 2009, **603**, 1926–1936.
33. F. Kollmer, *Appl. Surf. Sci.*, 2004, **231–232**, 153–158.
34. C. Prinz, F. Hook, J. Malm and P. Sjovall, *Langmuir*, 2007, **23**, 8035–8041.
35. N. Davies, D. E. Weibel, P. Blenkinsopp, N. Lockyer, R. Hill and J. C. Vickerman, *Appl. Surf. Sci.*, 2003, **203/204**, 223–227.
36. D. Weibel, S. Wong, N. Lockyer, P. Blenkinsopp, R. Hill and J. C. Vickerman, *Anal. Chem.*, 2003, **75**, 1754–1764.
37. A. Brunelle, D. Touboul and O. Laprevote, *J. Mass Spectrom.*, 2005, **40**, 985–999.
38. A. Delcorte and B. J. Garrison, *Nucl. Instrum. Methods Phys. Res., Sect. B*, 2007, **255**, 223–228.
39. B. J. Garrison and Z. Postawa, *Mass Spectrom. Rev.*, 2008, **27**, 289–315.
40. J. L. S. Lee, S. Ninomiya, J. Matsuo, I. S. Gilmore, M. P. Seah and A. G. Shard, *Anal. Chem.*, 2010, **82**, 98–105.
41. S. Ninomiya, K. Ichiki, H. Yamada, Y. Nakata, T. Seki, T. Aoki and J. Matsuo, *Rapid Commun. Mass Spectrom.*, 2009, **23**, 3264–3268.
42. A. Wucher and N. Winograd, *Anal. Bioanal. Chem.*, 2010, **396**, 105–114.
43. Z. Postawa, L. Rzeznik, R. Paruch, M. F. Russo, N. Winograd and B. J. Garrison, *Surf. Interface Anal.*, 2011, **43**, 12–15.
44. A. Wucher, J. Cheng, L. Zheng and N. Winograd, *Anal. Bioanal. Chem.*, 2009, **393**, 1835–1842.

45. M. Fartmann, C. Kriegeskotte, S. Dambach, A. Wittig, W. Sauerwein and H. F. Arlinghaus, *Appl. Surf. Sci.*, 2004, **231**, 428–431.

46. R. E. Peterson, A. Nair, S. Dambach, H. F. Arlinghaus and B. J. Tyler, *Appl. Surf. Sci.*, 2006, **252**, 7006–7009.

47. B. J. Tyler, S. Dambach, S. Galla, R. E. Peterson and H. F. Arlinghaus, *Anal. Chem.*, 2011, **84**, 76–82.

48. K. Chughtai and R. M. A. Heeren, *Chem. Rev.*, 2010, **110**, 3237–3277.

49. A. Gunnarsson, F. Kollmer, S. Sohn, F. Hook and P. Sjovall, *Anal. Chem.*, 2010, **82**, 2426–2433.

50. A. Carado, M. K. Passarelli, J. Kozole, J. E. Wingate, N. Winograd and A. V. Loboda, *Anal. Chem.*, 2008, **80**, 7921–7929.

51. R. Hill and P. W. M. Blenkinsopp, *Appl. Surf. Sci.*, 2004, **231**, 936–939.

52. J. L. S. Lee, I. S. Gilmore, M. P. Seah and I. W. Fletcher, *J. Am. Soc. Mass Spectrom.*, 2011, **22**, 1718–1728.

53. J. L. S. Lee, I. S. Gilmore, M. P. Seah, A. P. Levick and A. G. Shard, *Surf. Interface Anal.*, 2011, **44**, 238–245.

54. D. A. Brenes, Z. Postawa, A. Wucher, P. Blenkinsopp, B. J. Garrison and N. Winograd, *Surf. Interface Anal.*, 2012, **45**, 50–53.

55. P. Sjövall, D. Rading, S. Ray, L. Yang and A. G. Shard, *J. Phys. Chem. B*, 2010, **114**, 769–774.

56. D. Breitenstein, C. E. Rommel, R. Mollers, J. Wegener and B. Hagenhoff, *Angew. Chem., Int. Ed.*, 2007, **46**, 5332–5335.

57. J. S. Fletcher, N. P. Lockyer and J. C. Vickerman, *Mass Spectrom. Rev.*, 2011, **30**, 142–174.

58. M. E. Kurczy, P. D. Piehowsky, D. Willingham, K. A. Molyneaux, M. L. Heien, N. Winograd and A. G. Ewing, *J. Am. Soc. Mass Spectrom.*, 2010, **21**, 833–836.

59. S. Siljeström, J. Lausmaa, P. Sjövall, C. Broman, V. Thiel and T. Hode, *Geobiology*, 2010, **8**, 37–44.

60. P. Sjovall, D. Rading, S. Ray, L. Yang and A. G. Shard, *J. Phys. Chem. B*, 2010, **114**, 769–774.

61. E. Niehuis, R. Möllers, D. Rading, H. G. Cramer and R. Kersting, *Surf. Interface Anal.*, 2012, **45**, 158–162.

62. J. S. Fletcher, S. Rabbani, A. Henderson, N. P. Lockyer and J. C. Vickerman, *Rapid Commun. Mass Spectrom.*, 2011, **25**, 925–932.

63. C. Heim, J. Lausmaa, P. Sjövall, J. Toporski, T. Dieing, K. Simon, B. T. Hansen, A. Kronz, G. Arp and J. Reitner, *Geobiology*, 2012, **10**, 280–297.

64. A. Palmquist, L. Emanuelsson and P. Sjövall, *Appl. Surf. Sci.*, 2012, **17**, 6485–6494.

65. P. Echlin, *Low-Temperature Microscopy and Analysis*, Plenum Press, New York, 1992.

66. T. P. Roddy, D. M. Cannon, S. G. Ostrowski, N. Winograd and A. G. Ewing, *Anal. Chem.*, 2002, **74**, 4020–4026.

67. H. F. Arlinghaus, C. Kriegeskotte, M. Fartmann, A. Wittig, W. Sauerwein and D. Lipinsky, *Appl. Surf. Sci.*, 2006, **252**, 6941–6948.

68. M. L. Kraft, P. K. Weber, M. L. Longo, I. D. Hutcheon and S. G. Boxer, *Science*, 2006, **313**, 1948–1951.
69. J. Malm, D. Giannaras, M. O. Riehle, N. Gadegaard and P. Sjovall, *Anal. Chem.*, 2009, **81**, 7197–7205.
70. V. Thiel, C. Heim, G. Arp, U. Hahmann, P. Sjovall and J. Lausmaa, *Geobiology*, 2007, **5**, 413–421.
71. M. K. Passarelli and N. Winograd, *Biochim. Biophys. Acta, Mol. Cell Biol. Lipids*, 2011, **1811**, 976–990.
72. T. Leefmann, C. Heim, S. Siljeström, M. Blumenberg, P. Sjövall and V. Thiel, *Rapid Commun. Mass Spectrom.*, 2013, **27**, 565–581.
73. C. Heim, P. Sjövall, J. Lausmaa, T. Leefmann and V. Thiel, *Rapid Commun. Mass Spectrom.*, 2009, **23**, 2741–2753.
74. C. R. Anderton, B. Vaezian, K. Lou, J. F. Frisz and M. L. Kraft, *Surf. Interface Anal.*, 2011, **44**, 322–333.
75. P. Sjövall, J. Lausmaa, B.-L. Johansson and M. Andersson, *Anal. Chem.*, 2004, **76**, 1857–1864.
76. M. S. Wagner, D. J. Graham and D. G. Castner, *Appl. Surf. Sci.*, 2006, **252**, 6575–6581.
77. M. S. Wagner, D. J. Graham, B. D. Ratner and D. G. Castner, *Surf. Sci.*, 2004, **570**, 78–97.
78. S. Vaidyanathan, J. S. Fletcher, R. M. Jarvis, A. Henderson, N. P. Lockyer, R. Goodacre and J. C. Vickerman, *Analyst*, 2009, **134**, 2352–2360.
79. H. Jungnickel, E. A. Jones, N. P. Lockyer, S. G. Oliver, G. M. Stephens and J. C. Vickerman, *Anal. Chem.*, 2005, **77**, 1740–1745.
80. R. N. S. Sodhi, C. A. Mims, R. E. Goacher, B. McKague and A. P. Wolfe, *Surf. Interface Anal.*, 2012, **45**, 557–560.
81. R. E. Goacher, D. Jeremic and E. R. Master, *Anal. Chem.*, 2011, **83**, 804–812.
82. I. S. Gilmore, F. M. Green and M. P. Seah, *Surf. Interface Anal.*, 2007, **39**, 817–825.
83. E. A. Jones, N. P. Lockyer and J. C. Vickerman, *Appl. Surf. Sci.*, 2006, **252**, 6727–6730.
84. S. G. Ostrowski, C. Szakal, J. Kozole, T. P. Roddy, J. Xu, A. G. Ewing and N. Winograd, *Anal. Chem.*, 2005, **77**, 6190–6196.
85. A. M. Piwowar, N. P. Lockyer and J. C. Vickerman, *Anal. Chem.*, 2009, **81**, 1040–1048.
86. A. M. Piwowar, S. Keskin, M. O. Delgado, K. Shen, J. J. Hue, I. Lanekoff, A. G. Ewing and N. Winograd, *Surf. Interface Anal.*, 2013, **45**, 302–304.
87. I. Lanekoff, M. E. Kurczy, K. L. Adams, J. Malm, R. Karlsson, P. Sjövall and A. G. Ewing, *Surf. Interface Anal.*, 2011, **43**, 257–260.
88. A. G. Shard, A. Rafati, R. Ogaki, J. L. S. Lee, S. Hutton, G. Mishra, M. C. Davies and M. R. Alexander, *J. Phys. Chem. B*, 2009, **113**, 11574–11582.
89. M. E. Kurczy, P. D. Piehowski, C. T. Van Bell, M. L. Heien, N. Winograd and A. G. Ewing, *Proc. Natl. Acad. Sci. U. S. A.*, 2010, **107**, 2751–2756.

90. S. G. Ostrowski, C. T. Van Bell, N. Winograd and A. G. Ewing, *Science*, 2004, **305**, 71–73.
91. N. Tahallah, A. Brunelle, S. De La Porte and O. Laprévote, *J. Lipid Res.*, 2008, **49**, 438–454.
92. H. Nygren, K. Borner, B. Hagenhoff, P. Malmberg and J.-E. Mansson, *Biochim. Biophys. Acta, Mol. Cell Biol. Lipids*, 2005, **1737**, 102–110.
93. A. F. M. Altelaar, J. van Minnen, R. M. A. Heeren and S. R. Piersma, *Appl. Surf. Sci.*, 2006, **252**, 6702–6705.
94. E. A. Jones, N. P. Lockyer and J. C. Vickerman, *Int. J. Mass Spectrom.*, 2007, **260**, 146–157.
95. R. Jetter and R. Sodhi, *Surf. Interface Anal.*, 2011, **43**, 326–330.
96. A. M. Spool, *Surf. Interface Anal.*, 2004, **36**, 264–274.
97. J. Toporski and A. Steele, *Org. Geochem.*, 2004, **35**, 793–811.
98. N. Wehbe, A. Heile, H. F. Arlinghaus, P. Bertrand and A. Delcorte, *Anal. Chem.*, 2008, **80**, 6235–6244.
99. S. Siljeström, T. Hode, J. Lausmaa, P. Sjövall, J. Toporski and V. Thiel, *Org. Geochem.*, 2009, **40**, 135–143.
100. J. Levine, J. C. Billeter, U. Krull and R. Sodhi, *Surf. Interface Anal.*, 2011, **43**, 317–321.
101. A. Steele, J. K. W. Toporski, R. Avci, S. Guidry and D. S. McKay, *Org. Geochem.*, 2001, **32**, 905–911.
102. G. Wanger, D. Moser, M. Hay, S. Myneni, T. C. Onstott and G. Southam, *Geobiology*, 2012.
103. S. A. Sandford, J. Aléon, C. M. O. D. Alexander, T. Araki, S. Bajt, G. A. Baratta, J. Borg, J. P. Bradley, D. E. Brownlee and J. R. Brucato, *Science*, 2006, **314**, 1720–1724.
104. T. H. Stephan, K. J. Elmar, H. H. Christian and R. Detlef, *Meteorit. Planet. Sci.*, 2003, **38**, 109–116.
105. J. Lindgren, P. Uvdal, P. Sjövall, D. E. Nilsson, A. Engdahl, B. P. Schultz and V. Thiel, *Nat. Commun.*, 2012, **3**, 824.
106. A. Delcorte, B. G. Segda and P. Bertrand, *Surf. Sci.*, 1997, **381**, 18–32.
107. J. Toporski, A. Steele, F. Westall, R. Avci, M. Martill and D. S. McKay, *Geochim. Cosmochim. Acta*, 2002, **66**, 1773–1791.
108. D. E. Greenwalt, Y. S. Goreva, S. M. Siljeström, T. Rose and R. E. Harbach, *Proc. Natl. Acad. Sci. U. S. A.*, 2013, **110**(46), 18496–18500.
109. S. A. Guidry and H. S. Chafetz, *J. Sediment. Res.*, 2003, **73**, 806–823.
110. E. A. Mathez and D. M. Mogk, *Am. Mineral.*, 1998, **83**, 918–924.
111. G. Wanger, G. Southam and T. C. Onstott, *Geomicrobiol. J.*, 2006, **23**, 443–452.
112. F. Westall, B. Cavalazzi, L. Lemelle, Y. Marrocchi, J.-N. Rouzaud, A. Simionovici, M. Salomé, S. Mostefaoui, C. Andreazza, F. Foucher, J. Toporski, A. Jauss, V. Thiel, G. Southam, L. MacLean, S. Wirick, A. Hofmann, A. Meibom, F. Robert and C. Défarge, *Earth Planet. Sci. Lett.*, 2011, **310**, 468–479.
113. R. Li and J. Parnell, *J. Geochem. Explor.*, 2003, **78–79**, 377–384.

114. S. Siljeström, H. Volk, S. C. George, J. Lausmaa, P. Sjövall, A. Dutkiewicz and T. Hode, *Geochim. Cosmochim. Acta*, 2013, **122**, 448–463.
115. V. Thiel, J. Toporski, G. Schumann, P. Sjövall and J. Lausmaa, *Geobiology*, 2007, **5**, 75–83.
116. T. Leefmann, C. Heim, A. Kryvenda, S. Siljeström, P. Sjövall and V. Thiel, *Org. Geochem.*, 2013, **57**, 23–33.
117. M. C. Perkins, C. J. Roberts, D. Briggs, M. C. Davies, A. Friedmann, C. A. Hart and G. A. Bell, *Planta*, 2005, **221**, 123–134.
118. A. Seyer, J. Einhorn, A. Brunelle and O. Laprévote, *Anal. Chem.*, 2010, **82**, 2326–2333.
119. E. N. Tokareva, P. Fardim, A. V. Pranovich, H. P. Fagerholm, G. Daniel and B. Holmbom, *Appl. Surf. Sci.*, 2007, **253**, 7569–7577.
120. K. Saito, Y. Watanabe, M. Shirakawa, Y. Matsushita, T. Imai, T. Koike, Y. Sano, R. Funada, K. Fukazawa and K. Fukushima, *Plant J.*, 2012, **69**, 542–552.
121. R. E. Goacher, E. A. Edwards, A. F. Yakunin, C. A. Mims and E. R. Master, *Anal. Chem.*, 2012, **84**, 4443–4451.
122. J. F. C. De Brouwer, K. E. Cooksey, B. Wigglesworth-Cooksey, M. J. Staal, L. J. Stal and R. Avci, *J. Microbiol. Methods*, 2006, **65**, 562–572.
123. S. G. Ostrowski, C. T. V. Bell, N. Winograd and A. G. Ewing, *Science*, 2004, **305**, 71–73.
124. J. S. Fletcher, A. Henderson, R. M. Jarvis, N. P. Lockyer, J. C. Vickerman and R. Goodacre, *Appl. Surf. Sci.*, 2006, **252**, 6869–6874.
125. C. E. Thompson, J. Ellis, J. S. Fletcher, R. Goodacre, A. Henderson, N. P. Lockyer and J. C. Vickerman, *Appl. Surf. Sci.*, 2006, **252**, 6719–6722.
126. D. Debois, K. Hamze, V. Guérineau, J. P. Le Caër, I. B. Holland, P. Lopes, J. Ouazzani, S. J. Séror, A. Brunelle and O. Laprévote, *Proteomics*, 2008, **8**, 3682–3691.
127. S. Vaidyanathan, J. S. Fletcher, N. P. Lockyer and J. C. Vickerman, *Appl. Surf. Sci.*, 2008, **255**, 922–925.
128. V. Nihorimbere, H. Cawoy, A. Seyer, A. Brunelle, P. Thonart and M. Ongena, *FEMS Microbiol. Ecol.*, 2012, **79**, 176–191.
129. E. Kaivosoja, S. Virtanen, R. Rautemaa, R. Lappalainen and Y. T. Konttinen, *Eur. Cells Mater.*, 2012, **24**, 60–73.
130. A. Dutkiewicz, H. Volk, S. C. George, J. Ridley and R. Buick, *Geology*, 2006, **34**, 437–440.
131. T. J. Colla, R. Aderjan, R. Kissel and H. M. Urbassek, *Phys. Rev. B*, 2000, **62**, 8487.
132. M. Ivarsson, C. Broman, E. Sturkell, J. Ormö, S. Siljeström, M. v. Zuilen and S. Bengtson, *Sci. Rep.*, 2013, **3**, 3487.

Development and Use of Catalytic Hydropyrolysis (HyPy) as an Analytical Tool for Organic Geochemical Applications

WILL MEREDITH,*[a] COLIN E. SNAPE[a] AND GORDON D. LOVE[b]

[a] Department of Chemical and Environmental Engineering, Faculty of Engineering, University of Nottingham, NG7 2RD, UK; [b] Department of Earth Sciences, University of California Riverside, CA 92521, USA
*Email: William.Meredith@nottingham.ac.uk

6.1 Origins of Hydropyrolysis

Hydropyrolysis (HyPy) was developed originally in the 1950s as a route for the direct conversion of coal into mainly methane (synthetic natural gas) and light aromatic feedstocks, particularly benzene.[1] This required the combination of extremely high temperatures and pressures, ~ 800 °C and up to 300 bar.[2,3] During the late 1980s, it was found that by using dispersed catalysts, such as sulfided molybdenum, it was possible to achieve much higher overall conversions to liquid products at pressures no higher than 150 bar.[4,5] Further, there is the possibility of hydrotreating the primary oil vapours to further increase the yield of hydrocarbons by removing heteroatoms from polar species.[6] These developments laid the foundation to

RSC Detection Science Series No. 4
Principles and Practice of Analytical Techniques in Geosciences
Edited by Kliti Grice

develop HyPy as an analytical pyrolysis procedure based on the unique ability HyPy possesses of producing high yields of biomarkers from petroleum source rocks, together with achieving overall conversions close to 100% for macromolecular labile organic matter, while, at the same time, minimising structural alteration by isomerisation and cracking.

6.2   Methodology Development

6.2.1   First Biomarker Application and Optimisation of the Experimental Conditions

In a preliminary study[7] into the fate of biomarker hydrocarbons during HyPy, it was observed for a lignite that the hopanes released had predominately the biologically synthesised 17β(H),21β(H) stereochemistry. The first full description of HyPy being used as an analytical pyrolysis method for cleaving hydrocarbon biomarkers from kerogen was by Love *et al.*,[8] who used an immature type I kerogen (Göynük oil shale, of Miocene age) that had been exhaustively pre-extracted in dichloromethane (DCM) and pyridine to remove any free biomarkers. The sample was first impregnated with aqueous/methanol solution of the catalyst precursor ammonium dioxydithiomolybdate [$(NH4)_2MoO_2S_2$], to give a nominal molybdenum loading of 1 wt%. This would reductively decompose upon heating to yield catalytically active oxysulfide molybdenum species (MoS_2 at approximately 400 °C). The sample, held in a reactor tube, was heated resistively from ambient temperature to 100 °C at 10 °C min^{-1}, and then to 520 °C at 5 °C min^{-1}, under a hydrogen pressure of 15 MPa. A hydrogen flow of 10 dm^3 min^{-1}, measured at ambient temperature and pressure, through the reactor ensured that the overall conversion was not limited by mass transfer in the sample bed[9] and so, once formed, volatile products were quickly swept from the reactor. The generated oil was collected in the trap cooled with dry ice, and recovered in DCM for subsequent fractionation. A schematic of the HyPy rig similar to that used is shown in Figure 6.1.

The high yield of aliphatic hydrocarbons released by HyPy (222 mg g^{-1} TOC^{-1}) was consistent with the relatively low aromaticity of the shale. They were primarily composed of a bimodal distribution of n-alkanes/n-alk-1-ene doublets (maximum at C_{20} and submaximum at C_{28}), extending to homologues heavier than n-C_{35} (Figure 6.2). The triterpanes generated were dominated by hopanes up to C_{35}, with the biologically inherited but thermodynamically unstable 17β(H),21β(H) configuration, with yields of the C_{29}–C_{33} homologues found to be 3–10 times greater than those from DCM extraction, or from pyrolysis in an atmosphere of nitrogen (in which, as shown in Figure 6.2, the C_{34} and C_{35} homologues were entirely absent). The survival of these extended hopanes, together with n-alkanes >n-C_{35} which are not often observed in the products of online pyrolysis or chemical degradation, demonstrated the ability of HyPy to maximise the yields of covalently bound alkane biomarkers, without adversely affecting their

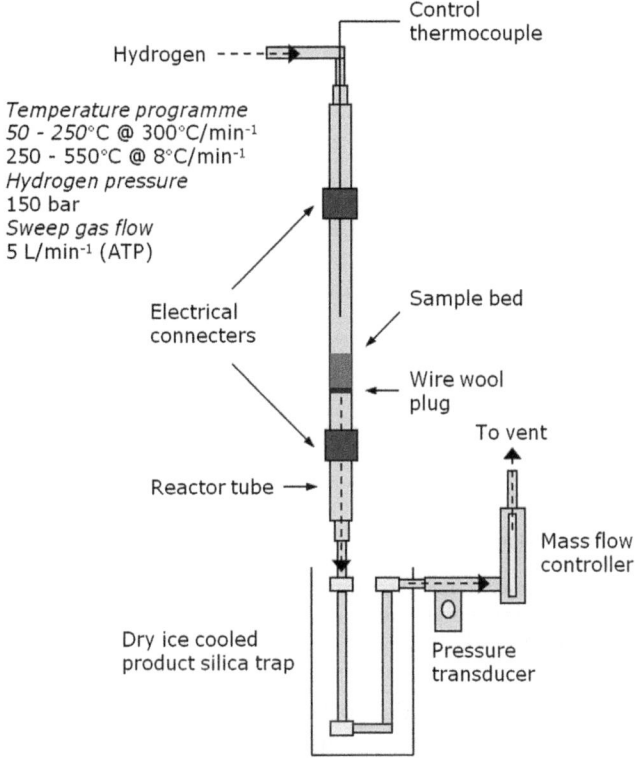

Figure 6.1 Schematic of HyPy apparatus.[77]

stereochemistries.[8] The results suggested that the bulk of the generated alkanes probably evolved before C–C bond scission occurred to a significant extent above 400 °C, and that the incorporation of biomarkers into kerogen, through functional groups present in the original biolipids, afforded steric protection for the hopanes and steranes.[8]

The bound biomarker distributions cleaved from the macromolecular matrix of this immature oil shale clearly exhibited well-preserved stereochemistries that were close to their original biological configurations, despite the apparent severity of the HyPy process.[8] The conditions used, including a combination of slow heating (5 °C min^{-1}), high hydrogen pressure (15 MPa), and a dispersed molybdenum catalyst, were employed intuitively, based on experience of the hydroliquefaction of coals and kerogens.[9] Therefore, it was necessary to systematically investigate the effect of altering key procedural variables on the product yield and molecular distributions obtained from the HyPy of selected kerogens, to ascertain the optimum conditions that should be employed for routine HyPy tests. Using the same type I oil shale as in the original study, together with a Tertiary lignite, heating temperature rates of 5, 20, and 300 °C min^{-1} and hydrogen pressures of 50 and 150 bar were tested. The efficacy of each set of conditions

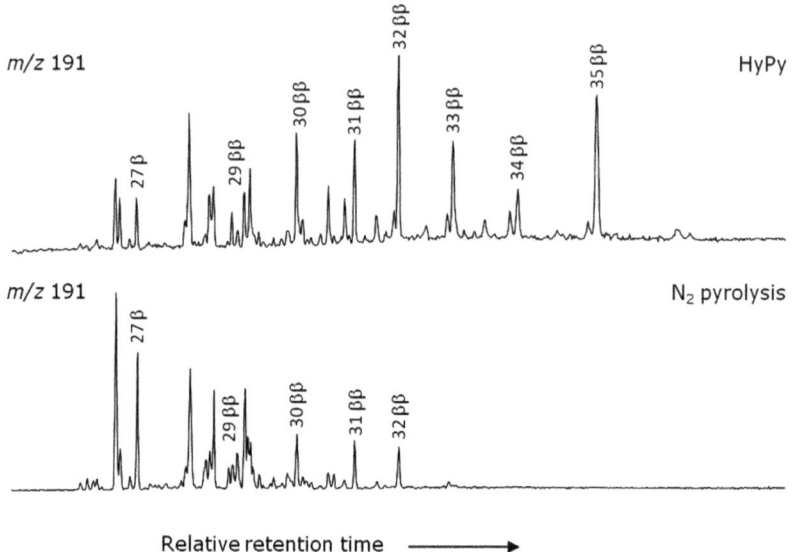

Figure 6.2   *m/z* 191 mass chromatograms from the HyPy and nitrogen pyrolysis of
                 solvent-extracted Göynük oil shale.
                 Samples from Love *et al.*[8]

was assessed by monitoring the total yields of soluble products, with the
distribution of the principal aliphatic products and biomarkers used to
measure the relative extent of thermal cracking.[10]

It was found that a heating rate of 5 °C min^{-1} to a maximum of 520 °C
under a hydrogen pressure of 150 bar represented the best regime for
achieving high conversions of kerogens to DCM-soluble oils. Using these
conditions, the generation of readily analysable products was maximised,
while structural rearrangement of biomarker species was minimal, with bulk
compositional changes in carbon aromaticity and long alkyl chain content
found to be relatively minor compared to traditional pyrolysis techniques.[11]
In order to save time, later tests confirmed that a fast heating rate of
300 °C min^{-1} to 250 °C followed by a slower rate of 8 °C min^{-1} to the final
temperature of 520 °C replicated the high product yields and structural
preservation, while reducing the total run time to 35 min. Even allowing for
this adaptation HyPy is more time-consuming than conventional analytical
pyrolysis methods such as flash pyrolysis, although it is clear that it yields
valuable biogeochemical information that is unlikely to be revealed *via*
other, more aggressive, degradation methods. In addition, replacing the
original coiled steel trap by one containing silica gel to adsorb the generated
oils[12] also reduced the time for product work-up, and minimised evaporative
losses in subsequent fractionation into aliphatics and aromatics. However, it
should be noted that unlike online pyrolytic methods and closed-system
microscale sealed-vessel pyrolysis (MSSVpy), HyPy does not routinely trap
the gaseous and low-molecular-weight liquid products (<n-C$_8$),[13] although

product recovery can be improved by replacing the dry ice in the trap with liquid nitrogen.[12]

6.2.2 Two-Stage HyPy

A subsequent development of HyPy has been the introduction of a second stage in the reactor, composed of a bed of hydrotreating (nickel/molybdenum or cobalt/molybdenum) catalyst below the sample. This was held isothermally at 320 °C which resulted in the effective catalytic defunctionalisation of the primary oil vapours to remove weaker heteroatomic functionalities (*e.g.* hydroxyls, carboxyls, and thiols), while preventing cracking and rearrangement of the hydrocarbons released.[14] This technique resulted in the doubling of the yield of gas chromatography (GC)-amenable aliphatic hydrocarbons, with a similar increase in the yield of hopane and sterane biomarkers observed, although the requirement for two different heating zones introduced too many complications into the HyPy procedure for routine operation. However similar results have also been found when replacing the second stage with an additional bed (250 mg) of the same molybdenum catalyst directly below the catalyst-impregnated sample.[15] This study also found that oil asphaltenes could be successfully treated when introduced into the reactor as solid flakes directly on top of the catalyst bed. It was assumed that the asphaltenes would soften on heating, and the flow of gas would then encourage them into intimate contact with the catalyst at the top of the bed, so removing the need for the time-consuming process of adsorbing each sample onto catalyst-impregnated silica. The remainder of the catalyst bed was then found to act as the second stage in the procedure to fully defunctionalise the primary oil vapours.

6.2.3 Model Compounds

In order to exploit HyPy to its full potential it was felt necessary to gain a better understanding of the efficiency and selectivity of the technique for releasing bound biomarkers from organic macromolecules such as kerogens and oil asphaltenes. Following a preliminary study[16], Meredith *et al.*[17] described a series of experiments with functionalised model compounds adsorbed to a silica substrate which included carboxylic acids (stearic acid, oleic acid, and 5β-cholanic acid), and sterols (cholesterol and cholestanol). This allowed for the selectivity of hydrogenation to their corresponding alkanes to be assessed, together with the extent of cracking and isomerisation undergone upon formation from specific functionalities.

The n-C_{18} acids were shown to be reduced to the n-C_{18} alkane, with a selectivity of more than 95% for stearic acid, although oleic acid was prone to cracking due to its unsaturated structure, with shorter-chained n-alkanes also being formed. The conversion of these compounds was relatively low, even at HyPy temperatures significantly above their boiling point, suggesting that interactions between the acids and the silica substrate led to the formation of stable entities (Si–O–C linkages) which significantly retarded volatilisation. Product yield was found to be increased when the model compound was placed directly onto a bed of catalyst, but for low-boiling compounds such as stearic acid this resulted in volatilisation and consequently cracking at temperatures below that of the activation point of the catalyst ($\sim 250\ ^\circ$C); however, this method produced improved yields of more than 95% pure product for higher-boiling compounds such as 5β-cholanic acid.[17] The presence of the functional group attached to the ring system of compounds such as 5α-cholestanol did not diminish the selectivity of the technique, although the double bond in cholesterol was found to induce rearrangements in the parent hydrocarbon. This resulted in the generation of cholestenes and diasteranes in addition to the expected isomers of 5α and 5β-cholestane.[18]

Studies employing model compounds have also sought to assess HyPy as a method for the treatment of biologically important functionalised compounds such as fatty acids and steroids, prior to compound-specific stable carbon isotope analysis by gas chromatography–combustion–isotope ratio mass spectrometry (GC-C-IRMS). The presence of functional groups that interact with the stationary phase of the GC column during the separation step hinders chromatographic resolution. Existing strategies to avoid such poor resolution involve derivatisation, which solves the chromatographic separation problem, but the addition of extra carbon atoms corrupts the carbon isotope signal of the target molecules. The absence of reactions which add or remove carbon prior to analysis makes HyPy an attractive alternative approach, which has been tested with both steroids[18] and fatty acids.[19] As noted above, while the results obtained for saturated steroids such as cholestanol were excellent, with the alkanes generated by HyPy seen to be isotopically faithful representatives of the starting materials, those for unsaturated compounds such as cholesterol were not satisfactory owing to rearrangements of the parent hydrocarbon induced by the presence of the carbon–carbon double bond. Later developments to the technique, including the use of a platinum catalyst, and a much lower reaction temperature (300 $^\circ$C), resulted in the same highly selective conversion of unsaturated cholesterol into cholestane, as previously found for saturated cholestanol when using the molybdenum catalyst, together with a far higher degree of conversion.[20]

A further direction to be explored in the future is the use of HyPy not only to defunctionalise steroids, but also to deconjugate them, as in biological systems such compounds are typically found conjugated to gluconoride or sulfide moieties.[21,22] Current analytical approaches include an unsatisfactory hydrolysis step followed by derivatisation, but with HyPy it may be possible to reduce biological conjugated steroids to a single hydrocarbon in one procedure without introducing significant isotopic fractionation.

6.3   Petroleum Geochemistry

The ability of HyPy to release high yields of biomarker hydrocarbons from macromolecular organic fractions, while minimising alteration to their isomeric distributions, gives the technique a wide variety of potential applications within petroleum geochemistry where traditional approaches, using the free, solvent-extractable biomarkers fail.

6.3.1 Assessment of Thermal Maturity

The assessment of thermal maturity that an oil or kerogen has undergone by using ratios of specific biomarkers in the free (bitumen) fraction is well documented. Typical ratios used include those of the ring system or side-chain isomerisation of cyclic alkanes, such as the hopanes and steranes, isomerisation of methylated aromatic hydrocarbons (*e.g.* the methylphenanthrene (MPI) and methyldibenzothiophene (MDR) indices), and the degree of aromatisation of the aromatic steroids.[23] Some of these ratios have been calibrated against vitrinite reflectance (% R_o) in order to identify the extent of thermal maturity in relation to the oil window that sample has reached.

These molecular indices are typically calculated from the biomarkers isolated from the free, solvent-extractable phase of oils or kerogens. For kerogens especially, this free phase is representative of only a small proportion of the total organic matter (typically <5%). Thus it does not follow that the distributions of the free biomarkers are necessarily representative of those covalently bound to the macromolecular structure.[24,25] Previous studies[26] have shown that, for hopanes and steranes bound into kerogen, preservation of biologically inherited structures is enhanced such that configurational isomerisation can be inhibited.[27] Therefore, compounds covalently bound to the macromolecular network may be less sensitive to thermal alteration, with isomerisation of kerogen-bound homohopanes ($22S/(S+R)$) proceeding at a slower rate during hydrous pyrolysis than that of the corresponding free homohopanes.[28]

A general trend of retarded maturity for bound biomarkers relative to their free counterparts has been established by HyPy for a series of vitrinite concentrates from six bituminous coals covering the rank range 0.46–1.32% R_o, which corresponded to a maturity range from just before to just after the oil-generating window for type III materials.[24] It was found that highly representative quantities of kerogen-bound hopanes and steranes released were generally less mature than their counterparts in the bitumen phase, both in terms of ring system and side-chain isomerisation, with the ratio of free to kerogen-bound aliphatic biomarkers increasing with maturation. In addition, while the solvent-extractable free biomarkers were found to be very sensitive to thermal stress at low maturity, this was not the case for bulk skeletal structural properties such as atomic H/C ratios and aromaticity as measured by ^{13}C nuclear magnetic resonance (NMR).[24]

Bishop *et al.*[29] compared the kerogen-bound hopanoids released by HyPy to their free bitumen counterparts for three unrelated source rock kerogens, ranging from below to within the oil-generating window, together with a Recent sediment. Maturation was found to control the composition of the

kerogen-bound hopanes, with excellent preservation of immature, biologically inherited 17β(H),21β(H) isomers found in the Recent sediment, and
the level of isomerisation apparent in the hopanes released from the source
rocks by HyPy found to be proportional to, but significantly lower than,
those observed in the bitumen extracts.[29] In a more systematic study for a
type II kerogen, Murray *et al.*[27] subjected six solvent-extracted Kimmeridge
Clay samples covering a range of vitrinite reflectance from 0.38–0.61% R_o
(equivalent to just before and into the oil-generating window) to HyPy. It was
found that the ratios of free to kerogen-bound aliphatic biomarkers increased markedly at relatively low maturity ($\sim$0.45–0.50% R_o), with large
reductions occurring in the concentrations of bound hopanes and steranes.
As this decrease was found to be much greater than that of the corresponding total aliphatics, it implied that the hopanes and steranes must be
bound by weaker covalent bonds (*i.e.* C–S and C–O) than the long-chain alkyl
moieties, which were the dominant structural components in the aliphatic
fraction as a whole.[27]

Figure 6.3 compares the hopane profile from the solvent extract and HyPy
products from the Kimmeridge Clay sample with a vitrinite reflectance of

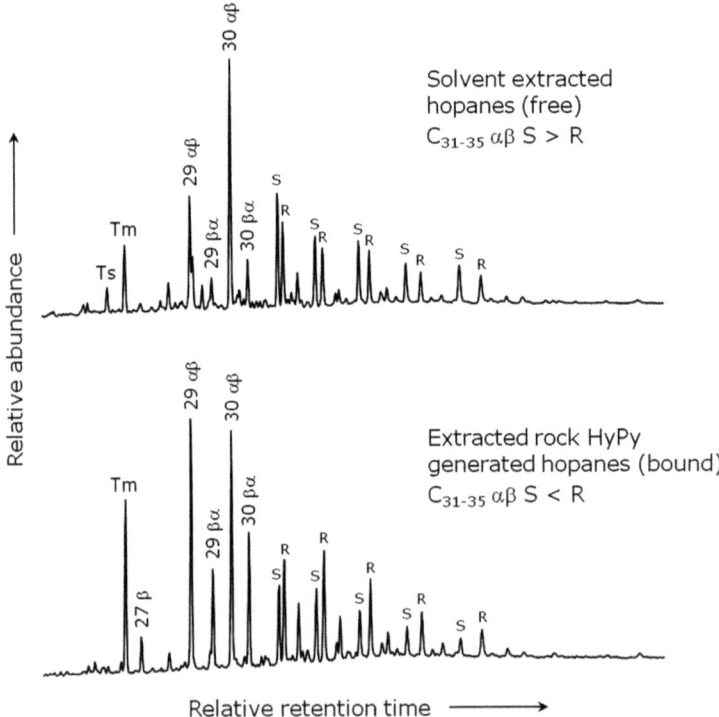

Figure 6.3 Comparison of the *m/z* 191 mass chromatograms showing the hopane
distributions from the free, solvent-extracted phase (top) and the bound,
HyPy-generated phase (bottom) for a sample of the Kimmeridge Clay
Formation (KCF).[27]

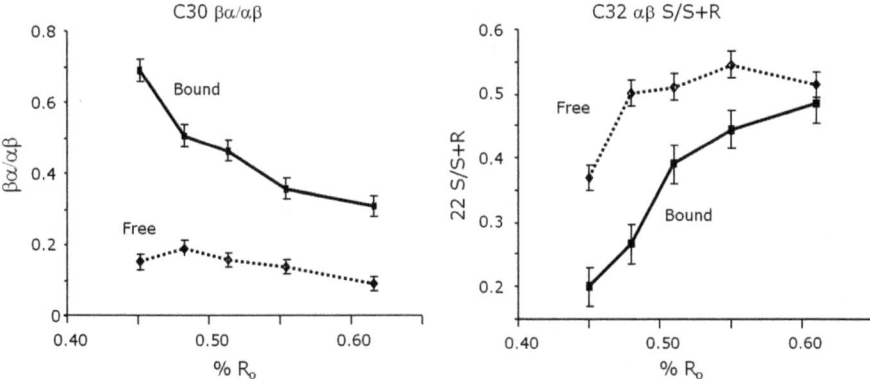

Figure 6.4 Variation in the hopane maturity parameters from the free and bound phases of the suite of KCF samples. (A) 17β,21α(H)-moretane/17α,21β(H)-hopane. (B) 17α,21β(H)-dihomohopane (22S)/[17α,21β(H)-dihomohopane (22S) + 17α,21β(H)-dihomohopane (22R)]. Samples from Murray *et al.*[27]

0.51% R_o. The most striking aspect of these chromatograms is the visual difference between the relative abundance of the 22S and 22R ratio isomers for the C_{31} to C_{35} 17α,21β(H)-homohopanes. The solvent extracts show a larger 22S epimer, as expected for a sample of this maturity, but the abundance is reversed for the bound phase with the 22R epimer being the larger. This is reflected in Figure 6.4 which shows the changes in the 22S/(S + R) ratio for the C_{32} 17α,21β(H)-hopanes, as well as the βα/αβ ratio (moretane/hopane index) for the C_{30} hopanes as a function of maturity for the solvent extracts and HyPy products. Other differences which are common for HyPy products include the higher abundance of the C_{29} αβ hopane relative to the C_{30} αβ homologue, and the absence of rearrangement products, such as trisnorneohopane (Ts) and the diasteranes.[30]

The study by Murray *et al.*[27] was the first to observe a systematic change in epimer ratios with increasing maturity for hopanes and steranes covalently bound in kerogens. These biomarkers were observed to undergo the same epimerisation reaction pathways as their free counterparts in the bitumen. As with previous studies they were shown to be generally less mature in terms of isomerisation at both ring system and side-chain chiral centres. Therefore, the range of ratios such as the moretane/hopane index for the HyPy products was found to be much larger than that for the bitumens (~5 times), indicating a much greater sensitivity to relatively small changes in maturity.[27] It should be noted that as with these ratios when determined for the free phase, the bound aliphatic biomarker maturation parameters cannot be accurately calibrated against vitrinite reflectance due to source facies variations. Bowden *et al.*[30] also quantified the biomarkers present in the bitumen and kerogen fractions from a series of organic-rich mudstones of up to near oil window thermal maturity. They found a similar abundance of biomarkers covalently bound into the kerogen as were present in the extractable aliphatic hydrocarbon fraction, and that the bound fraction therefore represented a potentially important source of molecular geochemical information.

These early studies were all conducted on samples from below, within, or just above the oil-generating window in terms of their maturity. Of great interest therefore, are kerogen-bound aliphatic biomarker distributions from samples at elevated, post-oil-window maturity, where those of the solvent-extractable free phase appear in much lower concentration, and have often long since reached equilibrium.[31] For such kerogens, the ability of HyPy to provide meaningful information at high thermal maturity was demonstrated by Lockhart *et al.*[32] for a suite of type II kerogen Kimmeridge Clay source rocks from the UK North Sea, with hydrogen index (HI) values as low as around 50 mg HC/g TOC. It was found that the isomerisation reactions of the kerogen-bound biomarkers were retarded with respect to the free phase until a HI value of approximately 300. Beyond this point ring-system isomerisation within the kerogen-bound phase is still retarded (until an HI of ~ 200), while the side-chain isomerisation for both hopanes and steranes is similar in the free and bound phases. For a corresponding series of kerogen samples, artificially matured *via* hydrous pyrolysis, both free and bound-phase side-chain isomerisation reactions were significantly retarded relative to the naturally occurring samples. This was consistent with earlier studies, but the extent of retardation for ring-system isomerisation in the bound phase was even more pronounced than for the free phase.[32]

Similar trends were apparent for a suite of type III kerogens consisting of 10 UK coals (0.50–3.29% R_o), with both ring-system and side-chain isomerisation reactions for the kerogen-bound steranes and ring-system reactions for the hopanes retarded until reaching apparent thermal equilibrium at a maturity of 0.85% R_o. These equilibrium values were therefore significantly lower than those for the free phase. At more elevated levels of maturity only the hopane ring-system parameters continued to show bound-phase retardation, not reaching thermal equilibrium, which was again at a reduced end-point until a maturity of 2.0% R_o. This suggested that the bound $C_{30}\beta\alpha/C_{30}\alpha\beta$ parameter may provide a particularly sensitive proxy for maturity applicable until late to post-oil-window maturity.[33]

Sequential (multiple temperature) HyPy can give an indication of the relative strengths of bonds being cleaved and, in the presence of deuterium, can distinguish a mixture of weak sulfidic and strong covalent linkages.[34] The finding of abundant kerogen-bound hopanes relative to the yields of extractable hopanoids in young sediments[34] highlights the fact that kerogen is formed from the earliest stages of diagenesis and a high proportion of kerogen is actually formed in the water column. Related to this, Pancost *et al.*[35] noted a bias against incorporation of C_{40} biphytane structures from glycerol dialkyl glycerol tetraether (GDGT) lipids from sedimentary methanogenic Archaea into kerogens in sediments from the Benguela Upwelling Region (Ocean Drilling Program (ODP) Site 1084) and a Mediterranean sapropel (S5). Compound-specific $\delta^{13}C$ analyses indicated that the bulk of

HyPy-released biphytane carbon isotopic signatures are relatively invariant, consistent with a dominant single source likely to be pelagic Crenarchaeota. These observations suggest that GDGTs can be incorporated into sedimentary geomacromolecules relatively rapidly, probably commencing in the water column *via* oxidative polymerisation or natural vulcanisation, with those derived from pelagic Crenarchaeota more likely to be incorporated than those derived from sedimentary methanogens living in anoxic sediment.

The potential of HyPy-generated aromatic hydrocarbons to characterise source rocks at elevated levels of maturity has also been investigated,[33] again using the suite of 10 UK coals (0.50–3.29% R_o). It was demonstrated the HyPy accessed a similar fraction of aromatic compounds, in terms of relative ring-size distribution as the solvent-extractable free phase, although at an order of magnitude greater in concentration from the diminishing pool of aromatic hydrocarbons available through to elevated levels of maturity. The methylphenanthrene index (MPI-1) was calibrated to type III kerogen-bound aromatic assemblages, and in common with the model of Radke and Welte[36] the bound methylphenanthrenes showed a positive correlation of vitrinite reflectance and MPI-1 at low and medium maturities (0.50–1.55% R_o), and a negative correlation at higher maturities (1.55–3.29% R_o). However, while consistent with the observed relationship for the aliphatic biomarkers of the kerogen-bound methylphenanthrenes appearing to undergo the same isomerisation reaction pathways to those in the free phase, as the bound MPI-1 ratios were consistently higher than those in the free phase, these aromatic parameters were found to be less sensitive to changes in maturity.[33] In addition it was found that alkyl–naphthalene ratios NMR and TMR-1 (N-modular and triple-modular redundancy) also displayed potential as maturity proxies at R_o of in excess of 1.0% and 1.5% respectively, although as with the MPI ratios the efficacy of these empirical maturity parameters appeared to be highly specific to the source characteristics of the kerogen type that they were calibrated against.[33]

6.3.2 Basin Filling History

One potential application for the subtle differences in thermal maturity exhibited by the bound biomarkers released by HyPy is to act as molecular tracers to aid the reconstruction of secondary migration pathways of oils in basins and reservoir filling history. The hypothesis tested by Russell *et al.*[37] was that bound biomarkers released from the adsorbed asphaltenes isolated from drilling core samples along a known migration pathway should reveal the maturity and source characteristics of the first oil charge that came into contact with the carrier substrate, even in cases where discrete molecular components of the oil had already been homogenised. This study was performed on six samples of reservoir cores and drill cuttings from the Blake Field (Outer Moray Firth, UK North Sea, blocks 13/24 and 13/29, Figure 6.5),

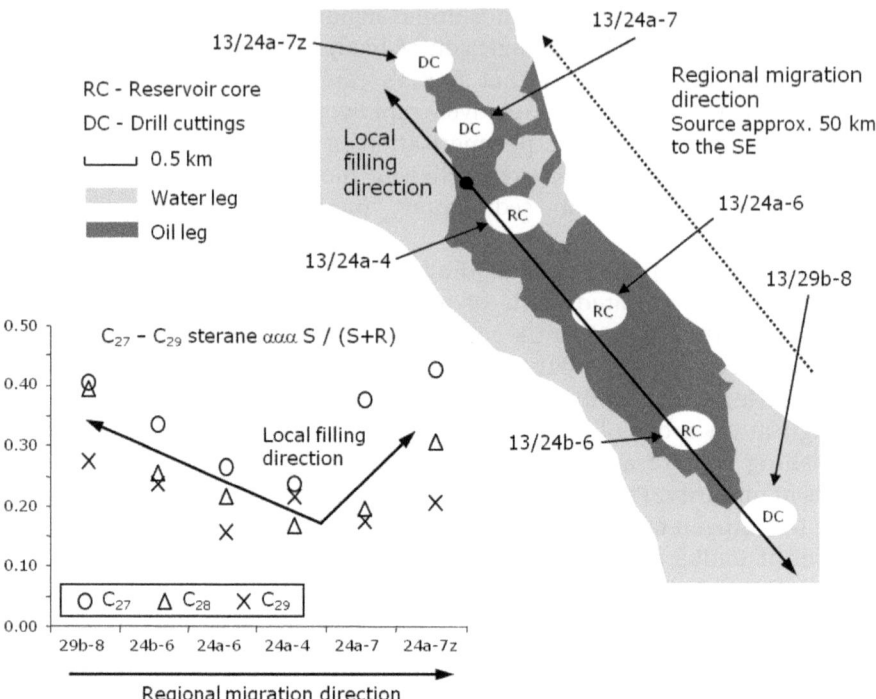

Figure 6.5 Schematic diagram of the samples from the Blake Field showing the regional and implied local filling direction based on maturity differences (as shown in graph of C_{27}–C_{29} sterane isomerisation ratios) of the biomarkers bound within the most strongly adsorbed asphaltenes.[37]

where the solvent-extractable biomarker hydrocarbon distributions exhibited no significant variation throughout the reservoir, offering little information as to reservoir filling events. However, the bound biomarkers obtained by the HyPy of the adsorbed asphaltene fractions revealed a discrete systematic increase towards the south-east of the reservoir from a point between samples 13/24a-4 and 13/24a-7 (Figure 6.5). This is potentially indicative of the location where oil first entered the reservoir, and a localised reservoir filling history that was opposite to the presumed regional migration direction.[37]

The hypothetical concept was also demonstrated in laboratory displacement experiments designed to examine asphaltene adsorption and displacement characteristics, where two oils of subtly different compositions (in terms of their bound biomarker profiles) were consecutively passed through core substrates under anhydrous and hydrous conditions. After elution with solvents of increasing polarity, it was revealed that the most strongly adsorbed asphaltenes exhibited bound biomarker distributions, indicative of approximately 70 wt.% of asphaltenes from the original oil having been retained on the core substrate.[37]

6.3.3 Heavily Biodegraded Oils

One aspect of petroleum geochemistry, where standard techniques that concentrate on the free, solvent-extractable phase cannot be usefully applied, is the characterisation of heavily biodegraded oils and bitumens. Many hydrocarbon classes such as the n-alkanes and small alkyl-substituted aromatic compounds are very susceptible to biodegradation, while more resistant compounds such as the steranes and hopanes, which are widely applied in source and maturity determinations as well as in correlation studies, are vulnerable to alteration at higher levels of biodegradation.[23] Once these compound classes have been irrevocably altered, the biomarkers, covalently bound into the macromolecular matrix of the asphaltene faction and not bioavailable to the degrading microbes, are thought to represent a remnant of the original oil.[38] Methods to access these biomarkers include chemical degradation[39] and pyrolytic approaches;[40] the former are time-consuming and release relatively low yields of biomarkers, but the latter can result in the cracking and molecular rearrangement of the biomarkers generated, which are at best therefore only partially representative of the bulk asphaltene structure.[24] More sensitive pyrolytic methods such as closed-system MSSVpy have been demonstrated to release biomarkers from petroleum asphaltenes with reduced structural alteration, and also allow for the analysis of the gaseous and low-molecular-weight products.[13]

The high yields of hydrocarbon biomarkers cleaved from asphaltenes, coupled with their excellent structural preservation, make HyPy an ideal approach for the characterisation of heavily biodegraded oils.[41] This was demonstrated in the study by Russell *et al.*[37] described above, in which it was found that one of the oils used in the laboratory displacement experiments was moderately biodegraded, with the n-alkanes completely removed and the isoprenoids greatly reduced in abundance. However, the HyPy of the asphaltenes isolated form this oil was shown to generate a pristine n-alkane distribution, with a chain length exceeding C_{35} implying excellent preservation of long-chain compounds. The HyPy of the asphaltenes from a more severely biodegraded oil, which contained no recognisable free biomarkers, was undertaken by Meredith *et al.*[42] who, in addition to the pristine n-alkanes, were also able to generate well-preserved sterane and hopane profiles.

A more systematic study of the free and bound biomarkers obtained from heavily biodegraded petroleum was undertaken by Sonibare *et al.*[43], who, following characterisation of the free, maltene fraction, subjected the asphaltenes isolated from three oil seeps from south-west Nigeria to HyPy. These seeps had been subjected to such severe biodegradation that the maltene fraction contained no n-alkanes, few identifiable aromatic hydrocarbons, and significantly altered sterane and hopane distributions. As shown in Figures 6.6 and 6.7 the HyPy of the asphaltenes again generated pristine n-alkane, sterane, and hopane profiles, allowing for source and maturity characterisation of the seeps to be undertaken. In this study geochemically significant aromatic hydrocarbons such as the alkylated

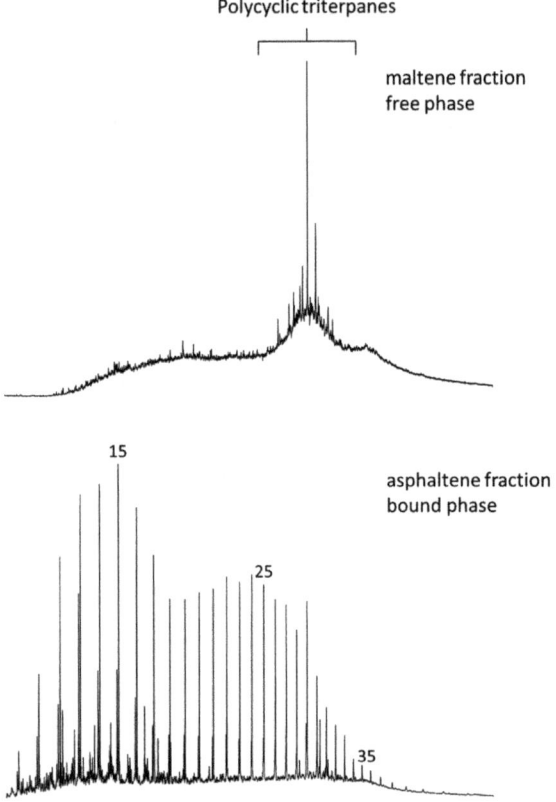

Polycyclic triterpanes

maltene fraction
free phase

15

asphaltene fraction
bound phase

25

35

Figure 6.6 Total ion current (TIC) chromatograms of the aliphatic hydrocarbons in
the maltene fractions (free phase) and asphaltene hydropyrolysates
(bound phase) from the Ilubinrin tar sand bitumen (numbers refer to
carbon chain lengths of n-alkanes).[43]

naphthalenes and phenanthrenes were also cleaved from the asphaltene
macromolecular matrix. Similar results were found by Lockhart *et al.*[44] who
used HyPy to characterise the asphaltene-bound biomarkers from a suite of
Australian crudes exhibiting a wide range in the extent of biodegradation.
The comparison between the distribution of the free and bound aliphatic
biomarkers was found to be consistent with those described above, while the
bound aromatic hydrocarbon fraction yielded a broader spectrum of compounds than present in the biodegraded free phase, particularly with respect
to the heavier components.[44]

6.3.4 Samples Contaminated with Drilling Fluids

A further example of where traditional geochemical approaches based on the
analysis of the free, solvent-extractable phase is compromised, is in instances of sample contamination with drilling fluids. Such fluids, both oil
and water based, are increasingly used in modern deep-water operations,

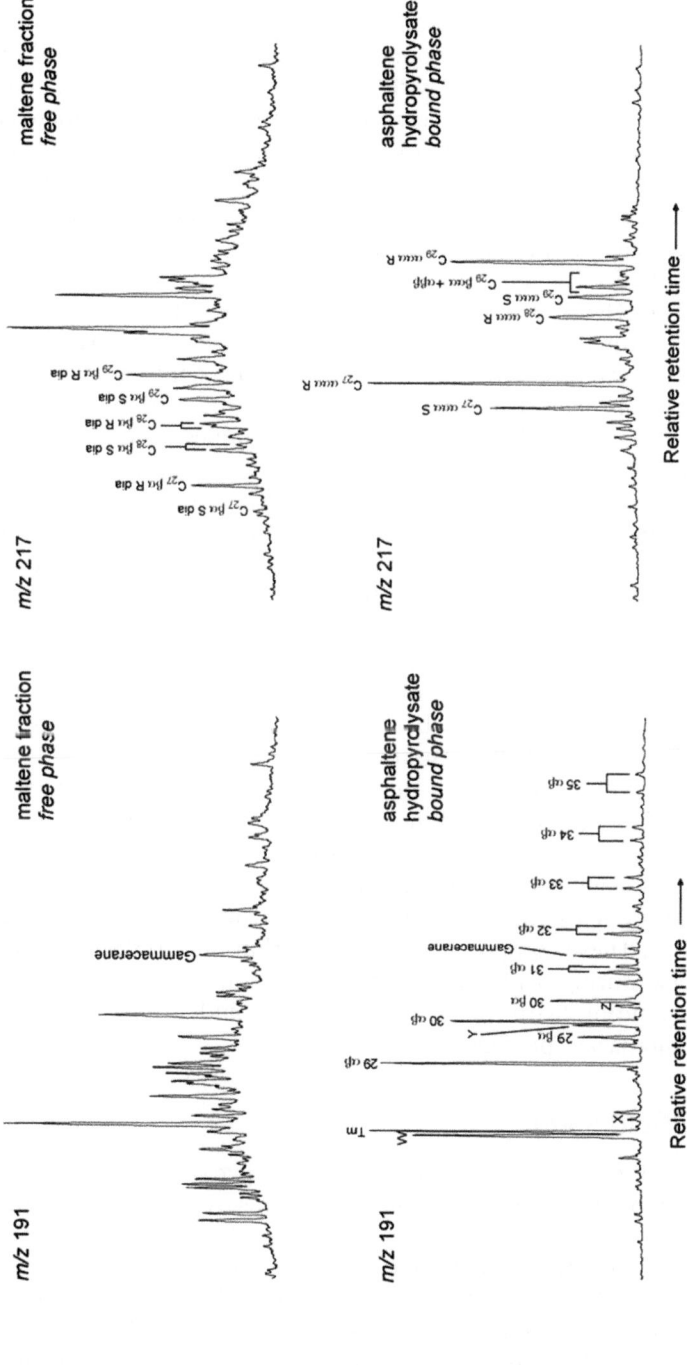

Figure 6.7 Comparison of the *m/z* 191 and *m/z* 217 mass chromatograms of the aliphatic hydrocarbons in the maltene fractions (free phase) and asphaltene hydropyrolysates (bound phase) from the Ilubinrin tar sand bitumen.[43]

and they potentially contain compounds which can interfere with the geochemical evaluation of produced oil, core, or drill-cutting samples.[45] Oil-based mud may be derived from diesel, and depending on the distillation-cut procedure may resemble unrefined crude oil (and so contain significant concentrations of biomarkers). Enhanced mineral oil (EMO) drilling fluids are refined diesels with a narrower boiling range, and reduced concentration of aromatic hydrocarbons. Synthetic muds are increasingly being used for environmental reasons, as they contain compounds such as olefins and esters which are designed to break down in the environment. Water-based muds may also contain diesel as an additive, together with lignites which may impart an immature signature to the biomarker distributions obtained.

The ability of HyPy-derived biomarkers to aid the maturity assessment of samples contaminated with drilling mud was first demonstrated for a UK North Sea (Central Graben) sample.[27] In addition, contamination of core samples with synthetic drilling mud was seen to be a problem in characterising the biomarkers from the drill-cutting samples used in the study of Russell *et al.*[37] on the Blake Field described above. High levels of the n-alkanes n-C_{16} and n-C_{18} were observed in the solvent extracts of the cuttings, and were thought to be artefacts from the drilling fluids used in this particular field (*e.g.* synthetic α-olefins). Such contamination would not be present in the bound phase, with the bound biomarkers generated from the HyPy of the asphaltenes isolated from the cutting extracts free of this material. However, potential contamination was also seen in the bound biomarkers, specifically the occurrence of immature hopanes with the ββ configuration, which may have been related to the presence of a lignite-derived additive in the drilling mud.

Investigations into the HyPy-derived bound biomarkers from commercially available synthetic drilling muds have revealed the presence of the high levels of n-C_{16} and n-C_{18} n-alkanes, together with the C29 sterane with the αααR stereochemical configuration.[46] These are probably derived from C_{16} and C_{18} fatty acids and a C_{29} sterol which are present in the drilling fluid as emulsifiers. These compounds would partition into to the n-heptane insoluble phase during asphaltene isolation, and are then hydrogenated to their corresponding hydrocarbons during HyPy, and so may interfere with the bound biomarker profiles.

6.3.5 Oilfield Solids

Deposits of oilfield solids such as tar mats and pyrobitumens have important implications for reservoir exploitation as they reduce porosity and permeability, causing significant reservoir heterogeneity.[47] An understanding of oilfield solids has important implications for the geological, engineering, and economic evaluation of affected reserves. Also, knowledge of their origin could help predict distributions in producing fields, which is essential for optimising field development and production. HyPy has considerable potential to address these issues as the pyrolysates produced possess similar bulk carbon skeletal characteristics as their parent kerogens, and so this relationship may also exist between HyPy products and other insoluble macromolecular materials such as pyrobitumens and tar mats.

HyPy tests on an extremely mature pyrobitumen (mean vitrinite reflectance greater than 2.0% R_o, and a carbon aromaticity of ∼90%) from the Kaybob South unit of the Swan Hills Formation, West Central Alberta, Canada, generated recognisable hopane and sterane profiles that were mature, and matched closely those obtained for the extremely small quantity of residual bitumen present.[48] In addition, the n-alkane profile generated, which extended to high-molecular-weight homologues was not what would have been expected for heavily cracked oil. Therefore, it was proposed that the small amount of residual bitumen present was largely the product of cracking of the macromolecular material in the pyrobitumen, as opposed to being residual trapped oil.

HyPy has also been applied to the characterisation of a tar mat from the Franklin Field (UK North Sea). The reservoir core from this field was heavily contaminated with an oil-based drilling mud and so any extractable aliphatic biomarker profiles would prove unreliable.[46] The bound aromatic fraction, believed to be contaminant free, was used to make a maturity assessment of the bitumen. This was accomplished by assessing the aromatic ring-size distribution, and its characterisation by ^{1}H-NMR. With increasing maturity the ring-size distribution of the bound aromatic fraction is seen to shift, with an increased abundance of alkylated 5/6-ring compounds seen in pyrobitumens. The Franklin bitumen by contrast was found to have a similar aromatic distribution to the North Sea oil asphaltene, one that was dominated by 3- and 4-ring compounds. In addition the ^{1}H-NMR spectra was also similar to that of the oil asphaltene with around 30% aromatic carbon, less than that typically seen for overmature pyrobitumens.[46]

6.4 Geobiology

6.4.1 Earliest Fossil Evidence of Animals

Hydrocarbon skeletons of lipids released by HyPy have been used as molecular fossils to track evolutionary and environmental change during the Precambrian. The parallel analyses of free and kerogen-bound biomarkers on the same rock samples affords much more confidence that we have correctly identified syngenetic compounds, which is a particularly important strategy when dealing with Precambrian sedimentary rocks.[49,50] Combining HyPy with lipid biomarker analysis using sophisticated multiple reaction monitoring–gas chromatography–mass spectrometry (MRM-GC-MS) allows highly detailed analysis of trace biomarker compounds (such as methylsteranes and methylhopanes) which are rarely reported from kerogen degradation products, and which are usually too low in abundance to be routinely detectable using conventional GC-MS.

This selective and highly sensitive MRM methodology for lipid biomarker analysis opens up the possibility of monitoring for a large suite of biomarker compounds for assessing ancient sedimentary biogenic inputs, and facilitating more accurate paleoenvironmental reconstructions. As with all lipid biomarker applications it is the level of thermal maturity of the host sedimentary organic matter, which can be gauged independently from Rock-Eval pyrolysis or elemental analysis (*e.g.* atomic H/C ratios), which is the principal control on whether the primary lipid biomarker signals will survive burial maturation. Thermally immature rocks (up to peak-oil-window maturity) are the best targets for lipid biomarker work, while mature rocks (particularly post-oil-window or overmature rocks) must be treated with particularly care and caution as indigenous biomarker yields fall precipitously at this high level of maturity, and the possibility of contamination influence from exogenous sources is more acute.

A diverse series of uncontaminated linear, branched and polycyclic alkanes (Figure 6.8) was released by HyPy from kerogens isolated from Neoproterozoic–Cambrian Huqf rocks from the South Oman Salt Basin (SOSB).[49,50] Rock-Eval pyrolysis parameters (*e.g.* T_{max} of 419–440 °C and

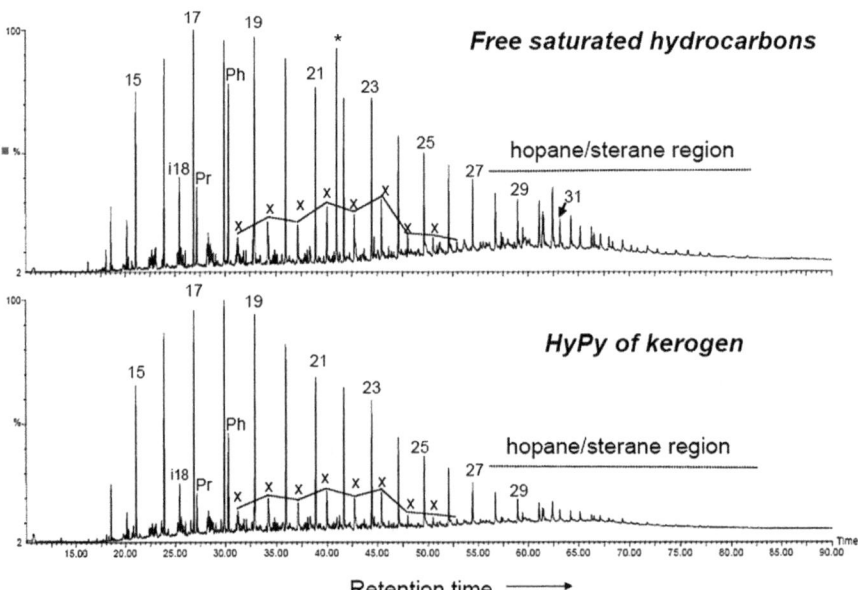

Figure 6.8   Total ion chromatograms (TIC) comparing the distributions of saturated hydrocarbons released by solvent extraction (top) in comparison with those generated from HyPy of the kerogen from an Ara Group carbonate rock (OMR 2229, from A1C interval). Numbers refer to carbon chain lengths of n-alkanes. X indicate a homologous series of C_{19}-C_{26+} mid-chain methylalkanes known as X peaks exhibiting a slight even over odd carbon number preference. X peaks elute immediately before n-alkanes for compounds with the same no. of carbon atoms. * = internal branched alkane standard, 3-methylheneicosane.[50]

hydrogen indices of 270–690 mg g^{-1} TOC) confirmed that the sedimentary organic matter was thermally immature (early- to mid-oil-window maturity)[51], and so ideal for lipid biomarker investigation. As for Phanerozoic rocks, the bound hopane and sterane biomarkers were protected against thermal alteration by their binding within a host polymeric matrix, and this explains why the HyPy products exhibit a slightly less mature isomeric distribution than the corresponding free biomarker hydrocarbons, confirming syngeneity of the lipid biomarker pools.[49,50] A 24-ethylcholestane (C$_{29}$ sterane) dominance amongst total steranes in Neoproterzoic rocks and oils from SOSB (constituting on average 70% of the C$_{27}$–C$_{29}$ steranes for all the Huqf rock samples[50,51]) and Eastern Siberian oils[52] reflect high inputs of chlorophyte (green) microalgae as major marine primary producers. Furthermore, the hopane/sterane ratios lie predominantly within the marine Phanerozoic range of 0.5–2.0,[50,51] indicating that by around 600–700 Ma eukaryotes were ecologically widespread and major contributors to total photosynthetic primary production in the Neoproterzoic ocean. In the Proterozoic rock record, microalgae are most likely a major source of these steroids. In contrast, HyPy of 1.64 Ga Barney Creek Formation and 1.4 Ga Roper Group kerogens yield no significant quantities of regular (4-desmethyl) steranes above detection limits (Love *et al.* unpublished results). As a way of characterising evolving eukaryotic microbial community structure, then, much interest lies in obtaining robust sterane records for Proterozoic rocks and detection of kerogen-bound steranes will be an important line of evidence in future investigations to rule out contamination influences from younger petroleum and other exogenous sources.

The biomarker investigation on strata from a number of sub-surface wells containing thermally well-preserved sedimentary rocks from the Huqf Supergroup, SOSB, also revealed a 100-million-year-long C$_{30}$ sterane biomarker (24-isopropylcholestane, or 24-ipc) record of demosponge radiation commencing in the late Cryogenian (between 635 and 713 Ma), and terminating in the Early Cambrian ($\sim$540 Ma).[50] This constituted the first robust evidence for animals predating the Marinoan glaciation (terminating at 635 Ma), and is supported by a close match with metazoan divergence age estimates from numerous molecular clocks[53,54] and the recent finding of 'sponge-grade metazoan' fossils from Cryogenian strata in South Australia.[55]

Demosponges are currently the only known extant organisms to produce 24-ipc sterols as their major sterol constituents, particularly demosponges from the order Halichondrida, and in some species the 24-ipc sterols can constitute up to 99% of the total monohydroxy sterol content.[56] Previous work[57] has established that a temporal pulse of elevated 24-iso/24-n-propylcholestane ratios (>0.5 and often around 1.0) existed in numerous sedimentary rocks from different basins of Neoproterozoic to Early Cambrian age (and later found in some Early Ordovician rocks), although the first occurrence of elevated amounts of this compound in the Neoproterozoic had not been well temporally constrained. To establish the stratigraphic range of the demosponge biomarkers beyond doubt in our South Oman rocks, kerogens were fragmented from various stratigraphic intervals using HyPy to release the

covalently bound forms of these molecules. We used the absolute uranium–
lead zircon age geochronological framework established for the host strata[58] to
better constrain the timing of the radiation of Porifera in the Neoproterozoic.
The parallel analysis of kerogen-bound products, containing abundant covalently bound 24-isopropylcholestanes alongside conventional solvent-
extractable biomarkers, was an important self-consistency check (see
Figure 6.9) which added significant confidence that the 24-ipc sterane
biomarkers were indigenous and syngenetic with the host sedimentary rock
and had not simply migrated from other, possibly younger, strata.

Quantification of biomarkers with internal standards showed that, on
average, C_{30} steranes made up 2.7% of the total steranes: 63% of these were
24-isopropylcholestanes and the remainder were 24-n-propylcholestanes
from marine pelagophyte algae[59], as identified in m/z 414–217 MRM-GC-MS
ion chromatograms. The ratio of 24-isopropylcholestane to 24-n-propylcho-
lestane (for all four regular geoisomers) was measured for extracts and
pyrolysates and found to be anomalously high (0.52–16.1, with an average

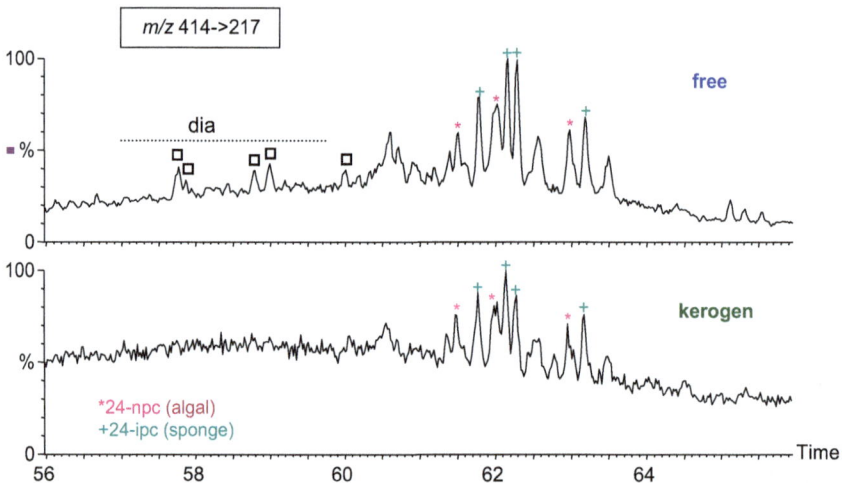

Figure 6.9   Multiple reaction monitoring (MRM)-GC-MS (from 414→217 Da transitions) chromatograms showing C_{30} steranes for the free (extractable)
hydrocarbons and the corresponding kerogen HyPy products from a
siltstone/shale from the Early Ediacaran Masirah Bay Formation shale
(JF-1) from the South Oman Salt Basin (SOSB).[50] Demosponge contributions are evident from abundant 24-isopropylcholestanes (+)
(24-ipc). 24-n-propylcholestanes (*) (24-npc) are markers of marine
pelagophyte algae and this confirms a marine depositional setting for
each formation in the SOSB. For both 24-ipc and 24-npc structures, four
regular sterane diasteriosomers are detected ($\alpha\alpha\alpha20R$, $\alpha\beta\beta20R$, $\alpha\beta\beta20S$,
$\alpha\alpha\alpha20R$) indicating a mature geoisomer distribution. Note the presence
of rearranged (diasteranes) products in the free extracts but not in the
kerogen HyPy products and a slightly less mature distribution of
diastereoisomers in the HyPy products, as should be observed for free
versus bound biomarker pools.

value of 1.51) in all Neoproterozoic–Cambrian SOSB samples in comparison with Phanerozoic oils and bitumens reported previously (typically <0.3). Since the absolute abundances of steranes varies widely, the 24-n-propylcholestane marker for marine pelagophyte algae[59] provides a molecular benchmark for inputs of sponge biomass to sediments through geological time. We proposed that only a significant input from demosponges could result in such a consistently high value for this ratio found in our Huqf rock bitumens and kerogen hydropyrolysates.

Oxygenation of oceans in the Neoproterzoic was apparently a long and protracted process, as gauged from compiled geochemical and stable isotopic records.[60–62] Modern sponges are able to live at relatively low oxygen conditions in benthic environments, as illustrated by the preferred modern habitat of hexactinellids on the continental slope, from 200 to 2000 m water depth. The shallow marine waters found in the Cryogenian may have allowed dissolved oxygen concentrations to be elevated for the first time above a critical threshold and would have likely been a major factor in sustaining the metabolic requirements of early metazoans. Benthic microbial mats could have become widespread in Neoproterozoic shallow marine environments from the Cryogenian onwards, and these benthic mats may have acted as environmental buffers consuming hydrogen sulfide and stabilising the redox conditions around the shallow seafloor for early benthic animals.[49] The 24-ipc biomarker was also more recently found as sterane constituent of Neoproterozoic oils from Eastern Siberia, with those from the Baykit Basin also possibly sourced from Cryogenian (850–635 Ma) parent source rocks.[52]

Moving forward in geological time, the ratio of 24-iso/n-propylcholestane drops significantly in the Late Cambrian–Early Ordovician and continues at lower values through the later Phanerozoic,[57] consistent with when the fossil record of sponges drops dramatically when the suspension-feeding eumatezaons (*e.g.* crinoids, brachiopods, corals) replaced these and sponges were more marginalised.[63] So, the temporal patterns of 24-ipc abundance are also consistent with predominantly a demosponge source for this compound preserved in Neoproterozoic–Ordovician sedimentary rocks.

6.4.2 HyPy Studies of Overmature 2.5–3.4 Ga Archean Kerogens

The main organic phase present in Archean rocks is largely aromatic-rich overmature kerogen.[64] Archean sedimentary organic matter exhibits a wide range of ^{13}C-depleted bulk carbon isotopic signatures (ranging from −30 to −55‰), not generally found in younger rocks, which reflects the balance of enigmatic autrotrophic and heterotrophic biological sources in the Archean ocean. Two previous investigations on overmature kerogens prepared from 2.5 Ga Mt McRae shales[65] and from 3.4–3.5 Ga Strelley Pool cherts[64] from the Pilbara Craton (Western Australia) have clearly demonstrated the proof-of-concept that genuine covalently bound hydrocarbon constituents can be released by HyPy and easily detected by GC-MS.

A complex distribution of molecular products, composed predominantly of aromatic hydrocarbons but with lower but detectable amounts of alkanes which tail off at n-C_{22}, was generated from Strelley Pool kerogens.[64] Being covalently bound, the structures are immobile and therefore most assuredly synsedimentary. The quantities of organics released by HyPy of the Strelley Pool kerogens are 10^4–10^6 times more abundant than the typical yield of polycyclic biomarkers reported from solvent extraction of Archaean rocks.[66] Importantly, in the Strelley Pool study we demonstrated for the first time a positive correlation between the degree of alkylation of aromatic products (related to thermal maturity) with bulk structural and molecular parameters (Figure 6.10) derived from elemental analysis and Raman spectroscopy.[64]

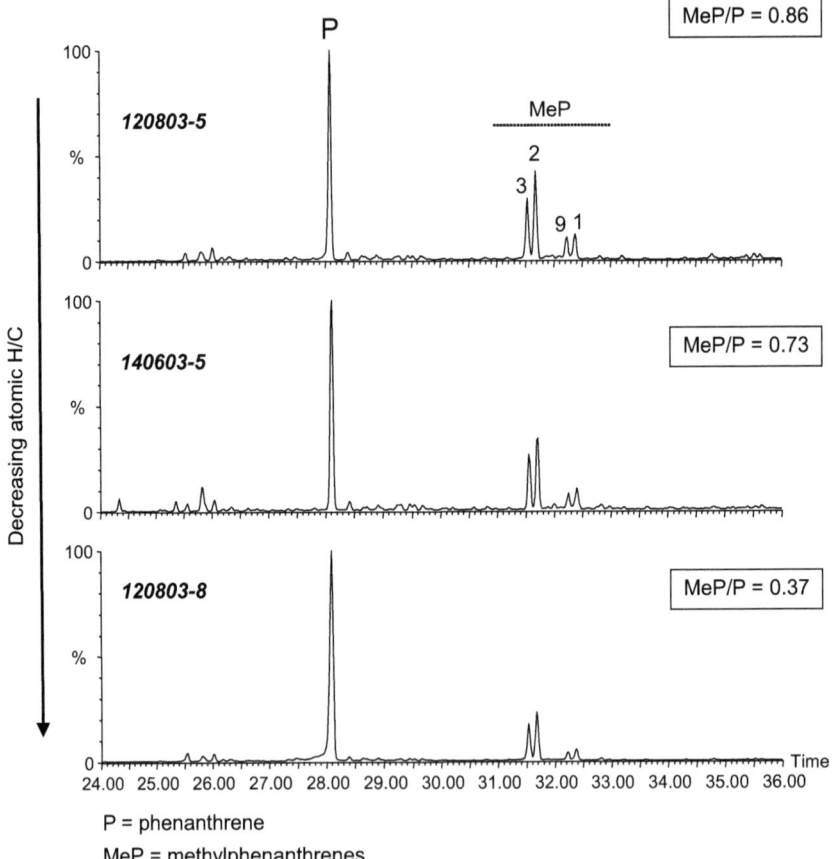

P = phenanthrene
MeP = methylphenanthrenes

Figure 6.10   Summed ion chromatograms (*m/z* 178 + 192) showing phenanthrene (P) and methylphenathrene (MeP) distributions generated from HyPy of three overmature 3.4 Ga Strelley Pool chert kerogens. Numbers (1, 2, 3, 9) refer to the position of the methyl substituent on phenanthrene for each MeP isomer. Note that the relative values of (summed MeP)/P abundance ratios are consistent with the maturity ordering determined independently from elemental analysis and Raman spectroscopy.[64]

Sufficient hydropyrolysate was generated for compound-specific $\delta^{13}C$ analyses to be performed on a small subset of the main hydrocarbon products (polycyclic aromatic hydrocarbons, PAHs) which showed a range of $\delta^{13}C$ signatures varying by 3–5‰ but with the $\delta^{13}C$ values for two of the most abundant GC-amenable parent PAH products (phenanthrene and pyrene) close to the stable carbon isotopic signature of the bulk kerogen.

The level of structural detail preserved in hydrocarbon products has yet to be fully explored for a larger set of Archean rocks but these results suggest that this is a promising way forward for analysing *in situ* Archean organics and that HyPy can partially fragment even the most recalcitrant and mature sedimentary kerogens (with atomic H/C ratios <0.1). It is important to point out that no robust polycyclic alkane biomarkers (*e.g.* hopanes, steranes) above detection limits have yet been reported from HyPy products of Archean kerogens. For overmature kerogens, as found in Archean and Paleoproterozoic strata, then the most robust biological source information may be preserved and encoded within the $\delta^{13}C$ signatures of the bound molecular constituents as measured by compound-specific $\delta^{13}C$ (CSIA) analyses on the least thermally transformed (*e.g.* alkanes) analytes—rather than from the structural features of organic molecules surviving metamorphism—though interpretation of CSIA data is unlikely to be very straightforward either.

6.4.3 Lipid Screening of Microbial Cultures and Extant Sponges

Understanding the information preserved in fossil biomarker lipid records requires detailed knowledge about the lipid content of extant microorganisms. Recently, a modified catalytic hydropyrolysis (HyPy) method has been devised as a rapid screening tool for intact or pre-extracted microbial cells to reveal carbon number patterns of linear, branched, and polycyclic aliphatic lipids and characterise any resistant aliphatic biopolymer constituents.[67] A revised experimental protocol, involving a modified catalyst-loading procedure, careful use of a silica support substrate, and a revised temperature programme was tested and optimised for handling biomass. Partial hydrogenation of double bonds inevitably occurred, although it was found that some unsaturation was preserved, particularly within branched and polycyclic hydrocarbon structures. Generation of hydrocarbons directly from biomass facilitates comparison of extant and fossil lipids in the same analytical window.

The key features of the HyPy methodology ensure that covalent bonds can be cleaved at the lowest possible temperatures in the heating cycle (typically between 250 and 450 °C). This combination of factors results in excellent preservation of structural and sterochemical features of hydrocarbon products in comparison to other analytical pyrolysis techniques which use inert gas atmospheres (*e.g.* He, N_2, or vacuum), since thermal cracking and isomerisation reactions are suppressed. The HyPy technique promotes reductive cleavage of carbon–oxygen and carbon–sulfur bonds and functional groups (including ether, sulfide, carboxyl, hydroxyl, thiols, and simple thiophenic groups) in discrete organic compounds and macromolecules, favouring release of hydrocarbon products.

The dominant aliphatic hydrocarbon products obtained from HyPy treatment of 23 microbial species of algae, bacteria, and Archaea were described previously,[67] and this demonstrated the efficacy of the technique as an effective lipid screening tool. A rapid and convenient way to assess the total distribution of the polymethylenic and branched alkyl contents of microorganisms is to isolate a total aliphatic hydrocarbon fraction produced from HyPy and then monitor m/z 85 ion chromatograms from routine GC-MS analyses (Figure 6.11). In general, it was observed that the bacteria and Archaea analysed biosynthesise a more restricted distribution of polymethylenic chain lengths, which are mainly n-C_{20} and shorter. In contrast, all of the algal cultures investigated generated significant quantities of waxy linear hydrocarbons with chain lengths greater than n-C_{20} and yielding a wider range of alkyl chain lengths than bacteria or Archaea.

Major steroidal products identified in hydropyrolysates of algal cultures included sterenes, steradienes, and diasterenes, as well as fully saturated steranes. The diasteroisomer distributions showed immature patterns as anticipated, with (20R)-5α,14α,17α(H) and (20R)-5β,14α,17α(H) being the dominant isomers. From GC-FID analyses, $\Delta^{13(17)}$-diasterenes and steranes were generally found to be the most abundant steroidal products. $\Delta^{13(17)}$-Diasterenes, which give a characteristic m/z 257 fragment ion, were generated as metastable products from rearrangement of sterenes and steradienes. These were, in turn, formed from elimination of the hydroxyl group at C-3 and subsequent mobilisation of double bonds.[68,69] The relative proportions of C_{27}–C_{29} steranes released by HyPy were in good agreement with published information about the steroid content of these algae (see[67] for references).

In certain instances, a mismatch between extractible and insoluble biomarker components was evident even for pure microbial biomass. HyPy of solvent-extracted insoluble cell residues of a cyanobacterium, *Phormidium luridum*, yielded a significantly higher relative proportion of 2-methylhopanols to non-methylated hopanes than that for whole cells of the species, implying a tighter association of the 2-methylated hopanoids in internal membranes. HyPy offers great potential to improve our understanding of the relationships between lipid biomarkers, their source cells, and secondary transformation processes as well as to expand the range of diagnostic lipid skeletons used to track source inputs or metabolic processes in the sedimentary record.

HyPy generation of sponge biomass is a useful and rapid means of assessing the diversity of lipid skeletons that these simple animals or their associated symbionts can biosynthesise.[50] HyPy of sponge biomass was used to discriminate the different possible C_{26} sterol (21-, 24-, and 27-norcholesterols) and C_{30} sterol (24-isopropylcholesterols and 24-n-propylcholesterols) structural isomers since the derivatised sterols (as trimethylsiliyl ethers) co-elute

m/z 85 ion chromatograms

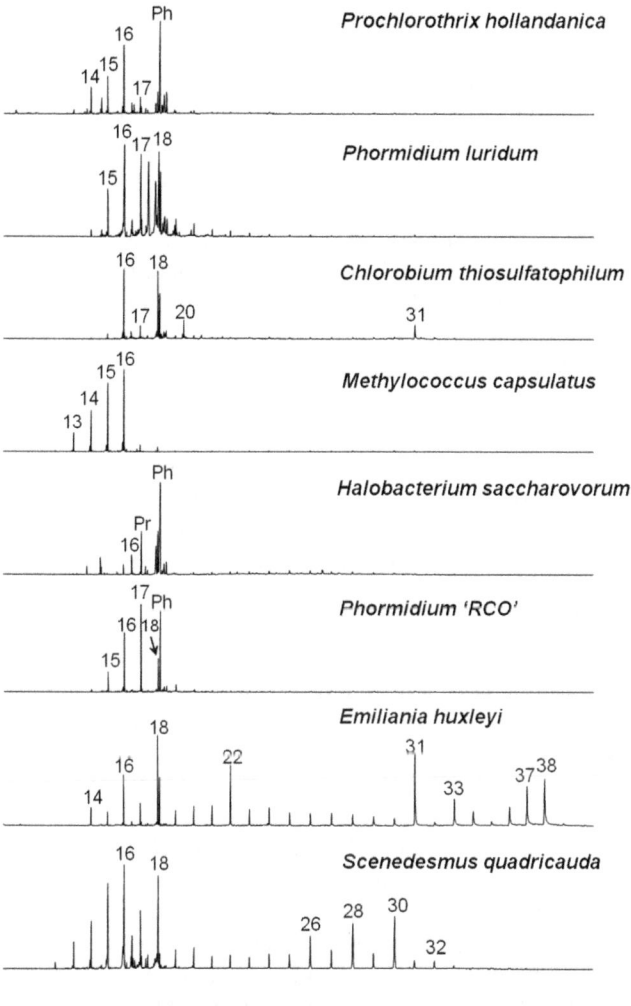

Figure 6.11 *m/z* 85 ion chromatograms showing the full distribution of n-alkanes and acyclic isoprenoids generated from HyPy of a selection of unextracted microbial cultures (Numbers refer to carbon chain lengths of n-alkane products; Ph = phytane).[50] *Prochlorothrix, Phormidium, Chlorobium, Methylococcus* are bacteria; *Halobacterium saccaharovorum* is an archaeon; *Scendedemus* and *Emiliania* are algae (eukaryotes).

on standard capillary GC columns while the sterane forms can be resolved and detected by MRM-GC-MS.

HyPy experiments on a small set of nine modern sponges (Table S4[50]) suggest that C_{26} sterols containing a 27-norcholestane hydrocarbon skeleton may actually be more commonly distributed lipids in modern demosponges than 24-isopropylcholesterols, although the abundance of these C_{26} compounds was low (<1% of total sterols) relative to the major sterols such as cholesterol and cholestanol. Intriguingly, as well as containing the 24-isopropylcholestane markers for demosponges, all our SOSB Neoproterozoic–Cambrian sedimentary rock samples[50] contained anomalously high amounts of another side-chain-modified fossil steroid, 27-norcholestane, in both solvent extracts and kerogen hydropyrolysates (Tables S1 and S2[50]) compared with other ancient rocks of similar thermal maturity. 27-Norcholestanes in SOSB rocks (total C_{26} steranes, typically dominated by 27-norcholestane) constituted on average 6.7% of total C_{26}–C_{29} steranes (Table S1[50]), in contrast to levels of less than 3% in most other oil-window-mature sediments and petroleum of any age. Although the enhanced 27-norcholestane content may reflect a fossil demosponge input, it is a much less robust molecular indicator than the elevated levels of 24-ipc steranes.

Cholesterol and cholestanol model compounds were both pyrolysed under standard HyPy conditions to monitor the level of side-chain cleavage produced by thermal cleavage during pyrolysis (by measuring the ratio of 27-norcholestane/cholestane products using MRM-GC-MS). The amount of side-chain scission which occurred was found to be negligible in comparison to the amounts of 27-norcholestanes produced by the sponges. While the common finding of low levels of 27-norsterols in modern demosponges is intriguing, at this stage we can only say that on balance, while there may have been a sponge contribution to the 27-norcholestanes in SOSB sedimentary rocks, we cannot rule out the possibility that these predominantly result from cleavage of the side chain of other steroid components of higher carbon number steroids (C_{27}–C_{30}) under unusual diagenetic and catagenetic conditions.

For C_{30} steroids, only *Dysidea fragilis* and *Axinella* sp. yielded detectable amounts of 24-ipc out of the nine sponge species screened with the immature $\beta\alpha\alpha(20R)$ and $\alpha\alpha\alpha(20R)$ diastereoisomers dominant from HyPy of sponge cells. The release of 24-ipc from HyPy of cells of *Dysidea fragilis* was consistent with a previous report of 24-ipc in another related species, *Dysidea herbacea*.[70] Thus, it appears that C_{30} sterols synthesised *de novo* by sponges containing the 24-ipc skeleton may actually be rarely found in modern sponges, but when they do occur they are produced only by certain demosponge genera, particularly from the order Halichondrida. In the case of *Pseudaxynissa* sp., 24-ipc and 22-dehydro-24-isopropylcholesterol together may make up more than 99% of total sterols,[56] suggesting a structural role for these sterols in the cell membrane. In contrast, 27-norcholesterols are usually minor sterol constituents (<1%) of total sterols in sponges.

Progress can be made in lipid analysis of extant organisms through rapid screening methods, such as HyPy of whole cells[67] used in conjunction with other modern analytical techniques, such as high-performance liquid chromatography (HPLC-MS), which allow us to directly probe the structure of complex, intact lipids.

6.5 Meteorites

Carbonaceous chondrites are a primitive class of meteorite that contain 2–5 wt% carbon, most of which is present as organic matter. The major organic component of carbonaceous chondrites is a solvent-insoluble, high-molecular-weight macromolecular phase that constitutes at least 70 wt% of the total organic content in these meteorites. HyPy of the insoluble Murchison macromolecular material released significant amounts of high-molecular-weight PAH, and heterocyclic aromatics up to 7-ring PAH including phenanthrene, carbazole, fluoranthene, pyrene, chrysene, perylene, benzo-perylene, and coronene with varying degrees of alkylation.[71] These are the highest molecular weight aromatic compounds released from meteoritic organic matter (>5 rings) by any analytical pyrolysis method, and approach the highest molecular weight of PAH compounds that can be analysed by conventional columns. Simple mass balance calculations, using elemental and isotopic constitution of pyrolysis residues, showed that HyPy more efficiently fragmented the insoluble meteoritic organic matrix than a water-pyrolysis method. An improved trapping regime using a silica trap allowed volatile components such as alkylbenzenes and alkylnaphthalenes (1- and 2-ring PAH) to be recovered more efficiently.[72]

Analysis of both products and residue from HyPy revealed that the insoluble meteoritic organic network contained both labile (pyrolysable) and refractory (nonpyrolysable) fractions, with the latter being dominated by large PAH structures with at least 5- or 6-ring (and mostly likely larger) PAH units cross-linked together. The refractory organic matter in the HyPy residue was significantly ^{13}C-depleted (having a bulk $\delta^{13}C$ signature of -20.7‰ compared with the pre-extracted and decalcified starting material with a bulk $\delta^{13}C$ signature of -12.7‰). The release of up to 7-ring PAH by HyPy, and the deduction that even larger entities must be present in the experimental residue, partly reconciles the apparent disharmony between the largest size of PAH detected in previous meteoritic organic analyses and the >20-ring PAH proposed for interstellar organic inventories. Two possible mechanisms for producing the range of bound PAH entities and the detailed $\delta^{13}C$ systematics that generally shows ^{13}C-depletion in the higher-molecular-weight components are gas-phase pyrolysis in circumstellar envelopes or cold irradiation of interstellar ices. The large carbon isotopic fractionation between sizeable PAH units suggests that a significant portion of the total macromolecular organic matter derived from complex interstellar organics that predate our solar system in age, with larger molecules built up from irradiation reactions of simple organics on mineral grains in the cold interstellar medium.[71]

6.6 Quantification and Characterisation of Black Carbon

Black carbon (BC), also known as pyrogenic carbon, is the carbon-rich (>60%), highly aromatic, and recalcitrant product of the incomplete combustion of biomass and fossil fuels.[73] BC encompasses a broad continuum, from combustion residues (*e.g.* charred biomass and charcoal) to combustion condensates (*e.g.* soot), reflecting different precursors and formation processes.[74] Currently a variety of thermal, chemical, spectroscopic, molecular marker, and optical methods are used to quantify BC, which inevitably give a wide range of results,[75] making it difficult to assess the occurrence and stability of BC in a range of environments. As BC is typically isolated *via* operational rather than chemical parameters, individual methodologies can only identify BC from a specific portion of the BC continuum, with no one method able to isolate or quantify BC across the whole range.[76]

HyPy facilitates the reductive removal of thermally labile organic matter, leaving behind any refractory, highly aromatic carbonaceous material as the residue. The high hydrogen pressure, slow heating rate, and sulfided molybdenum catalyst prevent the generation of secondary char, as often encountered with chemical or thermal oxidative methods. The fraction that remains after HyPy can be quantified by elemental analysis, represents the portion of the BC continuum that is stable under HyPy conditions, and can be defined as BC_{HyPy}.[77]

Ascough *et al.*[78] demonstrated for two soils that, following HyPy to 450 °C, and then in 25 °C increments to 600 °C, it was possible to identify a set of conditions under which lignocellulosic and other labile organic carbon material (*e.g.* lipids, proteins) are fully removed (by 550 °C), but at which degradation *via* hydrogasification to methane of the resistant BC component of the sample had not yet commenced (>575 °C). The resulting plateau in carbon content between 550 and 575 °C therefore represents the BC_{HyPy} content of the sample. Further tests[77] on a selection of 12 standard reference materials from the International BC Ring Trial,[75] containing a number of BC-rich samples, BC-containing environmental matrices, and BC-free potentially interfering materials, revealed that this plateau in the carbon content of the HyPy residues was again apparent between 550 °C and 575 °C in all of the soil and sediment samples studied (Figure 6.12). This suggested that HyPy isolates a carbonaceous fraction of consistent relative stability in samples from different environments.

The reported BC_{HyPy} values[77] for all of the reference materials fell within the range of BC contents reported in the Ring Trial. Determinations were highly reproducible ($\pm 2\%$), with strong evidence that at temperatures above 550 °C non-BC organic matter was effectively removed from soils and sediments by HyPy with minimal charring, and so the BC_{HyPy} values obtained did not show significant positive bias. BC_{HyPy} isolation also appeared to be largely independent of matrix effects, with the method providing good discrimination between BC and potentially interfering materials including a lignite, a BC-free oil shale, and the biopolymer melanoidin. However, in common with the other methods investigated in the Ring Trial, HyPy was not able to distinguish between true pyrogenic BC and the bituminous coal, which due to its high rank contained large aromatic ring clusters that are chemically indistinguishable from BC. HyPy also appeared able to

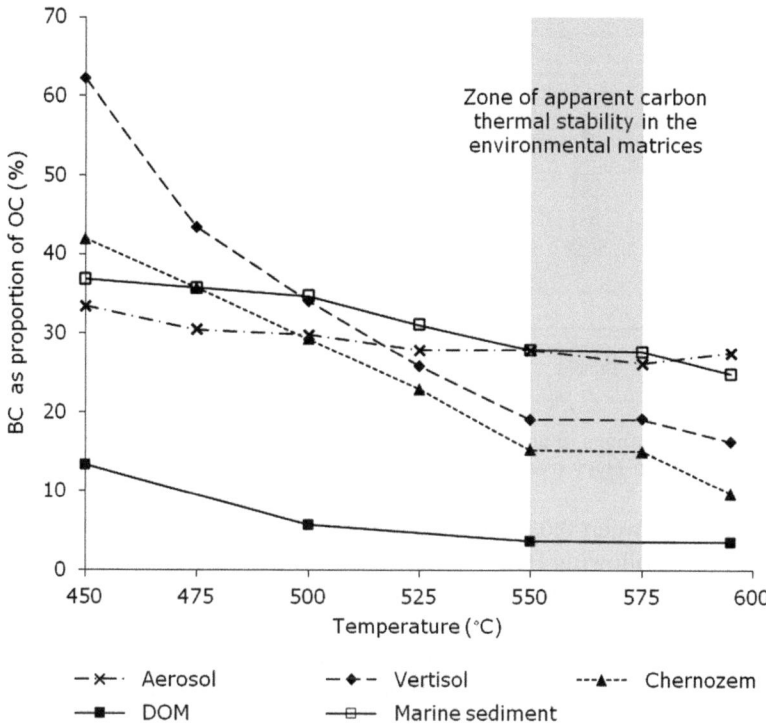

Figure 6.12 Black carbon (BC) as proportion of organic carbon (OC) as measured for the residues of the environmental matrices after HyPy at different temperatures.[77]

discriminate between relatively labile biochars reporting low BC_{HyPy} values and more refractory, high-BC_{HyPy} soot in pure samples, and between environmental samples from industrial sites with BC predominantly derived from the combustion of fossil fuels and agricultural sites dominated by the burning of vegetation.[77]

As the technique is non-destructive (with respect to the BC_{HyPy} fraction), in addition to quantifying the amount of BC which is stable under specific HyPy conditions (550 °C, 150 bar H_2), it is beneficial to characterise the isolated BC_{HyPy} fraction in order to identify what part of the BC continuum is being defined as BC. Ascough *et al.*[79] subjected five archaeological charcoals to HyPy, with the composition of the fresh and treated samples characterised by ^{13}C NMR and elemental analysis. An example of the NMR spectra produced is shown in Figure 6.13 which illustrates the effective removal of the resonances derived from cellulose and hemicellulose (60–105 ppm) from the fresh material, with the BC_{HyPy} fraction showing a dramatically reduced signal range that was restricted to the peaks from condensed aromatic structures (127 ppm and spinning side-bands). This was also reflected in the O/C atomic ratios of these samples, which, as shown in Figure 6.13, initially ranged from between 0.09 to 0.37 for the fresh charcoals, before being

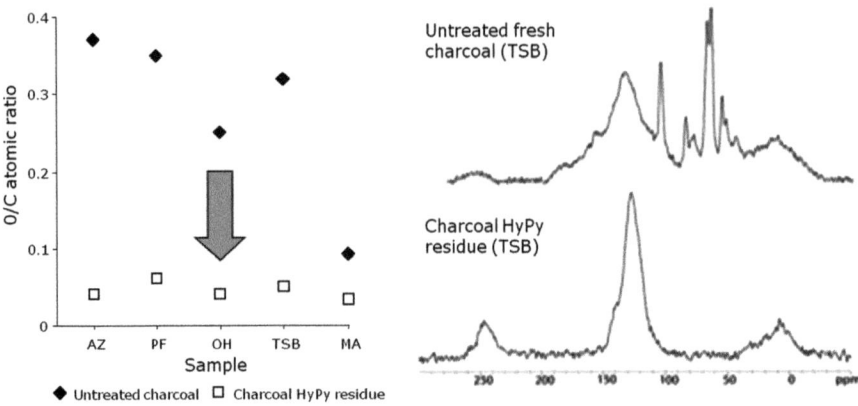

Figure 6.13   Changes in atomic O/C ratios and ^{13}C NMR spectra of the fresh (upper) and HyPy treated (lower) charcoal samples.[79]

reduced to between 0.03 and 0.07 following HyPy.[79] A similar reduction in the O/C ratio following HyPy was also seen for the BC-rich materials from the Ring Trial.[77] In addition, the H/C atomic ratios of each of these samples were observed to reduce from between 0.19 and 0.71 to a maximum of 0.5, which inferred an aromaticity of the BC_{HyPy} fraction of >7 peri-condensed ring PAHs (with up to 24 carbon atoms). Although without further detailed characterisation of the BC_{HyPy} residue itself, the selective discrimination of aromatic domains with less than 7 rings being entirely restricted to the non-BC_{HyPy} fraction remains uncertain.

As well as isolating the BC_{HyPy} fraction, HyPy also allows, following analysis by GC-MS, for the characterisation of the non-BC_{HyPy} material at a molecular level. While the non-BC_{HyPy} fraction from soils are likely to be dominated by the well-documented products of the thermal decomposition of labile organic matter such as lignocellulosic material and humic acids, the non-BC_{HyPy} fraction of charcoals has been shown to contain pyrogenic PAHs with between 2-ring compounds (naphthalene) and 7-ring compounds (coronene) cleaved from the macromolecular char structure.[79] Therefore, HyPy can be seen to discriminate against BC with a relatively low degree of aromatic condensation from lightly charred samples, such as biochars produced at relatively low temperatures.[77] Such pyrogenic PAHs with less than 7 rings have, however, been found as the degradation products of BC in soils, rivers, and marine systems[80], and so the BC_{HyPy} fraction, composed as it is of more highly condensed aromatic clusters, is likely to be less susceptible to degradation, and so more recalcitrant in the environment than the total BC fraction of soils and sediments.

The HyPy of BC fractions previously isolated by acidified dichromate oxidation has provided compelling evidence for the incomplete removal of non-BC material, and hence potential overestimation of BC by this popular method.[81] HyPy allowed for this erroneously assigned BC to be characterised, and for both a BC-rich soil and a BC-free oil shale, the non-BC material that survived oxidation was found to be largely paraffinic in structure.

This material was probably derived from compounds such as lipids and other waxy components containing long n-alkyl moieties, which are hydrophobic, and hence difficult to oxidise by wet chemistry.

6.7 Summary

Catalytic hydropyrolysis (HyPy) has been demonstrated to be a highly versatile technique with a range of different applications across organic geochemistry, encompassing petroleum and environmental geochemistry, natural product research, and geobiology. Despite the obviously attractive capabilities of HyPy for analysing covalently bound and highly functionalised lipid biomarkers, the technique has taken time to become more widely used because of the specialist nature of the equipment required to perform this high-pressure treatment safely and routinely. A commercially available system has been developed (Figure 6.14) and eight systems are currently installed worldwide in both academic and commercial institutions, which

Figure 6.14 Commercially available HyPy system, designed for non-specialist laboratory use.

are routinely used for the characterisation of a wide variety of organic macromolecules and other substrates. The efficacy of HyPy is founded upon decades of careful research which has optimised reactor design and analytical run conditions to allow excellent reproducibility of molecular profiles and retention of the major structural and stereochemical features of biomarker hydrocarbon products. The key empirical biomarker lipid systematics, as demonstrated for many Recent and ancient kerogens, show that the kerogen-bound hopane and sterane biomarkers released by HyPy exhibit a slightly less mature isomeric distribution than the corresponding free (extractable) biomarker hydrocarbons. The kerogen-bound biomarker pool is immobile and protected against alteration of structural and stereochemical features during thermal maturation reactions, since this pool is sequestered by binding within a high-molecular-weight macromolecular matrix. The ability to exhaustively remove soluble organic contaminants from sediments using solvent treatment while typically leaving no problematic residue has been demonstrated previously for oil-window-mature sediment cores contaminated with oil-based drilling mud. As well as work on geomacromolecules, progress in identifying the variety of hydrocarbon skeletons in lipids synthesised by different lineages of extant organisms, including microorganisms and sponges, can be made through rapid HyPy of whole cells used in conjunction with other modern analytical techniques which allow us to directly probe the structure of complex, intact lipids.

References

1. R. W. Hiteshu, R. B. Anderson and M. D. Schlesinger, Hydrogenating coal at 800 °C, *Ind. Eng. Chem.*, 1957, **49**, 2008–2010.
2. M. J. Finn, G. Fynes, W. R. Ladner and J. O. H. Newman, Light aromatics from coal hydropyrolysis, *Fuel*, 1980, **59**, 397–404.
3. G. Fynes, W. R. Ladner and J. O. H. Newman, The hydropyrolysis of coal to BTX, *Prog. Energy Combust. Sci.*, 1980, **6**, 223–232.
4. C. Bolton, C. Riemer, C. E. Snape, F. J. Derbyshire and M. T. Terrer, Effect of low temperature hydrogenation on pyrolysis and hydropyrolysis of a bituminous coal, *Fuel*, 1988, **67**, 901–905.
5. C. E. Snape, C. Bolton, R. G. Dosch and H. P. Stephens, High liquid yields from bituminous coal *via* hydropyrolysis with dispersed catalysts, *Energy Fuels*, 1989, **3**, 421–425.
6. H. P. Stephens, C. Bolton and C. E. Snape, Hydrocracking of hydropyrolysis tar with hydrous titanium oxide catalysts, *Fuel*, 1989, **68**, 161–167.
7. N. Robinson, G. Eglinton, C. J. Lafferty and C. E. Snape, Comparison of alkanes released from a bituminous coal *via* hydropyrolysis and low temperature hydrogenation, *Fuel*, **70**, 249–253.
8. G. D. Love, C. E. Snape, A. D. Carr and R. C. Houghton, Release of co-valently-bound alkane biomarkers in high yields from kerogen *via* catalytic hydropyrolysis, *Org. Geochem.*, 1995, **23**, 981–986.

9. M. J. Roberts, C. E. Snape and S. C. Mitchell, Hydropyrolysis: funda-
    mentals. Two-stage processing and PDU operation, in *Geochemistry,
    Characterisation and Conversion of Oil Shales, NATO ASI Series C, Vol.
    C455*, ed. C. E. Snape, Kluwer, Dordrecht, 1995, pp. 277–295.
10. G. D. Love, A. McAulay, C. E. Snape and A. N. Bishop, Effect of process
    variables in catalytic hydropyrolysis on the release of covalently-bound
    aliphatic hydrocarbons from sedimentary organic matter, *Energy Fuels*,
    1997, **11**, 522–531.
11. M. Maroto-Valer, G. D. Love and C. E. Snape, Close correspondence
    between carbon skeletal parameters of kerogens and their hydro-
    pyrolysis oils, *Energy Fuels*, 1997, **11**, 539–545.
12. W. Meredith, C. A. Russell, M. Cooper, C. E. Snape, G. D. Love, D. Fabbri
    and C. H. Vane, Trapping hydropyrolysates on silica and their sub-
    sequent thermal desorption to facilitate rapid fingerprinting by GC-MS,
    Org. Geochem., 2004, **35**, 73–89.
13. L. J. Berwick, P. F. Greenwood, W. Meredith, C. E. Snape and
    H. M. Talbot, Comparison of microscale sealed vessel pyrolysis (MSSVpy)
    and hydropyrolysis (Hypy) for the characterisation of extant and sedi-
    mentary organic matter, *J. Anal. App. Pyrol.*, 2010, **87**, 108–116.
14. G. D. Love, I. P. Murray and C. E. Snape, Maximising the yields of
    covalently-bound biomarkers from sedimentary organic matter, in,
    *Abstracts of the 19th International Meeting on Organic Geochemistry,
    Istanbul*, 1999, 135–136.
15. W. Meredith, C. E. Snape, C. Uguna and G. D. Love, Maximising the yield
    of bound aliphatic biomarkers *via* a convenient two-stage hydropyrolysis
    procedure, in *Petroleum Geochemistry and Exploration in the Afro-Asian
    Region*, ed. D. Liang, D. Wang and Z. Li, Taylor and Francis, London,
    2008, pp. 145–152.
16. O. E. Craig, G. D. Love, S. Isaksson, G. Taylor and C. E. Snape, Stable
    carbon isotopic characterisation of free and bound lipid constituents of
    archaeological ceramic vessels released by solvent extraction, alkaline
    hydrolysis and catalytic hydropyrolysis, *J. Anal. Appl. Pyrolysis*, 2004, **71**,
    613–634.
17. W. Meredith, C. Sun, C. E. Snape, M. A. Sephton and G. D. Love,
    The use of model compounds to investigate the release of covalently
    bound biomarkers in hydropyrolysis, *Org. Geochem.*, 2006, **37**, 1705–
    1714.
18. M. A. Sephton, W. Meredith, C. Sun and C. E. Snape, Hydropyrolysis as a
    preparative technique for compound-specific carbon isotope ratio
    measurement of endogenous steroids, *Rapid Comm. Mass Spectrom.*,
    2005, **19**, 3339–3342.
19. M. A. Sephton, W. Meredith, C. Sun and C. E. Snape, Hydropyrolysis as a
    preparative method for the compound-specific carbon isotope analysis
    of fatty acids, *Rapid Comm. Mass Spectrom.*, 2005, **19**, 323–325.
20. W. Meredith, R. L. Gomes, C. E. Snape, M. Cooper and M. A. Sephton,
    Hydropyrolysis over a platinum oxide catalyst as a preparation technique

for the compound-specific carbon isotope ratio measurement of C27 steroids, *Rapid Comm. Mass Spectrom.*, 2010, **24**, 501–505.

21. R. L. Gomes, W. Meredith, C. E. Snape and M. A. Sephton, Analysis of conjugated steroid androgens: preparative approaches and associated issues, *J. Pharm. Biomed. Anal.*, 2009, **49**, 1133–1140.

22. R. L. Gomes, W. Meredith, C. E. Snape and M. A. Sephton, Conjugated steroids: analytical approaches and applications, *Anal. Bioanal. Chem.*, 2009, **393**, 453–458.

23. K. E. Peters, C. C. Walters and J. M. Moldowan, *The Biomarker Guide. Volume 2: Biomarkers and isotopes in petroleum systems*, Cambridge University Press, Cambridge, 2004.

24. G. D. Love, C. E. Snape, A. D. Carr and R. C. Houghton, Changes in molecular biomarker and bulk carbon skeletal parameters of vitrinite concentrates as a function of rank, *Energy Fuels*, **10**, 149–157.

25. G. D. Love, C. E. Snape and A. E. Fallick, Differences in the mode of incorporation and biogenicity of the principal aliphatic constituents of a Type I oil shale, *Org. Geochem.*, 1998, **28**, 797–811.

26. B. Mykce and W. Michaelis, Molecular fossils from chemical degradation of macromolecular organic matter, *Org. Geochem.*, 1986, **10** 847–858.

27. I. P. Murray, G. D. Love, C. E. Snape and N. J. L. Bailey, Comparison of covalently-bound aliphatic biomarkers released *via* hydropyrolysis with their solvent-extractable counterparts for a suite of Kimmeridge clays, *Org. Geochem.*, 1998, **29**, 1487–1505.

28. K. E. Peters, J. M. Moldowan and P. Sundararaman, Effects of hydrous pyrolysis on biomarker thermal maturity parameters: monterey phosphatic and siliceous members, *Org. Geochem.*, 1980, **15**, 249–265.

29. A. N. Bishop, G. D. Love, C. E. Snape and P. Farrimond, Release of kerogen-bound hopanoids by hydropyrolysis, *Org. Geochem.*, 1998, **29**, 989–1001.

30. S. A. Bowden, P. Farrimond, C. E. Snape and G. D. Love, Compositional differences in biomarker constituents of the hydrocarbon, resin, asphaltene and kerogen fractions: An example from the Jet Rock (Yorkshire, UK), *Org. Geochem.*, 2006, **37**, 369–383.

31. G. W. Van, Graas, Biomarker maturity parameters for high maturities: calibration of the working range up to the oil/condensate threshold, *Org. Geochem.*, 1990, **16**, 1025–1032.

32. R. S. Lockhart, W. Meredith, G. D. Love and C. E. Snape, Release of bound aliphatic biomarkers *via* hydropyrolysis from Type II kerogen at high maturity, *Org. Geochem.*, 2008, **39**, 1119–1124.

33. R. S. Lockhart, *Behaviour of kerogen-bound biomarkers at elevated levels of thermal maturity*. Unpublished PhD thesis, University of Nottingham, 2010.

34. P. Farrimond, G. D. Love, A. N. Bishop, H. E. Innes, D. F. Watson and C. E. Snape, Evidence for the rapid incorporation of hopanoids into kerogen, *Geochim. Cosmochim. Acta*, 2003, **67**, 1383–1394.

35. R. D. Pancost, J. M. Coleman, G. D. Love, A. Chatzi, I. Bouloubassi and C. E. Snape, Kerogen-bound glycerol dialkyl tetraether lipids released by hydropyrolysis of marine sediments: A bias against incorporation of sedimentary organisms?, *Org. Geochem.*, 2008, **39**, 1359–1371.

36. M. Radke and D. H. Welte, The methylphenanthrene index (MPI): a maturity parameter based on aromatic hydrocarbons, in *Advances in Organic Geochemistry 1981*, eds. M. Bjorøy *et al.*, Wiley, Chichester, 1983, pp. 504–512.

37. C. A. Russell, W. Meredith, C. E. Snape, G. D. Love, E. Clarke and B. Moffatt, The potential of bound biomarker profiles released *via* catalytic hydropyrolysis to reconstruct basin charging history for oils, *Org. Geochem.*, 2004, **35**, 1441–1459.

38. C. M. Ekweozor, Tricyclic terpenoid derivatives from chemical degradation reactions of asphaltenes, *Org. Geochem.*, 1984, **6**, 51–61.

39. P. Peng, J. Fu and G. Sheng, Ruthenium-ions-catalysed oxidation of an immature asphaltene: structural features and biomarker distribution, *Energy Fuels*, 1999, **13**, 266–277.

40. D. M. Jones, A. G. Douglas and J. Connan, Hydrocarbon distributions in crude oil asphaltene pyrolysates. 1. Aliphatic compounds, *Energy Fuels*, 1987, **1**, 468–476.

41. I. P. Murray, C. E. Snape, G. D. Love and N. J. L. Bailey, Hydropyrolysis of heavy oils for source correlation studies, in *Abstracts of the 19th International Meeting on Organic Geochemistry. Istanbul*, 1999, pp. 341–342.

42. W. Meredith, C. E. Snape, A. D. Carr, H. P. Nytoft and G. D. Love, The occurrence of unusual hopenes in hydropyrolysates generated from severely biodegraded oil seep asphaltenes, *Org. Geochem.*, 2008, **39**, 1243–1248.

43. O. O. Sonibare, C. E. Snape, W. Meredith, C. N. Uguna and G. D. Love, Geochemical characterisation of heavily biodegraded oil sand bitumens by catalytic hydropyrolysis, *J. Anal. Appl. Pyrolysis*, 2009, **86**, 135–140.

44. R. Lockhart, M. T. Le, K. Grice and W. Meredith, Characteristics of biodegraded Australian oils *via* catalytic hydropyrolysis, in *Abstracts of the 25th International Meeting on Organic Geochemistry, Interlaken*, 2011, p. 214.

45. L. M. Wenger, C. L. Davis, J. M. Evensen, J. R. Gormly and P. J. Mankiewicz, Impact of modern deepwater drilling and testing fluids on geochemical evaluations, *Org. Geochem.*, 2004, **35**, 1527–1536.

46. A. D. Carr, I. C. Scotchman, W. Meredith and C. E. Snape, Bitumens in North Sea Jurassic sandstone reservoirs: Evidence for low temperature degradation rather than oil cracking, in *Abstracts of the 22nd International Meeting on Organic Geochemistry, Seville*, 2005, pp. 392–393.

47. A. Wilhelms and S. R. Larter, Overview of the geochemistry of some tar mats from the North Sea and U.S.A: implications for tar mat origin, in *The Geochemistry of Reservoirs*, eds. J. M. Cubitt, and W. A. England, Geological Society Special Publication 86, London, 1995, pp. 87–101.

48. W. Meredith, C. A. Russell, C. E. Snape, G. D. Love, E. Clarke, B. Moffatt, A. D. Carr and I. C. Scotchman, Potential of bound biomarkers released *via* hydropyrolysis for the characterisation of pyrobitumens and tar mats, *in Abstracts of the 21st International Meeting on Organic Geochemistry, Krakow*, 2003, pp. 305–306.
49. G. D. Love, C. Stalvies, E. Grosjean, W. Meredith and C. E. Snape, Analysis of molecular biomarkers covalently bound within Neoproterozoic sedimentary kerogen. In *From Evolution to Geobiology: Research Questions Driving Paleontology at the Start of a New Century*, eds. P. H. Kelley and R. K. Bambach, Paleontological Society Short Course, October 4, Paleontological Society Papers, Vol. **14**, 2008, pp. 67–83.
50. G. D. Love, E. Grosjean, C. Stalvies, D. A. Fike, J. P. Grotzinger, A. S. Bradley, A. E. Kelly, M. Bhatia, W. Meredith, C. E. Snape, S. A. Bowring, D. J. Condon and R. E. Summons, Fossil steroids record the appearance of Demosponges during the Cryogenian Period, *Nature*, 2009, **457**, 718–721.
51. E. Grosjean, G. D. Love, C. Stalvies, D. A. Fike and R. E. Summons, Origin of petroleum in the Neoproterozoic-Cambrian South Oman Salt Basin, *Org. Geochem.*, 2009, **40**, 87–110.
52. A. E. Kelly, G. D. Love, J. Zumberge and R. E. Summons, Hydrocarbon biomarkers of Neoproterozoic to lower Cambrian oils from eastern Siberia, *Org. Geochem.*, 2011, **42**, 640–654.
53. K. J. Peterson and N. J. Butterfield, Origin of the Eumetazoa: Testing ecological predictions of molecular clocks against the Proterozoic fossil record, *Proc. Natl. Acad. Sci. U. S. A.*, 2005, **102**, 9547–9552.
54. E. Sperling, J. M. Robinson, D. Pisani and K. J. Peterson, Where's the glass? Biomarkers, molecular clocks, and microRNAs suggest a 200-Myr missing Precambrian fossil record of siliceous sponge spicules, *Geobiology*, 2010, **8**, 23–46.
55. A. C. Maloof, C. V. Rose, R. Beach, B. M. Samuels, C. C. Calmet, D. H. Erwin, G. R. Poirier, N. Yao and F. J. Simons, Possible animal-body fossils in pre-Marinoan limestones from South Australia, *Nat. Geosci.*, 2010, **3**, 653–659.
56. W. Holfheinz and G. Oesterhelt, 24-Isopropylcholesterol and 22-Dehydro-24 isopropylchloesterol, novel sterols from a sponge, *Helv. Chim. Acta*, 1979, **62**, 1307–1309.
57. M. A. McCaffrey, J. M. Moldowan, P. A. Lipton, R. E. Summons, K. E. Peters, A. Jeganathan and D. S. Watt, Paleoenvironmental implications of novel C_{30} steranes in Precambrian to Cenozoic Age petroleum and bitumen, *Geochim. Cosmochim. Acta*, 1994, **58**, 529–532.
58. S. A. Bowring, J. P. Grotzinger, D. J. Condon, J. Ramezani, M. J. Newall and P. A. Allen, Geochronologic constraints on the chronostratigraphic framework of the Neoproterozoic Huqf Supergroup, Sultanate of Oman, *Am. J. Sci.*, 2007, **307**, 1097–1145.
59. J. M. Moldowan, F. J. Fago, C. Y. Lee, S. R. Jacobson, D. A. Watt, N. Slougui, A. Jeganthan and D. C. Young, Sedimentary

24-n-propylcholestanes, molecular fossils diagnostic of marine algae, *Science*, 1990, **247**, 309–312.

60. D. A. Fike, J. P. Grotzinger, L. M. Pratt and R. E. Summons, Oxidation of the Ediacaran ocean, *Nature*, 2006, **444**, 744–747.

61. C. Scott, T. W. Lyons, A. Bekker, Y. Shen, S. W. Poulton and A. D. Anbar, Tracing the stepwise oxygenation of the Proterozoic ocean, *Nature*, **452**, 456–459.

62. D. E. Canfield, S. W. Poulton, A. H. Knoll, G. M. Narbonne, G. Ross, T. Goldberg and H. Strauss, Ferruginous conditions dominated later Neoproterozoic deep-water chemistry, *Science*, 2008, **321**, 949–952.

63. K. Peterson, Macroevolutionary interplay between planktic larvae and benthic predators, *Geology*, 2005, **33**, 929–932.

64. C. P. Marshall, G. D. Love, C. E. Snape, A. C. Hill, A. C. Allwood, M. R. Walter, M. J. Van Kranendonk, S. A. Bowden, S. P. Sylva and R. E. Summons, Structural characterization of kerogen in 3.4 Ga Archaean cherts from the Pilbara Craton, Western Australia, *Precambrian Res.*, 2007, **155**, 1–23.

65. J. J. Brocks, G. D. Love, C. E. Snape, G. A. Logan, R. E. Summons and R. Buick, Release of bound aromatic hydrocarbons from late Archean and Mesoproterozoic kerogens *via* hydropyrolysis, *Geochim. Cosmochim. Acta*, 2003, **67**, 1521–1530.

66. J. J. Brocks, G. A. Logan, R. Buick and R. E. Summons, Archean molecular fossils and the early rise of eukaryotes, *Science*, 1999, **285**, 1033–1036.

67. G. D. Love, S. A. Bowden, R. E. Summons, L. L. Jahnke, C. E. Snape, C. N. Campbell and J. G. Day, An optimised catalytic hydropyrolysis method for the rapid screening of microbial cultures for lipid biomarkers, *Org. Geochem.*, 2005, **36**, 63–82.

68. T. M. Peakman and J. R. Maxwell, Early diagenetic pathways of steroid alkenes, *Org. Geochem.*, 1998, **13**, 583–592.

69. J. W. De Leeuw, H. C. Cox, F. W. Van Graas, F. W. Van de Meer, T. M. Peakman, J. M. A. Baas and B. Van de Graaf, Limited double bond isomerisation and selective hydrogenation of steranes during early diagenesis, *Geochim. Cosmochim. Acta*, 1989, **53**, 903–909.

70. M. RambaBu, N. S. Sarma and S. Nittala, Chemistry of herbacin and new unusual sterols from the marine sponge Dysidea herbacea, *Indian J. Chem.*, 1987, **26B**, 1156–1160.

71. M. A. Sephton, G. D. Love, J. S. Watson, A. B. Verchosky, I. P. Wright, C. E. Snape and I. Gilmour, Hydropyrolysis of insoluble carboanaceous organic matter in the Murchison meteorite: New insights into its macromolecular structure, *Geochim. Cosmochim. Acta*, 2004, **68**, 1385–1393.

72. M. A. Sephton, G. D. Love, W. Meredith, C. E. Snape, C. Sun and J. S. Watson, Hydropyrolysis: A new technique for the analysis of macromolecular material in meteorites, *Planet. Space Sci.*, 2005, **53**, 1280–1286.

73. E. D. Goldberg, *Black Carbon in the Environment. Properties and Distribution*, Wiley-Interscience, New York, 1985.

74. H. Schmid, L. Laskus, H. J. Abraham, U. Baltensperger, V. Lavanchy, M. Bizjak, P. Burba, H. Cachier, D. Crow, J. Chow, T. Gnauk, A. Even, H. M. T. Brink, K. P. Giesen, R. Hitzenberger, C. Hueglin, W. Maenhaut, C. Pio, A. Carvalho, J. P. Putaud, D. Toom-Sauntry and H. Puxbaum, Results of the 'carbon conference' international aerosol carbon round robin test Stage I, *Atmos. Environ.*, 2001, **35**, 2111–2121.

75. K. Hammes, M. W. I. Schmidt, R. J. Smernik, L. A. Currie, W. P. Ball, T. H. Nguyen, P. Louchouarn, S. Houel, O. Gustafsson, M. Elmquist, G. Cornelissen, J. O. Skjemstad, C. A. Masiello, J. Song, P. A. Peng, S. Mitra, J. C. Dunn, P. G. Hatcher, W. C. Hockaday, D. M. Smith, C. Hartkopf-Froeder, A. Boehmer, B. Lueer, B. J. Huebert, W. Amelung, S. Brodowski, L. Huang, W. Zhang, P. M. Gschwend, D. X. Flores-Cervantes, C. Largeau, J-N. Rouzaud, C. Rumpel, G. Guggenberger, K. Kaiser, A. Rodionov, F. J. Gonzalez-Vila, J. A. Gonzalez-Perez, J. M. de la Rosa, D. A. C. Manning, E. Lopez-Capel and L. Ding, Comparison of quantification methods to measure fire-derived (black/elemental) carbon in soils and sediments using reference materials from soil, water, sediment and the atmosphere, *Global Biogeochem. Cy.*, 2007, **21**.

76. L. A. Currie, B. A. Benner, J. D. Kessler, D. B. Klinedinst, G. A. Klouda, J. V. Marolf, J. F. Slater, S. A. Wise, H. Cachier, R. Cary, J. C. Chow, J. Watson, E. R. M. Druffel, C. A. Masiello, T. I. Eglinton, A. Pearson, C. M. Reddy, O. Gustafsson, J. G. Quinn, P. C. Hartmann, J. I. Hedges, K. M. Prentice, T. W. Kirchstetter, T. Novakov, H. Puxbaum and H. Schmid, A critical evaluation of interlaboratory data on total, elemental, and isotopic carbon in the carbonaceous particle reference material, NIST SRM 1649a, *J. Res. Natl. Inst. Stan. Tech.*, 2002, **107**, 279–298.

77. W. Meredith, P. L. Ascough, M. I. Bird, D. J. Large, C. E. Snape, Y. Sun and E. L. Tilston, Assessment of hydropyrolysis as a method for the quantification of black carbon using standard reference materials, *Geochim. Cosmochim. Acta*, 2012, **97**, 131–147.

78. P. L. Ascough, M. I. Bird, F. Brock, T. F. G. Higham, W. Meredith, C. E. Snape and C. H. Vane, Hydropyrolysis as a new tool for radiocarbon pre-treatment and the quantification of black carbon, *Quatern. Geochron.*, 2009, **4**, 140–147.

79. P. L. Ascough, M. I. Bird, W. Meredith, R. E. Wood, C. E. Snape, F. Brock, T. F. G. Higham, D. J. Large and D. C. Apperley, Hydropyrolysis: implication for radiocarbon pretreatment and characterization of black carbon, *Radiocarbon*, 2010, **52**, 1336–1350.

80. W. C. Hockaday, A. M. Grannas, S. Kim and P. G. Hatcher, Direct molecular evidence for the degradation and mobility of black carbon in soils from ultrahigh-resolution mass spectral analysis of dissolved organic matter from a fire-imp acted forest soil, *Org. Geochem.*, 2006, **37**, 501–510.

81. W. Meredith, P. L. Ascough, M. I. Bird, D. J. Large, C. E. Snape, J. Song, Y. Sun and E. L. Tilston, Direct evidence from hydropyrolysis for the retention of long alkyl moieties in soil black carbon fractions isolated by acid dichromate oxidation, *J. Anal. Appl. Pyrolysis*, 2013, **103**, 232–239.

CHAPTER 7

Microscale Sealed Vessel Pyrolysis

BRIAN HORSFIELD,*[a] FRANZ LEISTNER[b] AND KEITH HALL[c]

[a] German Research Centre for Geosciences, Potsdam, Germany;
[b] Forschungszentrum Jülich, Germany; [c] Hall Analytical, Manchester, U.K.
*Email: horsf@gfz-potsdam.de

7.1 Introduction

Pyrolysis is the process whereby solid, liquid, and gaseous materials are thermally degraded in the absence of oxygen into smaller molecular fragments. It has been defined as 'a chemical degradation reaction that is induced by thermal energy alone'.[1] In this chapter we discuss the methodology and applications of microscale sealed vessel pyrolysis (MSSV) which, while having a broad range of uses in the natural and earth sciences, was originally designed to replicate in the laboratory (10^{-1}–10^{-3} yr experimental duration) important petroleum-forming reactions taking place in nature over extremely long periods of time (10^6–10^8 yr).

7.1.1 Challenges

Pyrolysis has been widely used to examine the biological and diagenetic origins of complex, naturally occurring geopolymeric materials by breaking them down into more readily analysable, volatile fragments. Geopolymers or macromolecules make up the biggest proportion of the global carbon budget (>95%), with the mass amounting to 10^{16} tons, mainly in a dispersed form.[2] Kerogen, the fraction of sedimentary organic matter that is insoluble in

RSC Detection Science Series No. 4
Principles and Practice of Analytical Techniques in Geosciences
Edited by Kliti Grice
© The Royal Society of Chemistry 2015
Published by the Royal Society of Chemistry, www.rsc.org

common organic solvents, is the major macromolecular component of sedimentary organic matter in rocks. It forms from biological precursors *via* a combination of two pathways, one being the random 'repolymerisation' or 'humification' of the microbial breakdown products of proteins, poly-saccharides and lignins from plants, namely amino acids, sugars, and phenols,[3,4] and the other coming from the selective preservation of resistant, often morphologically structured, biopolymeric materials such as spores, pollen, and degraded cellular debris.[5-8] The proportion of kerogen that is attributed to either neoformation reactions in the so-called zone of dia-genesis (<100 °C) or direct preservation depends on the nature of the starting material and the depositional environment in question (as reviewed by Horsfield and Rullkötter[9]). Thus, a portion of the polymethylene com-ponents in aquatic autochthonous kerogen particles may originate by the random polymerisation of algal-derived polyunsaturated lipids[10-12] possibly *via* sulfur atoms,[13,14] whereas the rest could simply consist of preserved aliphatic cell wall material. Similarly, the polymethylene components in allochthonous kerogen particles may originate by the random grafting (initially esterification) of wax acids and wax alcohols from higher plant cuticles to a polycondensed nucleus[15] or represent preserved cuticular aliphatic biopolymeric materials. In the case of woody organic matter derived from land plant, the petrological habits of which indicate a variety of possible origins,[6] alkylphenolic and methoxyphenolic moieties may enter the kerogen *via* random repolymerisation of lignocellulosic degradation products or the preservation of lignin and similar materials.[16,17].

As kerogen is gradually exposed to progressively higher temperatures over millions to tens of millions of years, because its sediment host is buried beneath successively accumulating younger layers, its composition changes. Major aliphatic substituents of the kerogen structure are progressively cracked more or less in the order of bond strength, and there is concomitant structural rearrangement of the residues. A large proportion of the generated volatile and mobile components is expelled into surrounding strata, then migrates into geological traps to form conventional petroleum deposits. The retained portion can also constitute commercial shale gas and shale oil re-sources, if uplift has brought the source rock to within 2–3 km of the surface. The residual kerogen is retained in the rock as a non-volatile and immobile component. The zone of catagenesis, defined by the mean vitrinite reflect-ance range $R_m = 0.5$–2.0%[18] and paleotemperature range of approximately 100–200 °C,[19] is where oil and gas are generated and expelled, whereas the succeeding zone of metagenesis, with R_m greater than 2.0% and paleo-temperatures exceeding 200 °C,[18,19] is characterised by gas only, because residual oil is thermally unstable. The upper limit of gas generation appears to lie around $R_m = 4\%$ and a paleotemperature of 270 °C, according to measurements of late gas potential.[20]

Relating the structure and composition of macromolecules to their pre-cursor organisms and the diagenetic modification undergone since death, as well as modelling their progressive thermal degradation during progressive subsidence, are the fundamental lines of research being undertaken today in upstream (exploration and production) petroleum science. The bulk composition of the generated petroleum, which is highly variable, directly governs the petroleum's bulk physical properties such as wettability, PVT (pressure, volume, and temperature) characteristics, polarity, and bulk density (American Petroleum Institute (API) gravity), and thence the number of phases present, their volume(s), and buoyancy.

The tradition of using pyrolysis in all these contexts goes back 50 years or more, with humic acids and kerogen isolated from soils and coals[21,22] and sediments of Precambrian through to Recent age,[23-27] as well as oil shales, native bitumens, asphalts, and petroleum fractions (see e.g.[28-33,11]) coming under detailed scrutiny.

7.1.2 Pyrolysis Systems

Many different pyrolysis systems have been developed for researching geosystems, and reviews are available.[34-37] They can be divided into two basic categories, one being dynamic and open (anhydrous), and the other static and closed (hydrous or anhydrous). Dynamic, or as they are frequently termed analytical, pyrolysis systems often make use of extremely short temperature rise times, and are used to elucidate the chemical composition and structure of geomacromolecules, for example when evaluating the petroleum generating properties (organic richness, quality, thermal lability) of a given sample. An inert carrier gas is employed to sweep high-temperature pyrolysis products (400–600 °C range; heating time measured in minutes down to fractions of a second; extremely short temperature rise time) to a detection and/or analytical device. Pyrolysis–gas chromatography (Py-GC) is one such analytical system, using filaments, Curie-point wires, or furnaces for bringing about pyrolysis. Fundamental concerns regarding the reproducibility of results from these pyrolysers have been largely unsubstantiated, at least as far as the organic matter occurring in sedimentary rocks is concerned. This is because kerogens, being strongly depleted in heteroatoms during diagenesis and maturation, are much less sensitive or reactive to pyrolysis environment than are heteroatom-rich biopolymers, as long as adequate care is given to keep residence time in the pyrolyser low.[38] Importantly, compositional inferences from analytical pyrolysis are based on only a rather small mass fraction of resolved pyrolysates and an even smaller fraction of the total pyrolysate ('hump' underlying the resolved components in chromatograms, and condensed tarry materials at interfaces within the analytical plumbing system), but basic compositional attributes of the starting material (aromaticity, sulfur content, average alkyl chain length, and oxygen functionality), have all been reliably predicted using just the resolved pyrolysis products.[38-40] In other words, Py-GC provides representative insights into the geochemistry of macromolecules.

Static closed-system pyrolysis is commonly used to simulate the thermal degradation processes occurring during the progressive subsidence of sedimentary basins. For this exercise, samples of low maturity are pyrolysed over a temperature range (usually 250–500 °C) that is intermediate between that of analytical pyrolysis and that at which petroleum is generated in sedimentary basins (<200 °C). Heating times are measured in hours and days, and heating rates for non-isothermal experiments are nine to ten orders of magnitude faster than the geological system that is being simulated (1–5 K Ma^{-1}; 10^{-11}K min^{-1}).

In hydrous pyrolysis, tens or hundreds of grams of rock fragments are heated with excess water in a Parr bomb in order to generate and expel petroleum-like products. A liquid water phase is maintained during the experiments at subcritical water temperatures (<374 °C). Lewan[41] outlined how 'petroleum-like' products are collected in the floating pyrolysate, whereas polar compounds remain enriched in the non-expelled fraction. As far as instrumentation and application are concerned, the approach is well suited to studying bound biomarkers and inferring expulsion-related fractionations, but it is time consuming, especially when a full quantification of precursors and products is made.

Gold bag pyrolysis is a simulation approach where confining pressure is controlled. Typically, grams or tens of grams of starting material are used. Pyrolysis conditions may be selected to be hydrous or anhydrous. Full mass balances have been made using this approach, and compositional kinetic models of petroleum generation constructed from precursors, products and intermediates.[42,43] The method is robust but extremely time consuming.

MSSV pyrolysis[44,45] is a microanalytical technique developed for artificially maturing milligrams to tens of milligrams of kerogen, coal, asphaltenes, or whole rock powder and then quantifying the major GC-amenable organic components generated during this process (C$_{1+}$) in a single online analytical step. Liquid water is not employed. The degree to which the system is confined is determined by the ratio of sample loading, degree of conversion into products, and the dead volume within the vessel. MSSV has the same advantages as analytical pyrolysis in being online, rapid, and highly reproducible.

Implicit in simulated maturation experiments is the requirement that key reaction pathways in the laboratory and in nature are very similar, and that only reaction rates are increased during the simulation. Ideally, progressive changes in the compositions of products and residues from simulations must mimic those seen in nature. The nature of the pressuring medium can play a crucial role in the results obtained; pressure exerts only a minor influence in the case where water is essentially absent, but has a large influence in its presence.[46] Water is an additional hydrogen source for hydrocarbon formation during hydrous pyrolysis, whereas organic hydrogen alone is available in anhydrous systems.[41] Crucially, the pyrolysates from laboratory experiments, irrespective of whether they are conducted under hydrous or anhydrous conditions, are without exception rich in polar and aromatic components, and therefore fundamentally different from mature source rocks and undegraded petroleum which are hydrocarbon-rich.[38] Of the myriads of actual reactions taking place, it has been assumed that basically two sequential bulk reactions are active in petroleum formation (see Larter and Horsfield[38] and references therein): $K \rightarrow K' + B$ (kerogen to bitumen), followed by $B \rightarrow B' + O$ (bitumen to oil). Under laboratory conditions. where bitumen yields are high, the bitumen-to-oil reaction is the rate-controlling step, whereas under geological heating conditions, where hydrocarbon yields are high, the kerogen-to-bitumen reaction is the rate-controlling step. This phenomenon is manifested on an Arrhenius diagram by a cross-over of rate curves.[47] The divergence of the curves is small because the kinetic parameters determined for the $K \rightarrow K' + B$ reaction by open-system pyrolysis adequately predict the conversion of kerogen to petroleum (*i.e.* $K \rightarrow K' + B$, and $B \rightarrow B' + O$) in the subsurface.[48,49]

The difference in polarity has traditionally been attributed to a physical rather than chemical effect as far as hydrous pyrolysis is concerned, with expulsion fractionation leading to a hydrocarbon-rich product, irrespective of heating rate (laboratory *versus* natural system). As far as gold bag pyrolysis is concerned, a comprehensive first-order kinetic model, using parallel pseudo-reactions for *inter alia* polar compound formation and degradation, was devised by Béhar *et al.*[43] While the same fraction yields could also be readily determined using MSSV pyrolysis, only GC-amenable products are usually analysed and incorporated into compositional kinetic models. The approach is based on the concept of 'structural moieties' occurring in kerogen, polar products (bitumen), and hydrocarbons, and is described in the concluding section of this chapter.

7.2 Overview of the MSSV Technique

Here we describe the design, operational details, and reproducibility of both the MSSV system developed at Kernforschungsanlage (KFA) in the late 1980s, and its commercial successor (Quantum MSSV-2 Thermal Analysis System GC2 Chromatography, Manchester, UK). Because the crucial test of any simulation method is the extent to which it duplicates nature, we have chosen to focus on how MSSV can actually predict the extent of selected reactions occurring in real petroleum systems. In other words, the direct validation of predictions using geological data is key. Those interested in a detailed comparison of laboratory results from the respective systems are referred to the publications of Michels *et al.*,[46] Horsfield,[50] and Henry and Lewan.[51]

The MSSV technique consists of two sequential steps, namely simulated maturation to generate products followed by the online analysis of those products. The products can also be analysed offline if desired, as, for example in the case of gas isotope studies.[52,53]

7.2.1 Step 1: Simulated Maturation

7.2.1.1 *Preparation and Loading*

Pyrolysis is conducted using glass capillaries, or quartz for high-temperature experiments (>570 °C), without the addition of excess water as a reactant or pressurising medium. For each experiment that involves a whole rock or kerogen sample, a commercially produced capillary tube (100 μL) is bent (~150° angle), trimmed to approximately 40 μL volume, then sealed at one end using a hydrogen flame. Precleaned glass beads (80–120 mesh) are added up to the level of the elbow within the sealed arm of the tube. Heating at 350 °C for 12 h in air is used as a further pre-experimental cleaning step. After the tube has been cleaned it is loaded with 1–5 mg in the case of kerogen, or 10–40 mg in the case of whole rock. The internal volume is then reduced to about 10 μL by filling the remainder of the tube with thermally precleaned glass beads in order to confine the system as much as possible.

Asphaltenes are loaded differently because they can sometimes be highly viscous and difficult to handle. A 40 μL glass or quartz tube, open at both ends, bent as described above, and containing quartz wool is precleaned as outlined above. A load of 1–5 mg of asphaltenes is then introduced onto the glass or quartz wool in dichloromethane solution using a 10 μL syringe. Excess solvent is removed before sealing the tube by passing 100 mL of nitrogen across and through the sample and then by storage under vacuum at 120 °C for 12 h. Asphaltene separation methods vary, but that of Theuerkorn *et al.*,[54] which employs n-hexane as precipitating medium, and utilises thermovaporisation–GC to confirm the absence of maltenes (*e.g.* waxy paraffins) in the final asphaltene precipitate, has been demonstrated to be reproducible.

Sample aliquots are artificially matured in batches using a purpose-built high-precision oven whose maximum linear programming range extends to 700 °C. Programmed temperature heating is usually employed, but isothermal heating conditions can be selected. In the original paper,[44] isothermal pyrolysis was performed using a GC oven (operating temperature <350 °C). Unheated aliquots, also contained in tubes, can be prepared in order to measure the original concentrations of volatile organic species in each sample using thermovaporisation.

Changes in composition and yield are known to be precisely and accurately relatable to the heating conditions employed in the simulation because of (1) the low thermal mass of the tubes and (2) the excellent temperature control afforded by the pyrolysis ovens.

7.2.1.2 Oxidation Effects

The internal volume of an MSSV tube is approximately 40 μL. After filling with glass beads, the free volume is reduced, and the voids filled with air. One-fifth of the free volume is oxygen. A simple mass balance (Table 7.1) reveals that the proportion of carbon or hydrogen that can be consumed by oxidation to carbon dioxide or water from a 1–3 mg loading is 0.1%. This means that at least 99.9% of the carbon is in a chemical form which reflects the structure of the parent material and the thermal degradation (pyrolysis) reaction. Thus, the MSSV method can be considered essentially free of oxidation artefacts.

Table 7.1 Effect of air during MSSV pyrolysis—a simple mass balance calculation on carbon and hydrogen consumption.

Volume of oxygen

- 32 g O_2 occupies 22.4 L at STP
- 32 µg O_2 occupies 22.4 µL $\equiv$ tube volume
- 6.5 µg O_2 occupies tube (1/5 atmosphere $= O_2$)

Consumption of carbon

- Need 12 g C for every 32 g $O \rightarrow CO_2$
- Need $12/32 \times 6.5$ µg C $= 2.4$ µg C for 6.5 µg $O \rightarrow CO_2$
- Typically, we analyse 1–3 mg kerogen ($\sim 70\%$ C)
- Conversion $\sim 0.1\%$

Consumption of hydrogen

- Need 4 g H for every 32 g $O \rightarrow H_2O$
- Need $4/32 \times 6.5$ µg H $= 0.8$ µg H for 6.5 µg $O \rightarrow H_2O$
- Typically we analyse 1–3 mg kerogen ($\sim 5\%$ H) $\equiv 0.1$ mg H
- Conversion $\sim 0.1\%$

7.2.1.3 Partial Pressure Regime

While the pressure in gold bags can be regulated by an external hydraulic system during pyrolysis,[42] MSSV tubes are rigid, and thence pressures are directly controlled by the temperature of pyrolysis and the partial pressure of the volatiles inside. The highest temperatures (>600 °C), bringing about high primary and secondary gas yields, result in pressures of around 4 MPa, and the normal operating range for laboratory-based studies of oil generation is 2–4 MPa, equivalent to a maximum burial of 0.5 km in nature. Thus, MSSV is a closed-system low-to-intermediate pressure method, whose pressures are about one-tenth of those in generative petroleum systems (20–50 MPa).

7.2.2 Step 2: Analysis of Pyrolysis Products

7.2.2.1 Online Gas Chromatography

Gas chromatography is employed to quantify the major components in pyrolysates (*e.g.* hydrocarbon gases, resolved compound classes) and thence the major predicted components of petroleums. The MSSV interface consists of a purpose-built piston device/sample holder, a programmable pyrolysis furnace, a heated on–off split, a cryogenic trap packed with glass beads, a heated transfer zone, and an analytical device. Tubes containing the original unpyrolysed sample are heated at 300 °C for a few minutes and the released volatiles analysed in order to determine pre-existing compound abundances in sediments.

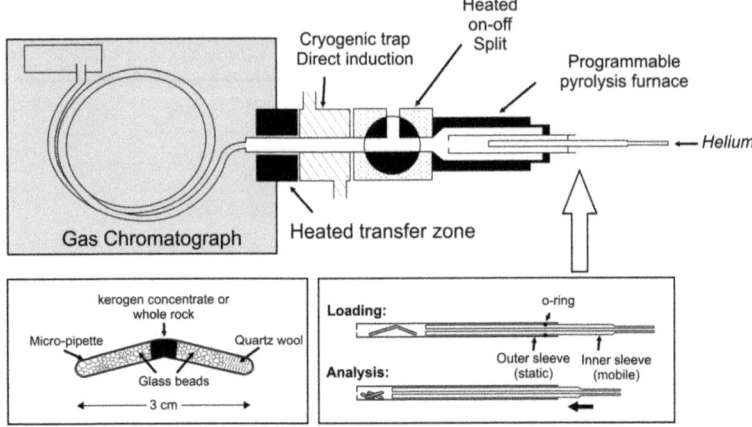

Figure 7.1 MSSV instrument: prototype with horizontal construction.

The original instrument (Figure 7.1) had a horizontal construction. The on–off split was open only when cleaning the outer surface of the MSSV tube. The split employed for injecting products on to the GC column was a different one, contained within the GC oven itself, *i.e.* after the trap. A sample was loaded into the system as shown in Figure 7.1. A 5 min purge cycle (300 °C) was used to clean the outer surfaces of the tube. During this time, effluents were vented *via* the on–off split. The tube was then cracked open by the piston action of the all-glass sample holder with the split in the off position. Products were swept from the heated zone (300 °C) by helium flow (32 mL min^{-1}) and condensed using the liquid nitrogen-cooled trap. Methane was only partially trapped by the configuration, and passed through to be quantified at the FID while the C2 + components were collecting in the trap. After ballistic heating of the trap using direct induction, GC is performed. For general purposes, with resolution of major components in the C_2–C_6 range as well as elution of n-alkanes up to C_{35}, a HP-1, 50 m length, internal diameter 0.32 mm, film thickness 0.52 μm will suffice. The GC oven temperature is programmed from 30 °C to 320 °C at 8 °C min^{-1}, providing baseline resolution in the C_2–C_8 range.

The commercial system (Quantum MSSV-2 Thermal Analysis System) is a miniaturised version of the above, built around the GC injector, and enjoys a vertical rather than horizontal space-saving design (Figure 7.2). It retains all of the original design features of the prototype, in that it has a piston device for breaking open the MSSV vessels, the interface to the gas chromatograph is held at 300 °C and the split for injection onto the GC column is located after the trap. The cryogenic trap is contained within the GC oven in this case. The reproducibility of the instrument is discussed below. In that regard, it should be noted that an earlier commercial MSSV system (Quantum MSSV-1 Thermal Analysis System) had its cryogenically cooled trap located after the split (GC column loop in liquid nitrogen), and as such proved suitable only for qualitative or semiquantitative studies.

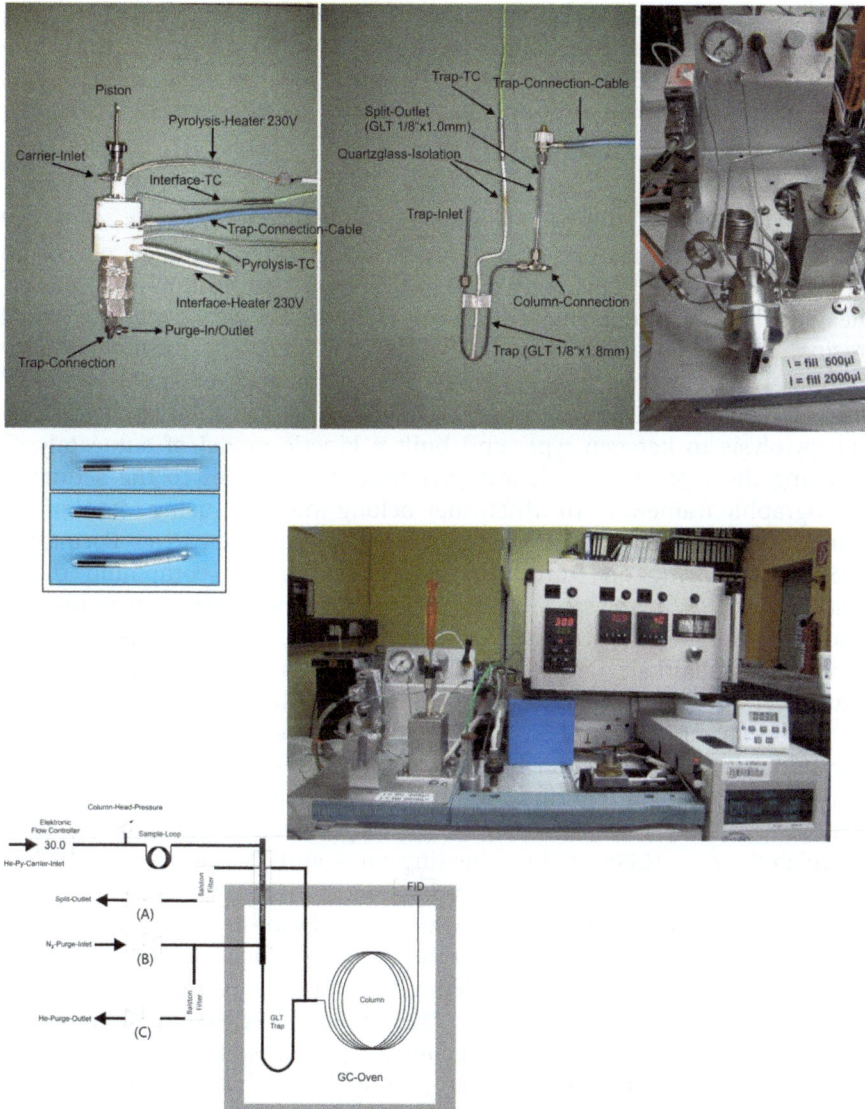

Figure 7.2 MSSV instrument: Quantum MSSV-2 Thermal Analyser.

Quantification. Total pyrolysate yield (C_{1+}), boiling ranges $(C_1, C_2–C_5, C_6–C_{14}, C_{15+})$, compound classes (n-alkenes, n-alkanes, alkylaromatic hydrocarbons, and alkylthiophenes) as well as individual compounds are routinely quantified by external standardisation using n-butane from a four-port sampling valve. Response factors for all compounds are assumed the same for the applications shown here, except that methane has a response factor of 1.1. Internal standards have also been tested over the years (phenylbutane, Horsfield *et al.*;[44] diamandoid, Acholla[55]) but the butane calibration can be performed quickly and is highly reproducible.

Analysis of Residues.   Upon completion of the GC cycle, pyrolysis–GC can be performed on the non-volatile residues of MSSV pyrolysis, using programmed heating (300–600 °C at 40 °C min^{-1}) of the same aliquot. This has enabled precursor–product mass balance calculations to be readily made.[45]

Quick Overview of Research Applications.   The GC analysis of MSSV pyrolysis products has provided data for use in kinetic studies, oil correlation, and reservoir geochemistry. Tegelaar *et al.*[56] and van Aarssen *et al.*[57] heated aliphatic biopolymer extracted from *Agave americana* and fossil resins, respectively, and found that the n-alkane and cadinane distributions were close to those of crude oils. Düppenbecker and Horsfield[58] and Horsfield and Düppenbecker[45] related changes in product yields during pyrolysis to kerogen type, and built a kinetic model of autocatalysis. Relating the types of petroleums generated from shales to the sequence stratigraphic framework in which they belong was done using MSSV of the Laney Shale and Luman Tongue members of the Green River Formation (USA).[59] Following on from preliminary studies,[60–62] compound distributions in MSSV pyrolysates of reservoir asphaltenes were used to identify sources that had contributed to the biodegraded petroleum in the Holzener Asphaltkalk, and to quantify mass loss during biodegradation. Muscio *et al.*[63] and Horsfield *et al.*[64] found that the average chain lengths in kerogens of the Bakken Shale (North America) and Alum Shales (Scandinavia) are anomalously short as compared to most type II kerogens, thus drawing attention to the need for molecular typing of kerogens. Kinetic parameters for the conversion of reservoired oil into gas were calculated using MSSV at three heating rates and 75 sample aliquots per sample,[64,65] and comparisons made with in-source secondary gas formation.[19] Thus, in-source oil cracking was shown to take place at lower levels of thermal stress ($R_m = 1.2\%$) than required for in-reservoir cracking ($R_m = 2\%$). The kinetics of petroleum formation were studied by Ritter *et al.*[66] using a Quantum MSSV-1 Thermal Analysis System followed by deconvolution of primary from secondary products. Late gas formation ($R_m > 2\%$) was documented in source rocks,[67,68] with the potential shown to increase as maturation proceeded.[69] Hydrocarbon and nitrogen gas generation was studied using MSSV and open-system methods by Idiz *et al.*[70] The progressive neoformation of dead carbon (non-volatile under pyrolysis conditions) from live carbon (originally released from the precursor as volatile) in the Bakken and Alum Shales was documented by the sequential analysis of MSSV products and residues,[64,71] thereby contrasting sharply with the Posidonia Shale (Germany) and Green River Shale (USA) from which neoformation of dead carbon did not occur. A comparison of open *versus* closed-system pyrolysis and the overlap of primary and secondary reactions was made by Schenk and Horsfield[72] and Dieckmann *et al.*[73] The MSSV interface has been employed for thermovaporisation of adsorbed or occluded volatiles. The pore space of sedimentary rocks can be occluded by diagenetic cements such as calcite or quartz, and contained in the latter cement are both aqueous fluid inclusions that are a trapped portion of the brine from which the calcite crystals grew, and hydrocarbon-bearing inclusions, representing a trapped coexisting free petroleum phase. The release and analysis of free gas and/or oil from fluid inclusions in reservoir cements and ore minerals has been done using the original KFA prototype[74,75] or the Quantum MSSV-2 set-up[76,77] using the liner as a sample holder and the modified piston as crusher, with the interface and GC system remaining standard.

7.2.2.2   Online Gas Chromatography(–IR)–Mass Spectrometry

GC-MS has been employed to identify trace components released from macromolecules in rocks, air particulates and water samples *via* thermovaporisation and stepwise MSSV pyrolysis. The Quantum MSSV-2 Thermal Analysis System has been interfaced to a variety of instruments.

Quick Overview of Research Applications. Hall *et al.*[78] and Waterman *et al.*[79,80] quantified polycyclic aromatic hydrocarbons in the NIST Standard Reference Material (SRM1649A) Urban Dust using the Quantum MSSV-2 system interfaced to GC-MS (Fisons MD800 mass spectrometer run in EI mode and scanned from m/z 40 to 520). The GC used a Phenomenex ZB-5 column, 25 m, 0.25 mm internal diameter, 0.25 μm film thickness). The GC oven was set at 40 °C for 13 min and then ramped at 5 °C min^{-1} to 300 °C where it was held for 25 min (total analysis time 90 min).

Hall *et al.*[81] were the first to analyse the stable carbon isotopic compositions of gases, n-alkanes, and alkylbenzenes generated from oil-to-gas cracking by online GC-IR-MS analysis of MSSV products, and the same system was later used to analyse light hydrocarbons released by thermovaporisation from the Bakken Shale.[82] GC-IR-MS was performed in triplicate using a VG Isochrom II system interfaced to a Dani 8510 gas chromatograph and Quantum MSSV-1 Thermal Analysis System. The GC was fitted with a 50 m OV-1 fused silica column. Reproducibility was usually ± 0.3‰. Mycke *et al.*[83] developed and extended the method to study the maturation of source rocks, and Wilhelms *et al.*[84] to study asphaltenes. More recently the stable hydrogen isotopes of pyrolysates have been studied using a similar analytical configuration.[85]

Berwick *et al.*[86] compared MSSV with high-pressure hydrogen pyrolysis (HyPy) as regards releasing intact biomarkers from numerous matrices. This comparison revealed many product similarities, but also several important features unique to each. Greenwood *et al.*[87] studied biomarkers in dissolved organic matter using MSSV-GC-MS analysis. A Hewlett-Packard (HP) 5890 Series II GC was interfaced to an Autospec (UltimaQ) double-focusing mass spectrometer. A 25 m × 0.32 mm DB5 capillary column, film thickness 0.52 μm, was used with helium carrier gas at a constant pressure of 8 psi. Numerous oven temperature programmes were used for analysis. Full scan

(FS), selected ion recording (SIR), and multiple (metastable) reaction monitoring (MRM) data were separately acquired. FS analyses were performed over the range 50–550 Da and at 3 scans per second. Follow-up integrated studies using similar GC-MS configurations have demonstrated that the qualitative speciation provided by MSSV can make a significant contribution to the structural characterisation and source recognition of indigenous aquatic organic matter and pollutants in wastewater.[88–90]

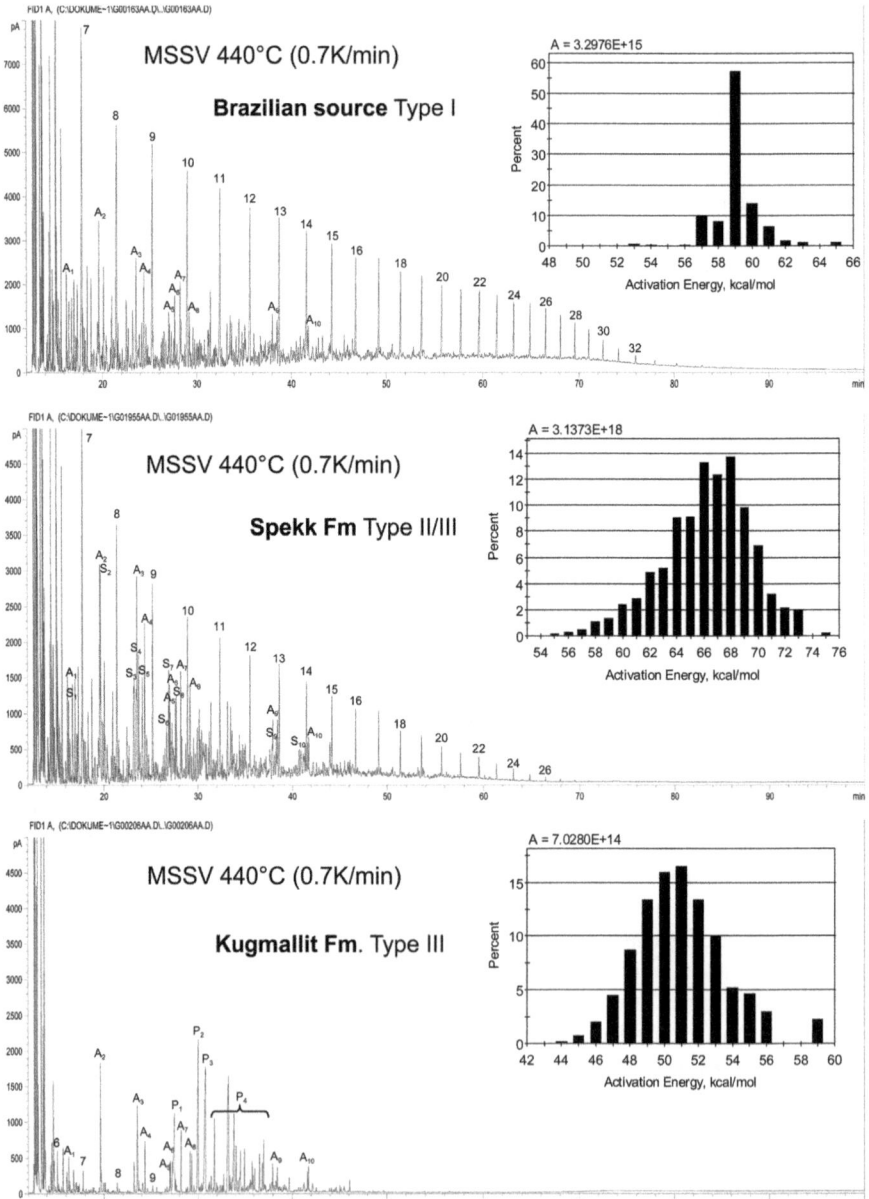

7.3 Resolution and Range of Products from MSSV Simulations

Here we focus on the simulation of petroleum generation reactions. Figure 7.3[91] shows a selection of MSSV gas chromatograms for selected petroleum source rocks. Each example portrays the simulated generated products at peak oil generation (transformation ratio (TR) = 0.4–0.8; TR is the fraction of the total petroleum potential that has been realised), and reveals differences in wax content, aromaticity, and gas *versus* oil potential.

Figure 7.3 MSSV gas chromatograms of selected source rocks at simulated peak generation. The unit for the frequency factors given in the insets is s^{-1}. Gas chromatographic peaks are labelled as follows:

Numbers	n-Alkanes
Pr	Pristane
Ph	Phytane
A	Aromatics
A1	Benzene
A2	Toluene
A3	*m/p*-Xylene
A4	*o*-Xylene
A5	(Alkyl)-Benzene
A6	1,3,5-Trimethylbenzene
A7	1,2,4-Trimethylbenzene
A8	1,2,3-Trimethylbenzene
AN	Naphthalene
A9	Methylnaphthalenes
A10	Dimethylnaphthalenes
S	Sulfur-bearing compounds
S1	Thiophene
S2	Methylthiophenes
S3	2,5-Dimethylthiophene
S4	2,4-Dimethylthiophene
S5	2,3-Dimethylthiophene
S6	2-Propylthiophene
S7	2-Ethyl-5-Methylthiophene
S8	2,3,5-Trimethylthiophene
S9	Methylbenzothiophene
S10	Dimethylbenzothiophene
P	Phenolic compounds
P1	Phenol
P2	*o*-Cresol
P3	*m/p*-Cresol
P4	(Alkyl)-phenols

The Brazilian type I source rock, deposited in a lacustrine setting, consisting largely, at the microscopic scale, of alginite A, and originating from *Botryococcus,* generates mainly paraffinic products on MSSV pyrolysis, extending from low molecular weight up to n-C$_{32}$. Based on open-system pyrolysis-GC the sample belongs to the High Wax Paraffinic Petroleum Organofacies of Horsfield.[39] Its narrow activation energy distribution (inset), calculated from a first-order cracking model that was populated by open-system pyrolysis data, is typical of a structurally homogeneous kerogen.[65,92] The relative abundance of each homologue is controlled to variable degrees by the chain length of the precursor moiety in the kerogen and the secondary reactions that occur during pyrolytic cleavage. Bimolecular combination reactions are not prevalent, and therefore n-alkyl chain lengths for pyrolysates represent minimum values of those present in the kerogen. Crude oils generated from these lacustrine source rocks in nature are known to contain an abundance of long-chain hydrocarbons and are thence termed waxy.

Marine type I kerogens which contain the residues of *Tasmanites* or *Gloeocapsamorpha* also generate extremely paraffinic products,[44,93] but with a shorter average chain length. Marine type II kerogens, like the Spekk Formation (prolific source rock, mid-Norway) generate an MSSV pyrolysate with a higher proportion of aromatic and branched hydrocarbons in the C$_4$–C$_{15}$ range than do type I kerogens. Toluene and the xylenes are prominent, with amounts roughly equal to those of n-hydrocarbons in the same boiling range. The Spekk belongs to the Low Wax Paraffinic-Naphthenic-Aromatic Petroleum Type Organofacies of Horsfield.[39]

While type II kerogens generate higher absolute quantities of aromatic hydrocarbons on pyrolysis, type III vitrinitic kerogens yield the highest relative amounts so that their pyrolysates are dominated by alkylbenzenes, alkylnaphthalenes, and alkylphenols. These products originate from the pyrolytic degradation of lignin, sporopollenin, and polycarboxylic acids.[94] The Kugmallit Formation of western Canada is such a kerogen, generating high proportions of gas in addition to the aforementioned aromatic compounds.

Figure 7.4 shows a broader simulated maturity range for the Spekk sample and for another lacustrine type I kerogen, from the Green River Formation, using programmed heating (1 K min^{-1}). The latter contains the remains of *Botryococcus* algal bodies. Mid-chain length normal alkanes are generated initially (350 °C), and thereafter hydrocarbons of lower molecular weight increase in relative abundance. These components are initially paraffinic (450 °C), but are then succeeded by aromatic hydrocarbons at highest conversions (500 °C). The gas content increases progressively, and is high at advanced levels of maturation. The Spekk sample also reveals changes in chain length distribution of n-alkanes, followed by aromatisation at the highest simulated maturity level.

MSSV data such as these have been used to populate compositional kinetic models of oil generation, oil cracking, and gas formation from both primary (kerogen source) and secondary (oil source) gas, as described below. Strict quality control must be applied if the results are to be used in kinetic modelling.

7.4 Data Quality Requirements for Kinetic Models

The transformations of sedimentary organic matter upon burial leading from the high-molecular-weight kerogen to oil and gas are generally accepted to proceed through a multitude of parallel and consecutive reactions which are unknown in detail, but which are recognized to be quasi-irreversible and controlled by chemical kinetics. Kinetic laws are therefore routinely considered as the mathematical link between high-temperature/short time and low temperature–long time configurations, allowing extrapolations from laboratory to natural heating conditions. However, the extremely complex structure of the organic components involved necessitates the application of gross kinetic concepts whereby individual product precursors are replaced by bulk potentials which are fractions of the total product yield. Predicting the composition of natural petroleum requires kinetic investigations of both the primary generation of individual components from kerogen and the secondary transformations among the primary products.[65]

MSSV is well suited to delivering this information, as seen by the gas chromatograms referred to in the preceding discussion. But the caveat is that data precision must be exceedingly high. Schenk and Horsfield[95] stressed how geological predictions are extremely sensitive to the frequency factor, and that this in turn is fixed by the shift in the temperature maxima of generation rates as a function of heating rate. To convert the cumulative MSSV data (Figure 7.5a) into rates, smoothing by appropriate spline functions is followed by a numerical differentiation of the evolution profiles of each compound group of interest (Figure 7.5b). It is the shift in the T_{max} of these rate curves that is critical for the calculation of frequency factor,[96] followed by a least squares iteration method that determines an activation energy distribution.

7.5 Data Quality Requirements for MSSV Simulations

Changes in total yield (C_{1+}) and boiling range yields (C_1–C_5, C_6–C_{14}, and C_{15+}) as a function of temperature (26 stages, 0.7 K min^{-1}) are shown in Figure 7.6 for the pyrolysis products of an immature sample from the Nordegg Formation (western Canada). Increasing gas yields are linked initially with primary generation and later secondary breakdown of higher-boiling compounds. Boiling range data, measured at three or more heating rates, has been used to calculate the kinetic parameters of bulk oil formation and destruction using data like this.[19,64,65,67,69,97] Because data scatter is low, best-fit splines have a low error factor and are reproducible, facilitating accurate frequency factor calculation.[98] For triplicate analysis of a boghead

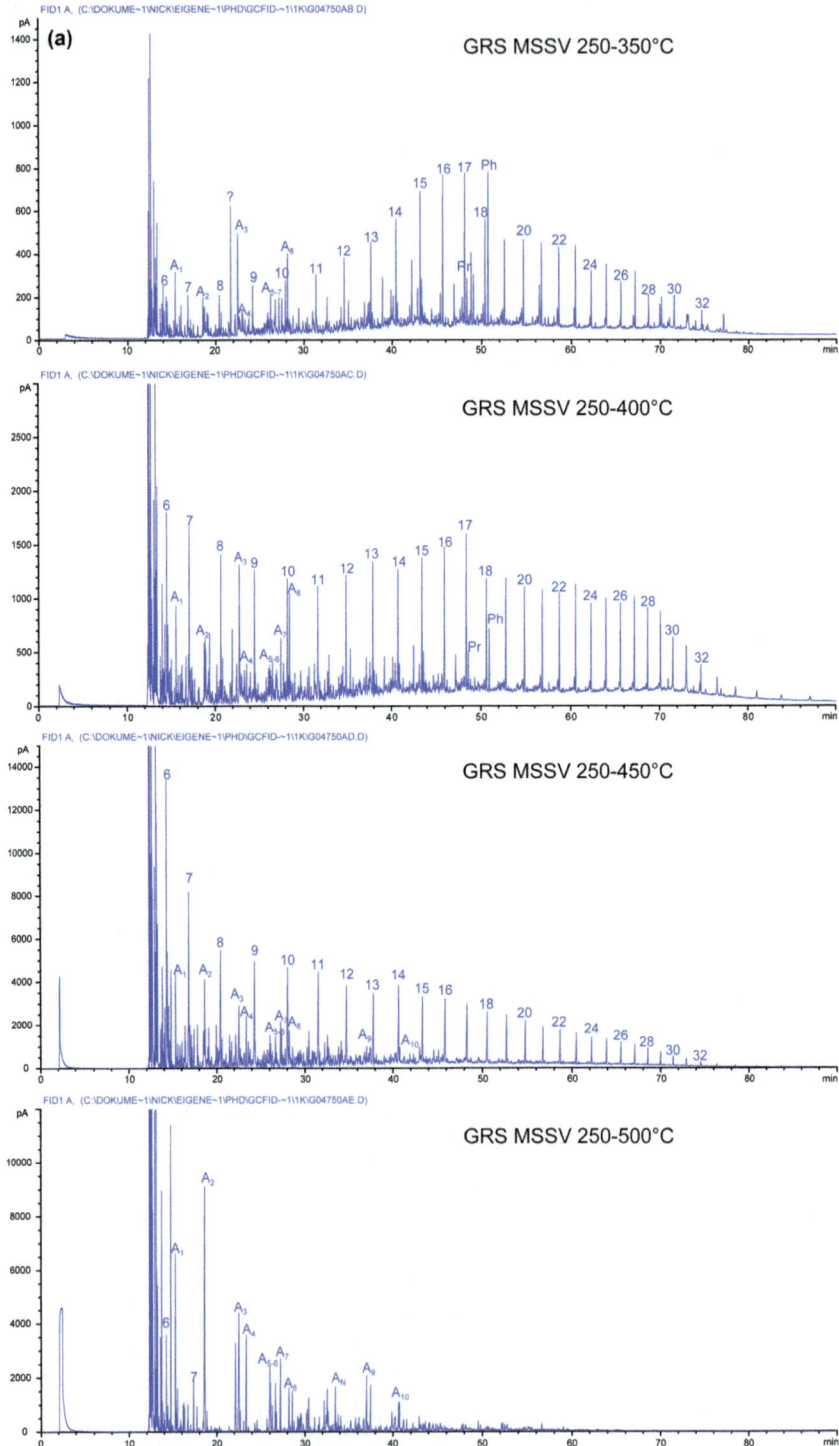

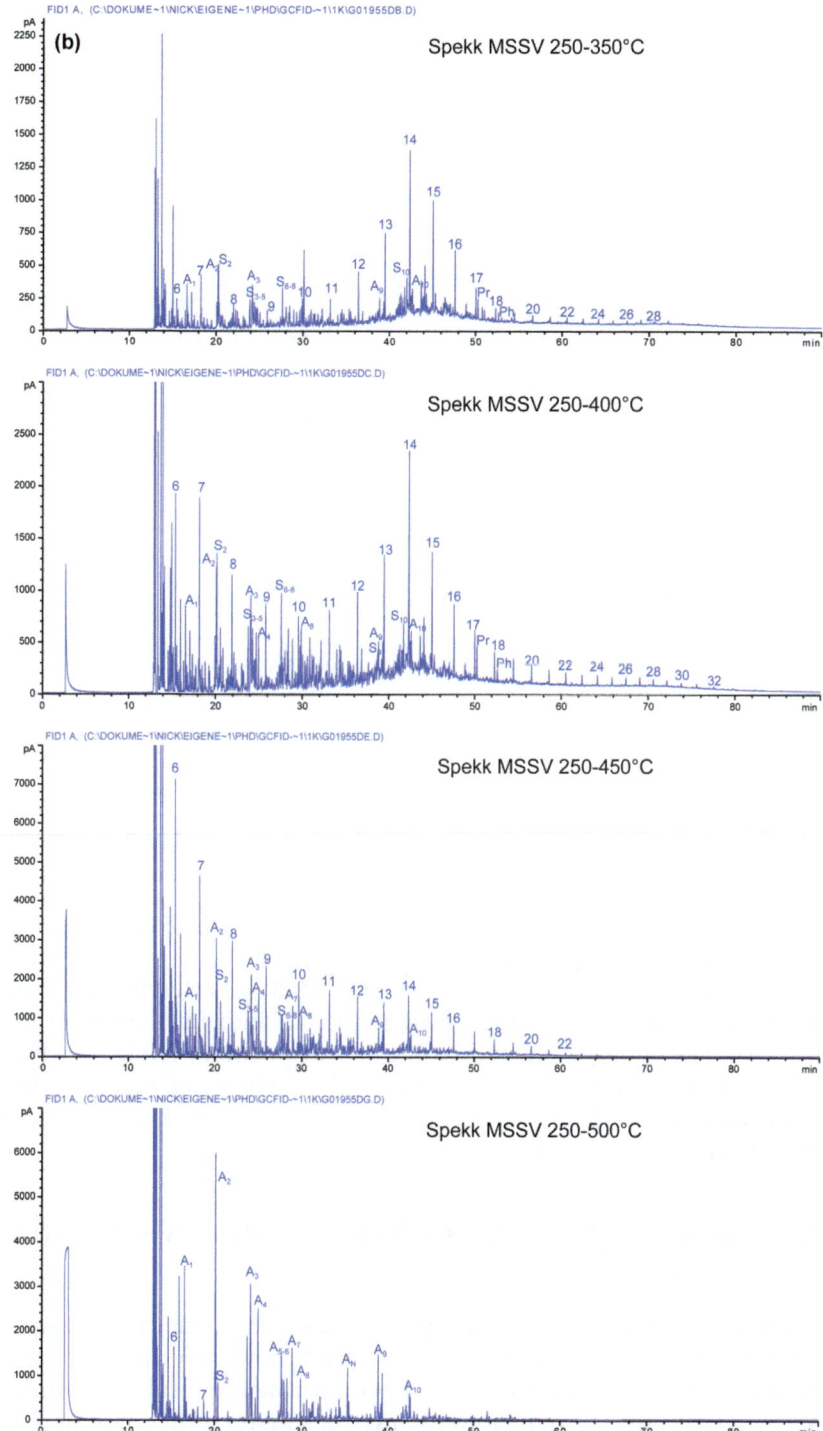

Figure 7.4   MSSV gas chromatograms: Spekk and Green River Formation over an extended simulated maturity range. Peaks labelled as in Figure 7.3.

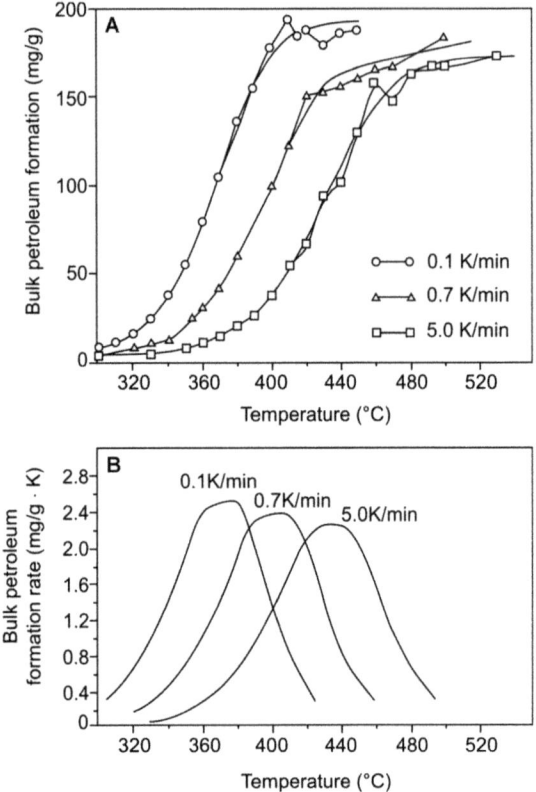

Figure 7.5 Cumulative gas yields with spline function, and the rate curve produced by differentiation of the spline function.

coal (type I kerogen), errors were as follows: C_1–C_5, $\pm 2.0\%$; C_6–C_{14}, $\pm 4.8\%$; C_{15}–C_{19}, $\pm 2.7\%$; C_{20}–C_{24}, $\pm 3.2\%$; C_{25}–C_{32}, $\pm 6.1\%$.

The high precision of the method using the MSSV-2 is clearly evident from the curves which, though derived from numerous individual analyses, have the appearance of open-system dynamic data acquisition. While the MSSV-2 system provides excellent reproducibility, it should be noted that the MSSV-1 system was devised mainly as a qualitative tool, and any kinetic data generated by that device are often of questionable accuracy.

7.6 Simulating Nature by MSSV Pyrolysis—a Reality Check

The MSSV technique has a high analytical performance, but the key question is whether the reactions taking place in the tubes actually approximate nature. We have addressed the question by comparing the results of simulated maturation with those from natural source rock maturation series, and with the bulk composition and phase characteristics of reservoired petroleum.

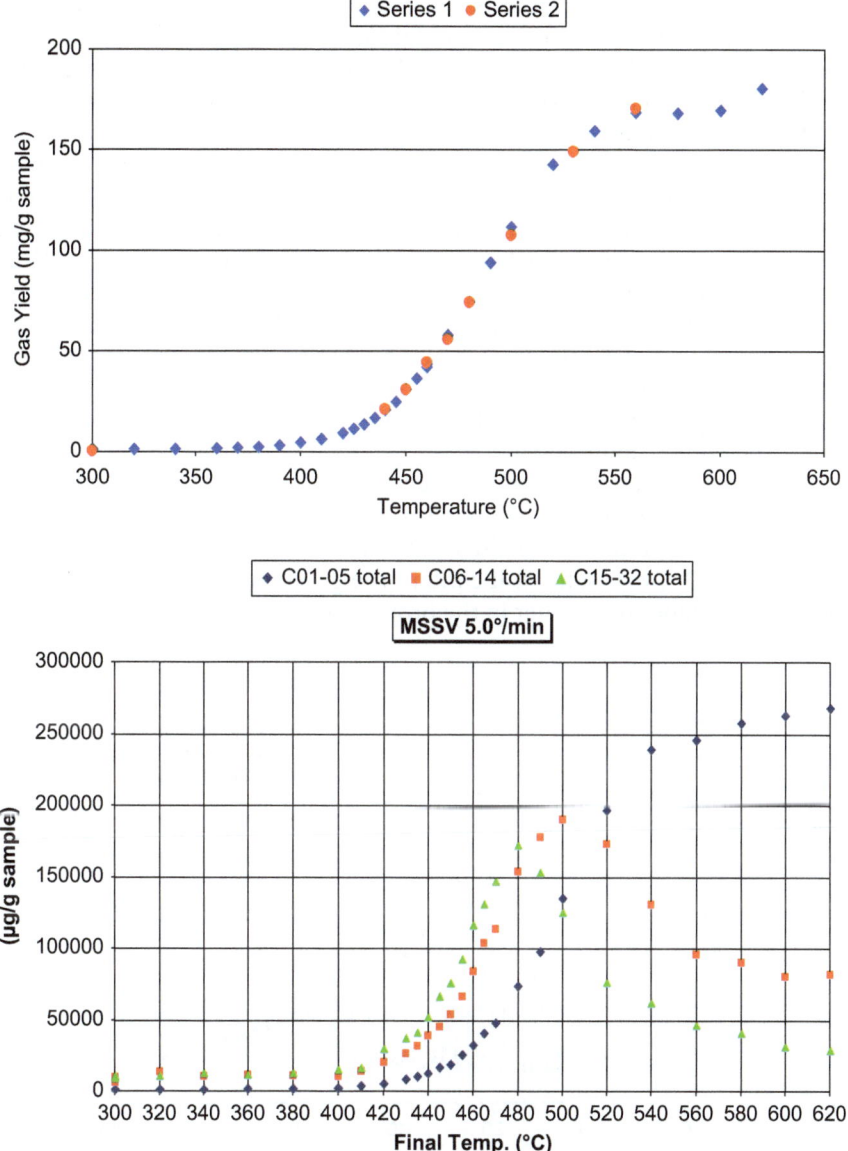

Figure 7.6 Data quality assessment: Nordegg kerogen, Canada. The blue and red symbols refer to two series run 2 months apart (5 K min^{-1}). In the top box, gas yields are defined as C_1–C_5 mg g^{-1} sample. In the bottom box, blue, red and green symbols refer to C_1–C_5, C_6–C_{14}, and C_{15+} boiling ranges in μg g^{-1} sample

7.6.1 Absence of Alkenes

The individual peaks, representing aliphatic and alkylaromatic compounds, show a more or less general decrease in abundance with increasing retention time and carbon number, especially for higher-temperature pyrolysates. Linear n-alkyl moieties in macromolecular structures, which crack to give n-alkene and n-alkane doublets during open-system pyrolysis, generate n-alkanes under closed-system conditions. This is because the free radicals on first-formed volatile products readily abstract hydrogen from the kerogen structure so that n-alkenes are not a major product of the closed-system pyrolysis of kerogens. Contributions from alkenes are generally restricted to the low C_2–C_4 range, for example very minor concentrations of but-1-ene, *cis*-but-2-ene, and *trans*-but-2-ene were reported by Horsfield *et al.*[44] and shown again in Figure 7.7. It should nevertheless be noted that the abundance of alkenes for a given kerogen is enhanced if low-temperature experiments (*e.g.* <300 °C) are done using sample weights that are at the low end of the recommended range (*e.g.* 1 mg). Similarly, whole rock samples whose pyrolytic yield is low (S2 < 50 mgHC g^{-1} rock) may generate and preserve alkenes upon MSSV.

The chromatograms shown in Figures 7.3, 7.4, and 7.7 can be considered to be 'simulated whole petroleum chromatograms'. Unlike natural whole oil samples, collected at the oilfield separator (*e.g.* whole oil chromatograms in Illich[99]), MSSV pyrolysates still contain the gaseous (<C_5) as well as higher-boiling components together in the proportions in which they are generated.

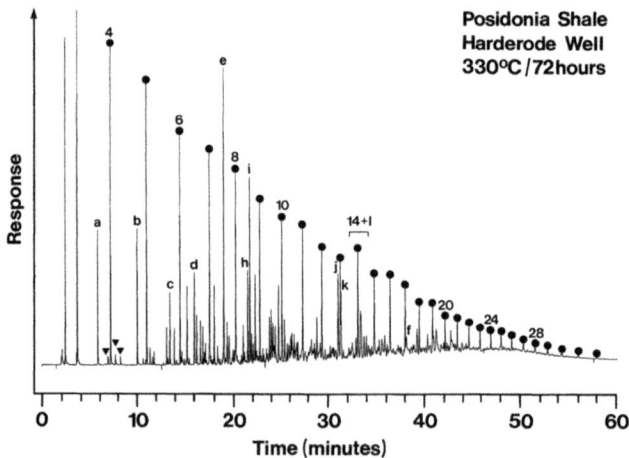

Figure 7.7 Whole petroleum chromatogram displaying enhanced resolution in the low-molecular-weight range of an MSSV pyrolysate (330 °C/3 days) of Posidonia Shale kerogen,[44] illustrating the low concentration of alkenes. The n-alkane peaks have been highlighted and selectively numbered. 2-methylbutane (a), 2-methylpentane (b), 2-methylhexane (c), benzene (d), toluene (e), ethylbenzene (h), meta- and para-xylenes (i), 2-methylnaphthalne (j), 1-methylnaphthalene (k), dimethylnaphthalenes (l) and pristine (f) are marked. The peaks marked with triangles are, in order of increasing retention time, *cis*-but-2-ene and *trans*-but-2-ene.

7.6.2 Bulk Kinetic Parameters—Open *versus* Closed Pyrolysis

Schenk and Horsfield[95] were the first to make a direct comparison of open *versus* closed (<5 MPa pressure) system pyrolysis, independent of extraneous variables (*e.g.* calibration standard, pneumatic design). The programmed heating of Posidonia Shale showed that the pre-exponential factors and activation energy distributions for open and closed (MSSV) pyrolysis were similar (Figure 7.8). The result pointed to pyrolysis scission reactions having the same heating rate-dependence under both laboratory and geological heating conditions.[59] The validity of the cracking model to the natural system was verified a short time later using a maturation series from the Hils Syncline (Germany).[72] Other investigations of compositional kinetics using MSSV have followed.[19,64–67,69]

Interestingly, total cumulative MSSV pyrolysate yields are appreciably lower in the closed system. The 40% lower yield for the closed-system experiment is noteworthy, and points to enhanced tar and coke formation taking place under the confined system, and equally for each potential. The results signal that petroleum yield predictions are dependent on system configuration. The fundamental question as to how best to assess total petroleum potential remains unanswered to this day; does open-system pyrolysis (*e.g.* Hydrogen Index or C_R/C_T)[100,101] overestimate petroleum yields in sedimentary basins?

7.6.3 Compositional Mass Balance of Expelled Petroleum

Mass balancing emphasises the utility of MSSV predictions. Santamaria and Horsfield[97] presented a compositional mass-balance model for petroleum formation based on pyrolysis-GC and Rock-Eval data for a source rock maturation series from the Western Canada Basin (upper Devonian Duvernay Formation, type II) and the Sonda de Campeche (offshore Mexico Tithonian source rock sequence, type IIS), respectively. Briefly, the model had the following sequential elements: calculation of TR from Rock-Eval data using the algebraic scheme of Pelet;[102] gathering of quantitative Py-GC data as being representative of the macromolecular structure *in toto*;[39] normalisation of Py-GC data to original total organic content; calculation of component yields (boiling ranges, compound classes, and individual compounds) by subtracting normalised yields from that of the least mature sample in the series; presentation of component yields as a function of TR. Compound class and individual component yields *versus* TR calculated from the natural maturation series were very similar to those measured directly from MSSV of the least mature sample in the series, up to TR = 80%.

Some of the results from the compositional mass balance for the Duverney Formation are shown here, and results compared with the predictions from MSSV of the least mature member of the maturity series. The correlation for resolved products is generally excellent, even down to individual compounds. The match for C_1–C_5 and C_6–C_{14} total products (including small

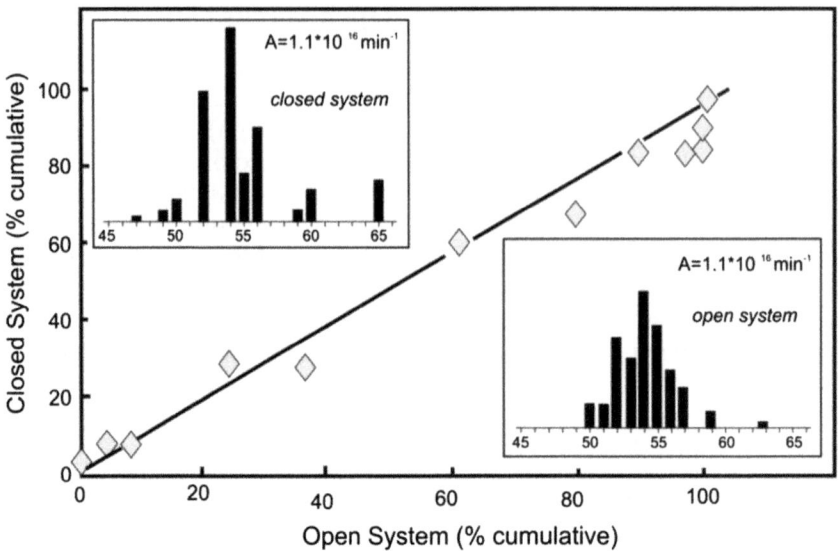

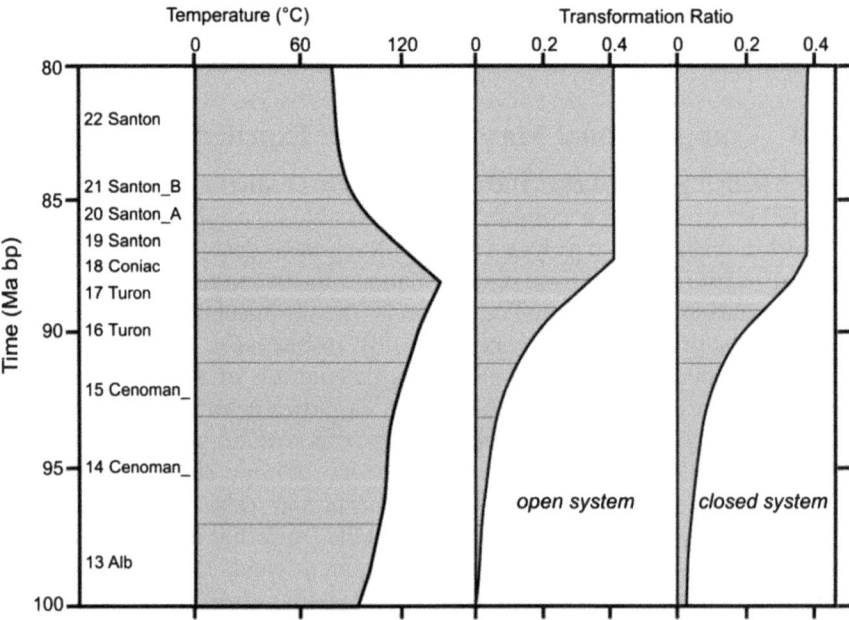

Figure 7.8 Kinetic parameters of bulk petroleum generation calculated from open and closed-system pyrolysis data from Posidonia Shale.[62]

hump) are shown in Figure 7.9. The only mismatch, in the case of both the Tithonian and Duverney examples, was for C_{15+} total products, which was strongly influenced by an unresolved Py-GC hump. That the problem lies

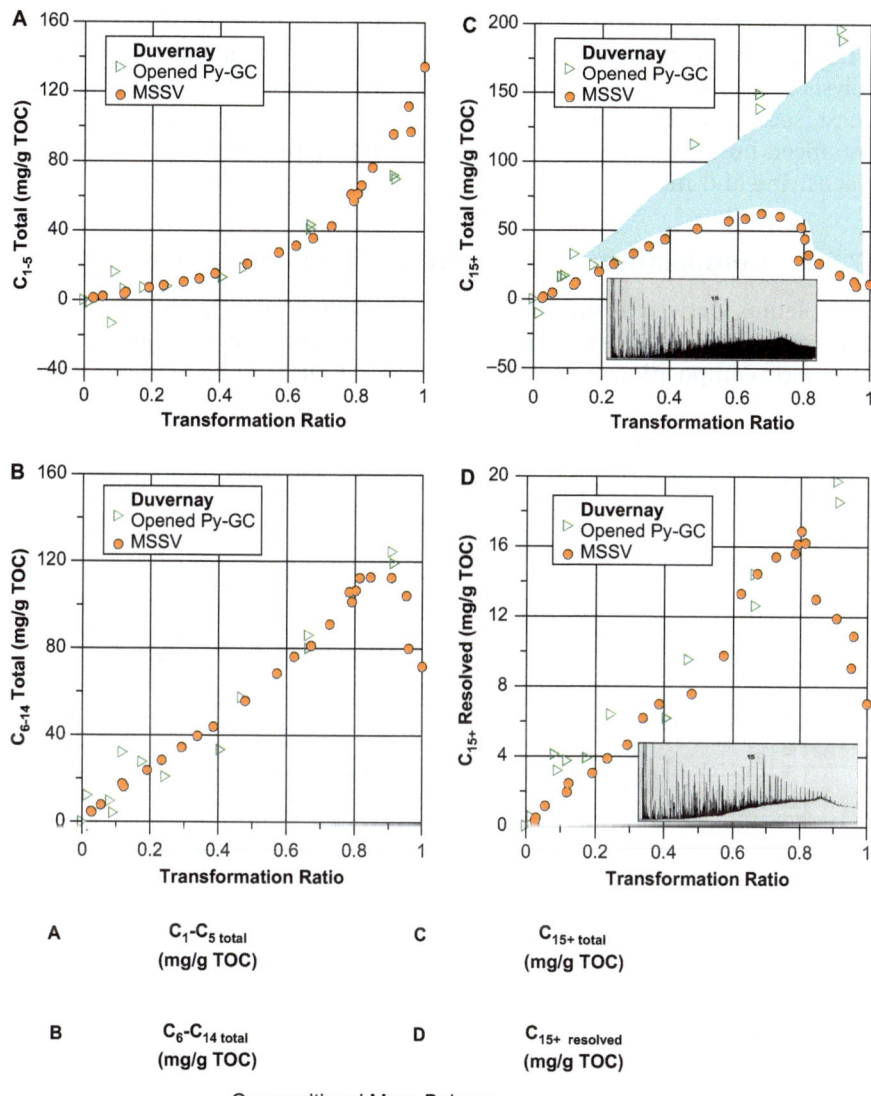

Figure 7.9 Compositional mass balance, Duvernay Formation, Canada.

with the large C_{15+} hump is clearly demonstrated by the excellent fit between C_{15+} resolved components from the mass balance as compared to MSSV. Thus, gas–oil ratio (GOR, $(C_1–C_5)/C_{6+totals}$) of MSSV products from a given TR (simulation) are higher than the Py-GC products produced by rapidly heating a sample of the same TR (natural) under open-system conditions. GOR calculated from the resolved components $((C_1–C_5)/C_{6+resolved})$ of MSSV products are identical.

A detailed discussion on whether the hump might consist of oligomeric fragments derived from kerogen breakdown or aggregates of reactive pyrolysis monomers is given by Horsfield.[50] The important point made in the next section is that the GOR predictions from MSSV (without the pronounced hump) match those observed in liquid petroleum provinces, both lacustrine and marine.

7.6.4 Composition of Reservoired Petroleum, Including GOR

Petroleums derived from marine source rocks containing type II organic matter have a low wax content and are 'mixed base' (paraffinic–naphthenic–aromatic composition of Tissot and Welte.[18] GOR is linked strongly to level

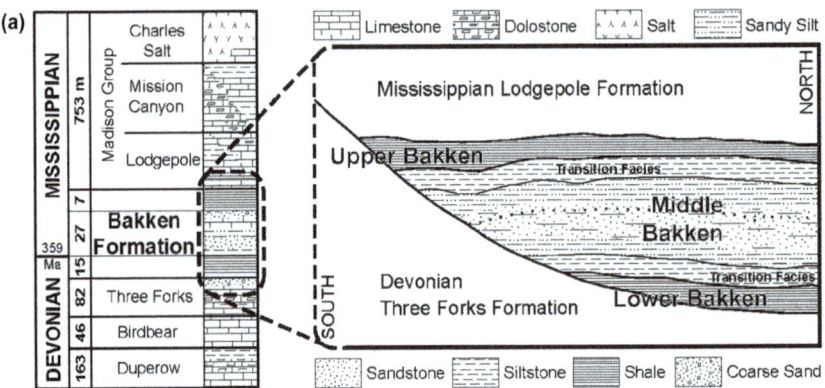

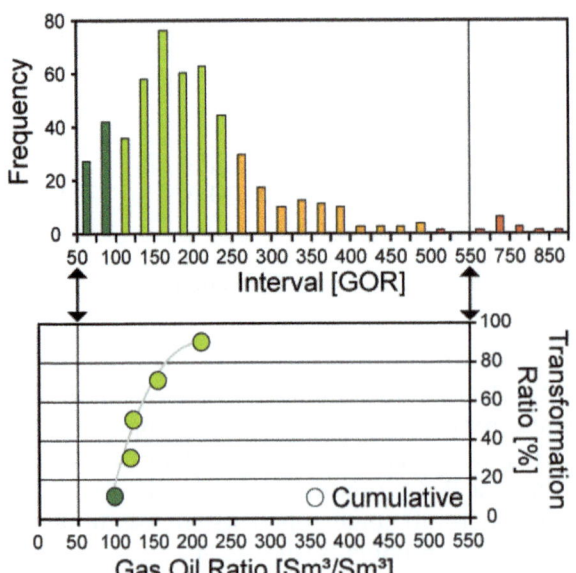

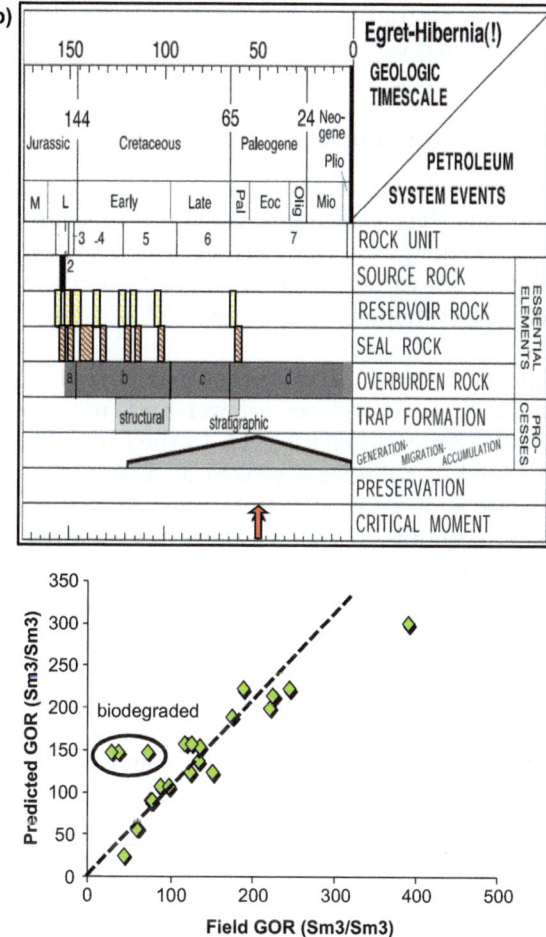

Figure 7.10 Gas–oil ratio predictions compared with field data for (a) the Jeanne d'Arc Basin, Canada, and (b) the Williston Basin, USA.

of maturity, with values increasing as a function of thermal stress,[97,103,104] though phase behaviour plays an overriding role. In high-pressure–high-temperature (HPHT) reservoirs of the North Sea, which can be considered closed systems, black to light oil GOR distributions in the North Sea Viking Graben closely matched the predictions of MSSV pyrolysis experiments performed on the Draupne Formation source rock.[105,cf. 106] di Primio and Neumann,[107] in a similar study of the Jade and Judy Fields in the Central Graben (North Sea) that included pressure prediction, reported that GOR predictions from MSSV pyrolysis bore a close resemblance to the natural HPHT system. Other examples of excellent GOR predictive capability are provided by modelling of the Egret Shale and its generated petroleum of the Jeanne d'Arc Basin, Canada,[108] and the Bakken Shale and its petroleums of the Williston Basin, USA,[109,110] as shown in Figure 7.10. The selective loss of light ends during biodegradation[111] accounts for the discrepancy between predicted and measured values in Figure 7.10b.

7.6.5 Fluid (Gas and Liquid) Compositions

Whereas MSSV pyrolysis can accurately reconstruct hydrocarbon GOR, it is inherently incapable of correctly reproducing the gas composition of natural fluids, generating wetter gases. High gas wetness is common to all pyrolysis methods—hydrous, anhydrous, open or closed systems—and isothermal or non-isothermal experimental conditions,[19,112–115] and independent of source rock type.

di Primio and Horsfield[116] showed that it is actually the higher ethane content in pyrolysis gases, and not a lower methane content, that makes them wetter, as illustrated in Figure 7.11. In the same figure it is clear that catalytically induced distributions, evoked as a potential cause of the discrepancy,[117–120] are a long way away from gas compositions associated with natural petroleum, ruling that process out as an explanation for compositional discrepancies between natural petroleum and pyrolysates.

Because gas composition dominantly controls the phase behaviour of petroleum,[121] the direct use of pyrolysis-derived compositional information, at least for the gas range compounds, would be highly misleading. The PhaseKinetics compositional kinetic model (Figure 7.12) therefore uses corrected gas compositions from MSSV pyrolysates, as well as pseudo-boiling ranges in the C_{6+} range, to populate bulk activation energy potentials. Five MSSV experiments are performed using a heating rate of 0.7 °C min^{-1}

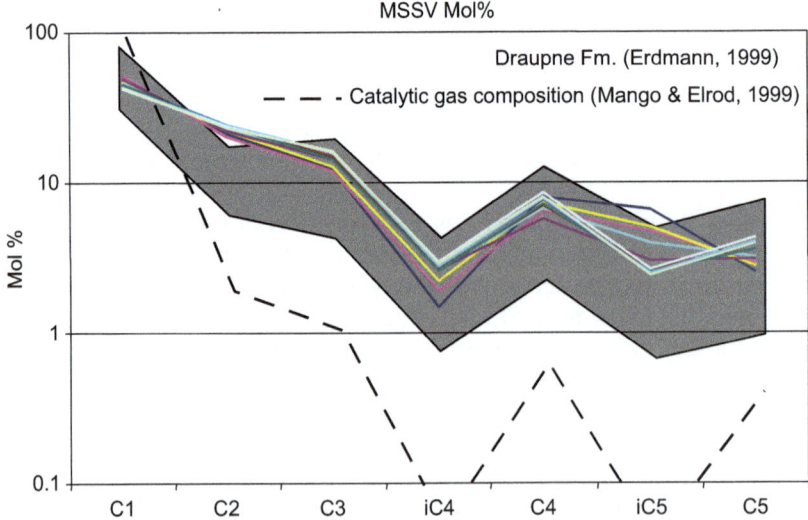

Figure 7.11 Comparison of light hydrocarbon distributions for natural petroleums (shaded area), MSSV pyrolysates, and the compositions reflecting catalysis.[117]

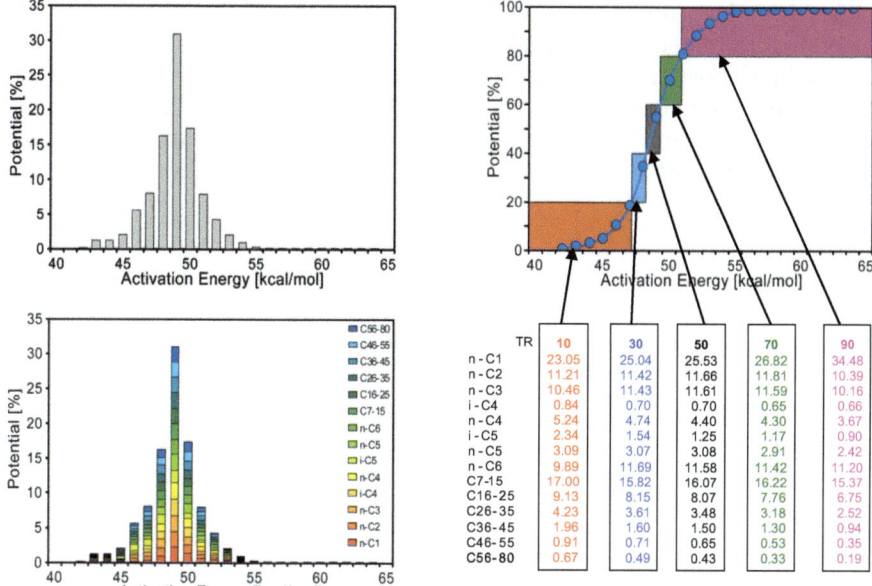

Figure 7.12 Procedural steps used in the building of PhaseKinetics models.[116] The bulk kinetic model (top left) is populated (bottom left) using compositions from MSSV (0.7 K min^{-1}) to temperatures corresponding to selected levels of conversion (TR = fraction of conversion as normalised to maximum MSSV product yield).

up to temperatures representing 10, 30, 50, 70, and 90% transformation to do this. Output from the model are molar proportions of components in the gas and liquid phases, as well as phase behaviour and fluid properties (saturation pressure, formation volume factor, API gravity).

Examples from the Norwegian North Sea, Brazil, and Mexico have demonstrated the close correspondence of the tuned compositional predictions with field data.[116]

7.6.6 A Word of Caution—Aromatic Moieties

As far as aliphatic components are concerned the compositions predicted from MSSV are representative of the expelled fluid phase in nature. This is not the case if the C_{6+} fraction consists largely of aromatic hydrocarbons, cracked from terrigenous kerogen of lignocellulosic origin. These aromatic compounds, while labile during laboratory pyrolysis, are likely in nature to be incorporated into aromatic structures *via* condensation and aromatisation reactions.[72] The point is illustrated here using coaly source rocks from Australia and condensates purportedly generated from them during natural maturation (Figure 7.13). MSSV pyrolysates are aromatic at all stages of simulated maturation, though aliphatic and alicyclic components are also readily discernable. The pseudocomponent distribution in the

PhaseKinetics model (boiling ranges, as defined by di Primio and Horsfield[116]) is strongly influenced by mono- and diaromatic components extending above an otherwise smooth envelope of decreasing abundance with increasing elution time. Yields of trimethylbenzenes (eluted before and after n-C_{10}), tetramethylbenzenes (before n-C_{12}), trimethylnaphthalenes, and tetramethylnaphthalenes $(C_{15} - C_{17})$ are exceptionally high. The

(a)

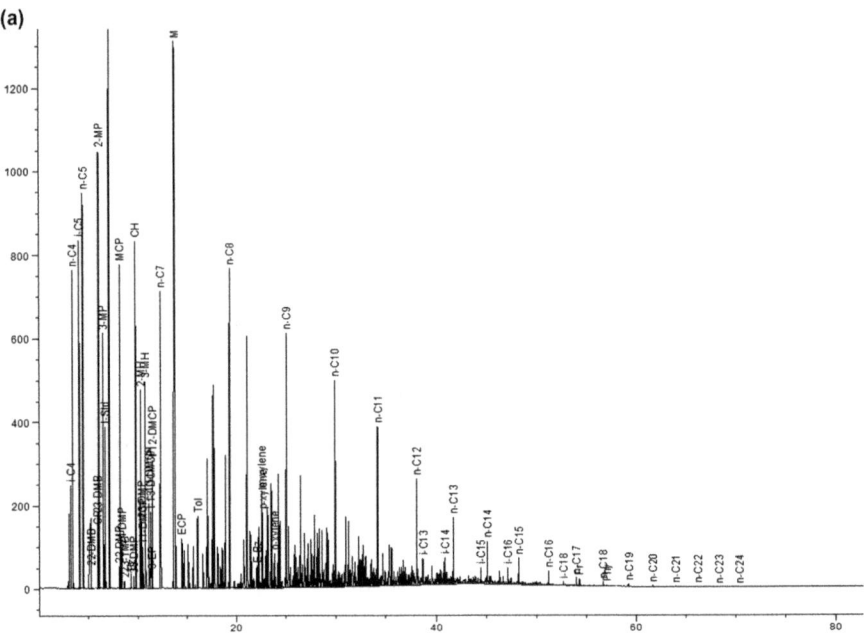

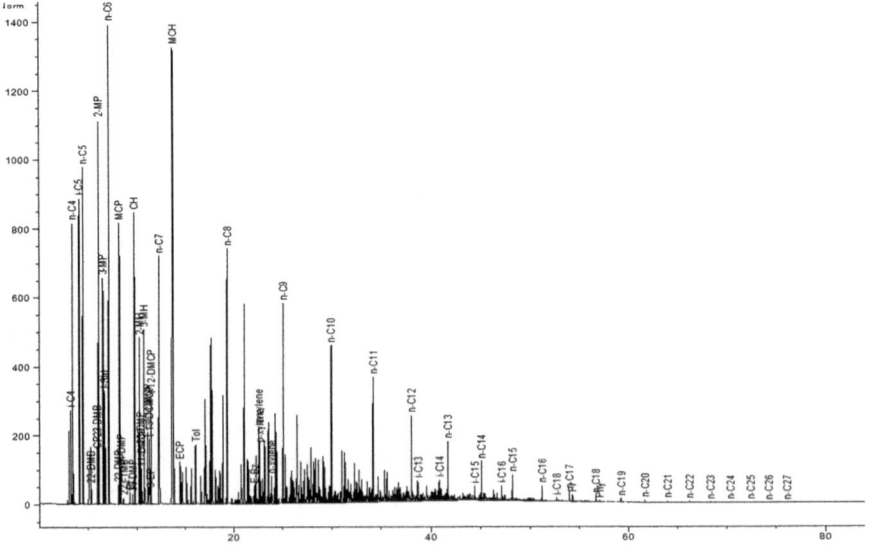

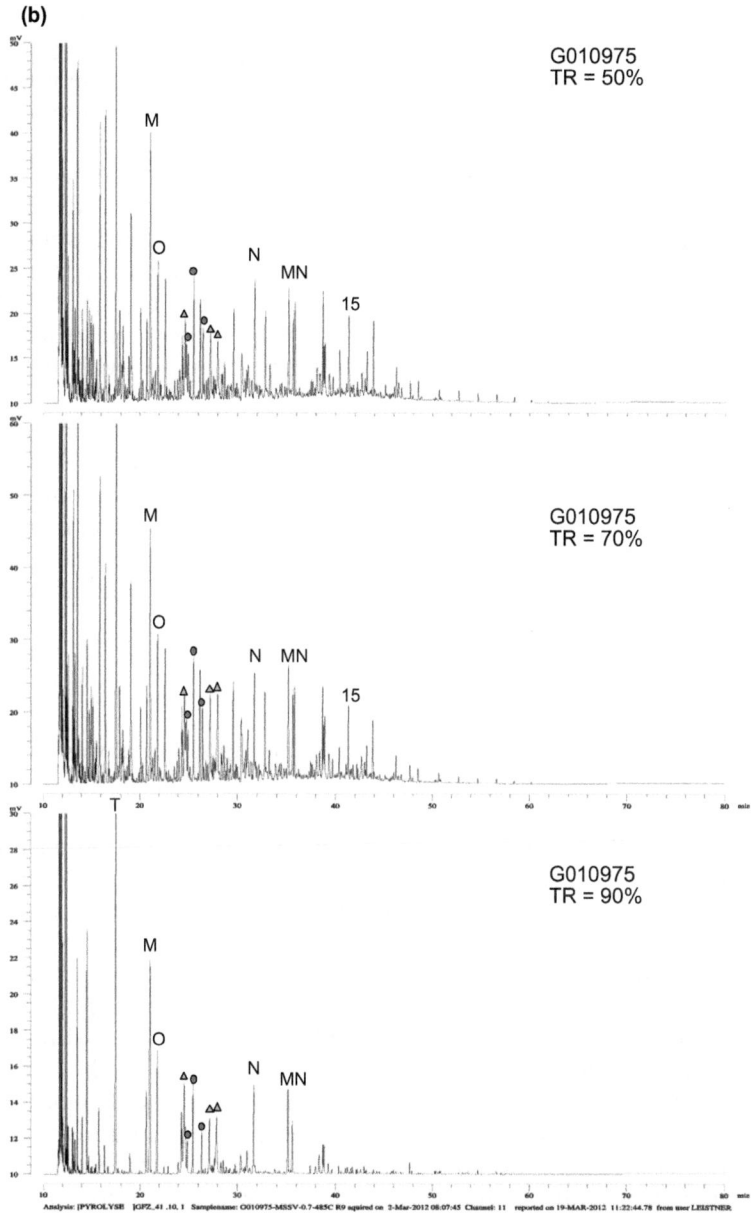

Figure 7.13   (a) Whole oil gas chromatogram of the condensates. Selected normal alkanes are marked according to carbon number. 2-Mp, 2-methylpentane; MCH, methylcyclohexane; ECP, ethylcyclopentane; I-C13, I-C14, I-C15, I-C16, acyclic isoprenoid alkanes; PR, pristine; PH, phytane. (b) MSSV pyrolysates of Australian coals, the purported source of the condensates. Numbers refer to chain length of n-alkanes. T, toluene; M, *meta*-plus *para*-xylenes; O, *ortho*-xylene; N, naphthalene; MN, methylnaphthalenes; red circles, trimethylbenzenes, green triangles, phenol + cresols.

condensates, on the other hand, are predominantly aliphatic/alicyclic with high abundances of normal alkanes and alkylcyclohexanes (Figure 7.13). Their pseudo-compound groups, calculated in exactly the same way as for MSSV pyrolysates, decrease smoothly in abundance with retention time; the aromatic peaks seen for the pyrolysates are absent.

An engineering approach for eliminating what can be considered cracking artefacts is to simply smooth the MSSV data, as depicted by the red line in Figure 7.14. The smoothing procedure results in a C_{7+} compositional distribution that resembles those of the condensates, though it is not exactly the same. The smoothed MSSV data does not match exactly that of the condensates, but it follows the natural fluid composition rule, whereby the molar percentage decreases while the density increases with increasing carbon number.

The PhaseKinetics model built using the skimmed MSSV data predicts GOR in the range 784–3019 Sm^3/Sm^3 at simulated maturity levels from TR 10% to 70%. At the highest simulated maturity level, no saturation point is found. Simply stated, the prediction is for gas and condensate, followed by gas with increasing maturity. While this prediction is essentially correct, actual predicted GOR values are too low, possibly because adsorption of C_{6+} components in nature[122] is significant.

7.7   Gas Generation

The thermal alteration of reservoired petroleum upon burial was simulated by Horsfield et al.[64] using MSSV of a medium-gravity oil from the Norwegian North Sea Central Graben. Kinetic modelling of the oil to gas conversion resulted in a narrow gas potential versus activation energy distribution (66–70 kcal mol^{-1}) and a pre-exponential factor of 1.1×10^{16} s^{-1}). By extrapolation to natural maturation conditions the onset of gas generation was predicted to occur between around 190 °C ($\sim 2\% R_m$) for geological heating rates between 0.53 and 5.3 K Ma^{-1}. Crucially, the predictions from the model were in accordance with the observed preservation of liquid hydrocarbons in a deep, hot (165 °C) closed system (sand body within Upper Jurassic) petroleum reservoir from the Saga 2/4–14 well, Norwegian Continental Shelf. Subsequently, Schenk et al.[65] used the same approach for four classic crude oil types and demonstrated that all had a high inherent stability. Oils with high wax contents were slightly more stable, the onset of cracking occurring 10 °C higher than for low-wax oils under geological heating rates. Conducting the same types of experiments using source rocks, Dieckmann[123] and Dieckmann et al.[19] deconvoluted primary from secondary gas reactions and found that the secondary cracking of oil into gas (isolated kerogens from the Duvernay and Posidonia Shales, respectively) occurred at 150 °C (1.2% R_m) under geological conditions. Combining these finding from in-source and in-reservoir simulations, it could be inferred that reservoired crude oils are likely to be flushed by overmature gas coming from their source rock prior to the reservoired oil ever becoming thermally

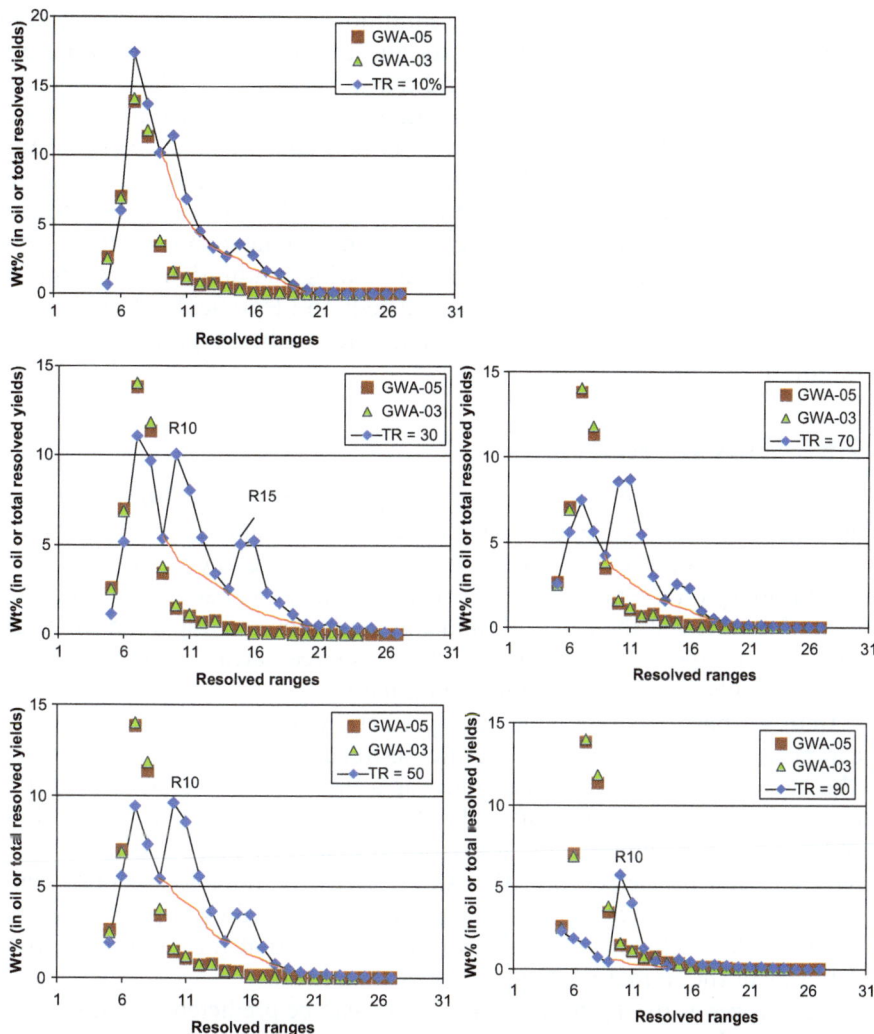

Figure 7.14 Smoothed pseudocomponent boiling range distributions (see Michels *et al.*[116] for definitions) aimed at reducing aromatic hydrocarbon concentrations so as to be similar to the condensate.

unstable, and that this controls the oil floor in open source-reservoir systems. As far as gas shales are concerned (closed system, unconventional exploration targets), the kinetic calculations of Hill *et al.*,[124] based on sealed gold-tube pyrolysis experiments, closely resemble those of Dieckmann *et al.*[19] and Schenk *et al.*[65]

Asphaltenes extracted from crude oils are proposed to possess structural features of the related source rock kerogen. Lehne *et al.*[125] used MSSV and combustion isotope ratio mass spectrometry (GC-C-IRMS) to show that secondary gas formation from whole rock covers a much broader temperature range under geological conditions than that from the asphaltene products, but that both the onset and the maximum temperature are nearly identical under geological conditions.

MSSV experiments on immature Heather (Norway) and Taglu Formations (Canada) inferred an additional late gas potential in mixed marine and terrestrial sources,[67,68] signalling the generation of gas above $R_m = 2\%$, according to kinetic modelling. Originally proposed to be generated from a high-molecular-weight bitumen precursor, the gas now appears to be generated from methyl groups concentrated during maturation, according to the MSSV experiments reported by Mahlstedt and Horsfield.[69] Importantly, natural maturation series demonstrated the predicted growth and realisation of potential.

7.8   Closing Remarks

Petroleum consists of an exceedingly complex mixture of hydrocarbons and non-hydrocarbons, ranging from methane to macromolecular aggregates. The relative proportions of these components are highly variable and depend initially on the nature of the kerogen in the parent source rock and its level of maturity at the time of expulsion, and subsequently upon the pressure and temperature conditions of the source–carrier–reservoir system during expulsion, migration and accumulation.

Here we have demonstrated that predictions of first-formed petroleum compositions from MSSV, for prolific petroleum source rocks, are correct. The key in coming to this assertion was a direct comparison with natural petroleum systems. Thus, GOR and bulk fluid compositions, as well as fluid stabilities and generation characteristics, for lacustrine and marine systems have all been predicted correctly using MSSV. The notable exception concerns petroleum correlated with organic-rich but hydrogen-poor (type III) source rocks such as coals; here we have shown that the prediction from pyrolysis of the immature sample is less than ideal.

The answer as to why some compositions can be predicted and others not lies in the fact that MSSV takes place under confined, low-pressure conditions. It has strong similarities to open-system analytical pyrolysis in that the products are dispersed in a relatively large dead volume. We noted earlier in the chapter that major resolved species in high-temperature pyrolysates can actually give compositional information on the kerogen as a whole rather than on atypical part-structures.[38–40] In other words, Py-GC provides representative insights into the structure of organic macromolecules. In a similar way, MSSV strips away representative structural moieties in the order of thermal lability or bond strength. On the premise that kerogen composition directly controls the types and yields of volatile products generated by cracking reactions during natural maturation, the abundances and distributions of pyrolysis products resemble the bulk compositions of natural petroleum. The condensation and aromatisation reactions taking place during coalification are simply not simulated properly by the MSSV technique, leading to compositional discrepancies. Importantly, using structural moieties in MSSV protocols does not require a description of products according to polarity,[43] and is extremely rapid and cost effective to perform on suites of samples, rather than single sample.

From the standpoint of instrumentation, the MSSV interface is now established as a flexible means for conducting thermovaporisation, open-system pyrolysis, and MSSV analyses, as well as the mechanical crushing of fluid inclusion hosted samples, with the same online configuration. We are currently exploring its use for microscale derivatisation and selective chemical degradation[126] coupled with the online analysis of C_{1+} products; an exciting new field in the making.

References

1. I. Ericsson and R. P. Lattimer, Pyrolysis nomenclature, *J. Anal. Appl. Pyrol.*, 1988, **14**, 19–221.
2. S. D. Killops and V. J. Killops, *An Introduction to Organic Geochemistry*, John Wiley & Sons, New York, 1993, pp. 217–238.
3. A. Nissenbaum and I. R. Kaplan, Chemical and isotopic evidence for the *in situ* origin of marine humic substances, *Limnol. Oceangr.*, 1972, **17**, 570–582.
4. F. J. Stevenson, Non biological transformations of amino acids in soils and sediments, in: *Advances in Organic Geochemistry*, ed. B. Tissot and F. Bienner, Editions Technip., Paris, 1973, pp. 701–714.
5. R. P. Philp and M. Calvin, Possible origin for insoluble organic (kerogen) debris in sediments from soluble cell-wall materials of algae and bacteria, *Nature*, 1976, **262**, 134–136.
6. E. Stach, M.-T. Mackowsky, M. Teichmüller, R. Teichmüller, G. H. Taylor, D. Chandra, D. G. Murchinson and F. Zierke (ed.) *Stach's Textbook of Coal Petrology*, Gebrüder Bornträger, Berlin, 1982.
7. C. Largeau, E. Casadevall, A. Kadouri and P. Metzger, Formation of Botryococcus-derived kerogens-comparative study of immature torbanites and of the extant alga *Botryococcus braunii*, *Org. Geochem.*, 1984, **6**, 327–332.
8. E. W. Tegelaar, J. W. de Leeuw, S. Derenne and C. Largau, A reappraisal of kerogen formation, *Geochim. Cosmochim. Acta*, 1989a, **53**(11), 3103–3106.
9. B. Horsfield and J. Rullkötter, Diagenesis, catagenesis and metagenesis, in *The Petroleum System from Source to Trap*, ed. L. Magoon and W. G. Dow, AAPG Memoir **60**, 1994, pp. 189–199.
10. B. A. Knights, A. C. Brown, E. Conway and B. S. Middleditch, Hydrocarbons from the green form of the freshwater alga *Botryococcus braunii*, *Phytochemistry*, 1970, **9**, 1317–1324.
11. R. F. Cane and P. R. Albion, The organic geochemistry of torbanite precursors, *Geochim. Cosmochim. Acta*, 1973, **37**, 1543–1549.

12. J. D. Saxby, Thermogravimetric analysis of oil shales, *Thermochim. Acta*, 1981, **47**, 121–123.

13. J. S. Sinninghe Damsté and J. W. de Leeuw, Analysis, structure and geochemical significance of organically-bound sulphur in the geosphere: state of the art and future research, *Org. Geochem.*, 1990, **16**, 1077–1101.

14. P. C. Adam, B. Schmid, B. Mycke, C. Strazielle, J. Connan, A. Huc, A. Riva and P. Albrecht, Structural investigation of non-polar sulfur cross-linked macromolecules in petroleum, *Geochim. Cosmochim. Acta*, 1993, **57**, 3395–3419.

15. S. R. Larter, H. Solli and A. G. Douglas, Phytol containing melanoidins and their bearing on the fate of isoprenoid structures in sediments, in *Advances in Organic Geochemistry 1981*, ed. M. Bjoroy *et al.*, John Wiley, Chichester, 1983, pp. 513–523.

16. B. Mycke and W. Michaelis, Lingnin-derived molecular fossils from geological materials, *Naturwissenschaften*, 1986, **73**, 731–734.

17. J. W. de Leeuw and C. Largeau, A review of macromolecular organic compounds that comprise living organisms and their role in kerogen, coal, and petroleum formation, in *Organic Geochemistry*, ed. M. H. Engel and S. A. Macko, Plenum Press, New York, 1993, pp. 23–72.

18. B. P. Tissot and D. H. Welte, *Petroleum Formation and Occurrence*, Springer-Verlag, New York, 1984, 699 p.

19. V. Dieckmann, B. Horsfield, H. J. Schenk and D. H. Welte, Kinetics of petroleum generation and cracking by programmed-temperature closed-system pyrolysis of Posidonia Shale, *Fuel*, 1998, **77**, 23–31.

20. N. Mahlstedt, *Evaluating the late gas potential of source rocks stemming from different sedimentary environments*, PhD thesis, Technical University of Berlin, 2012.

21. G. W. Girling, Evolution of volatile hydrocarbons from coal, *J. Appl. Chem.*, 1963, **13**, 77–91.

22. R. W. L. Kimber and P. L. Searle, Pyrolysis gas chromatography of soil organic matter. 1. Introduction and methodology, *Geoderma*, 1970, **4**(1), 47–55.

23. W. M. Scott, V. E. Modzelezki and B. Nagy, Pyrolysis of early Precambrian Onverwacht organic matter ($>3 \times 10(9)$ yr old), *Nature*, 1970, **225**(5238), 1129–1130.

24. G. Dungworth, and A. W. Schwartz, Kerogen isolates from the Precambrian of South Africa and Australia, in *Advances in Organic Geochemistry 1971*, ed. H. R. V. Gaertner and H. Wehner, Pergamon Press, Oxford, 1972, pp. 699–706.

25. J. Leventhal, S. E. Suess and P. Cloud, Non-prevalence of biochemical fossils in kerogen from pre-Phanerozoic sediments, *Proc. Natl. Acad. Sci. U. S. A.*, 1975, **72**, 4706–4710.

26. A. G. Douglas, R. C. Coates, B. F. J. Bowler, and K. Hall, Alkanes from pyrolysis of recent sediments, in *Advances in Organic Geochemistry 1975*,

ed. J. A. Gomez-Angulo and R. Campos, Enadimsa, Madrid, 1977, pp. 357–374.

27. A. Giraud, Application of pyrolysis and gas chromatography to the geochemical characterization of kerogen in sedimentary rocks, *Bull. AAPG*, 1970, **54**, 439–455.

28. J. Connan, Laboratory simulation and natural diagenesis, 1. Thermal evolution of asphalts from the Aquitaine basin (SW France), *Bull. Cent. Rech. Pau-SNPA*, 1972, **6**, 195–214.

29. J. Connan and B. M. van der Weide, Diagenetic alteration of natural asphalts, in *Oil Sands: Fuel of the Future*, CSPG Memoir **3**, 1974, pp. 134–147.

30. P. Le, Plat, Application of pyrolysis-gas chromatography to the study of the non-volatile petroleum fractions, *J. Gas Chromatogr.*, 1967, **5**, 128–135.

31. D. W. Poxon and R. G. Wright, The characterization of bitumens using pyrolysis gas chromatography, *J. Chromatogr.*, 1971, **61**, 142–144.

32. A. E. George, R. C. Banerjee, G. T. Smiley and H. Sawatzky, Simulated geothermal maturation of Athabasca bitumen, *Bull. Can. Petrol. Geol.*, 1977, **25**, 1085–1096.

33. S. E. Moschopedis, S. Parkash and J. G. Speight, Thermal decomposition of asphaltenes, *Fuel*, 1978, **57**, 431–434.

34. H. L. C. Meuzelaar, J. Haverkamp and F. D. Hileman, *Pyrolysis Mass Spectrometry of Recent and Fossil Biomaterials*, Elsevier, Amsterdam, 1982.

35. R. P. Philp, Application of pyrolysis-gas chromatography and pyrolysis-gas chromatography-mass spectrometry to fossil fuel research, *Trends Anal. Chem.*, 1982, **1**, 237–241.

36. B. Horsfield, Pyrolysis studies and petroleum exploration, in *Advances in Petroleum Geochemistry*, ed. J. Brooks and D. H. Welte, Academic Press, New York, 1984, pp. 247–298.

37. S. R. Larter, Application of analytical pyrolysis techniques to kerogen characterization and fossil fuel exploration/exploitation, in *Analytical Pyrolysis, Methods and Application*, ed. K. Voorhees, Butterworth, London, 1984, pp. 212–275.

38. S. R. Larter and B. Horsfield, Determination of structural components of kerogen using analytical pyrolysis methods, in *Organic Geochemistry*, ed. M. Engel and S. Macko, Plenum Publishing, New York, 1993, pp. 271–287.

39. B. Horsfield, Practical criteria for classifying kerogens: Some observations from pyrolysis-gas chromatography, *Geochim. Cosmochim. Acta*, 1989, **53**, 891–901.

40. T. I. Eglinton, J. S. Sinninghe-Damsté, M. E. L. Kohnen and J. W. De, Leeuw, Rapid estimation of the organic sulfur content of kerogens, coals and asphaltenes by pyrolysis-gas chromatography, *Fuel*, 1990, **69**, 1394–1404.

41. M. D. Lewan, Evaluation of petroleum generation by hydrous pyrolysis experimentation, *Philos. Trans. R. Soc. London, Series A*, 1985, **315**, 123–134.

42. M. Monthioux, P. Landais and J.-C. Monin, Comparison between natural and artificial maturation series of humic coals from the Mahakam delta, Indonesia, *Org. Geochem.*, 1985, **8**, 275–292.

43. F. Béhar, S. Roy and D. Jarvie, Artificial maturation of a Type I kerogen in closed system: mass balance and kinetic modelling, *Org. Geochem.*, 2010, **41**, 1235–1247.

44. B. Horsfield, U. Disko and F. Leistner, The micro-scale simulation of maturation: outline of a new technique and its potential applications, *Geol. Rundsch.*, 1989, 78/1, 361–374.

45. B. Horsfield and S. J. Düppenbecker, The decomposition of Posidonia Shale and Green River Shale kerogens using Microscale Sealed Vessel (MSSV) pyrolysis, *J. Anal. Appl. Pyrol.*, 1991, **20**, 107–123.

46. R. Michels, P. Landais, B. E. Torkelson and R. P. Philp, Effects of effluents and water pressure on oil generation during confined pyrolysis and high-pressure hydrous pyrolysis, *Geochim. Cosmochim. Acta*, 1995, **59**, 1589–1604.

47. R. L. Braun and A. J. Rothman, Oil-shale pyrolysis: kinetics and mechanism of oil production, *Fuel*, 1975, **54**, 129–131.

48. P. Ungerer and R. Pelet, Extrapolation of the kinetics of oil and gas formation from laboratory experiments to sedimentary basins, *Nature*, 1987, **327**, 52–54.

49. A. K. Burnham and R. L. Braun, Development of a detailed model of petroleum formation, destruction, and expulsion from lacustrine and marine source rocks, *Org. Geochem.*, 1990, **16**, 27–39.

50. B. Horsfield, The bulk composition of first-formed petroleum in source rocks, in *Petroleum and Basin Evolution*, ed. D. H. Welte, B. Horsfield, and D. R. Baker, Springer Verlag, Heidelberg, 1997, pp. 335–402.

51. A. A. Henry and M. D. Lewan, Comparison of kinetic model predictions of deep gas generation, in *Geologic Studies of Deep Natural Gas Resources*, Chapter D, ed. T. S. Dyman and V. A. Kuuskraa, U.S. Geological Survey Digital Data Series, 2001.

52. A. Vieth, M. Gabriel, F. Perssen, V. Dieckmann and B. Horsfield, Offline coupling of MSSV and GC-IRMS: a new approach to determine the isotopic composition of generated gases, in *Book of Abstracts, 24th International Meeting on Organic Geochemistry (IMOG)*, Bremen, 2009, p. 395.

53. M. Ladjavardi, L. J. Berwick, K. Grice, Ch. J. Boreham and B. Horsfield, Rapid offline isotopic characterisation of hydrocarbon gases generated by micro scale sealed vessel pyrolysis, *Org. Geochem.*, 2013, **58**, 121–124.

54. K. Theuerkorn, B. Horsfield, H. Wilkes, R. di Primio and E. Lehne, A reproducible and linear method for separating asphaltenes from crude oil, *Org. Geochem.*, 2008, **39**, 929–934.

55. F. V. Acholla, *Quantitative pyrolysis-gas chromatography using diamondoid compounds*, U.S. Patent 5394733, 1995.

56. E. W. Tegelaar, R. M. Matthezing, B. H. Jansen, B. Horsfield and J. W. de Leeuw, Possible origin of n-alkanes in high-wax crude oils., *Nature*, 1989, **342**(6249), 529–531.

57. B. G. K. van Aarssen, J. W. de Leeuw and B. Horsfield, A comparative study of three different pyrolysis methods used to characterise a biopolymer isolated from fossil and extant Dammar resins, *J. Anal. Appl. Pyrol.*, 1991, **20**, 125–139.

58. S. Düppenbecker and B. Horsfield, Compositional information for kinetic modelling and petroleum type prediction, *Org. Geochem.*, 1990, **16**, 259–266.

59. B. Horsfield, D. J. Curry, K. Bohacs, R. Littke, J. Rullkötter, H. J. Schenk, M. Radke, R. G. Schaefer, A. R. Carroll, G. Isaksen and E. G. Witte, Organic geochemistry of freshwater and alkaline lacustrine environments, Green River Shale, Wyoming, *Org. Geochem.*, 1994, **22**, 415–440.

60. G. P. A. Muscio, B. Horsfield and D. H. Welte, Compositional changes in the macromolecular organic matter (kerogens, asphaltenes and resins) of a naturally matured source rock sequence from northern Germany as revealed by pyrolysis methods, in *Organic Geochemistry: Advances and Applications in the Natural Environment*, ed. D. A. C. Manning, Manchester University Press, Manchester, 1991, pp. 447–449.

61. B. Horsfield, J. Heckers, D. Leythaeuser, R. Littke and U. Mann, A study of the Holzener Asphaltkalk, Northern Germany: observations regarding the distribution, composition and origin of organic matter in an exhumed petroleum reservoir, *Mar. Petrol. Geol.*, 1991, **8**, 198–211.

62. B. Horsfield, S. J. Düppenbecker, H. J. Schenk and R. G. Schaefer, Kerogen typing concepts designed for the quantitative geochemical evaluation of petroleum potential, in *Basin Modelling; Advances and Applications*, ed. A. G. Doré, J. H. Augustson, C. Hermanrud, D. J. Steward and O. Sylta, Norwegian Petroleum Society Special Publication, 1993, **3**, 243–249.

63. G. Muscio, B. Horsfield and D. H. Welte, Occurrence of thermogenic gas in the immature zone—implications from the Bakken in-source reservoir system, *Org. Geochem.*, 1994, **22**, 461–476.

64. B. Horsfield, H. J. Schenk, N. Mills and D. H. Welte, Investigation of the in-reservoir conversion of oil to gas: compositional and kinetic findings from closed-system programmed-temperature pyrolysis, *Org. Geochem.*, 1992, **19**, 191–204.

65. H. J. Schenk, R. di Primio and B. Horsfield, The conversion of oil into gas. Part 1: Comparative kinetic investigation of gas generation from crude oils of lacustrine, marine and fluviodeltaic origin by programmed-temperature closed-system pyrolysis, *Org. Geochem.*, 1997, **26**, 467–481.

66. U. Ritter, M. B. Myhr, T. Vinge and K. Aareskjold, Experimental heating and kinetic models of source rocks: comparison of different methods., *Org. Geochem.*, 1995, **23**, 1–9.

67. M. Erdmann and B. Horsfield, Enhanced late gas generation potential of petroleum source rocks via condensation reactions: Evidence from the Norwegian North Sea, *Geochim. Cosmochim. Acta*, 2006, **70**, 3943–3956.

68. V. Dieckmann, R. Ondrak, B. Cramer and B. Horsfield, Deep basin gas: new insights from kinetic modelling and isotopic fractionation in deep-formed gas precursors, *Mar. Petrol. Geol.*, 2006, **23**, 183–199.

69. N. Mahlstedt and B. Horsfield, Metagenetic methane generation in gas shales I. Screening protocols using immature samples, *Mar. Petrol. Geol.*, 2012, **31**, 27–42.

70. E. Idiz, B. M. Krooß, B. Horsfield, R. Littke and B. Müller, Generation of hydrocarbon gases and molecular nitrogen from coals, in *Organic Geochemistry: Developments and Applications to Energy, Climate, Environment and Human History*, ed. J. O. Grimalt and C. Dorronsoro, AIGOA, Donostia-San Sebastian, 1995, pp. 1089–1091.

71. G. P. A. Muscio and B. Horsfield, Neoformation of inert carbon during the natural maturation of a marine source rock; Bakken Shale, Williston Basin, *Energy Fuels*, 1996, **10**, 10–18.

72. H. J. Schenk and B. Horsfield, Using natural maturation series to evaluate the utility of parallel reaction kinetics models: an investigation of Toarcian shales and Carboniferous coals, Germany, *Org. Geochem.*, 1998, **29**, 137–154.

73. V. Dieckmann, B. Horsfield and H. J. Schenk, Heating rate dependency of petroleum-forming reactions: implications for compositional kinetic predictions, *Org. Geochem.*, 2000, **31**, 1333–1348.

74. Jochum, G. Friedrich, A. Germann, B. Horsfield, F. Leistner and W. Pickel, Occurrence of hydrocarbons in ore minerals of the Triassic sandstone-hosted lead-zinc deposits Maubach and Mechernich, Germany, in *Economic Geology in Europe and Beyond II: Models for Mineral Deposits in Sedimentary Basins*, ed. R. P. Foster, BGS/IMM, Nottingham and London, 1994, pp. 25–27.

75. J. Jochum, A. Germann, G. Friedrich, B. Horsfield and W. Pickel, Mechanical decrepitation coupled with gas chromatography—a new method for the determination of hydrocarbons in ore minerals, in *3rd Biennial SGA Meeting (Society for Geology Applied to Mineral Deposits). Mineral Deposits: From their Origin to their Environmental Impacts*. Prague, Czech Republic, 28–31 August, 1995.

76. H. Volk, U. Mann, O. Burde, B. Horsfield and V. Suchú, Petroleum inclusions and residual oils; constraints for deciphering petroleum migration, *J. Geochem. Explor.*, 2000, **69–70**, 595–599.

77. H. Volk, B. Horsfield, U. Mann and V. Suchý, Variability of petroleum inclusions in vein, fossil and vug cements – a geochemical study in the Barrandian Basin (Lower Palaeozoic, Czech Republic), *Org. Geochem.*, 2002, **33**, 1319–1341.

78. P. A. Hall, A. F. R. Watson, G. V. Garver, K. Hall, S. Smith, D. Waterman and B. Horsfield, An investigation of micro-scale sealed vessel thermal extraction-gas chromatography-mass spectrometry (MSSV-GC-MS) and micro-scale sealed vessel pyrolysis-gas chromatography-mass spectrometry applied to a standard reference material of an urban dust/organics, *Sci. Total Environ.*, 1999, **235**, 269–276.

79. D. Waterman, B. Horsfield, F. Leistner, K. Hall and S. Smith, Quantification of polycyclic aromatic hydrocarbons in the NIST standard reference material (SRM1649A) urban dust using thermal desorption GC/MS, *Anal. Chem.*, 2000, **72**, 3563–3567.

80. D. Waterman, B. Horsfield, K. Hall and S. Smith, Application of microscale sealed vessel thermal desorption gas chromatography-mass spectrometry for the organic analysis of airborne particulate matter: linearity, reproducibility and quantification, *J. Chromatogr. A*, 2001, **912**, 143–150.

81. K. Hall, B. Horsfield and N. Mills, On the feasibility of coupling microscale sealed vessel (MSSV) pyrolysis, gas chromatography and isotope ratio mass spectrometry for studying the thermal degradation of crude oil into natural gas, in *Poster sessions from the 16th International Meeting on Organic Geochemistry, Stavanger 1993*, ed. K. Oygard, Falch Hurtigtrykk, Oslo, 1993, pp. 794–797.

82. G. Muscio, B. Horsfield and D. H. Welte, The Bakken in-source reservoir system—new findings on organofacies and petroleum generation, *Org. Geochem.*, 1994, **22**, 461–476.

83. B. Mycke, K. Hall and P. Leplat, Carbon isotopic composition of individual hydrocarbons and associated gases evolved from micro-scale sealed vessel (MSSV) pyrolysis of high molecular weight organic material, *Org. Geochem.*, 1994, **21**, 787–800.

84. A. Wilhelms, S. R. Larter and K. Hall, A comparative study of the stable isotopic composition of crude oil alkanes and associated crude oil asphaltene pyrolysate alkanes, *Org. Geochem.*, 1994, **21**, 751–759.

85. J. R. Cerqueira and E. Vaz dos Santos Neto, Distinction between lacustrine and marine thermogenic gases based on hydrogen and carbon isotopic compositions, in *AAPG Hedberg Research Conference 'Natural Gas Geochemistry: Recent Developments, Applications and Technologies'*, Beijing, China, 9–12 May, 2011.

86. L. Berwick, P. F. Greenwood, W. Meredith, C. E. Snape and H. M. Talbot, Comparison of microscale sealed vessel pyrolysis (MSSVpy) and hydropyrolysis (HyPy) for the characterisation of extant and sedimentary organic matter, *J. Anal. Appl. Pyrol.*, 2010, 7, 108–116.

87. P. F. Greenwood, J. A. Leenheer, C. McIntyre, L. Berwick and P. D. Franzmann, Bacterial biomarkers thermally released from dissolved organic matter, *Org. Geochem.*, 2006, **37**, 597–609.

88. L. Berwick, P. F. Greenwood, R. Kagi and J.-P. Croué, Thermal release of nitrogen organics from natural organic matter using microscale sealed vessel pyrolysis, *Org. Geochem.*, 2007, **38**, 1073–1090.

89. L. Berwick, P. F. Greenwood and R. J. Smernik, The use of MSSV pyrolysis to assist in the molecular characterisation of aquatic natural organic matter, *Water Resources*, 2010, **44**, 3039–3054.

90. P. F. Greenwood, L. J. Berwick and J.-P. Croué, Molecular characterisation of the dissolved organic matter of wastewater effluents by MSSV pyrolysis GC–MS and search for source markers., *Chemosphere*, 2012, **87**, 504–512.

91. N. Mahlstadt, *Die Rolle von Reaktionen zweiter Ordnung bei der Gasgenese in sedimentären Systemen*, Diploma Thesis, Technical University Berlin, 2006.

92. B. P. Tissot, R. Pelet and P. Ungerer, Thermal history of sedimentary basins, maturation indices, and kinetics of oil and gas generation., *Bull. AAPG*, 1987, **71**, 1445–1466.

93. P. A. Hall, D. M. McKirdy, G. P. Halverson, J. B. Jago and J. G. Gehling, Biomarker and isotopic signatures of an early Cambrian Lagerstätte in the Stansbury Basin, South Australia, *Org. Geochem.*, 2011, **42**, 1324–1330.

94. S. A. Stout and J. J. Boon, Structural characterization of the organic polymers comprising a lignite's matrix and megafossils, *Org. Geochem.*, 1994, **21**, 953–970.

95. H. J. Schenk and B. Horsfield, Kinetics of Petroleum generation by programmed-temperature closed-*versus* open-system pyrolysis, *Geochim. Cosmochim. Acta*, 1993, **57**, 623–630.

96. K. H. van Heek and H. Jüntgen, Bestimmung der reaktionskinetischen Parameter aus nichtisothermen Messungen., *Ber. Bunsengesellschaft Phys. Chem.*, 1968, **72**, 1223–1231.

97. D. Santamaria-Orozco and B. Horsfield, Gas generation potential of Upper Jurassic (Tithonian) source rocks in the Sonda de Campeche, Mexico, in *The Circum-Gulf of Mexico and the Caribbean: Hydrocarbon habitats, Basin Formation and Plate Tectonics*, ed. C. Bartolini, R. T. Buffler and R. F. Blickwede, Memoir 79, AAPG, Tulsa, OK, 2003, pp. 349–363.

98. H. J. Schenk and B. Horsfield, Simulating the conversion of oil into gas in reservoirs: the influence of frequency factors on kinetic predictions, in *Organic Geochemistry: Developments and Applications to Energy, Climate, Environment and Human History*, ed. J. O. Grimalt and C. Dorronsoro, AIGOA, Donostia-San Sebastian, 1995, pp. 1102–1103.

99. H. A. Illich, F. R. Haney and T. J. Jackson, Hydrocarbon geochemistry of oils from Maranon Basin, Peru., *Bull. AAPG*, 1977, **61**, 2103–2114.

100. J. Espitalié, M. Madec, B. Tissot, J. J. Menning and P. Leplat, Source rock characterization methods for petroleum exploration, in, *Proceedings Offshore Technology Conference*, 1977, **3**, 439–444.

101. J. A. Gransch and E. Eisma, Characterisation of the insoluble organic matter of sediments by pyrolysis, in *Advances in Organic Geochemistry 1966*, ed. G. P. Hobson and G. C. Speers, Pergamon, Oxford, 1970, pp. 407–426.

102. R. Pelet, Evaluation quantitative des produits formés lors de l'évolution geochemique de la matiere organique, *Rev. Inst. Français Pétrole*, 1985, **40**, 551–561.

103. C. Cornford, J. A. Morrow, A. Turrington, J. A. Miles and J. Brooks, Some geological controls on oil composition in the UK North Sea, in *Petroleum Geochemistry and Exploration of Europe*, ed. J. Brooks, Special Publication 12, Geological Society, London, 1983, pp. 175–194.

104. W. A. England, A. S. Mackenzie, D. M. Mann and T. M. Quigley, The movement and entrapement of petroleum fluids in the subsurface, *J. Geol. Soc.*, 1987, **144**, 327–347.

105. R. di Primio and J. E. Skeie, Development of a compositional kinetic model for hydrocarbon generation and phase equilibria modelling: a case study from Snorre Field, Norwegian North Sea, in *Understanding Petroleum Reservoirs: Towards an Integrated Reservoir Engineering and Geochemical Approach*, ed. J. M. Cubitt, W. A. England, S. R. Larter, Geological Society, London, 2004, pp. 157–174.

106. M. Vandenbroucke, F. Béhar and J. L. Rudkiewicz, Kinetic modelling of petroleum formation and cracking: implications from the high pressure/high temperature Elgin Field (UK, North Sea), *Org. Geochem.*, 1999, **30**, 1105–1125.

107. R. di Primio and V. Neumann, HPHT reservoir evolution: a case study from Jade and Judy fields, Central Graben, UK North Sea, *Int. J. Earth Sci.*, 2008, **97**, 1101–1114.

108. F. Baur, R. Littke, H. Wielens, C. Lampe and Th. Fuchs, Basin modeling meets rift analysis—A numerical modeling study from the Jeanne d'Arc basin, offshore Newfoundland, Canada., *Mar. Petrol. Geol.*, 2010, 27, 585–599.

109. P. Kuhn, R. di Primio and B. Horsfield, Bulk composition and phase behaviour of petroleum sourced by the Bakken Formation of the Williston Basin, in *Petroleum Geology: From Mature Basins to New Frontiers—Proceedings of the 7th Petroleum Geology Conference*, ed. B. A. Vining and S. C. Pickering, 2010, pp. 1065–1077.

110. P. Kuhn, R. di Primio, R. Hill, J. R. Lawrence and B. Horsfield, Three-dimensional modeling study of the low-permeability petroleum system of the Bakken Formation, *Bull. AAPG*, 2012, **96**, 1867–1897.

111. N. J. L. Bailey, H. R. Krouse, C. R. Evans and M. A. Rogers, Alteration of crude oils by waters and bacteria: evidence from geochemical and isotopic studies, *Bull. AAPG*, 1973, **57**, 1276–1290.

112. F. Béhar, S. Kressmann, J. L. Rudkiewicz and M. Vandenbroucke, Experimental simulation in a confined system and kinetic modelling of kerogen and oil cracking, *Org. Geochem.*, 1992, **19**, 173–189.

113. P. Andresen, N. Mills, H. J. Schenk and B. Horsfield, The importance of kinetic parameters in modelling oil and gas generation - a case study in 1D from well 2/4–14, in *Basin Modelling; Advances and Applications*, ed. A. G. Doré, J. H. Augustson, C. Hermanrud, D. J. Steward and O. Sylta, Special Publication 3, Norwegian Petroleum Society, 1993, pp. 563–571.

114. U. Berner, E. Faber, G. Scheeder and D. Panten, Primary cracking of algal and land plant kerogens: kinetic models of isotope variations in methane, ethane and propane, *Chem. Geol.*, 1995, **126**, 233–245.

115. R. Michels, N. Enjelvin-Raoult, M. Elie, L. Mansuy, P. Faure and J. L. Oudin, Understanding of reservoir gas compositions in a natural case using stepwise semi-open artificial maturation, *Mar. Petrol. Geol.*, 2002, **19**, 589–599.

116. R. di Primio and B. Horsfield, From petroleum-type organofacies to hydrocarbon phase prediction, , *Bull. AAPG*, 2006, **90**, http://dx.doi.org/ 1031–1058.

117. F. D. Mango, Transition metal catalysis in the generation of natural gas, *Org. Geochem.*, 1996, **24**, 977–984.

118. F. D. Mango, The origin of light hydrocarbons, *Geochim. Cosmochim. Acta*, 2000, **64**, 1265–1277.

119. F. D. Mango, Methane concentrations in natural gas: the genetic implications, *Org. Geochem.*, 2001, **32**, 1283–1287.

120. F. D. Mango and L. W. Elrod, The carbon isotopic composition of catalytic gas: a comparative analysis with natural gas, *Geochim. Cosmochim. Acta*, 1999, **63**, 1097–1106.

121. R. di Primio, V. Dieckmann and N. Mills, PVT and phase behaviour analysis in petroleum exploration, *Org. Geochem.*, 1998, **29**, 207–222.

122. A. S. Pepper and P. J. Corvi, Simple kinetic models of petroleum formation. Part I: oil and gas generation from kerogen, *Mar. Petrol. Geol.*, 1995, **12**, 291–319.

123. V. Dieckmann, *Zur Vorhersage der Erdöl- und Erdgaszusammensetzungen durch die Integration von Labor- und Fallstudien.* PhD Thesis, Lehrstuhl für Geologie, Geochemie und Lagerstätten des Erdöls und der Kohle, Rheinisch-Westfälische Technische Hochschule Aachen, 1998.

124. R. J. Hill, E. Zhang, B. J. Katz and Y. Tang, Modeling of gas generation from the Barnett Shale, Fort Worth Basin, Texas, *Bull. AAPG*, 2007, **91**, 501–521.

125. E. Lehne, V. Dieckmann, R. di Primio, A. Fuhrmann and B. Horsfield, Changes in gas composition during simulated maturation of sulfur rich type II-S source rock and related petroleum asphaltenes, *Org. Geochem.*, 2009, **40**, 604–616.

126. C. Glombitza, K. Mangelsdorf and B. Horsfield, A novel procedure to detect low molecular weight compounds released by alkaline ester cleavage from low maturity coals to assess its feedstock potential for deep microbial life., *Org. Geochem.*, 2009, **40**, 175–183.

CHAPTER 8

High-Precision MC-ICP-MS Measurements of $\delta^{11}B$: Matrix Effects in Direct Injection and Spray Chamber Sample Introduction Systems

MICHAEL HOLCOMB,* KAI RANKENBURG AND MALCOLM MCCULLOCH

ARC Centre of Excellence for Coral Reef Studies, School of Earth and Environment & Oceans Institute, The University of Western Australia, 35 Stirling Highway M004, Crawley, WA 6009 Australia
*Email: mholcomb3051@gmail.com

8.1 Introduction

As a consequence of the growing importance of CO_2 induced climate change and the associated phenomenon of ocean acidification, high-precision measurements of boron isotopic compositions, especially in marine carbonates, is of growing importance. The isotopic composition and elemental abundance of boron is of special interest because it provides one of the few means to track the evolution of pH in seawater,[1-3] as well as being a probe into the processes of biomineralization.[4-6] Central to the interpretation of boron isotopic and elemental data is ensuring that the veracity of the measurements are not compromised by artefacts from, for example, memory

RSC Detection Science Series No. 4
Principles and Practice of Analytical Techniques in Geosciences
Edited by Kliti Grice
© The Royal Society of Chemistry 2015
Published by the Royal Society of Chemistry, www.rsc.org

or matrix effects. This is especially the case in the analysis of modern samples where shifts in $^{11}B/^{10}B$ ratios due to changing seawater pH are still relatively subtle[7] and hence require accurate, precise measurements.

Although protocols for sample preparation and isotopic analysis of boron by thermal ionization mass spectrometry (TIMS) are now relatively well established,[8-10] there is an increasing demand for multi-collector inductively coupled plasma mass spectrometer (MC-ICP-MS) measurements, due to the inherently greater sample throughput, but also importantly because of the potential for more precise measurements. This latter possibility arises mainly from the greater ability to correct for instrumental mass bias through standard–sample–standard bracketing. The commonly used approach of standard–sample bracketing, however, implicitly assumes a constant mass bias response between the standard *versus* the (unknown) sample and hence is subject to a range of matrix-induced effects depending on the nature of the sample. The thresholds at which matrix-induced effects become critical in undertaking accurate isotopic fractionation measurements has not, however, been extensively investigated for boron and hence represents a source of potential uncertainty. Furthermore, simplified extraction protocols[11] are now being more commonly used in MC-ICPMS approaches,[12] with the potential to be more subject to sample matrix effects and hence require specific investigation.

A major factor limiting the wider application of MC-ICPMS to boron measurements is the relatively large and persistent memory or blank effects, requiring prolonged washout to avoid cross-contamination. This poses a significant issue as boron tends to be retained at high levels for extended periods of time by the glass spray chambers typically used for ICPMS. Several different approaches are being used to reduce the effects of boron retention. These include the use of ammonia[3,13] to make the spray chamber more basic and hence reduce boron volatility, additives such as mannitol,[14,15] and hydrofluoric acid[16,17] used to complex the boron, among others.[18-20] Additionally, perfluoroalkoxy alkane (PFA) spray chambers have been employed,[17,21] but under acidic conditions they still require extended washout periods. An alternative approach is to eliminate the spray chamber altogether, as in the case of direct injection,[22,23] where the sample is directly introduced into the plasma.

Here we compare results obtained with both direct injection and spray chamber introduction systems. We also investigate the effects of anions, in particular SO_4^{2-} and Cl^- which are commonly found in samples such as seawater and marine carbonates at concentrations much higher than those of boron, yet are not necessarily fully removed by some of the commonly used protocols,[3,7,8,11] and thus represent a source of potential matrix interferences.

8.2   Methods

8.2.1   Mass Spectrometry

All measurements were carried out using a Neptune Plus MC-ICPMS (ThermoFisher Scientific) equipped with nickel sample and skimmer cones

(ThermoFisher part numbers 1044530 and 1067600 respectively), a Spetec Perimax peristaltic pump, and ESI SC2-DX autosampler using either 0.2 or 0.25 mm internal diameter Teflon-coated carbon fibre autosampler probes (ESI). An extraction voltage of −2000 V and an acceleration voltage of 10 kV were used for all measurements. Boron isotopes were collected in Faraday cups H4 and L4.

8.2.1.1 Direct Injection

Boron measurements were carried out using direct injection MC-ICPMS as described by Louvat *et al.*[23] Settings are detailed in Table 8.1. The setup was modified slightly from that described by Louvat *et al.*[23] Instead of interrupting flow during sample changes, we used a continuous flow system. To maintain continuous flow through the direct injector, a three-way solenoid pinch valve (Biochem Valve) was installed, but with the supplied Y-connector being replaced by a ~0.5 mm internal diameter PEEK Y-fitting (Idex Health and Science). This allowed the input solution to alternate between the sample line and a rinse solution such that uptake from the sample line could be stopped when the autosampler moved between samples, with a rinse solution being taken up in its place. To further reduce problems with bubble formation, a glass debubbler (Seal Analytical) was installed before the direct injector to remove bubbles which formed periodically. A dilute (~0.2%) Triton X-100 solution was passed through the sample line occasionally to reduce bubble growth. Sample measurements were bracketed by the SRM-951 standard.

8.2.1.2 Spray Chamber

Protocols for boron measurements using a spray chamber introduction system follow standard methods,[16,17,24,26] with instrument settings as detailed in Table 8.1. A 100 μL min^{-1} nebulizer (PFA-ST, ESI) was used to introduce the sample into a quartz cyclonic spray chamber (ESI), and a

Table 8.1 Neptune settings for each introduction system and type of measurement. Where two values are listed for a given setting, the first was used for sample measurements, the second for blank measurements.

Instrument settings	Direct injection	Spray chamber	Sulfur
RF power	1200 W	1300 W	1300 W
Ar cooling gas flow	15 L min^{-1}	16 L min^{-1}	16 L min^{-1}
Ar auxiliary gas flow[a]	1.24 L min^{-1}	0.8 L min^{-1}	0.8 L min^{-1}
Ar sample gas flow[a]	0.205 L min^{-1}	1.21 L min^{-1}	1.113 L min^{-1}
Sample uptake rate	~50 μL min^{-1}	~100 μL min^{-1}	~100 μL min^{-1}
Resolution	Low	Low	Medium
Integration time	4.194 s	4.194/2.097 s	1.049 s
Number of cycles	60	60/30	25

[a]Typical settings are given, but the rates were optimized each run and thus values for an individual run may differ slightly.

sapphire injector (1.8 mm, ESI) used to introduce the sample into the plasma. Samples were bracketed with a standard. Sample and standard measurements were bracketed by a rinse solution consisting of 0.01 M HF in 0.15 M HNO_3, which was passed through the system for ~ 4 min followed by a ~ 1 min measurement cycle to quantify the residual blank concentration and its isotopic ratio. The sample or standard was then aspirated for several minutes followed by ~ 5 min measurement of isotopic values. Blank corrections were applied to both samples and standards.[24]

8.2.2 Reagents

All resins were cleaned before use. Amberlite IRA 743 resin was washed with 1 N HNO_3 followed by 0.075 N HNO_3 and then stored in 0.075 N HNO_3 until use. Cation (Bio-Rad AG50W-X8 200–400 mesh) and anion (Bio-Rad AG1-X8 100–200 mesh) exchange resins were rinsed $3\times$ with 7 N HCl followed by $1\times$ with H_2O. Cation and anion resins were then loaded into columns and further cleaned (see below). Before use all columns were checked for air bubbles, and, if present, these were removed.

All acids used were distilled in a Teflon subboiling distillation system (Savillex DST-1000); water was from a Milli-Q unit (Integral 5, Millipore). SRM-951 was dissolved in water and then diluted in the desired acid to prepare standards and spike solutions. All plastic ware was rinsed repeatedly before use with water and dilute acid.

8.2.3 Testing Matrix Effects on MC-ICPMS

Since any variations in solution composition between samples and standards have the potential to change the mass bias in ICP measurements and thus alter the measured values,[23,25] it is best to measure samples and standards with identical boron concentrations and matrices. However, when separating boron from environmental samples, complete separation of boron from the sample matrix is not necessarily achieved, thus it is important to check the sensitivity of measurements to likely variations in composition. Samples (*e.g.* marine carbonates) often vary in their boron content, thus it is important to establish measurement sensitivity to variations in boron concentration. Acid concentration will vary among samples as well, either due to impurities which may contribute to the sample mass but not react with the acid added to dissolve the sample, or due to alteration of the sample chemistry by the boron extraction procedure used (*e.g.* ion exchange—see section 8.2.4). A wide range of cations are present in seawater and marine carbonates at concentrations known to cause matrix effects for boron measurements,[3,15,23,24] and thus must be removed. Anions are present as well, with the two most abundant in seawater and many marine carbonates being Cl^- (which has previously been suggested to interfere with ICPMS measurements[15]) and SO_4^{2-} (CO_3^{2-} is present as well, but this is lost as CO_2 when samples are acidified and thus not considered), though their influence on $\delta^{11}B$ measurements *via* ICPMS has not been extensively tested.

8.2.3.1 Effects of Variations in Nitric Acid and Boron Concentrations

To check for the dependence of boron isotope ratios on nitric acid concentration, a range of acid concentrations were prepared with equivalent concentrations of SRM 951 and measurements made using both introduction systems. The effect of sample boron concentration was tested by preparing different concentrations of SRM 951 or a 24.7‰ $\delta^{11}B$ laboratory standard in nitric acid. Concentrations and bracketing standards are specified in the legends of Figures 8.1 and 8.2.

8.2.3.2 Chloride

Since many samples of potential interest (*e.g.* seawater) contain Cl⁻, the effects of Cl⁻ were tested by adding various amounts of hydrochloric acid to a nitric acid matrix. Sets of experiments were run with direct injection, and later using spray chamber introduction. For direct injection: to test for the effects of Cl⁻, solutions of SRM 951 (355 ppb B) in 0.075 N acid were prepared by mixing 0.075 N HCl and HNO_3 with a SRM 951 stock solution.

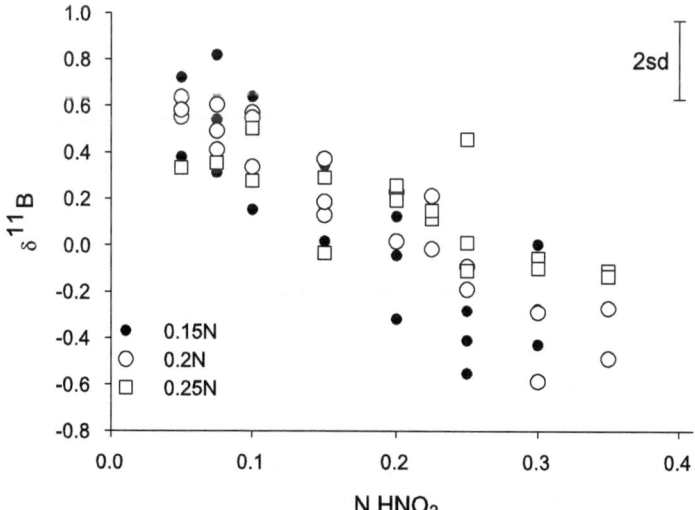

Figure 8.1 Effect of nitric acid concentration on $\delta^{11}B$ (‰) values of SRM 951 measured using a spray chamber introduction system. All solutions contained ~300 ppb boron, and measurements were normalized against a bracketing SRM solution in either 0.15, 0.2, or 0.25 N HNO_3 as indicated in the legend. Symbols represent individual measurements. For reference, the average measurement error on the bracketing standards (2 standard deviations) is shown in the upper right.

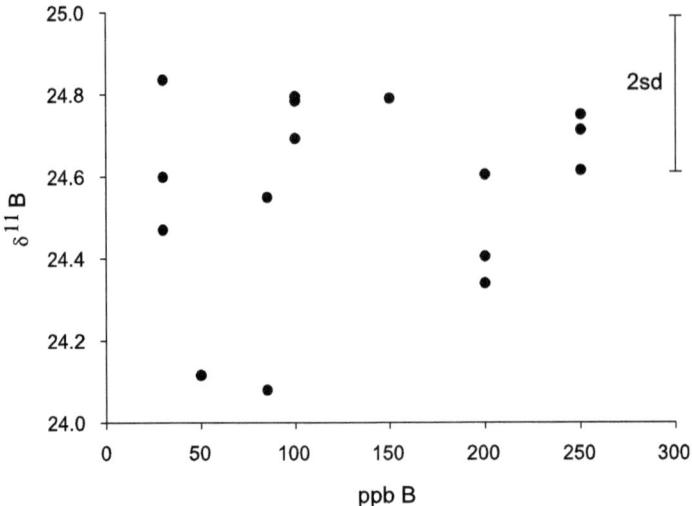

Figure 8.2   Effect of boron concentration on $\delta^{11}B$ values of a 24.7‰ solution in 0.15 N HNO_3 measured using a spray chamber. Points are individual measurements; all measurements were made against a 300 ppb solution. For reference, the measurement error on the bracketing standard (2 standard deviations) is shown in the upper right.

For spray chamber tests, a 300 ppb boron solution of SRM 951 was used in 0.15 N acid with various ratios of $HCl:HNO_3$ (Figure 8.3a). Since the presence of Cl^- in a nitric acid matrix appeared to affect measurements we also tested to see if variations in chloride in a hydrochloric acid matrix would have less of an effect. To check for the dependence of boron isotope ratios on hydrochloric acid concentration, a range of acid concentrations were prepared with equivalent concentrations of SRM 951 and measured; hydrochloric acid was used in the rinse solution run between samples. $\delta^{11}B$ values are plotted *versus* hydrochloric acid concentration in Figure 8.3b,c. Additional tests are described in the Appendix.

8.2.3.3   Sulfate (SO_4^{2-})

To test the effects of SO_4^{2-} on measurements made *via* direct injection, SRM 951 was prepared in nitric acid both with and without H_2SO_4 (~ 6 ppm final SO_4^{2-} concentration, similar to a typical aragonite extract). Solutions were processed through various extraction protocols (described below) which either removed SO_4^{2-} or left levels unchanged to verify the effects were likely due to SO_4^{2-} (Figure 8.4a).

To test the effects of SO_4^{2-} on measurements made using a spray chamber, measurements were made on a range of samples prepared (as described below) with both cation and anion resin (and low SO_4^{2-} concentrations)[12] *versus* cation resin alone (Figure 8.4b).

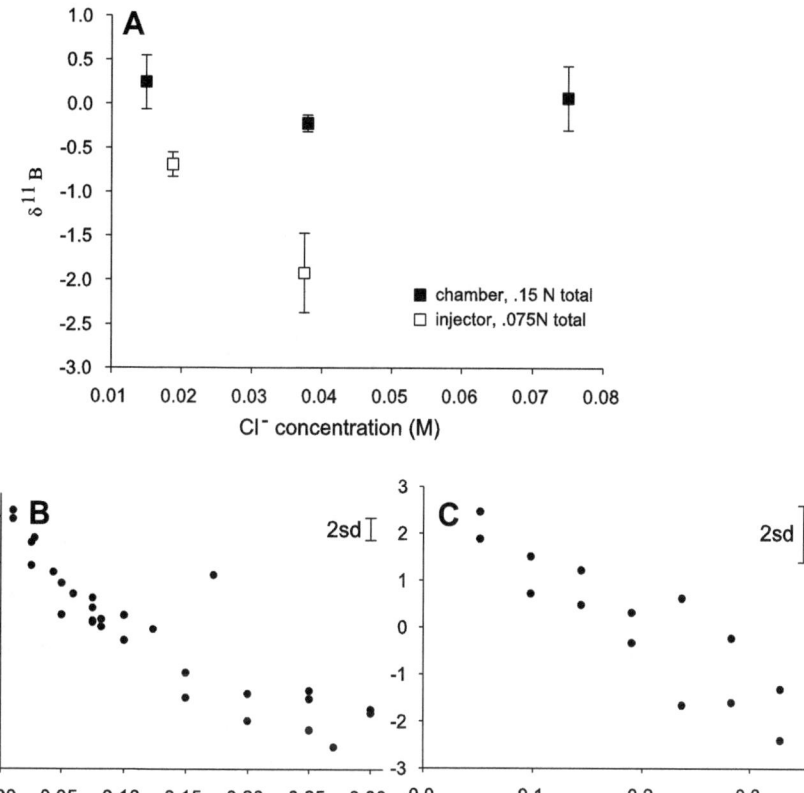

Figure 8.3 Effect of hydrochloric acid concentration on $\delta^{11}B$ values of SRM 951. (A) Measurements made in an nitric acid background *versus* a standard in nitric acid. Total acid concentration was kept constant. Measurements were made either with a direct injector (open symbols) or spray chamber (filled symbols). Symbols are means, error bars are 1 standard error. (B) Hydrochloric acid concentration dependence for direct injection measurements of SRM 951 (measured against SRM in 0.075 N HCl). Symbols are individual measurements. (C) As for (B) except samples were introduced using a spray chamber and measured against SRM in 0.2 N HCl. In (B) and (C), the measurement error on the bracketing standard (2 standard deviations) is shown in the upper right.

8.2.4 Boron Extraction Methods

A number of approaches for boron extraction were evaluated: (1) absorption of boron onto a boron-specific resin (Amberlite IRA 743)[8,26] as well as the addition of a cation ion exchange column procedure;[10,27] the latter is used due to the possibility of cations present in the samples forming hydroxides with low solubilities (*e.g.* Mg^{2+} in seawater) during boron absorption at high pH, and (2) a cation and/or cation + anion ion exchange resin column.[11,12,28]

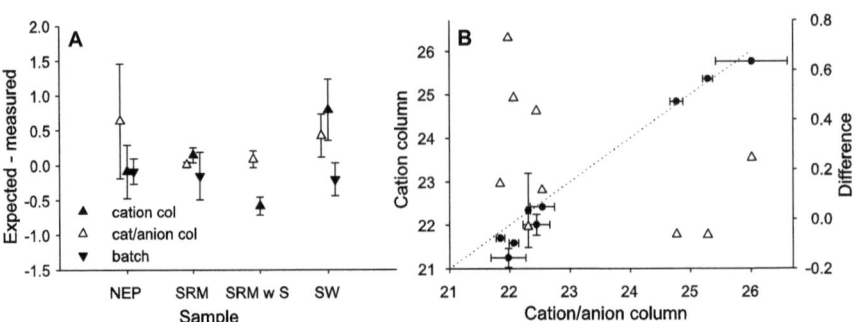

Figure 8.4 Comparison of different sample preparation approaches and the effect of $SO_4{}^{2-}$. (A) Expected – measured $\delta^{11}B$ values measured *via* direct injection for a laboratory coral standard (NEP, expected value 26‰), SRM-951 (expected value 0‰), and seawater (SW, expected value 39.8‰) purified with either a cation column only (filled up-triangle), or a cation column plus an anion column (open up-triangle), or batch chemistry with boron-specific resin (down-triangle). SRM samples with ~6 ppm $SO_4{}^{2-}$ added were also run (SRM w S) to test the effects of $SO_4{}^{2-}$. Values are averages of at least three replicate extractions of the given sample type, error bars are 2 standard error of the mean. (B) $\delta^{11}B$ (‰) values for coral powders prepared with cation + anion resin *versus* cation resin alone (measured using a spray chamber; black circles represent means, error bars are 2 standard error for repeated measurements of a given sample, not all measurements were repeated), and the difference between the average values (cation/anion value – cation alone value, up triangles). The 1:1 line is shown for reference.

8.2.4.1 Boron-Specific Resin (Batch Chemistry)

Purification using Amberlite IRA 743 resin followed a batch chemistry protocol based on existing protocols.[8,10,26,29] Samples (≤0.5 mL of NEP (a laboratory coral standard), seawater, or SRM-951 in 0.075 N HNO_3) were first passed through a 0.8 mL cation column (Bio-Rad Micro bio-spin column) and boron eluted with 2.5 mL of 0.075N HNO_3. 50 µl of boron-specific resin was added to the extract, 1 drop of phenol red (~1 mg/mL) added and 5.2–7.2 mL of 0.075 N NaOH (Rowe CS12613, stored with boron-specific resin added to it) added (to reach a basic pH for boron absorption, targeting pH 8–10). Tubes were placed on a shaker table and incubated overnight. The supernatant was discarded and the resin rinsed 5× with water. Boron was then eluted with 5 × 0.5 mL of 0.175 N HNO_3, with ~8 h incubations for each volume. Final boron concentrations were generally 200–300 ppb. To test accuracy, four samples were extracted as above with 12.2 mg coral powder (NEP) in 0.075 N HNO_3 mixed with varying amounts of SRM 951 and measured against SRM 951 in ~0.175 N HNO_3 (see Appendix and Figure 8.5). This protocol was also tested without the use of the initial cation column. However, without the initial cation column high concentrations of magnesium were present in some samples.

8.2.4.2 Cation and Anion Resins

Matrix sensitivity tests were also conducted for boron elutions using various combinations of cation and anion exchange columns. Resins were washed with 6 N HNO_3 and then equilibrated with 0.075 N HNO_3. Samples for purification on cation resin alone were either loaded onto 2 mL cation columns (Bio-Rad Poly-prep chromatography columns) or 0.8 mL cation columns and eluted with 0.075 N HNO_3. Cation plus anion resin purification used 0.8 mL cation columns followed immediately by 0.8 mL anion columns as well as procedures that are described in detail elsewhere.[12] The results for seawater, NEP, SRM 951, and SO_4^{2-} spiked SRM 951 processed with these procedures are given for direct injection measurements in Figure 8.4a, and samples measured using spray chamber introduction are presented in Figure 8.4b.

8.2.5 Spray Chamber and Direct Injection Inter-Comparison

To verify that different measurement methods yield equivalent results, a range of samples were prepared and measured using both direct injection and spray chamber introduction approaches. The samples included SRM 951; corals *Stylophora pistillata* (grown at the Centre Scientifique de Monaco), *Pocillopora damicornis* (grown at the Centre Scientifique de Monaco), and *Porites* sp. (from the Great Barrier Reef, Australia and Papua New Guinea); synthetic aragonite (precipitated at Woods Hole

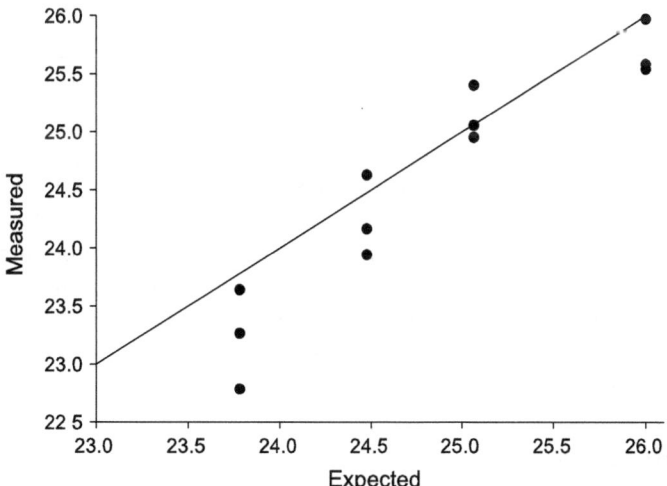

Figure 8.5 Expected *versus* measured $\delta^{11}B$ values for mixtures of SRM and NEP in nitric acid extracted with cation resin followed by batch chemistry. Samples contain equivalent quantities of NEP and variable quantities of SRM. Expected values were calculated assuming $\delta^{11}B$ of NEP = 26 and that the 12.2 mg of powder used contained 705 ng boron. Points represent individual measurements; the 1 : 1 line is shown for reference.

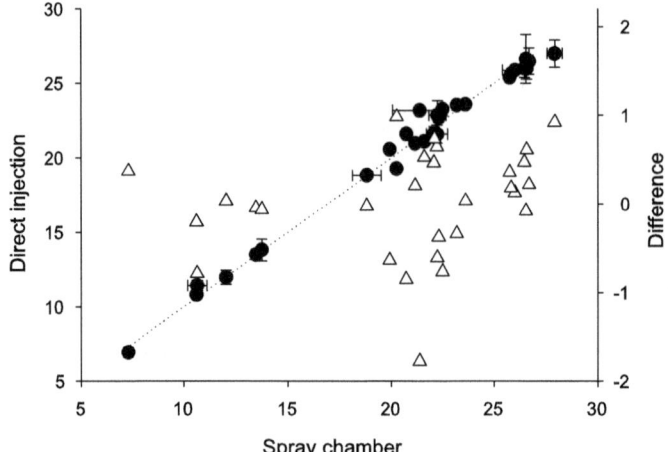

Figure 8.6 $\delta^{11}B$ values for samples measured *via* both direct injection and spray chamber techniques. Black circles are means of replicate measurements for a given sample, error bars are 2 standard errors, except for samples which were only measured once, for which error bars are omitted. The difference (spray chamber–direct injection value) between the average value obtained with each introduction system is show as open triangles. The 1:1 line is shown for reference.

Oceanographic Inst.); and *Marginopora vertebralis* (a high-magnesium calcite foraminifera cultured at the University of South Pacific, Fiji) The results are compared in Figure 8.6.

8.2.6 Contaminants Remaining in Extracted Samples

Likely cation contaminants (sodium. magnesium. aluminium, calcium), as well as boron, remaining after extraction were measured on residual boron samples using Q-ICPMS (X-series 2, ThermoFisher Scientific).

Sulfur (mass 32 and 34) was measured as well *via* MC-ICPMS (Neptune Plus). Samples were introduced as described for spray chamber introduction except that a quartz dual cyclonic spray chamber (ThermoFisher Scientific) was used. Instrument settings were as listed in Table 8.1. Nitric acid containing known amounts of sulfuric acid was run periodically to calibrate measurements.

8.3 Results

8.3.1 Effects of Nitric Acid Concentration on $\delta^{11}B$ Measurements

Measurements made using a spray chamber showed a decline in $\delta^{11}B$ of $\sim 0.6\%$ for an increase in normality from 0.075 to 0.3 N HNO_3, though values appear to plateau between 0.25 N and 0.35 N (Figure 8.1). This is similar to results described previously for direct injection[23] with a similar sensitivity being found for the dependence of $\delta^{11}B$ on changing concentrations of nitric acid; however, in this case $\delta^{11}B$ values increased with increasing acid concentration.

8.3.2 Effects of Boron Concentration on $\delta^{11}B$ Measurements

Consistent with previous findings[23] for direct injection, little concentration dependence was found for $\delta^{11}B$ values with boron concentrations in the ~190–350 ppb range. Below 190 ppb the dependency on boron concentration became more pronounced, with declines in $\delta^{11}B$ approaching 1‰ at ~90 ppb. Data obtained using a spray chamber showed a similar decline in $\delta^{11}B$ at low boron concentrations when data were not corrected for the acid blank. For blank-corrected data, using the spray chamber, no systematic concentration effect was observed over the 30–300 ppb boron range (Figure 8.2).

8.3.3 Chloride (Hydrochloric Acid)

For both direct injection and spray chamber measurements the addition of Cl^- to a nitric acid background was associated with reduced precision, and in the case of direct injection, values tended to be lower (Figure 8.3a). Samples containing ~0.019 M Cl^- had $\delta^{11}B = -0.69 \pm 0.28$, and samples containing ~0.038 M Cl^- had $\delta^{11}B = -1.92 \pm 0.9$ (mean $\pm$ 2 standard errors) when measured *via* direct injection against SRM 951 in nitric acid. In contrast to nitric acid the use of hydrochloric acid was associated with a strong effect of acid concentration on $\delta^{11}B$ values (Figure 8.3b,c).

8.3.4 Sulfate

With the direct injection method, repeated extractions of the SRM-951 standard yielded values within error of the expected value regardless of the extraction chemistry used (Figure 8.4a). However, when SRM was spiked with SO_4^{2-} (~6 ppm final SO_4^{2-} concentration in the boron extract), $\delta^{11}B$ increased by ~0.5‰. Passing the SO_4^{2-}-spiked SRM through anion resin reduced sulfur concentrations to near background levels and $\delta^{11}B$ returned to the expected value.

NEP coral samples passed through cation resin alone had nearly constant concentrations of ~7 ppm sulfur, while those passed through cation plus a 0.8 mL anion resin column still had sulfur concentrations of 2–3 ppm and were associated with higher measurement variability when measured *via* direct injection (Figure 8.4a).

8.3.5 Spray Chamber *versus* Direct Injector

Measurements were made on a range of carbonate samples with ppb levels of SO_4^{2-} and boron concentrations ranging from ~50 to 250 ppb using both spray chamber and direct injection sample introduction techniques. Comparable values were generally obtained with both measurement approaches (Figure 8.6) with sample-standard matching of acid strength and boron concentration.

8.3.6 Other Interferences

Boron extracts analysed for potentially interfering cations (sodium, magnesium, aluminium, calcium) generally had count rates near background levels. Background subtracted count rates normalized to boron were almost always less than 0.6 for Na/B, 0.006 for Mg/B, 0.02 for Al/B extracted *via* column chemistry, 0.3 for Al/B extracted *via* batch chemistry, and 0.01 for Ca/B. The major exception was seawater samples extracted with boron-specific resin without the use of a cation column, for which Mg/B was as high as 0.3.

8.4 Discussion

8.4.1 Acid Concentrations

Sample matrices are critically important for measurement of $\delta^{11}B$ with any contaminating elements or variations in sample composition having the potential to create problems for sample measurement.[23,24,30] For measurements made *via* direct injection, Louvat *et al.*[23] have shown $\delta^{11}B$ values to increase by $\sim 1‰$ over the nitric acid range we tested (0.05–0.3 N HNO_3) which is consistent with our observations (data not shown). Similarly, $\delta^{11}B$ measurements made using a spray chamber introduction system exhibited effects, but in contrast showed a decline of $\sim 1‰$ with an increase of acid concentration from 0.05 to 0.35 N HNO_3 (Figure 8.1). Regardless of the sample introduction method, nitric acid concentration influenced $\delta^{11}B$ measurements, showing the importance of closely matching acid concentrations for samples and standards.

Since Cl^- in an nitric acid background was associated with high measurement variability (Figure 8.3a), and many samples of interest (such as seawater) contain high concentration of Cl^-, the possibility of masking the effects of Cl^- variations by using high background levels of Cl^- was tested. When hydrochloric acid was used for the carrier solution a pronounced $\sim 1‰$ (spray chamber) to 4‰ (direct injection) decline in $\delta^{11}B$ was observed with increasing acid concentration (0.1–0.2 N HCl, Figure 8.3b,c), indicating a strong sensitivity to Cl^-. Thus the use of hydrochloric acid is unlikely to provide an effective means of reducing Cl^- matrix effects from samples containing high concentrations of Cl^- such as seawater.

8.4.2 Boron Concentrations

Variations in boron concentration have been associated with shifts in $\delta^{11}B$ values.[23,24] For direct injection, our measurements confirm this, with a decline in $\delta^{11}B$ ($\sim 1\permil$ for a change from 190 to 90 ppb boron) with decreasing boron concentration as observed by Louvat *et al.*[23] For spray chamber measurements, blank corrections are necessary especially when standards and samples differ in boron concentration. With the application of blank corrections, no concentration dependence is observed for spray chamber measurements (Figure 8.2). This contrasts with the findings of Guerrot *et al.*[24] who reported a decline of $\sim 1\permil$ with an increase in boron concentration from 20 ppb to 500 ppb. Although the source of such differences is not known, there are a number of differences in how the spray chamber introduction system was implemented in the study of Guerrot *et al.*[24] relative to the current study which could contribute to such different behaviours.

The $^{11}B/^{10}B$ ratio for the blank was consistently lower than that measured in either samples or standards (Table 8.2). Thus when samples and standards have different boron concentrations, and no blank correction is made, the residual boron will tend to reduce the isotopic ratio of the lower concentration solution to a greater extent than the higher concentration solution, thus giving rise to an apparent concentration dependency. Although the blank was not routinely measured between samples for direct injection measurements as no correction is required, when measured, the blank also had a lower isotopic ratio (Table 8.2). This could account for the observed decline in $\delta^{11}B$ when low-concentration standards were measured against high-concentration standards. Concentration matching of samples and standards thus greatly reduces the influence of residual boron on measurements since it will affect both values to the same extent. Furthermore, the use of standards with a similar isotopic composition to the samples can reduce the influence of the residual blank, and the corresponding corrections.[12] However, in the event that concentration matching of standards and samples is impractical, then the use of a blank correction procedure can help compensate for variations in boron concentration and blank isotopic compositions.

Table 8.2 Typical $^{11}B/^{10}B$ ratios and signal intensities (^{11}B) for the SRM standard (200 ppb), blanks (hydrochloric acid or nitric acid), and standard bracketing blanks as measured with both direct injection and spray chamber introduction.

Sample	Direct injection		Spray chamber[a]	
	Intensity (mV)	Ratio	Intensity (mV)	Ratio
HCl (blank)	2.8–3.3	3.767–4.234	n.d.	n.d.
SRM in HCl	1550	4.653–4.668	1300	4.61–4.62
HCl (bracket)	n.d.[b]	n.d.[b]	18–22	4.55–4.59
HNO$_3$ (blank)	1–4	3.3–4.41	4–7	4.5–4.6
SRM in HNO$_3$	1300–2000	4.61–4.73	1300	4.620–4.655
HNO$_3$ (bracket)	n.d.[b]	n.d.[b]	16–36	4.57–4.65

[a]Spray chamber blank values are for blanks measured after extensive rinsing of the spray chamber with clean acid. Note that values are measured ratios, not corrected for background.
[b]n.d., not determined.

8.4.3 Sample Preparation

Since a wide range of cations,[23,24] as well as anions[15] such as SO_4^{2-} (Figure 8.4a) and Cl^- (Figure 8.3a) are potential sources of interference, they must be taken into account when selecting a boron purification protocol. The protocols tested here are to remove many of these interfering ions while retaining the boron, but all suffer certain limitations.

8.4.3.1 Boron-Specific Resin

Boron-specific resin (Amberlite 743) has been widely used for the purification of boron.[8,26,31] A wide range of protocols have been used both in column[3,31] and batch[26,32] format. The use of boron-specific resin has many advantages, especially for samples with unusual matrices, imprecisely known quantities of sample, or low boron concentrations. Boron-specific resin is used to absorb boron at high pH, then boron is eluted at low pH;[30] this allows boron to be concentrated from low-concentration samples. However, when using boron-specific resin, maintaining the pH within the desired range is important for both boron adsorption and preventing the precipitation of cations, which can occur if the pH is too high. Thus purification with boron-specific resin requires the use of buffers,[3] or the monitoring of pH during adsorption, or the use of additional purification steps (*e.g.* cation resin).[26]

Even with the use of boron-specific resin, not all interfering elements are necessarily removed. For instance, ppm levels of sulfur were still present in some samples prepared with boron-specific resin which may account for the slightly lower than expected $\delta^{11}B$ values for seawater (Figure 8.4a) measured *via* direct injection. What percentage of this sulfur is from the original sample *versus* the pH indicator used in the extraction is not known, but sulfur concentrations were consistently higher for seawater and coral powders than for SRM 951 or blank controls, suggesting the sample contributes much of the sulfur present in the final extract. Such data suggest that passage through anion exchange resin could yield a cleaner extract when used in combination with boron-specific resin, or alternatively rinsing with sodium nitrate.[31] Cl^- too can be absorbed by the resin,[33] thus Cl^- contamination has the potential to be an issue even with the use of boron-specific resin. The high concentrations of cations observed in some samples may reflect the precipitation of cations present in the sample at basic pH during boron adsorption. The use of a cation column reduces this problem, though this problem can be minimized by careful control of pH for boron absorption.[26,31]

Thus depending on the matrix, purification with boron-specific resin alone may be adequate, or a combination of boron-specific resin plus cation and anion exchange resins may be needed to purify samples for boron isotope measurements. Such protocols are generally robust and suitable for a diverse range of samples as indicated by their widespread adoption for purifying boron for TIMS.[7,10]

8.4.3.2 Cation/Anion Resins

Cation exchange resins appear effective for the removal of cations which are well established as being sources of interference.[11,23,24] Since both Cl^- and SO_4^{2-} were found to interfere with $\delta^{11}B$ measurements, and SO_4^{2-} can be removed with anion exchange resin, the use of an anion exchange resin in addition to cation resin was tested. The anion exchange resin can be equilibrated with a wide range of anions. For the resin tested, the use of OH^- or acetate could increase the effectiveness of the resin for removal of Cl^- and SO_4^{2-}, but these would in turn create potential problems with acid strength (and potential conversion of boric acid to borate) or replace one matrix anion with another, creating other potential interferences (as seen with hydrochloric acid; Figure 8.3, Appendix). Equilibrating the anion resin with NO_3^- allows the combined cation and anion chemistry to increase the nitric acid concentration without adding additional ions, and by testing the dependence of $\delta^{11}B$ on nitric acid concentration, a region can be chosen over which small variations in acid normality have little effect on the measured values. However, in the nitrate form, the anion resin has little capacity for the removal of Cl^- (based on the manufacturer's data). The capacity to absorb SO_4^{2-} is also limited, as indicated by the high levels of SO_4^{2-} present in some extracts. Thus care must be taken not to overload the anion column with SO_4^{2-}, otherwise variable elevations of SO_4^{2-} in samples could lead to increased errors and reduced precision as suggested by the high variability for direct injection measurements of the NEP coral standard post anion column (Figure 8.4a). For seawater extractions, both Cl^- and SO_4^{2-} are present at concentrations likely to affect $\delta^{11}B$ measurements. However, the lower than expected values (Figure 8.4a) suggest that the effect of Cl^- dominated the matrix effect (*e.g.* Figure 8.3a). Thus, so long as nitric acid is used as the carrier, the use of a combined anion/cation resin purification is likely to be of limited applicability, being restricted to samples with relatively low concentrations of interfering anions (*e.g.* Cl^-). The use of other specialized resins, or acids such as acetic acid which are more readily displaced from the anion resin, may be more generally applicable to a wider range of samples; however, such chemistries are unlikely to achieve the same robustness seen with boron-specific resin purification.

8.4.4 Comparison of Direct Injection with Spray Chamber Sample Introduction

Although both direct injection and spray chamber introduction systems can be used successfully for boron isotope measurements, and the two methods yield similar values[32] (Figure 8.6), each has certain advantages and disadvantages.

Direct injection exhibits shorter washout times, allowing rapid switching between samples with little risk of carry-over, thus allowing a higher sample throughput. With the setup described, 6 min was required for signal intensities to stabilize upon switching from a 350 to <1 ppb boron solution. Most of this time was due to the delay between uptake of the solution from the sample vial and its introduction into the plasma, and thus could be reduced further by reducing tubing lengths and volumes. In our system the debubbler (1 mm internal diameter) added considerably ($\sim$2 min) to the uptake time as did the tubing (0.6 mm diameter) associated with the three-way valve; the use of smaller-diameter tubes or reducing their length could have further decreased uptake time and increased throughput. The short washout time and the direct introduction of the sample into the plasma allows measurements to be made on relatively small sample volumes (0.5 mL in the current setup, but this could easily be reduced). For high sample throughput and low volume samples, direct injection offers distinct advantages. However, this is partially offset by increased matrix effects for both cations[23] and anions (Figures 8.3 and 8.4), thus requiring greater care in sample purification. The introduction system itself is less stable than the spray chamber. Retuning is required regularly during the first day of operation to compensate for drift, and bubbles can destabilize measurements, if not extinguish the plasma.

The use of a spray chamber provides a more robust sample introduction system. It is generally less sensitive to both matrix effects and air introduction, as well as stabilizing more rapidly (1–2 h). However, substantially more time is required for the boron concentration to return to blank values following a sample ($\sim$1 h). Although it may be possible to reduce the washout time,[15,17] spray chamber introduction will still likely require longer washout times than direct injection because of the inherent limitation of the large surface area represented by the spray chamber. A much shorter washout time can be used if sample and standard measurements are bracketed by blank measurements and the blank value used to correct the sample and standard values.[24] Although doing a blank correction reduces the necessity of extended washout periods between samples, it adds an additional measurement which reduces sample throughput relative to direct injection. With the use of a spray chamber the sample is not introduced directly into the plasma, and much of the sample is lost during the introduction process, reducing the efficiency and hence increasing the sample volume required. Thus while a spray chamber offers greater stability than the direct injector, it is at the cost of needing larger volumes of sample per measurement and lower sample throughput.

8.5   Summary

Established protocols using boron-specific resin for the separation of boron from sample matrices appear to be robust and likely to give the most satisfactory results for a wide range of samples for MC-ICPMS. Although there has been recent interest in the use of cation exchange resins for purifying samples for MC-ICPMS, the presence of anions which can also interfere with measurements will pose a significant challenge for such chemistries, though there may be specific cases for which such chemistries are advantageous.[12] Sample introduction *via* either the low-blank direct injection or spray chamber generally yields equivalent values for a given sample, and each approach can be advantageous depending on sample size and potential matrix interferences.

Appendix

A.1 Column Optimizations

Both cation resin (Bio-Rad AG50W-X8) alone, and cation followed by anion resin (Bio-Rad AG1-X8) were tested for their potential to separate boron from other elements present in typical samples. All columns were optimized for sample and elution volumes to ensure complete recovery of the boron while minimizing the elution of other elements. Sulfur was generally the first contaminating element measured to elute from the columns, thus the effects of sulfur contamination were specifically tested (see text).

A.2 Column Chemistry Using Hydrochloric Acid

Samples dissolved and eluted in hydrochloric acid were run to test the reproducibility of this method using direct injection. However, the strong effect of hydrochloric acid concentration (Figure 8.3) led us to abandon further testing; the data collected are included here for reference.

Initial tests consisted of three replicate extractions of SRM 951 (0.5 mL of a 1500 ppb boron solution in 0.075 N HCl), NEP (12.2 mg of coral powder in 0.5 mL 0.075 N HCl), and acidified seawater (170 µL seawater mixed with 0.075 N HCl, 0.5 mL total volume), as well as a blank (0.075 N HCl) control. Samples were passed through cation and anion exchange resin. Data are presented as differences from expected values (Figure 8.7). The large offsets for NEP and seawater are likely due to the increased acid concentration following column chemistry.

To test the linearity of the offset from expected values, four columns were loaded (as above) with 12.2 mg coral powder (NEP) in 0.12 N HCl mixed with varying amounts of SRM 951. Samples were eluted with 0.12 N HCl, and measured against SRM 951 in $\sim$0.2 N HCl; a higher acid concentration was used to reduce the effect of acid concentration variations on the measured values. Measured values were offset by a nearly constant value from the expected value (Figure 8.8), suggesting the effect of hydrochloric acid concentration differences are constant and not dependent on the isotopic composition of the sample.

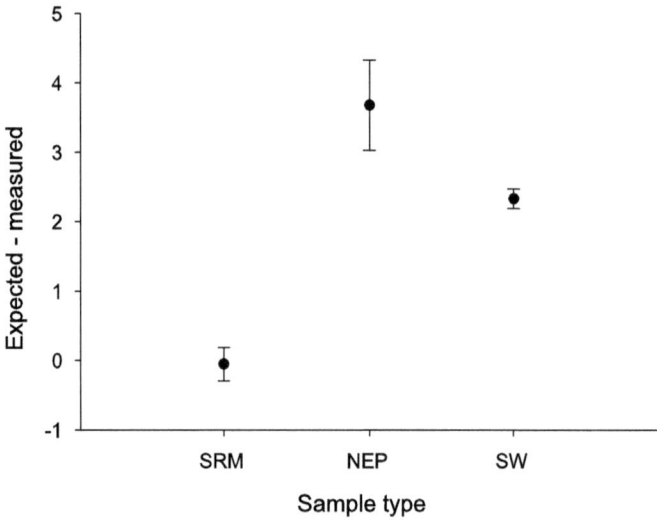

Figure 8.7 Expected – measured $\delta^{11}B$ values for SRM 951 (expected value = 0), NEP (expected value = 26), and seawater (SW, expected value = 39.8) extracted in hydrochloric acid. Symbols represent the mean of three replicate extractions, error bars are 2 standard errors. Measurements were made against SRM in 0.075 N HCl.

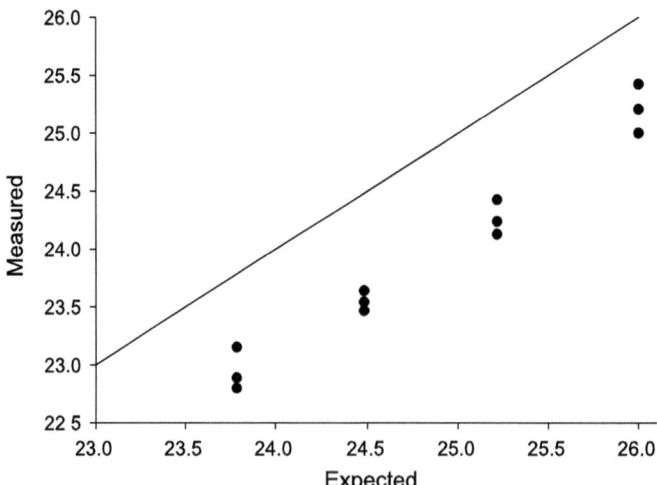

Figure 8.8 Expected *versus* measured $\delta^{11}B$ values for mixtures of SRM and NEP in hydrochloric acid. Samples contain equivalent quantities of NEP and variable quantities of SRM. Expected values were calculated assuming $\delta^{11}B$ of NEP = 26 and that the 12.2 mg of powder used contained 705 ng boron. Points represent individual measurements; the 1 : 1 line is shown for reference.

Acknowledgements

The authors are grateful for assistance provided by J. Trotter, D. Dissard, and L. Georgio, and for samples provided by D. Dissard; R. Naidu (University of the South Pacific); A. Venn, E. Tambutté, S. Tambutté, and D. Allemand at the Centre Scientifique de Monaco; G. Gaetani and A. Cohen at the Woods Hole Oceanographic Institution. This work was supported by the Australian Research Council (ARC) Centre of Excellence in Coral Reef Studies. MH was supported by an ARC Super Science Fellowship and MM by a Western Australian Premiers Fellowship and the ARC Centre of Excellence in Coral Reef Studies.

References

1. P. N. Pearson and M. R. Palmer, *Science*, 1999, **284**, 1824–1826.
2. C. Pelejero, E. Calvo, M. T. McCulloch, J. F. Marshall, M. K. Gagan, J. M. Lough and B. N. Opdyke, *Science*, 2005, **309**, 2204–2207.
3. G. L. Foster, *Earth Planet. Sci. Lett.*, 2008, **271**, 254–266.
4. N. Allison, A. A. Finch and EIMF, *Geochim. Cosmochim. Acta*, 2010, 74, 5537–5548.
5. A. C. Gagnon, *Proc. Natl. Acad. Sci. U. S. A.*, 2013, **110**, 1567–1568.
6. M. McCulloch, J. Falter, J. Trotter and P. Montagna, *Nature Clim. Change*, 2012, **2**, 623–627.
7. G. Wei, M. T. McCulloch, G. Mortimer, W. Deng and L. Xie, *Geochim. Cosmochim. Acta*, 2009, **73**, 2332–2346.
8. E. Kiss, *Anal. Chim. Acta*, 1988, **211**, 243–256.
9. G. L. Foster, Y. Ni, B. Haley and T. Elliott, *Chem. Geol.*, 2006, **230** 161–174.
10. J. Trotter, P. Montagna, M. McCulloch, S. Silenzi, S. Reynaud, G. Mortimer, S. Martin, C. Ferrier-Pagès, J.-P. Gattuso and R. Rodolfo-Metalpa, *Earth Planet. Sci. Lett.*, 2011, **303**, 163–173.
11. D. C. Gregoire, *Anal. Chem.*, 1987, **59**, 2479–2484.
12. McCulloch, *et al.*, In Prep.
13. A. Al-Ammar, R. K. Gupta and R. M. Barnes, *Spectrochim. Acta, Part B.*, 1999, **54**, 1077–1084.
14. D. H. Sun, R. L. Ma, C. W. McLeod, X. R. Wang and A. G. Cox, *J. Anal. At. Spectrom.*, 2000, **15**, 257–261.
15. J. K. Aggarwal, D. Sheppard, K. Mezger and E. Pernicka, *Chem. Geol.*, 2003, **199**, 331–342.
16. K. Nagaishi and T. Ishikawa, *Geochem. J.*, 2009, **43**, 133–141.
17. W. Gangjian, W. Jingxian, L. Ying, K. Ting, R. Zhongyuan, M. Jinlong and X. Yigang, *J. Anal. At. Spectrom.*, 2013, **28**, 606.
18. T. U. Probst, N. G. Berryman, P. Lemmen, L. Weissfloch, T. Auberger, D. Gabel, J. Carlsson and B. Larsson, *J. Anal. At. Spectrom.*, 1997, **12**, 1115–1122.
19. C. Wright, F. Fryer and G. Woods, *Agilent ICP-MS J.*, 2008, 4–5.

20. B.-S. Wang, C.-F. You, K.-F. Huang, S.-F. Wu, S. K. Aggarwal, C.-H. Chung and P.-Y. Lin, *Talanta*, 2010, **82**, 1378–1384.
21. C. J. Park, K. J. Kim, M. J. Cha and D. S. Lee, *Analyst*, 2000, **125**, 493–497.
22. F. G. Smith, D. R. Wiederin, R. S. Houk, C. B. Egan and R. E. Serfass, *Anal. Chim. Acta*, 1991, **248**, 229–234.
23. P. Louvat, J. Bouchez and G. Paris, *Geostandards and Geoanalytical Research*, 2011, **35**, 75–88.
24. C. Guerrot, R. Millot, M. Robert and P. Négrel, *Geostand. Geoanal. Res.*, 2011, **35**, 275–284.
25. I. I. Stewart and J. W. Olesik, *J. Anal. At. Spectrom.*, 1998, **13**, 843–854.
26. C. Lécuyer, P. Grandjean, B. Reynard, F. Albarède and P. Telouk, *Chem. Geol.*, 2002, **186**, 45–55.
27. H. Wei, Y. Xiao, A. Sun, C. Zhang and S. Li, *Int. J. Mass Spectrom.*, 2004, **235**, 187–195.
28. A. J. Spivack and J. M. Edmond, *Anal. Chem.*, 1986, **58**, 31–35.
29. D. Dissard, E. Douville, S. Reynaud, A. Juillet-Leclerc, P. Montagna, P. Louvat and M. McCulloch, *Biogeosci. Discuss.*, 2012, **9**, 5969–6014.
30. J. K. Aggarwal and M. R. Palmer, *Analyst*, 1995, **120**, 1301–1307.
31. D. Lemarchand, J. Gaillardet, C. Göpel and G. Manhès, *Chem. Geol.*, 2002, **182**, 323–334.
32. E. Douville, M. Paterne, G. Cabioch, P. Louvat, J. Gaillardet, A. Juillet-Leclerc and L. Ayliffe, *Biogeosciences*, 2010, 7, 2445–2459.
33. Y.-K. Xiao, B.-Y. Liao, W.-G. Liu, Y. Xiao and G. H. Swihart, *Chin. J. Chem.*, 2003, **21**, 1073–1079.

CHAPTER 9

Radioactive Carbon in Environmental Science

JOHN DODSON

Institute for Environmental Research, Australian Nuclear Science and Technology Organisation, Kirrawee, NSW 2232, Australia
Email: jdd@ansto.gov.au

9.1 Radioactive Carbon

Carbon is a very abundant element in biological and Earth materials and it is widely distributed in the atmosphere, biosphere, hydrosphere, and in soil and rock systems. Its ubiquity means its study lends itself to a wide range of applications in the geosciences and biosciences. The main isotopes of carbon have nuclear masses of 12, 13, and 14 and by virtue of these masses they have slightly different reaction rates in chemical processes. The isotope ^{14}C is unstable and decays at a known rate: it is called radiocarbon and the amount of this isotope in any carbon-bearing material or compound can give an indication of the time since it was in equilibrium with the ^{14}C content through exchange with its environment. This chapter describes the background to and measurement of radiocarbon, along with some applications in the geo- and biosciences.

9.2 Basis and History of Radiocarbon Dating

Carbon-14 was discovered by Kamen in 1940, and he found it had a half-life of about 5700 years. It was found that ^{14}C is formed by the interaction of ^{14}N in the atmosphere with neutrons in incoming cosmic rays. The amount of

RSC Detection Science Series No. 4
Principles and Practice of Analytical Techniques in Geosciences
Edited by Kliti Grice
© The Royal Society of Chemistry 2015
Published by the Royal Society of Chemistry, www.rsc.org

^{14}C formed is thus a function of the cosmic ray influx. The radioactive carbon rapidly oxidises to carbon dioxide and becomes a component of the global carbon cycle, and is cycled through the biosphere and geosphere. In the long term the production rate is matched by the decay rate back to ^{14}N, otherwise nitrogen would gradually disappear from the atmosphere. The abundance of ^{14}C is less than about 0.1% of carbon in the biosphere.

In 1947 Willard Libby (Figure 9.1) worked out that plants would absorb some of the trace ^{14}CO$_2$ during photosynthesis and this would go through the biosphere through eating relationships (Figure 9.2). In this way most plants and animals would be in approximate equilibrium with the ^{14}CO$_2$ in the atmosphere.[1] Once the living organism died and could not absorb carbon, the gradually declining ^{14}C would provide the basis for a chronometer to work out the time since death. This could be done by measuring the relative concentration of ^{14}C to ^{12}C. If this could be done over about 10 half-lives then ages up to about 50 000–60 000 years could be estimated from samples such as wood, plant macrofossils, wooden artefacts, peat, textiles, bones, animal shells, and more. In 1960 Libby received the Nobel Prize in Chemistry for this work.

Radiocarbon dating, therefore, is essentially any method that measures the residual radioactivity from ^{14}C. In short, the radiocarbon age indicates when the organism was last alive and in equilibrium with the environment in which it lived, and also the ^{14}C production rate at the time when it was alive.

Figure 9.1   Willard Libby, founding father for radiocarbon dating. Professor Libby received the Nobel Prize in Chemistry in 1960 for his pioneering work in this field. University of Chicago Photograph Archive (apf1–03871) Special Collection Research Center, University of Chicago Library.

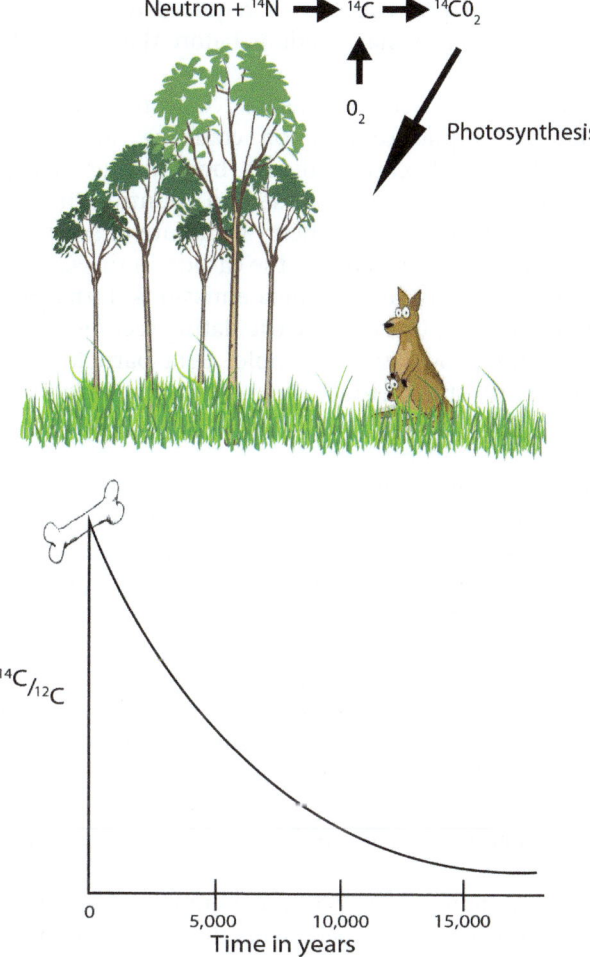

Figure 9.2 Formation of radiocarbon by neutron capture in ^{14}N. The resulting ^{14}C oxidises to carbon dioxide and is taken up by plants through photosynthesis. Herbivores and carnivores then convert this to animal tissue and into the detrital food chains. These processes result in living organisms having a ^{14}C content approximately like that of the atmosphere.

9.3 Measurement

Probably all organic compounds can be used for radiocarbon dating, as well as some inorganic materials like shell aragonite, coral, and speleothem carbonate. Inorganic carbon sources are valid where the carbonate, oxalate, or other mineral formation involves carbon capture where the ^{14}C was in equilibrium with the atmosphere. Materials such as water containing dissolved carbon or carbon compounds in suspension can also be dated. Since carbon is abundant in the biosphere and geosphere a certain amount of

physical and chemical pretreatment in required to remove possible contaminants or to select particular fractions before they are analysed for their
radiocarbon content.

The main techniques used to measure ^{14}C from a sample are gas proportional counting, liquid scintillation counting, and accelerator mass
spectrometry (AMS). Radiocarbon decays by emitting beta particles and
conventional radiocarbon dating is based on counting the number of beta
particle emissions which have the energy signature of a beta emission from
carbon. For gas proportional counting the carbon in the sample is converted
to pure carbon dioxide gas, and the beta emissions in the gas are counted.

Conventional counting is a radiometric dating technique that counts the
beta particles emitted by a given sample. Beta particles are products of
radiocarbon decay. In this method, the carbon sample is first converted to
carbon dioxide gas before measurement in gas proportional counters
takes place.

Liquid scintillation counting[2] is another radiocarbon dating technique
that was popular in the 1960s. In this method, the sample is in liquid form
and a scintillator is added. This scintillator produces a flash of light when it
interacts with a beta particle. A vial with a sample is passed between two
photomultipliers, and it is only when both devices register the flash of light
that a count is made.

AMS is the modern radiocarbon dating method and is considered to be
the more efficient way to measure radiocarbon content of a sample.[3] In this
method, the ^{14}C content is directly measured relative to the ^{12}C and ^{13}C
present. The method does not count beta emissions but the number of
carbon atoms present in the sample and the proportion of the isotopes
(Figure 9.3) This method typically measures atoms of a given mass to about 1
in 10^{15}.

The estimated amount of ^{14}C needs a small correction. Heavier isotopes
tend to have lower reaction rate coefficients than light isotopes. Thus ^{12}C will
tend to be a bit more abundant in materials than ^{14}C, leading to an
underestimate of the actual ^{14}C in a sample, and hence its age. Since this
depends on the material itself the ratio of the two stable isotopes of carbon,
^{12}C and ^{13}C, is measured in the sample. This gives an estimate of the discrimination factor applying to the ^{13}C and from this for the ^{14}C in the
sample, and the direct correction that needs to be applied.

Finally one more correction for the radiocarbon measurement is needed.
This arises because of the 'reservoir effect'. It was noted above that in the
normal sort of circumstances an organism has a ^{14}C concentration which is
in equilibrium with the biosphere where this is set by the cosmic ray flux and
radiocarbon production. If an organism is growing in an environment where
the ^{14}C is set by other parameters then the age will appear to be different.
Ocean carbon has a considerable residence time, and corrections of centuries or more need to be applied. Organisms which feed on ocean fish will
likewise incorporate relatively more ancient carbon in their bodies. Biota

Figure 9.3 The STAR Accelerator at the Australian Nuclear Science and Technology Organisation in Sydney. A major part of the time of this machine is used for AMS radiocarbon dating.

Figure 9.4 A reservoir effect arises when the radiocarbon content along food chains is in part driven by the starting value of radiocarbon concentration. In the oceans, for example, the modern radiocarbon concentration is diluted by 'old' carbon, so that a measured age on living material looks much older, perhaps by several centuries. A diet based largely on marine fish will have this reservoir effect evident in a fish eating bird or human. The map is based on one from Matsumoto and Key[18] for deep waters. The ages represent carbon run-off from the continents, upwelling, and current velocities across the ocean floor.

which process ancient radiocarbon may have no apparent age at all, and those which live in environments fed by ancient groundwater will also return apparent anomalous ages. Each case needs to be assessed independently (Figure 9.4).

9.4 Why Radiocarbon Measurements are Not True Calendar Ages

Radiocarbon measurements are always reported, by convention, as years 'before present' (BP); and as you will see below, in calibrated calendar ages. The conventional age value is directly based on the proportion of radiocarbon found in the sample. It is calculated on the assumption that the atmospheric radiocarbon concentration has always been the same as it was in 1950 and that the half-life of radiocarbon is 5568 years. For this purpose 'present' refers to 1950, so you do not have to make adjustments for the year in which the measurement was made.

To illustrate this we will refer to an example. If a sample is found to have a radiocarbon concentration exactly half of that for material which was modern in 1950, the radiocarbon measurement would be reported as 5568 BP (plus or minus an error term). This does not mean that the sample comes from 3619 BC (plus or minus the error term) for two very important reasons. Firstly, the half-life of radiocarbon is 5730 years,[4] and not the originally estimated value of 5568 years, but we still use the earlier value so that each age is reported on the same time scale. Secondly, the proportion of radiocarbon in the atmosphere has varied by a few per cent over time. We know this since radiocarbon measurements on materials of known age reveal this variability. Thus, in order to see what a radiocarbon determination means in terms of a true age, we need to know how the atmospheric concentration has changed with time.

Tree rings have been one of the means to test the variability in the radiocarbon record. In certain situations many tree species reliably lay down one tree ring every year. The cellulose and lignin in the wood in these rings stays relatively unchanged during the life of the tree, and thus reflects the record of the radiocarbon concentration in the year in which it was formed. If we have a tree that is 200 years old we can measure the radiocarbon in the individual 200 rings and observe what the radiocarbon concentration is for each calendar year across this time span. The longest-living trees, such as *Pinus longaeva* (Bristlecone pine) in the mountains of the south-western USA, and *Lagarostrobus franklinii* (Huon pine) in western Tasmania can live for over 2000 years and up to 4700 years. From these trees it is possible to calibrate radiocarbon measurements back to a few thousand years ago. Reimer *et al.*[5] describe this in some detail.

Since tree rings tend to vary in width from year to year with changing weather patterns during the growing seasons, this provides a cross-checking procedure to growing trees with dead and fossil wood, to extend radiocarbon dating calibration back much further in time (see Figure 9.5) This has been done with Bristlecone pines, Huon pines (from wood found in river sediments), and fossil oak wood (*Quercus petraea*) from peats in Ireland and Germany. These have provided records of known calendar ages that extend beyond 11 000 years. New Zealand kauri (*Agathis australis*), from northern

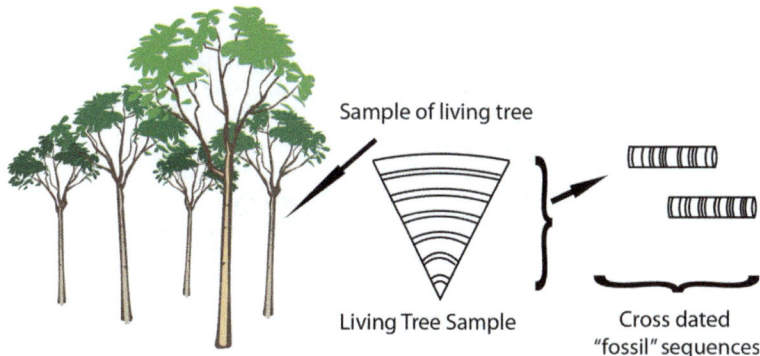

Sample of living tree

Living Tree Sample

Cross dated "fossil" sequences

Figure 9.5 Wood from a living tree, and an illustration of how this may be cross-matched to old and fossil ring-width sequences. Radiocarbon dating of individual rings provides a method of comparing radiocarbon with calendar ages.

New Zealand, has fossil wood preserved in peats that extend beyond the radiocarbon time scale and this may ultimately provide a year-by-year sequence for 50 000 years or more.[6] In the meantime, radiocarbon calibration with calendar ages from corals has been used to extend the time scale from the Holocene into the late Pleistocene.[7]

9.5 Calibration of the Radiocarbon Time Scale

In principle, the calibration of radiocarbon determinations is simple. If you have a radiocarbon measurement on a sample, you can try to find a tree ring with the same proportion of radiocarbon, then the ring wood age will be the same as the sample age. Of course this is limited to some extent by the level of measurement precision, and this means that there will be a range of possible calendar year dates for any measurement. The other complication, which is scientifically more interesting, is that the atmospheric radiocarbon concentration has varied in the past, and there might be several possible age ranges for a sample measurement. One of the complications has arisen from the large injection of old carbon from carbon dioxide released into the atmosphere from burning fossil fuels.

Figure 9.6 shows how calibration of radiocarbon works. The diagram shows how a radiocarbon measurement of 3000 ± 30 BP can be calibrated. The measurement is shown as a distribution on the right-hand side (with mean and normalised plot of age range). The line across the middle of the graph shows ages based on tree ring data with the range at any point representing ± 1 standard deviation. The distribution at the bottom shows the calendar age of the sample, which is derived from the intersection of the radiocarbon and tree ring ages. Here we can say that the true age has a 95% probability of being between 1375 cal BC and 1129 cal BC, and by estimating

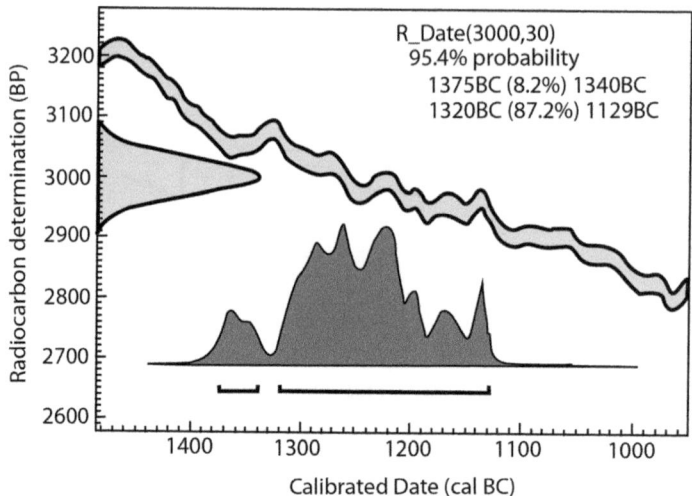

Figure 9.6   A demonstration of how a radiocarbon date $(3000 \pm 30 \text{ BP})$ can be calibrated to a calendar age (between 1375 and 1129 BC). Redrawn from Oxford University Radiocarbon web site (http://c14.arch.ox.ac.uk/embed.php?File=calibration.html).

the ages from the two distributions a 8.2% probability of being between 1375 and 1340 cal BC, and an 87.2% probability of being between 1320 and 1129 cal BC.

There are several calibration programs available online which enable easy conversion of radiocarbon ages to calendar ages,[5] and IntCal09, and the University of Cologne package available at http://www.calpal-online.de).

9.6   Small Carbon Measurements

When Libby developed the radiocarbon technique, about 5 g of pure carbon (within the radiocarbon time scale range) was needed to obtain an age estimation within generally acceptable error limits. As scintillation counters improved the sample size was progressively reduced to 1 g or less. The development of atom counting by AMS, rather than decay counting, reduced the necessary sample size to milligrams and less. This immediately provided approaches to new kinds of questions. For example, if an age for valuable materials such as the Turin Shroud was asked for it could be done on less than a square centimetre or a few fibres of material, compared to a substantial portion of the artefact as required previously. Residues in ancient pottery could be dated, valuable cultural items and even seeds or small shells could be considered without destroying irreplaceable items or significant parts of excavated materials. The other problem which now became more obvious was the potential for contamination or mixed materials to give erroneous ages. For example, it was routine to take bulk sediment samples with their various components of reworked and directly deposited organics to build an age profile for a section. If the older reworked material was a very small fraction of the sample it had no major effect on the age. However, younger carbon has proportionally more ^{14}C per gram, and large amounts of variously aged subcomponents could substantially alter the age of what the researcher thought they were dating. Very small samples are more prone to cross-contamination either in the sedimentary setting or even during sample collection.

Small-sample analyses opened the possibility of independently dating fractions within a sample. If they all return the same age then this suggests the assemblage was formed at the same time. If the ages vary then this demonstrates a variety of sources of material in the assemblage. With such data the researcher can now make a judgement on what is the reliable component(s) to give the age that is relevant for their research problem.

Samples as small as 50–100 μg can now be prepared and measured by many radiocarbon laboratories that have AMS facilities. At the Australian Nuclear Science and Technology Organisation in Sydney a new technique involving a microfurnace and laser ablation can sputter samples as small as 2–5 μg of carbon into the accelerator for measurement of a radiocarbon age. Thus even smaller samples can be analysed, and new types of questions can be addressed. For example, the possibility of dating specific compounds within samples has arrived. These questions can now be addressed at the molecular scale instead of the microscopic scale. Thus the separation of compounds from complex organic samples using gas or liquid chromatography can be accessible. The potential to expand the range of questions that may be addressed has risen by an order of magnitude or more.

For example, Pessenda *et al.*[8] dated total organic matter, humin fraction, and charcoal from soils in Brazil. They showed that total soil organic matter returned ages that were significantly younger than the humin and charcoal, with the latter returning similar ages. The total soil organic matter was contaminated by younger carbon while the humin fraction ages could be assumed as the minimum age for carbon in soils.

Compound-specific radiocarbon dating has proved to be useful in defining chronologies in challenging environments. For example, Ohkouchi and Eglinton[9] found that solvent-extractable short-chain fatty acids gave consistent radiocarbon ages while bulk sediment ages in Ross Sea sediments gave complex and unreliable chronologies around the margin of Antarctica.

9.7 Bomb Pulse

As described above, the amount of ^{14}C in the atmosphere varied by only a few per cent but after nuclear-bomb testing in the atmosphere it rose by 50% or more. The Test Ban Treaty in 1963 has led to the rapid decline of atmospheric ^{14}C as it diffused into the oceans. By about 2010 the values of ^{14}C had almost returned to the pretesting levels. Figure 9.7 shows the measured trends, and the peak centred around the mid-1960s is referred to as the

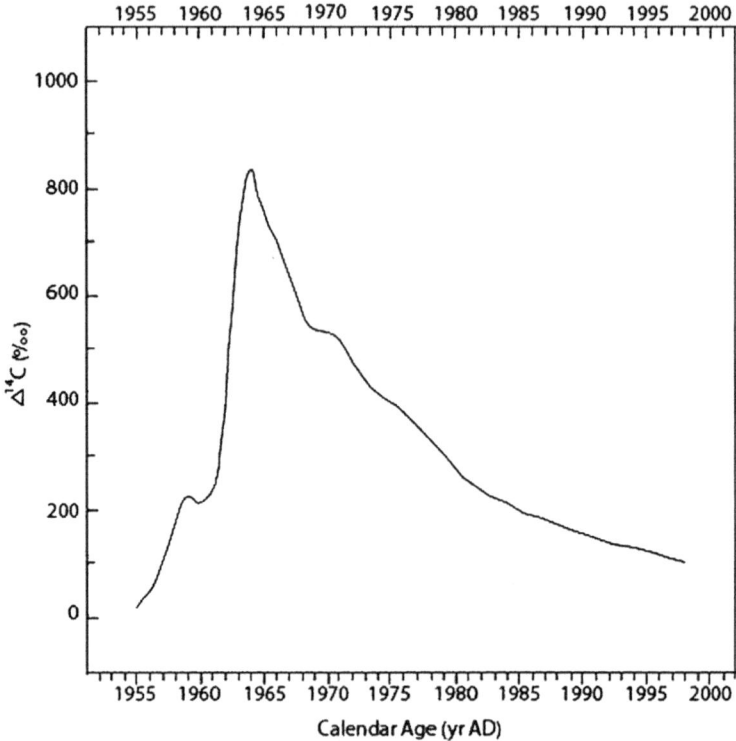

Figure 9.7 The bomb pulse radiocarbon profile. Redrawn from data shown in Hua.[19]

bomb pulse signature. This has provided a window for very accurate dating of samples with ages in this time frame.

There have been several applications of the bomb pulse signature, such as checking the vintages of wines, brain cells, fat cells,[10] and even the formation of eye lens crystallines. Comparing the amount of ^{14}C in eye lens has made it possible to investigate the timing of the formation of the crystals. These are formed at the time of birth, and people born after 1955 have the bomb pulse signature present in the crystals of their eye lens.[11]

9.8   Case Studies

9.8.1   Environmental History at the Local and Regional Scale

From the 1960s to 1990s radiocarbon was applied to sediment sequences to provide chronologies for proxy analyses of climate and environmental changes. The proxies were usually were based on fossil pollen, sediment facies, geochemistry, or grain size analyses. From the 1940s to about the 1960s many of these sequences were compared regionally by cross-referencing of pollen zones or sediment records, and it was assumed that these were driven by large-scale climate processes. The application of radiocarbon to many of these showed that while some of these assumptions were fine, in many cases it could be demonstrated that individual sites had offset chronologies for changes that were apparent. It became clear that understanding these differences was crucial for recognising the response times of sediment systems, vegetation changes, and the role of climate in driving some of these. It was later recognised that human impact was a key element in the response of many systems. Without radiocarbon most of these records were floating free in time.[12] From about the 1990s several parts of the world had multiple records of climate change built on robust proxy analyses and sound radiocarbon-based chronologies. It then became possible to look beyond national boundaries and within regions to identify large-scale climatic patterns. The COHMAP project,[13,14] for example, brought scientists together to compile regional record syntheses and compare these with climate model simulations. This was a major step in identifying some of the global and synoptic scale processes that drive climate changes.

9.8.2   Radiocarbon and Ecology

Fire is a significant factor in the ecology of many vegetation systems, yet little is understood about the spatial and temporal variability in the recent past, or what drives these. Mooney *et al.*[15] recently compiled over 220 sedimentary charcoal records from Australasia in order to address these questions. It is absolutely crucial that these are well dated so that regional patterns can be examined. The data reveals that on orbital time scales, fire in Australasia predominantly reflects climate, with colder periods characterized by less biomass burning, and warmer intervals by more. Furthermore, within the limits of dating uncertainties of individual records, it is apparent that the composite charcoal record is similar to the form, number, and timing of Dansgaard–Oeschger cycles as observed in Greenland ice cores; and surprisingly, less like the variability expressed in the Antarctic ice-core record. Millennial-scale variability is characteristic of the composite record of the subtropical high-pressure belt during the past 21 ka, but the tropics show a somewhat simpler pattern of variability with major peaks in biomass burning around 15 ka and 8 ka. There is no distinct change in fire regime corresponding to the arrival of humans in Australia at 50 ± 10 ka and no correlation between archaeological evidence of increased human activity during the past 40 ka and the history of biomass burning. However, changes in biomass burning in the last 200 years may have been exacerbated or influenced by humans. These kinds of studies are useful for policy-makers who set practices for the protection and environmental management aspects of fire.

9.8.3   Radiocarbon and Groundwater

Groundwater systems contain many times the amount of water in rivers, lakes, and dams. As global warming increases its grip on climate systems, most especially in the mid-latitude regions of both hemispheres, many societies are looking to exploit the groundwater resources for agriculture, industry, and potable water. There are now increasingly questions about the sustainability of use of groundwater systems since radiocarbon ages on dissolved inorganic carbon in these systems show that the water is in fact quite often thousands of years old. This raises the possibility that the recharge rates might be very low, and if exploitation exceeds the recharge then groundwater usage may be analogous to mining a finite resource.

The Pilbara region of Western Australia is one of Australia's most important mining areas. The State Government of Western Australia has proposed that the population of Port Hedland should expand considerably to meet the infrastructure and service needs of the region; however, with little surface water this will depend on the ability to abstract considerable amounts of groundwater which will possibly have impacts from climate variation, coastal discharge, and ecological water requirements in the overall mix of water allocations for the region. The Wallal Sandstone has a large groundwater storage of high-quality water.[16] Dissolved inorganic carbon from water samples was processed into carbon dioxide by acidifying the samples with H_3PO_4 and extracting the liberated carbon dioxide. Radiocarbon ages were corrected and samples collected from the recharge areas were found to be only a few decades old or slightly older ('submodern'). These waters are apparently recharged each time tropical cyclones are nearby, since the water has $\delta^{18}O$ and δ^3H signatures typical of tropical rainfall, and this coast experiences more cyclones than any other coast in Australia. Groundwaters from the north-eastern section of the aquifer were calculated to be about 5000–8000 years old, while those from the north-western section had measured ages between about 21 000 and 46 000 years old. In short, the radiocarbon data shows that the groundwater flow rates are very low, only about ~ 2 m yr^{-1}, and abstraction from the older waters could have significant negative impact on sustainability of the resource.

9.8.4 Applications in Archaeology

Accurate chronology for ancient societies and their activities is an essential element in tracing the history and interactions of ancient cultures, and their place in history. The following example comes from a study in south-east Spain. It has long been thought that early copper metallurgy is a barometer of social complexity, and that this was barely developed in south-east Spain compared to other areas in the region. Eight settlements near modern Valencia were studied in detail through the application of 66 radiocarbon dates on slag charcoal, furnace charcoal, and bone of domesticated sheep and goats.[17] Careful selection of sites which covered households, workshops, agricultural and mining sites, and smelting quarters along the Guadalquivir valley enabled a precise analysis of social behaviours and the development of social complexity in the region. First sites settled included the south-western Iberian pyrite belt at about 3100 BC, then expanded activity in the lower

Guadalquivir valley from about 2900 BC, and the upper Guadalquivir valley after about 2700 BC. The dates on activities also show the time of abandonment of sites after about 1900 BC. The uninterrupted records over 500 years and the rise of large regional centres is pivotal in understanding the beginnings of landscape transformation in the region.

9.9 Conclusion

In summary Libby's radiocarbon methodology has come a long way and is now a frequently used method in a wide range of sciences. The improvements in the technique have evolved largely as a result of a better understanding of the carbon cycle, cosmogenic flux rates, and the dynamics of Earth systems in general. Improvements in measurement technology have resulted in the use of smaller samples, more accurate atom counting, and faster turnaround times. Calibration of the radiocarbon time scale and the quantification of reservoir effects have resulted in the routine reporting of ages in calendar years. The reduction of sample size requirements from grams to micrograms has opened up the world of dating compound-specific materials at molecular level. Every one of these advances has opened up new kinds of questions which can be approached by the application of radiocarbon in the worlds of archaeology, environmental and geosciences, biosciences, forensics, and medicine.

Acknowledgements

Many thanks for Rhiannon Still for preparing the figures.

References

1. W. F. Libby, *Radiocarbon Dating*. University of Chicago Press, Chicago, 1952.
2. D. L. Horrocks, *Applications of Liquid Scintillation Counting*. North-Holland, Amsterdam, 1974.
3. A. E. Litherland, Ultrasensitive mass spectrometry with accelerators., *Annu. Rev. Nucl. Particle Sci.*, 1980, **30**, 437–473.
4. H. Godwin, Half-life of radiocarbon., *Nature*, 1962, **195**, 984.
5. P. J. Reimer, M. G. L. Baillie, E. Bard, A. Bayliss, J. W. Beck, P. G. Blackwell, C. Bronk Ramsey, C. E. Buck, G. S. Burr, R. L. Edwards, M. Friedrich, P. M. Grootes, T. P. Guilderson, I. Hajdas, T. J. Heaton, A. G. Hogg, K. A. Hughen, K. F. Kaiser, B. Kromer, F. G. McCormac, S. W. Manning, R. W. Reimer, D. A. Richards, J. R. Southon, S. Talamo, C. S. M. Turney, J. van der Plicht and C. E. Weyhenmeyer, IntCal09 and Marine09 radiocarbon age calibration curves, 0–50,000 years cal BP., *Radiocarbon*, 2009, **51**, 1111–1150.

6. S. W. Leavitt and B. Bannister, *Dendrochronology and Radiocarbon Dating: The Laboratory of Tree Ring Research Connection*. University of Arizona, Tucson, AZ, 2004.

7. T.-C. Chiu, R. G. Fairbanks, Li Cao and R. A. Mortlock, Analysis of the atmospheric ^{14}C record spanning the past 50,000 years derived from high-precision [230]Th/[234]U/[238]U and [231]Pa/[235]U and ^{14}C dates on fossil corals., *Quat. Sci. Rev.*, 2007, **26**, 18–36.

8. L. R. Pessenda, S. M. Gouveia and R. Aravena, Radiocarbon dating of total soil organic matter and humin fraction and its comparison with (super 14)C ages of fossil charcoal., *Radiocarbon*, 2001, **43**, 595–601.

9. N. Ohkouchi and T. I. Eglinton, Compound-specific radiocarbon dating of Ross Sea sediments: A prospect for constructing chronologies in high-latitude oceanic sediments., *Quat. Geochronol.*, 2008, **3**, 235–243.

10. D. Grimm, The mushroom cloud's silver lining., *Science*, 2008, **321**, 1434–1437.

11. N. Lynnerup, H. Kjeldsen, S. Heegaard, C. Jacobsen and J. Heinemeier, Radiocarbon dating of the human eye lens crystallines reveal proteins without carbon turnover throughout life., *PLoS ONE*, 2008, 3(1), e1529.

12. N. Roberts, *The Holocene: An Environmental History*. Blackwell Publishers, Oxford, 1998.

13. COHMAP, Project Members, Climatic changes of the last 18,000 years: observations and model simulations., *Science*, 1988, **241**, 1043–1052.

14. H. E. Wright Jr, J. E. Kutzbach, T. Webb III, W. F. Ruddiman, F. A. Street-Perrott and P. J. Bartlein (ed), *Global Climates since the Last Glacial Maximum*. University of Minnesota Press, Minneapolis, MN, 1993.

15. S. D. Mooney, S. P. Harrison, P. J. Bartlein, A.-L. Daniau, J. Stevenson, K. C. Brownlie, S. Buckman, M. Cupper, J. Luly, M. Black, E. Colhoun, D. D'Costa, J. Dodson, S. Haberle, G. S. Hope, P. Kershaw, C. Kenyon, M. McKenzie and N. Williams, Late Quaternary fire regimes of Australasia., *Quat. Sci. Rev.*, 2011, **30**, 28–46.

16. K. Meredith, *Radiocarbon Age Dating Groundwaters of the West Canning Basin, Western Australia*. ANSTO Report C-1038 prepared for the Government of Western Australia, Perth, 2009.

17. F. Nocete, R. Sáez, M. R. Bayona, A. Peramo, N. Inacio and A. Abril, Direct chronometry (^{14}C AMS) of the earliest copper metallurgy in the Guadalquivir Basin (Spain) during the third millennium BC: first regional database., *J. Archaeol. Sci.*, 2011, **38**, 3278–3295.

18. K. Matsumoto and R. M. Key, Natural radiocarbon distribution in the deep ocean, in *Global Environmental Change in the Ocean and on Land*, ed. M. Shiyomi *et al.*, Terra Publishing Company, Tokyo, Japan, 2004, pp. 45–58.

19. Q. Hua, Radiocarbon: a chronological tool for the recent past., *Quat. Geochronol.*, 2009, **4**, 378–390.

CHAPTER 10

Development and Initial Biogeochemical Applications of Compound-Specific Sulfur Isotope Analysis

P. F. GREENWOOD,*[a,b,c] A. AMRANI,[d] A. SESSIONS,[e]
M. R. RAVEN,[e] A. HOLMAN,[c] G. DROR,[d] K. GRICE,[c]
M. T. MCCULLOCH[b] AND J. F. ADKINS[e]

[a] Centre for Exploration Targeting, The University of Western Australia, M006, 35 Stirling Highway, Crawley, WA 6009, Australia; [b] School of Earth and Environment, The University of Western Australia, M006, 35 Stirling Highway, Crawley, WA 6009, Australia; [c] Western Australian Organic and Isotope Geochemistry Centre, Curtin University, Perth, WA 6845, Australia; [d] The Institute of Earth Sciences, The Hebrew University, Edmond J. Safra Campus, Givat Ram Jerusalem, 91904, Israel; [e] Division of Geological and Planetary Sciences, California Institute of Technology, 1200 East California Boulevard, Pasadena, CA 91125, USA
*Email: paul.greenwood@uwa.edu.au

10.1 Introduction

Compound-specific isotope analysis (CSIA) has previously been applied to the stable isotopes of hydrogen, carbon, nitrogen, and oxygen, but has only recently been extended to sulfur. The $\delta^{34}S$ values of chromatographically resolved organic analytes can now be reliably measured by modern

RSC Detection Science Series No. 4
Principles and Practice of Analytical Techniques in Geosciences
Edited by Kliti Grice
© The Royal Society of Chemistry 2015
Published by the Royal Society of Chemistry, www.rsc.org

commercial multi-collector inductively coupled plasma mass spectrometry (MC-ICPMS) instruments providing high mass resolution. The m/z 32 chromatogram and $\delta^{34}S$ values of resolved hydrocarbons from the gas chromatography (GC)-ICPMS analysis of a Caspian Sea oil is shown in Figure 10.1 and was similar to the data reported from the same oil in the first practical application of this technology to a natural sample.[1]

Gas-source isotope ratio mass spectrometry (irMS), traditionally used for stable isotope measurements, is unable to support the analysis of sulfur isotopes in organic sulfur compounds (OSCs) isolated by GC. The main hurdle in this pursuit is the need for continuous oxidation of OSCs to sulfur dioxiode and the separation of other combustion products (*e.g.* water, carbon dioxide). ICPMS overcomes this problem by atomising and ionising OSCs in the plasma source enabling the direct measurement of monoatomic sulfur ions (*i.e.* $^{32}S^+$ and $^{34}S^+$). A new problem arises in that the isotopologues of oxygen interfere with those of sulfur at both m/z 32 and 34. A theoretical mass-resolving power ($m/\Delta m$) of 1808 is required to distinguish $^{32}S^+$ (31.9721 Da) from $^{16}O_2^+$ (31.9898 Da), and of 1221 to distinguish $^{34}S^+$ (33.9678 Da) from $^{16}O^{18}O^+$ (33.9940 Da). In practice, mass resolution of several thousand is needed to achieve stable isotope ratios,[2–4] but this is well within the capabilities of most modern MC-ICPMS instruments. The use of plasma ionisation to produce monoatomic ions yields several other benefits, including the lack of oxidative combustion reactors and catalysts that must be regenerated, and eliminating the need for oxygen isotope corrections in calculating $\delta^{34}S$. A substantial drawback, however, is the complexity and cost of MC-ICPMS instrumentation.

Early steps to use organic sulfur fractions included extraction methods to gain at least similar functional groups (*e.g.* humic acids).[5,6] Just before the emergence of the GC-ICPMS procedure for continuous-flow sulfur-CSIA,

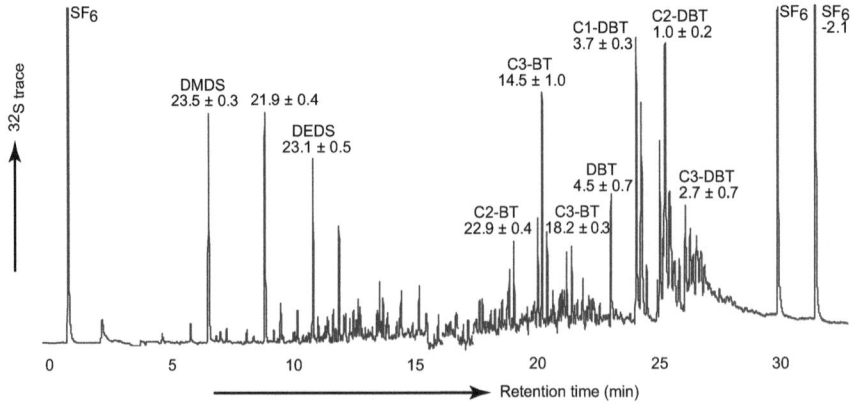

Figure 10.1   *M/z* 32 chromatogram from GC-ICPMS of a low-sulfur Caspian Sea crude oil. $\delta^{34}S$ values are shown atop peaks corresponding to major organic sulfur compounds.
(Figure modified from Amrani *et al.*[1])

measurement of the δ^{34}S values of individual OSC was performed by EA-irMS analysis following their isolation using preparative high-performance liquid chromatography (HPLC).[7] These were very labour-intensive and time-consuming analyses, but they demonstrated the concept of sulfur-CSIA which represents a major step forward for sulfur biogeochemical studies. The prototype GC-ICPMS technology for sulfur-CSIA was developed in 2009[1] and two other similar facilities have since been established.[8,9]

10.1.1 Sulfur Biogeochemistry and δ^{34}S

Organic sulfur (OS) is the second major pool of reduced sulfur in sediments after pyrite, and has significant influence on the coupled global biogeochemical cycles of carbon, sulfur, and oxygen. Sulfurisation is thought to play a key role in the preservation of organic matter,[10] including functionalised organic compounds which retain a structural link to their biological source (*i.e.* biomarkers). Whereas biomolecules with abundant functional groups (*e.g.* carbohydrates)[11,12] are highly labile and subject to rapid microbial mineralisation,[13–15] their reaction with reduced sulfur species during diagenesis leads to cross-linking of molecules and a highly polymerised, stable molecular structure that is more resistant to low-temperature degradation.[16–19] Sulfurisation can thus sequester certain labile compounds otherwise vulnerable to diagenetic processes. Reduced sulfur can also participate in catalytic reduction of unsaturated compounds.[20]

The sulfur incorporated into sedimentary organic matter may potentially derive from biotic as well as abiotic pathways. OS species (*e.g.* the amino acid cysteine) are synthesized through direct reduction and assimilation of dissolved sulfate. These OS compounds are quite labile, however, and seem unlikely to survive diagenesis and contribute significantly to sedimentary OS.[13,21] Nevertheless, the contribution of biotic sulfur to sedimentary organic matter remains debated. Abiotic sulfur derives from the incorporation of reduced inorganic sulfur species, mainly HS^- and S_x^{2-}. Although the reduced sulfur species are themselves derived from biotic processes (*e.g.* microbial sulfate reduction), they are thought to be incorporated into OS by abiotic reactions during diagenesis.[22,23]

The occurrences and stable isotopic relationships of reduced inorganic and OS species in sediments have helped us to understand the timing and pathways of OS formation in a variety of environments, including modern oceans,[24,25] hypersaline basins,[26] freshwater environments,[27] and ancient systems such as the Miocene Monterey Formation.[28] The distribution of sulfur isotopes in the environment is controlled to first order by fractionations imparted during abiotic sulfate reduction. Many studies have shown this process strongly favours the lighter isotope, yielding under open system conditions sulfides that are significantly depleted in ^{34}S relative to the source sulfate,[22,29–35] sometimes by more than 70‰.[34,35] In contrast, OS derived from biotic pathways typically has a δ^{34}S value similar to that of seawater sulfate (present-day value of 21.0‰).[36,37] A second control on δ^{34}S is the microbial cycle of sulfide oxidation and subsequent disproportionation of intermediate oxidation phases of sulfur, such as elemental sulfur, thio-sulfate, or polysulfides.[31,32,38–40] These forms of sulfur are generally enriched in [34]S relative to co-occurring sulfide.[41] While the fractionations associated with sulfide oxidation are generally small,[23] those associated with microbial disproportionation can be up to 37‰.[32,39] It has been assumed that repeated cycles of reduction, oxidation, and disproportionation may lead to a large offset between the δ^{34}S values of sulfate and sulfide,[31] thereby also im-pacting the OS fractions.

Sedimentary OS often has δ^{34}S values between those of biotic and abiotic end members, suggesting possible contributions from both processes.[42–45] However, without the ability to measure δ^{34}S in individual compounds, the occurrence of significant isotope effects accompanying the incorporation of reduced sulfur species has not yet been carefully evaluated. It remains possible that most OS is derived from pore water sulfide or polysulfide (*i.e. via* abiotic reactions) but with fractionations that lead to greater [34]S enrichments in OSCs.

10.2   GC-ICPMS Instrumentation and Analytical Performance

Continuous-flow compound-specific sulfur isotope analysis was achieved by interfacing GC with MC-ICPMS. The coupling of GC with MC-ICPMS for isotope ratio measurements was first applied to other elements[46–49] and later adapted at Caltech for the study of sulfur isotopes in organic compounds.[1] Two similar instruments have recently been established at the Hebrew University, Israel[8] and The University of Western Australia (UWA).[9] The Caltech and UWA facilities both use an Agilent GC while the Hebrew University has a Perkin Elmer (Clarus 580) GC, but essentially any commercially available GC could be used with minor modifications. All three instruments employ a ThermoFisher Scientific Neptune MC-ICPMS which has the im-portant characteristic of having a grounded interface. The high mass reso-lution ($m/\Delta m$) it can provide is required to clearly separate (by $\sim$15 mDa) sulfur and oxygen isotopes, such as the minor isobaric interference of $^{16}O_2$ from the ^{32}S signal.

The UWA instrument is shown in Figure 10.2, along with a schematic layout of the instrument currently in use at Caltech.

It should be noted that GC peak resolution remains constrained by the complexity of the sample mixture. Coeluting peaks, noisy backgrounds, and unresolved complex mixtures all create difficulties and thus can reduce the accuracy and precision of δ^{34}S measurements. The same limitation impacts CSIA for other elements. In all of these cases, chromatographic complexity can sometimes be overcome with chemical, thermal, or other separation procedures to reduce organic extracts into less complex fractions more amenable to uncompromised GC resolution of individual compounds.[1,7]

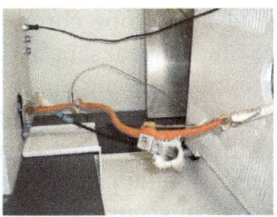

Transfer line from GC (right) and ICPMS (left). High Temp. maintained by low-V resistive heating

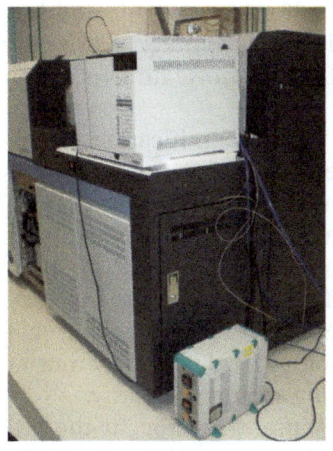

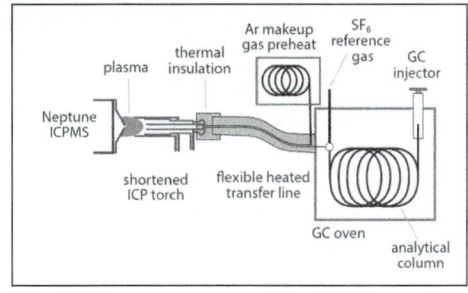

GC (top-front) -ICPMS

Schematic prototype instrument (Amrani et al 2009). GC carrier gas and Ar sample gas flow coaxially through the transfer line and into the injector, where they mix. SF6 reference gas (50 ppm S) is introduced at end of analytical column.

Figure 10.2 GC-ICPMS facility at the University of Western Australia and schematic layout of first GC-IPMS instrument at Caltech.

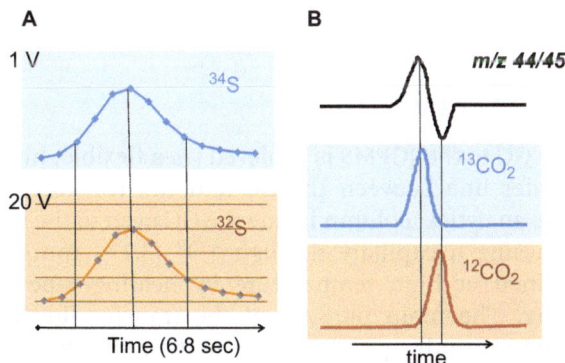

Figure 10.3 Comparison of the temporal profiles typical of (A) ^{32}S and ^{34}S signals from GC-ICPMS measurement of an organic sulfur compound; and (B) $^{13}CO_2$ and $^{12}CO_2$ (and corresponding *m/z* 44/45) signals from GC-irMS analysis of an organic analyte.

Unlike for $\delta^{13}C$ and δD analysis, however, there seems to be little chromatographic separation of the sulfur isotopologues (*i.e.* molecules having a ^{34}S substitution). The temporal profiles of ^{32}S and ^{34}S measured from GC-ICPMS analysis of a representative OSC are shown in Figure 10.3, and are compared to the $^{13}CO_2$ and $^{12}CO_2$ signals from the GC-irMS analysis of an organic analyte. In the absence of an isotope chromatographic effect, partial integration of the sides of co-eluting peaks, or of the central most intense part of a peak, may still yield accurate δ^{34}S measurements. Amrani *et al.*[1] compared the integration of whole *versus* centre parts of sulfur hexafluoride and hexylthiophene peaks and found the latter partial peak strategy provided a slight improvement in precision (approaching the shot-noise limit) and only a slight difference ($\sim 0.2‰$) in accuracy.

A related, ongoing challenge of sulfur-CSIA is the identification of analytes. Because the ICPMS operates as an element-specific detector, lone organosulfur peaks eluting in the midst of many other (non-sulfur-bearing) hydrocarbons can still be accurately measured with only minor reduction in accuracy and precision.[1] However, the molecular mass spectra (*e.g.* obtained by a GC-MS) of these coeluting mixtures are sufficiently complex that analyte identification can be difficult or impossible. S-CSIA thus presents a challenge that is somewhat unique in the stable isotope world, namely that we can measure the δ^{34}S values of many peaks that we cannot identify.

There have been other analytical challenges more specific to the GC-ICPMS technology, perhaps most significantly the efficient transfer of GC-resolved analytes (particularly less volatile high molecular weight OSCs) to the ICP torch. Fortunately, this issue has been recently addressed by more effective heating of the transfer line and, particularly the connection to the ICP torch, such that analysis of the traditional GC range of organic analytes can now be supported.

Several key aspects of the analytical technology are separately described below.

10.2.1   GC-ICPMS Interface

Coupling of the GC to the ICPMS is achieved *via* a flexible, high-temperature (>300 °C) transfer line between the GC and the ICP torch (Figure 10.2). Extension of the analytical column inside the GC oven to the GC torch can be achieved by passing a capillary through 1/8″ (3.175 mm) stainless steel tubing maintained at high temperature by heating tape or low-voltage resistive heating. The main purpose of the transfer line is to maintain GC-resolved analytes in the gas phase during their transfer to the ICP torch. Flexibility of the transfer line is critical to preserve the ability of the ICPMS to 'tune' torch position relative to the cone/skimmer orifice.

Argon makeup gas is required at relatively high flow (1–2 L min^{-1}) to operate the inductively coupled plasma, and should be preheated to minimise the condensation of analytes as they elute from the transfer capillary into the torch. This was initially achieved by diverting the argon flow through the GC oven (so as to track with oven temperature) and then flowing coaxially with the GC carrier gas through the heated transfer line.[1] However, it was subsequently discovered that such a setup leads to variable gas flow rates and associated ICPMS tuning problems, probably due to an imperfect response of mass flow controllers to changing flow resistance in the GC oven. Currently, the argon sample gas supply is heated by passage through a separate, heated box maintained at 300 °C.

10.2.2 Sulfur Hexafluoride Reference Gas

Sulfur hexafluoride, an inert, odourless, and inexpensive gas, has proved convenient for instrument tuning and isotopic calibration. A simple gas inlet system can be constructed to provide either continuous or pulsed introduction of sulfur hexafluoride diluted in helium to the mass spectrometer.[1] A continuous supply of sulfur hexafluoride is required for instrument tuning and this exercise also benefits from a capacity to control the sulfur hexafluoride concentration *via* the level of helium dilution. Microvolume capillary loops can introduce known volumes of sulfur hexafluoride–helium to the GC carrier gas stream as discrete peaks. This produces sulfur hexafluoride peaks during sample analysis, useful for $\delta^{34}S$ calibration. Calibration can alternatively be achieved by reference to coinjected internal standards, as is often done for $\delta^{13}C$ and δD CSIA, although the requirement for a tuning gas still remains.

10.2.3 Data Processing and Calibration

As for other CSIA methods, an MS scan cycle of sufficiently short duration (<200 ms) is required to provide adequate temporal resolution of chromatographic peaks. Current Neptune software does not support the data processing of transient (time-varying) signals, so data must be exported and processed separately. Current solutions include Microsoft Excel, MatLAB, or Isodat (Thermo IRMS software). Analyte isotope ratios ($^{34}S/^{32}S$) are measured from background-subtracted peak areas. Various algorithms could be implemented to speed the processing of the exported data, especially for complex chromatograms. However, manual peak and baseline interval identification in Isodat generally produces more accurate data than simple peak detection algorithms.

The results are expressed in conventional $\delta^{34}S$ notation as a ‰ deviation from the international standard Vienna Canyon Diablo Troilite (V-CDT) according to the following equation:

$$\delta^{34}S(‰) = (R_a/R_s) - 1$$

where R_a is the $^{34}S/^{32}S$ ratio of analyte and R_s is the $^{34}S/^{32}S$ ratio of V-CDT standard ($\equiv 0.044151$).[50]

The ICPMS data can then be converted into $\delta^{34}S$ values through reference to a normalisation curve constructed using the $\delta^{34}S$ values of known standards (*e.g.* measured in isolation by EA-irMS or ICPMS) covering a broad range of $\delta^{34}S$ values, as demonstrated in Figure 10.4. The ICPMS-determined $\delta^{34}S_{analyte}$ values are further calibrated against $\delta^{34}S_{SF6}$ values of the same chromatogram.

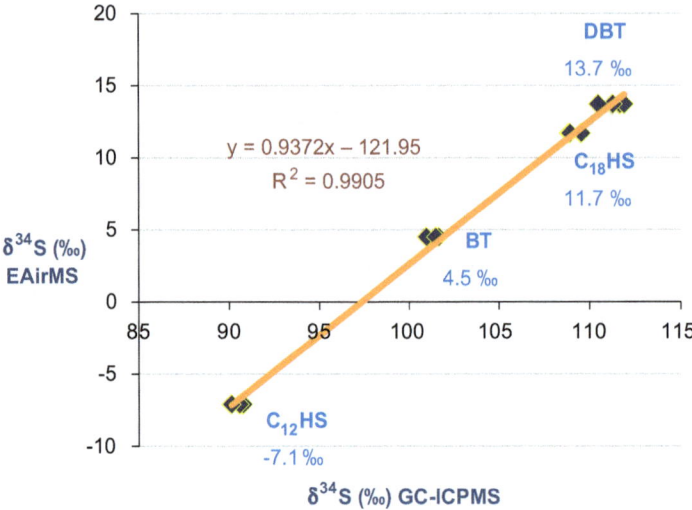

Figure 10.4 Normalisation curve from EA-irMS *versus* GC-ICPMS measured δ^{34}S values of several commercial standards.

10.2.4 Sensitivity and Precision

The typically low relative concentrations of many OSCs in natural samples such as oils or sedimentary organic matter can be a challenge to the precision of δ^{34}S analysis. GC-ICPMS measurements of commercial standards have demonstrated impressive precision (<0.2‰) and accuracy (<0.3‰) for peaks containing <100 pmol sulfur per compound, yielding ^{34}S peak areas of 0.5–2 V s in the medium resolution ($\sim$4000, $m/\Delta m$, 5–95%) mode of the MC-ICPMS system.[1] Based solely on counting statistics, *i.e.* assuming no other sources of error, roughly 5 billion sulfur ions (0.2 billion ^{34}S ions; $\sim$8 fmol of sulfur) must be counted to achieve a standard deviation for δ^{34}S of 0.1‰. Assuming a useful ion yield in the ICPMS of 10 ppm (which is typical at medium resolution), this corresponds to a requirement for injecting $\sim$800 pmol of sulfur. A 10-fold reduction in the demand for precision (*i.e.* to 1.0‰ standard deviation, still sufficient for many biogeochemical problems of interest) leads to a 100-fold reduction in sample requirement (to 8 pmol of sulfur). In real-world analyses, additional sources of error mean that results are typically a factor of 2 or 3 above this 'shot-noise' limit.[1] Nevertheless, sulfur-CSIA can exceed the sensitivity of carbon and hydrogen CSIA, primarily because of the relatively high abundance of the rare isotope ^{34}S (4.21%, *versus* 1.11% for ^{13}C and 0.015% for ^{2}H).

The recent analysis of seawater dimethylsulfide (DMS) and dimethylsulfoniopropionate (DMSP) on a separate GC-ICPMS facility with a purge and trap GC inlet and operating in high mass resolution mode (n.b., sensitivity reduced by $\sim$2-fold relative to medium resolution) on just 23–179 pmol synthetic DMS or DMSP was conducted with precision better than 0.3‰.[8] A sub-ppb level of sensitivity is comparable to standard GC-MS.

10.3 δ^{34}S Model of a Sedimentary Sulfur Cycle

As an example of sedimentary sulfur cycling and associated δ^{34}S behaviour, a simple flowchart of predicted mechanism and δ^{34}S values associated with the Here's Your Chance (HYC) sediment-hosted metal sulfide deposit is shown in Figure 10.5. Sedimentary recycling processes are critical in the formation of reduced sulfur species crucial to the deposition of many mineral types (*e.g.* transportation and precipitation of metal species from hydrothermal fluids).

The HYC ore body, the largest Palaeoproterozoic Pb–Zn–Ag deposit of north-east Australia, is hosted in the 1640 ± 3 Ma[51] Barney Creek Formation (BCF) of the McArthur Group. The deposit contains eight separate ore bodies,[52,53] and several different sulfide phases with distinctive δ^{34}S values reflecting multiple sulfur sources.[54] For instance, the ^{34}S enriched nature of one major pyrite phase (δ^{34}S up to +45‰) relative to an earlier diagenetic pyrite phase (−13 to +15‰) was attributed to closed-system reduction of sulfate limited within the sediment pile.[54]

The unmineralised organic matter at HYC was only marginally mature with H/C >1.6 and Rock-Eval T_{max} values as low as 435 °C, reflecting exceptional preservation for its age.[55] Hydrocarbon biomarkers from several different microbial sources have been detected, including aliphatic biomarkers of sulfide-oxidising bacteria.[56] Dibenzothiophenes (DBTs) and

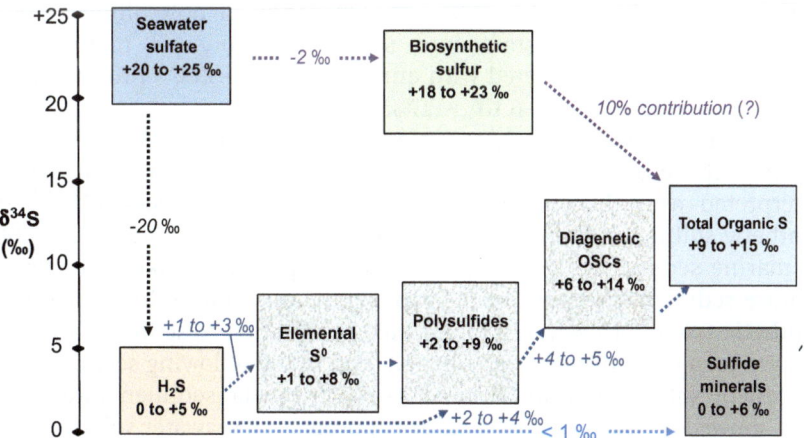

Figure 10.5 Proposed sulfur cycle model and associated δ^{34}S behaviour at the HYC mine (McArthur River Station, northern Australia), a sediment-hosted massive sulfide deposit. From Holman *et al.*[64] with δ^{34}S values and fractionations predicted from several previous studies.[33,67,69]

other OSCs have also been detected in the HYC organic matter.[57] A number of studies have proposed that organic sulfur are sourced from poly-sulfides[7,24,44] or elemental sulfur,[58] and DBTs can be produced by the re-action of reduced sulfur with unsaturated hydrocarbons[59,60] or from thermochemical sulfate reduction (TSR).[61] Other sedimentary studies have similarly shown that metal sulfide mineralisation and organic matter sul-furisation can occur simultaneously.[44,62] This is true even with pyrite for-mation, despite the kinetic preference for pyrite,[63] presumably reflecting a subset of organic compounds that are more highly reactive than the bulk organic matter.

The sulfur cycle and δ^{34}S model shown in Figure 10.5 represents an ex-tension of a previous scheme[33] for sulfur isotopic fractionation in nature.[64] Briefly, bacterial reduction of seawater sulfate, estimated from measurement of evaporite deposits to have Paleoproterozoic δ^{34}S value of $+20$ to $+25$‰[65] produced ^{34}S-depleted hydrogen sulfide with a δ^{34}S value of $\sim +2$‰ (~ 20‰ depletion)—based on δ^{34}S values of Wollogorang pyrite, underlying the BCF and through which mineralising fluid is believed to have flowed.[66] The δ^{34}S fractionation associated with biotic sulfate reduction will be comparatively negligible (< -2‰).[33] Sulfide minerals in the HYC (PbS, ZnS, AgS) have δ^{34}S values of 0 to $+8$‰,[54] consistent with very little fractionation ($< +1$‰) of dissolved hydrogen sulfide driven precipitation.[33] Elemental sulfur and polysulfides, if produced by phototrophic sulfur-oxidising bacteria would be expected to show a small (~ 1 to 3‰) enrichment in ^{34}S compared to hydrogen sulfide.[67] Laboratory experiments have shown that polysulfides in chemical equilibrium with elemental sulfur and dissolved sulfide are typically 2–4‰ enriched compared to sulfide, although as high as 6‰ enrichment is also possible.[68] The incorporation of polysulfide anions of intermediate oxidation state into sedimentary organic matter during diagenesis might be expected with enrichment of 4–5‰,[69] producing OSCs with a predicted δ^{34}S of $+6$ to $+11$‰.

Sedimentary OS, however, commonly exhibits bulk δ^{34}S values that lie between those of biotic and abiotic end members, a result that has been interpreted as reflecting contributions from both processes.[42,44,45] Bio-synthetic sulfur has been historically estimated to contribute 10–25% of OS in marine sediments,[58] although a lower proportion is likely in thermally mature sediments because of the rapid mineralisation of highly labile bio-synthetic OSCs.[44] However, a biological contribution to sedimentary OS has not yet been unequivocally established, and in the following section on early diagenetic sulfurisation in Cariaco Basin (Venezuela) sediments we note the data showed no OSCs with δ^{34}S values approaching seawater sulfate, thus no evidence for biotic-sourced OS. Despite these uncertainties, the total OS range of the present model is based on a 10% contribution of biologically sourced sulfur.

δ^{34}S values for elemental sulfur ($+6$ to $+8$‰), OS ($+5$ to $+8$‰) and early phase sulfide minerals (-5 to $+6$‰) recently measured in HYC sediments were in the ranges predicted by this model.[64] Furthermore, preliminary δ^{34}S values of several DBTs isolated from the OM of HYC sediments were +10 to +11‰ (within the +6 to +14‰ range predicted for diagenetic OSCs, Figure 10.5), although this sulfur-CSIA data is considered tentative because of the low concentrations and precision of analysis.

10.4   Early Applications of Sulfur-CSIA

Several contemporary research questions in sulfur biogeochemistry that have attracted initial attention include the mechanism and timeframes of diagenetic organic sulfurisation, the characterisation of ocean derived sulfur aerosols and both oil and mineral (as discussed above for the HYC model) exploration applications. Results and analytical insights from these initial sulfur-CSIA studies are summarised below.

10.4.1   Diagenetic Formation of Organosulfur Compounds

The concentrations and sulfur isotopic compositions of sedimentary sulfur species have been used to investigate organic matter sulfurisation in a variety of environments, including modern oceans,[24,25] hypersaline basins,[26] freshwater environments,[27] and ancient systems such as the Miocene Monterey Formation.[28] Most sulfur is incorporated into kerogen on timescales of thousands of years or less, so kerogen $\delta^{34}S$ will reflect early diagenetic and sedimentary processes. The heterogeneity of sedimentary organic matter, however, necessitates a compound-specific approach to interpreting the ancient kerogen $\delta^{34}S$ record. Sulfur isotopic signals generated by the organic matter sulfurisation process may make it possible to use sulfur-CSIA to inform studies of the intertwined sulfur, carbon, and oxygen cycles in Earth's history. Additionally, sulfur-CSIA itself is a powerful tool for improving our understanding of organic sulfurisation mechanisms. Sulfur isotopic compositions of individual OSC can indicate the timescale and location of their formation because inorganic sulfur species exhibit strong isotopic gradients in many anoxic sediments.

Case Study: Early Diagenetic Sulfurisation in the Cariaco Basin

Werne and colleagues conducted the first compound-specific $\delta^{34}S$ analyses on OSC from the sediments of the Cariaco Basin.[7] Individual sulfurised isoprenoids were isolated from large (200–250 g) samples by preparative chromatography (prep-HPLC) prior to EA-irMS analysis. The $\delta^{34}S$ values of these OSCs were typically around −15‰, substantially [34]S-enriched relative to other coexisting sulfur pools, including both kerogen sulfur and dissolved sulfide. The sulfur isotopic enrichment of roughly 10‰ could be only partially explained by an equilibrium isotope effect during formation identified in previous laboratory experiments.[69]

Sulfur-CSIA by GC-ICPMS requires much smaller sediment samples (~ 5 g) which eliminates the need for laborious manual separation. By this method, the same OSCs from the same core (ODP Core 1002B) yielded different results[70] to the previous EA-irMS based measurements.[7] The GC-ICPMS data on previously investigated isoprenoids and other OSCs were consistently [34]S-depleted relative to kerogen and dissolved sulfide. The δ^{34}S profiles of selected OSCs are shown in Figure 10.6, and reflect progressive [34]S enrichment with depth, consistent with continuous formation from inorganic sulfur species, which also become increasingly [34]S-enriched with depth. The δ^{34}S discrepancy of these studies[7,70] is likely due to the presence of elemental sulfur in the fractions isolated by HPLC. Elemental sulfur in Cariaco Basin sediments is abundant in non-polar lipid extracts and was relatively [34]S-enriched, with preliminary δ^{34}S measurements, ranging from roughly −19‰ to +5‰,[66] that are capable of explaining the more enriched values previously reported.[7] Sulfur-CSIA by GC-ICPMS avoids this problem caused by homogenisation of an isolated, imperfectly purified fraction.

The [34]S-depleted nature of volatile OSCs was similar to that of coexisting pyrite. Like pyrite, these OSCs appear to have a combination of syngenetic (water column) and diagenetic (sediment) sources, the relative importance of which varies for different individual compounds. For example, one of the isoprenoid compounds that was measured in the earlier study and also shown to accumulate in Cariaco Basin sediments may have an exclusively diagenetic source.[7] Accumulation and isotopic patterns for the C$_{20}$ isoprenoid thiophene, another dominant OSC, indicate a substantial syngenetic source and do not conclusively establish diagenetic formation.[70]

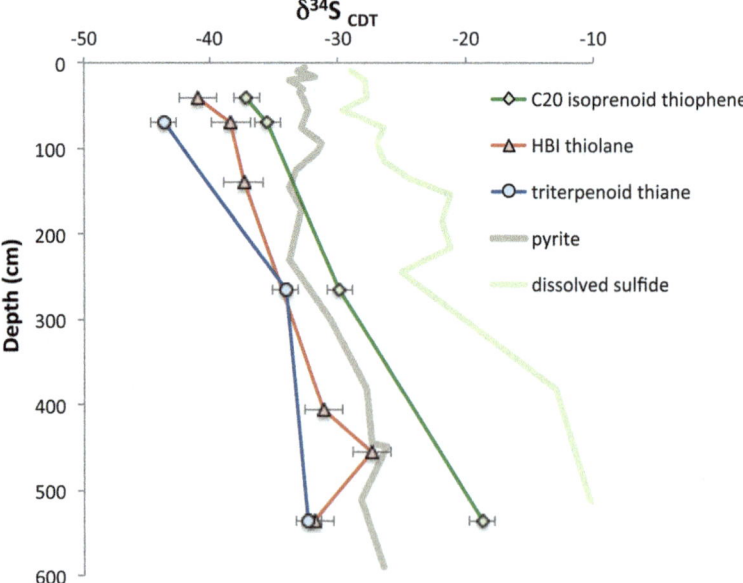

Figure 10.6 δ^{34}S profile of selected organic sulfur compounds detected in Cariaco Basin sediments and also for co-occurring pyrite and dissolved sulfide at different water column depths.

Syngenetic and diagenetic sulfurisation may also have different reaction mechanisms and isotope effects. The ^{34}S-depleted volatile OSC in Cariaco Basin sediments appear to record a kinetic isotope effect with respect to their sulfur source, which is likely dissolved bisulfide (HS$^-$). Syngenetic organic sulfurisation, in contrast, may have a more reversible reaction mechanism and record equilibrium isotope effects like those seen for OSC polymers in experimental studies.[69] Bulk kerogen δ^{34}S values would then reflect a complex mixture of OSCs, each potentially recording distinct information about the sedimentary sulfur cycle.

Thus far, no evidence of individual OSC with δ^{34}S values approaching that of seawater sulfate, and that might be candidates for the postulated addition of biosynthetic sulfur.

10.4.2 Oceanic Emissions of DMS

Sulfur-CSIA has also been extended to target gaseous-range OS analytes such as DMS. DMS is a degradation product of DMSP, a metabolite of marine phytoplankton, and represents the largest natural source of biogenic sulfur to the global atmosphere. DMS is emitted to the atmosphere and rapidly oxidises to sulfate, a significant constituent of submicrometre aerosols. These aerosols scatter incoming solar radiation and may act as cloud condensation nuclei (CCN) that change the albedo of clouds. There are multiple sources of sulfur to the atmosphere including biogenic, sea-spray, anthropogenics, volcanic emissions, dust, and biomass burning. Sulfur isotope ratios may offer a way to calculate the contribution of ocean-derived DMS to global sulfur cycling and aerosol budgets. However, the δ^{34}S of DMS emitted to the atmosphere from the ocean has been difficult to measure because of the low natural concentration of DMS in oceanic water (typically few nM) and a high volatility and reactivity that make such analysis very difficult. Recently, a multistep extraction procedure coupled with Raney nickel dehydrosulfurisation, and then fluorination (to sulfur hexafluoride) for sulfur isotope analysis by irMS was used to obtain δ^{34}S values for DMSP from intertidal macroalgae (+17.3 to +19.3‰) and an estuarine phytoplankton bloom (+19 to +20‰).[71] These labour-intensive and time-consuming analyses of DMSP required processing of large quantities of algae and seawater (50 L) during blooms, where DMSP is at high concentrations, to obtain sufficient sulfur (>6 µmol) for isotope analysis.

The current coupling of GC and ICPMS technology offers the required sensitivity (pmol rather than µmol level) and a significantly simpler way for the sulfur isotope analysis of trace level compounds such as DMS, thereby enabling its measurement even at oligotrophic regions. Preliminary measurements using this technique on water samples from San Pedro, CA, Tasmania, and Pacific Costa Rica were the first δ^{34}S analysis of DMS in oceanic water and showed relatively uniform values of about +18 to +21‰.[72] However, DMS recoveries were low (<50%) due to storage and transfer problems and uncertainties were thus relatively large. An improved purge and trap method recently developed was able to concentrate the small amounts of DMS and DMSP with more than 98% recovery.[8] A few millilitres of ocean water (> 23 pmol sulfur) were sufficient for precise and accurate (<0.3‰) $\delta^{34}S$ analysis by GC-ICPMS. This purge and trap GC-ICPMS method was then used to measure the $\delta^{34}S$ of DMS and DMSP in a range of geographically distinct marine waters from around the globe.[73] These DMSP samples showed a remarkable $\delta^{34}S$ consistency of +18.9 to +20.3‰ in the surface ocean. Different seasons, ocean basins, phytoplankton blooms or non-blooms, did not seem to affect the $\delta^{34}S$ of DMSP at the ocean surface. Depth profiles (Figure 10.7) of DMS and DMSP in Eilat (Northern Red Sea, Israel) down to 140 m reveal less than 1‰ difference between DMS and its precursor DMSP.[73] Very little sulfur isotope fractionation (<0.5‰) was observed during evaporation experiments of DMS from aqueous solution. This shows that the $\delta^{34}S$ of DMS emitted to the atmosphere is similar to that of DMS in the surface water which in turn is similar to that of DMSP. These data indicated that the $\delta^{34}S$ of oceanic DMSP is relatively homogenous, contributing to DMS with a quite distinctive $\delta^{34}S$ signature. Hence, $\delta^{34}S$ analysis might help measure the oceanic contribution of atmospheric DMS from anthropogenic sources of atmospheric sulfate, thereby enabling estimation of the DMS contribution to aerosols.

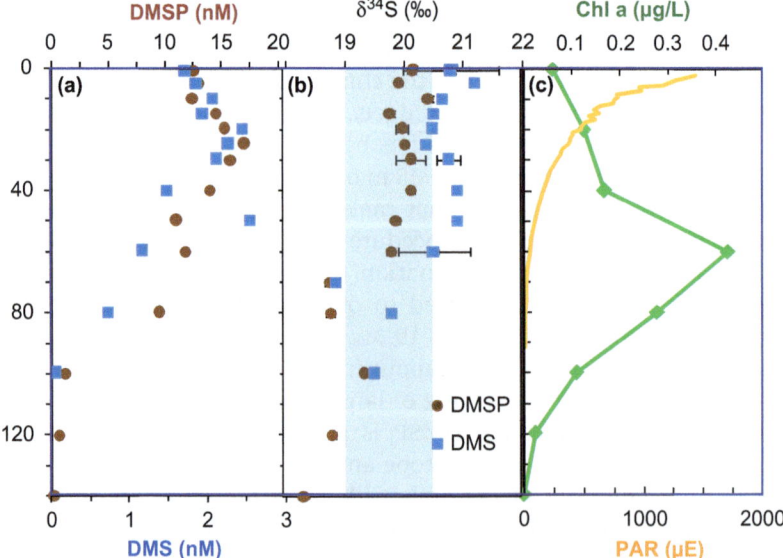

Figure 10.7 Depth distribution (0–140 m) in open water at Eilat, Red Sea of (A) DMSP and DMS concentrations; (B) $\delta^{34}S$ values of DMSP and DMS. The shaded area represents the range of $\delta^{34}S$ values of DMSP obtained for surface water (≤5 nm depth) from other locations around the globe; and (C) chlorophyll a concentration (green line) and photosynthetically active solar radiation (PAR, yellow line) between 400 and 700 nm (summer conditions of September 2012). Figure modified from Amrani *et al.*[73]

10.4.3 Oil Analysis

The largest petroleum systems in the world are carbonate/evaporite sequences, which are typically high in sulfur.[74] However, crude oils with sulfur concentrations of 4% or more are often excluded from industrial processing because of the interference of OSCs during the refinery process[41,75] and the effect relatively weak S–S bonds have on petroleum kinetics.

Unlike the large range of hydrocarbon biomarkers that have been detected in oils and source rocks, OSCs provide quite limited information about fossil fuel sources. Sulfur is usually incorporated into organic matter *via* secondary processes,[76] most significantly by TSR. Although TSR is a major control on the $\delta^{34}S$ of organic sulfur in oils, other factors such as depositional environment and thermal maturity may also influence the $\delta^{34}S$ of OSCs. Bulk $\delta^{34}S$ values of petroleum can vary over a wide range (−8 to 32‰)[77] and have proved quite useful for oil–oil correlations (see *e.g.* Gaffney *et al.*[78]).

10.4.3.1 *Thermochemical Sulfate Reduction and the $\delta^{34}S$ of Organic Sulfur*

TSR is a high-temperature redox process in which sulfates such as gypsum or anhydrite are reduced and organic matter oxidised.[79,80] It can have a drastically negative impact on oil quality since normal and branched alkanes, particularly in the gasoline range, are easily oxidised by TSR,[81–83] and the high concentrations of hydrogen sulfide and carbon dioxide generated can result in sour gas reservoirs of low economic value.

Several studies of TSR-impacted organic matter have shown this process can lead to variances in the distribution and $\delta^{13}C$ of hydrocarbon products,[84,85] from which certain trends were identified to reflect the extent of TSR.[76] For instance, loss of n-alkanes in preference to aromatics, and kinetic control over greater ^{12}C loss, leads to a saturate hydrocarbon fraction with a heavier $\delta^{13}C$ value. Likewise, the $\delta^{34}S$ of whole oils and their fractions can change with TSR.[80] The hydrogen sulfide produced has a value close to the source sulfate and its secondary reaction with hydrocarbons may lead to OS with similar $\delta^{34}S$ values.

The TSR process can lead to the formation of aromatic sulfur compounds such as benzothiophenes (BTs) and DBTs. The investigation of TSR and $\delta^{34}S$ was extended to the molecular level by separately measuring the $\delta^{34}S$ of parent and methyl BTs and DBTs in oils impacted by TSR.[61] The sample set comprised a number of Upper Jurassic oils including from the Smackover Formation (southern USA) where they had been varyingly impacted by TSR. These geological samples were complimented by a set of gold tube pyrolysis experiments conducted with three calcium sulfate spiked oils to simulate subsurface TSR occurrences.

Significantly, the oil and pyrolysis sample sets both showed the BT and methylated analogues were quick to adopt the $\delta^{34}S$ of the sulfate source at relatively low levels of TSR. A variance in the ^{34}S fractionation of BT and the

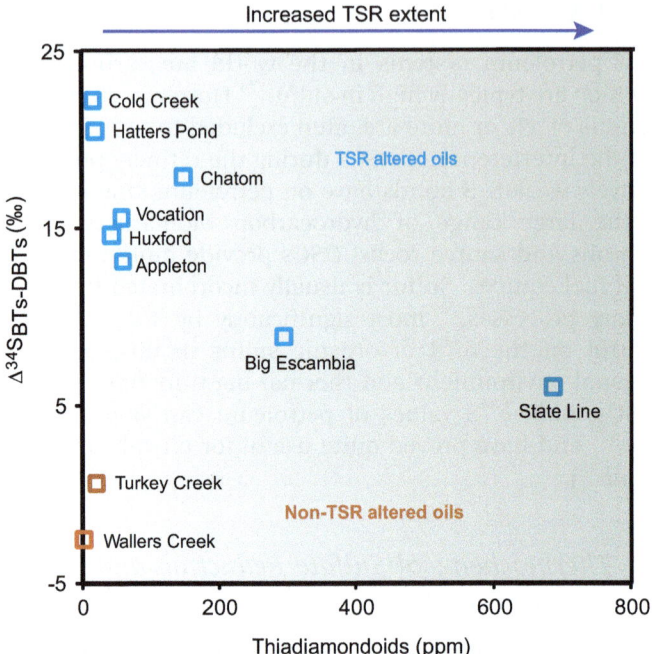

Figure 10.8 $\Delta\delta^{34}S_{BT\text{-}DBT}$ *versus* thiadiamondoid concentrations from a suite of Smackover Formation oils impacted to different degrees by thermochemical sulfate reduction (TSR). The concentrations of thiadiamondoids has previously been correlated against extent of TSR.[95] $\Delta\delta^{34}S_{BT\text{-}DBT}$ was measured using average $\delta^{34}S_{BT}$ and $\delta^{34}S_{DBT}$ values of all measured isomers.
(Figure modified from Amrani *et al.*[61])

higher molecular weight and more stable DBT, formed less rapidly by TSR, was shown to provide a good measure of the extent of TSR. As an example of the relationship between $\Delta\delta^{34}S_{BT\text{-}DBT}$ and TSR, GC-ICPMS measured values from several Smackover Formation oils impacted by TSR to different degrees are shown in Figure 10.8.[61]

$\Delta\delta^{34}S_{BT\text{-}DBT}$ was subsequently proposed as a sensitive indicator of the extent of early TSR and this new parameter should be of practical benefit in oil exploration studies. Further investigation of the $\delta^{34}S$ of, and $\Delta\delta^{34}S$ between, OSC products of TSR may also help illuminate the mechanistic complexities of this process which typically involves major alteration of the hydrocarbon composition of oils.[61]

10.4.3.2 *Other Controls on the $\delta^{34}S$ of Petroleum OSCs*

Thermal maturity. $\delta^{34}S$ values of particular OSCs may vary modestly with thermal maturity. Maturity has generally been observed to have only a small influence on the bulk $\delta^{34}S$ of oils and source rocks.[76,86] Pyrolysis experiments of source rocks and kerogens support this observation, showing relatively small sulfur isotope fractionation of ~2‰ between the kerogen, and the free hydrocarbons (*i.e.* oil/bitumen) and hydrogen sulfide thermally produced from it.[87,88]

Laboratory-controlled maturation experiments of an immature source rock were conducted to further investigate the impact of thermal maturity on the δ^{34}S of individual OSCs in fossil fuels.[89] Semi-batch pyrolysis experiments were operated at low pressures (up to 50 psi), at a slow heating rate of 4 °C day^{-1} (200–400 °C). The source rock used in the study was a Ghareb Formation oil shale attained from core samples from the Shfela Basin (Aderet 1 drillhole, Israel) exhibiting both high TOC (17.3 wt%) and high OS content (10 wt% from the kerogen).

The unheated bitumen showed a large distribution of OSCs with δ^{34}S values which ranged by about 15‰. The δ^{34}S value of pyrite in the source rock was −27‰. The OSCs consisted mainly of alkyl thiophenes ($>C_{10}$), and to a lesser extent sulfides and alkylbenzothiophenes. The bitumen released from the residual rock on heating to 400 °C consisted mainly of DBTs and other more condensed aromatic sulfur species. These OSCs exhibited a very narrow δ^{34}S range (−0.9 to +0.9‰), centred on the bulk kerogen δ^{34}S value of +0.6‰. The δ^{34}S value of a several OSCs identified as major products of the analytical pyrolysis as a function of temperature are shown in Figure 10.9.

Throughout these experiments the pyrolysates were condensed at 4 °C and collected in intervals of approximately 25 °C. Preliminary results show that the OSCs (*i.e.* thiolanes, alkylated thiophenes, benzothiophenes, and dibenzothiophenes) have a unique thermally dependent sequential molecular distribution. The δ^{34}S values generally increased with temperature by up to 3–4‰ for alkylthiophenes and sulfides relative to the initial samples that were collected at 250 °C and the kerogen. Benzothiophenes and dibenzothiophenes were consistently ^{34}S depleted compared with alkylated

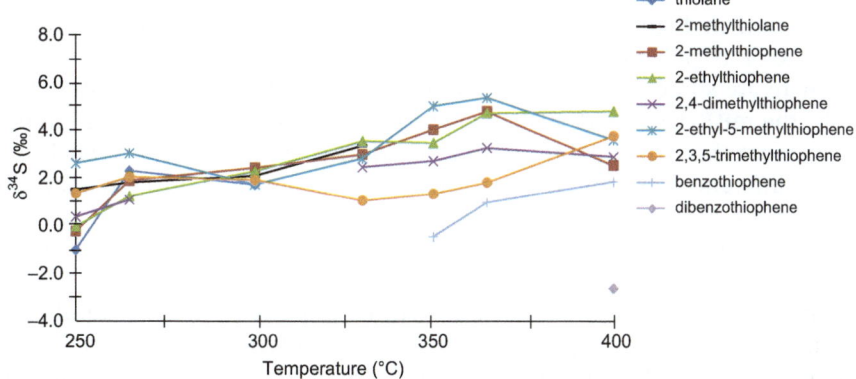

Figure 10.9 δ^{34}S profiles of selected organic sulfur compounds from oils that collected at different pyrolysis temperatures of Gahreb Formation source rock.

thiophenes and were closer to those in the raw kerogen. Differences in the $\delta^{34}S$ values for the different OSC families may be due to their formation from different types of sulfur bonding in the kerogen controlled by the thermal stability of the C–S bonds. Since BTs and DBTs form at relatively higher thermal maturities they probably form by cleavage of the more stable C–S bonds in the kerogen.

In general, the condensed oil fraction exhibits significantly less sulfur isotope variability than that of the unheated bitumen. This observation suggests that sulfur isotope mixing during thermal maturation may be responsible for the relative homogeneity of unaltered oils (*e.g.* by TSR or bacterial sulfate reduction (BSR)). Moreover, the 3–4‰ difference evident between some OSCs in the generated oils and the source rock further support the applicability of sulfur isotopes in oil–source rock correlations.

These preliminary results indicate that the molecular and $\delta^{34}S$ distribution of OSCs show a promising potential for indicating the thermal maturation stage of the source rock and its structural transformations.

Oil Source/Deposition Environment.   Despite the fact that OSCs are not as diagnostic of organic matter inputs as hydrocarbon biomarkers, sulfur-CSIA may still extend the oil source correlations of bulk $\delta^{34}S$. Bulk $\delta^{34}S$ values of petroleum can vary over a wide range (−8 to +32‰)[77] and have proved quite useful for oil–oil correlations.[78]

The $\delta^{34}S$ of BTs and DBTs from a small range of oils not impacted by TSR have now been measured across several GC-ICPMS studies. These include oils from the Smackover Formation (Gulf of Mexico; n.b. these are not TSR affected),[61] Oman,[61] Tarim Basin (north-west China),[9] and Jinxian Sag (northern China)[9] although the $\delta^{34}S$ values of the latter two oils were considered with some caution due to low analyte concentrations and the limited oil volumes available for analysis. Characteristic features of these oils are briefly given in Table 10.1.

The $\delta^{34}S$ values of parent and methyl BT and DBTs from the different oils are shown in Figure 10.10. The values for the methyl analogues are the average of all polymethyl (C_1–C_3) isomers analysed.

There are some clear differences in the $\delta^{34}S$ data of these oils. Most notably, the mBT, DBT, and mDBT values of the two Smackover Formation oils were more than 20‰ lighter than those of five Tarim Basin oils. Such isotopic differences are much larger than could be attributed to any maturity differences. Different palaeo-depositional environments might be a more plausible explanation for the large deviation in the $\delta^{34}S$ values of the OSCs in these oils.

The $\delta^{34}S_{OSC}$ data of the Smackover Formation oils were consistently less than 5‰, which was significantly lighter than the same products from a larger suite of TSR-affected oils from the Smackover Formation,[61] and is likely more indicative of the $\delta^{34}S$ of the source kerogen.

The sulfur-CSIA data for the Jinxian Sag oil was between the relatively high $\delta^{34}S_{OSC}$ values of the Tarim Basin oils and low $\delta^{34}S_{OSC}$ values in the

Table 10.1   Oils for which sulfur-CSIA data has been measured.

Oil name	Brief description
Smackover Formation oils, Gulf of Mexico	Upper Jurassic age; many Smackover oils have been impacted by TSR, data is presented here for the Turkeys Creek and Wallers Creek oils which were not thought to be TSR affected[61]
Oman oil	High-sulfur (3 wt%), relatively immature oil
Tarim Basin oils (NW China)	Numerous episodes of hydrocarbon generation, migration and accumulation. Multiple source rocks, but mainly of Cambrian or Ordovician age. Ordovician oils characterised by high DBT/phenanthrene ratio[96].
Jinxian Sag oil (China)	Tertiary age, Lacustrine; hydrocarbon reservoirs typically contain high sour gas and hydrogen sulfide concentrations, with much speculation about whether these are due to TSR or BSR[86]

BSR, bacterial sulfate reduction; DBT, dibenzothiophene; TSR, thermochemical sulfate reduction.

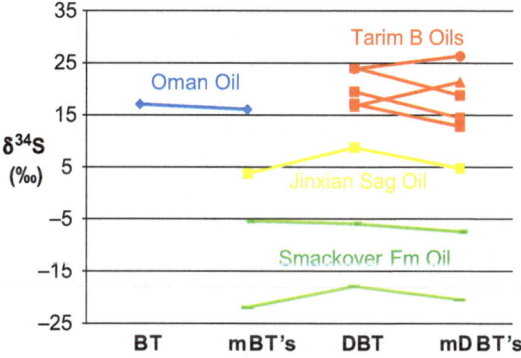

Figure 10.10 $\delta^{34}S$ data of benzothiophene (BT), dibenzothiophene (DBT), and C_1–C_3 polymethylated analogues (average of all measured isomers) from the GC-ICPMS analysis of several different oils unaffected by thermochemical sulfate reduction.[9,61]

Smackover Formation oils. Significantly, the $\delta^{34}S$ of the mBTs are similar to that of co-occurring DBT and mDBTs which suggests they had not been exposed to TSR, and implies instead a BSR influence on this oil. There has been much conjecture about the control on the sour gas and hydrogen sulfide-rich nature of Jinxian Sag hydrocarbon reservoirs.[90]

The concentrations of BTs and DBTs in the Oman oil were below detection limits, but the $\delta^{34}S$ of a range of BTs could be measured following their concentration by sealed gold tube pyrolysis. The data reported here corresponds to pyrolysis conditions of 360 °C applied for 40 or 89 h in the absence of calcium sulfate (n.b. added to other pyrolysis experiments to promote TSR). These $\delta^{34}S$ values were in the lower $\delta^{34}S$ range of the Tarim Basin oil analytes.

10.5   Conclusions and Future Application

Compound-specific $\delta^{34}S$ represents an entirely new analytical capability which could be applied to much of the OS system, with applications ranging from atmospheric and ocean chemistry to sediment diagenesis and ancient proxies. Already impressive precision and reproducibility with $\delta^{34}S$ of less than 0.5‰ has been demonstrated for the $\delta^{34}S$ analysis of volatile OSCs in seawater and associated aerosols, sediment extracts, crude oils and rock extracts. Following the development and subsequent commercial availability of continuous-flow CSIA technology for other light stable isotopes, particularly carbon and hydrogen, there was a rapid transition from measuring the isotopic compositions of bulk organic fractions to measuring individual molecular species. A similar trend might now be anticipated following the recent invention of sulfur-CSIA.

The $\delta^{34}S$ value of bulk sedimentary OS largely reflects that of the major sulfur source (*e.g.* porewater H_2S) and lacks the resolution to provide detail about the factors influencing $\delta^{34}S$ at the molecular level. The $\delta^{34}S$ values of specific OSCs can vary substantially and cover a wide range of values in different environments reflecting differences in (1) inorganic sulfur source; (2) pathway of OSC formation; (3) environmental conditions such as redox state; (4) timing of diagenetic sulfurisation of OM; and (5) diagenetic overprinting, including that from microbial sulfur cycling. Over two decades ago, the sedimentary occurrences of more than 1500 novel OSCs was reported,[10] but their geochemical significance was limited by a lack of knowledge about their formations mechanism(s) and reactivity with sulfur in the natural environment. These remain critical areas of study and will be well served by a capacity to measure the $\delta^{34}S$ of individual OSCs in ancient sediments or extant precursors.

Measurement of the $\delta^{34}S$ values of OSCs by GC-ICPMS has enormous analytical potential extending from biogeochemical studies to characterisation of synthetic chemicals and environmental contaminants and possibly even source correlation of chemical warfare agents. Sulfur-CSIA may help address a wide variety of questions in biogeoscience, including reconstruction of biomarker taphonomy, microbial sulfur cycling, and frontier petroleum basins. However, careful evaluation of the controls on sulfurisation and connection of the major pathways of sulfur incorporation during sedimentary diagenesis to patterns of $\delta^{34}S$ in individual molecules will be required to fully exploit this unique capability. This will require the acquisition of a substantial amount of compound-specific $\delta^{34}S$ data in modern and palaeoenvironments, as well as supporting data on inorganic sulfur species, from different compounds, depths, and environments. Measurement of the $\delta^{34}S$ of OSCs produced by reaction of organic matter with reduced inorganic sulfur and preserved in the sedimentary record for long time periods at natural abundance levels will help to greatly improve our understanding of past organic sulfur systems.

An example of one geoscience sulfur-CSIA application worthy of further exploration relates to the behaviour of organic and inorganic sulfur species during mass extinctions and the recovery periods that followed. The δ^{34}S values of sedimentary OSCs could help reconstruct the sulfur cycles associated with these key evolutionary periods. Some of the most significant pulses of evolution throughout Earth's history have coincided with extinction events. Fossil and geochemical evidence tends to suggest the five major mass extinctions—end-Ordovician, Frasnian–Famennian (F/F), Permian–Triassic (P/Tr), Triassic–Jurassic (Tr/J) and Cretaceous–Tertiary (K/T)—were prolonged periods of biotic stress triggered by a combination of tectonically induced hydrothermal and volcanic processes, leading to eutrophic oceans, global warming, sea-level rise, and global anoxia.[91] Perturbations in the marine sulfur cycle and thus the redox state of the ancient seas during mass extinctions have been reflected in the δ^{34}S of pyrite, showing a general positive shift when conditions are highly euxinic (rapid burial of pyrite) and a negative shift when hydrogen sulfide is rapidly oxidised.[92,93] A number of carotenoid biomarkers (*e.g.* isorenieratane) specific to Chlorobi (biological proxies of photic zone euxinic conditions), strict anaerobes which can photosynthesise using hydrogen sulfide as an electron donor have been identified within P/Tr, Tr/J, and end-Devonian sections.[91,94] GC-ICPMS measurement of these biomarkers and other OSCs in sediments spanning these boundaries should help reveal the extent and consequences of euxinic conditions during the mass extinctions periods and subsequent recovery phases.

Acknowledgements

PG, KG, AH, and MM acknowledge funding support from CSIRO Flagship Collaboration Fund Cluster for Organic Geochemistry of Mineral Systems and KG also from an ARC Discovery Outstanding Research Award. AA thanks Shimon Feinstein, Itay Reznik, and IEI Ltd for the oil shale and oil samples from Aderet 1 drillhole and the support of ISF grant 1269/12. ALS and MRR acknowledge the support of NSF EAR-1024919. Our valued instrument technicians, Guillaume Paris (Caltech), Kai Rankenburg (UWA), and Ward Said-Ahmad (HUJI) are thanked for extensive help in developing, maintaining, and implementing the respective sulfur-CSIA systems. Michael Böttcher is thanked for an insightful peer review which helped improve this manuscript.

References

1. A. Amrani, A. L. Sessions and J. F. Adkins, Compound-specific delta(34)S analysis of volatile organics by coupled GC/multicollector-ICPMS., *Anal. Chem.*, 2009, **81**, 9027–9034.
2. T. Prohaska, C. Latkoczy and G. Stingeder, Precise sulfur isotope ratio measurements in trace concentration of sulfur by inductively coupled

plasma double focusing sector field mass spectrometry., *J. Anal. At. Spectrom.*, 1999, **14**, 1501–1504.

3. R. Clough, P. Evans, T. Catterick and E. H. Evans, Delta34S measurements of sulfur by multicollector inductively coupled plasma mass spectrometry., *Anal. Chem.*, 2006, **78**, 6126–6132.

4. P. R. Craddock, O. J. Rouxel, L. A. Ball and W. Bach, Sulfur isotope measurement of sulfate and sulfide by high-resolution MC-ICP-MS., *Chem. Geol.*, 2008, **253**, 102–113.

5. A. Nissenbaum, M. J. Baedecker and I. R. Kaplan, Organic geochemistry of Dead Sea sediments., *Geochim. Cosmochim. Acta*, 1972, **36**, 709–727.

6. R. Francois, A study of sulphur enrichment in the humic fraction of marine sediments during early diagenesis., *Geochim. Cosmochim. Acta*, 1987, **51**, 17–27.

7. J. P. Werne, T. W. Lyons, D. J. Hollander, S. Schouten, E. C. Hopmans and J. S. Sinninghe-Damsté, Investigating pathways of diagenetic organic matter sulfurization using compound-specific sulfur isotope analysis., *Geochim. Cosmochim. Acta*, 2008, **72**, 3489–3502.

8. A. Said-Ahmad and A. Amrani, A sensitive method for the sulfur isotope analysis of dimethylsulfide and dimethylsulfoniopropionate in seawater., *Rapid Commun. Mass Spectrom.*, 2013, **27**, 2789–2796.

9. P. F. Greenwood, M. McCulloch, K. Grice, A. Holman, L. Hong, H. Ling and S. Jin, Compound specific $\delta34S$ analysis—development and applications, in *26th International Meeting of Organic Geochemistry, Book of Abstracts*, Tenerife, Spain, 2013, vol. **2**, pp. 146–147.

10. J. S. Sinninghe-Damsté and J. W. de Leeuw, Analysis, structure and geochemical significance of organically-bound sulphur in the geosphere: State of the art and future research., *Org. Geochem.*, 1990, **16**, 1077–1101.

11. J. S. Sinninghe-Damsté, M. Kok, J. Köster and S. Schouten, Sulfurized carbohydrates: an important sedimentary sink for organic carbon?, *Earth Planet. Sci. Lett.*, 1998, **164**, 7–13.

12. M. D. Kok, S. Schouten and J. S. Sinninghe-Damsté, Formation of insoluble, nonhydrolyzable, sulfur-rich macromolecules *via* incorporation of inorganic sulfur species into algal carbohydrates., *Geochim. Cosmochim. Acta*, 2000, **64**, 2689–2699.

13. J. I. Hedges and R. G. Keil, Sedimentary organic matter preservation—an assessment and speculative synthesis., *Mar. Chem.*, 1995, **49**, 81–115.

14. D. J. Burdige, Preservation of organic matter in marine sediments: Controls, mechanisms, and an imbalance in sediment organic carbon budgets?, *Chem. Rev.*, 2007, **107**, 467–485.

15. M. Vandenbroucke and C. Largeau, Kerogen origin, evolution and structure., *Org. Geochem.*, 2007, **38**, 719–833.

16. Z. Aizenshtat, A. Stoler, Y. Cohen and H. Nielsen, The geochemical sulfur enrichment of recent organic matter by polysulfides in the Solar-Lake, in *Advances in Organic Geochemistry 1981*, ed. M. Bjorøy, John Wiley & Sons, New York, 1983, pp. 279–288.

17. M. E. L. Kohnen, J. S. Sinninghe-Damsté and J. W. de Leeuw, Biases from natural sulphurization in palaeoenvironmental reconstruction based on hydrocarbon biomarker distributions., *Nature*, 1991, **349**, 775–778.

18. P. Adam, J. C. Schmid, B. Mycke, C. Strazielle, J. Connan, A. Huc, A. Riva and P. Albrecht, Structural investigation of nonpolar sulfur cross-linked macromolecules in petroleum., *Geochim. Cosmochim. Acta*, 1993, **57**, 3395–3419.

19. P. Adam, P. Schneckenburger, P. Schaeffer and P. Albrecht, Clues to early diagenetic sulfurization processes from mild chemical cleavage of labile sulfur-rich geomacromolecules., *Geochim. Cosmochim. Acta*, 2000, **64**, 3485–3503.

20. Y. Hebting, P. Schaeffer, A. Behrens, P. Adam, G. Schmitt, P. Schneckenburger, S. Bernasconi and P. Albrecht, Biomarker evidence for a major preservation pathway of sedimentary organic carbon., *Science*, 2006, **312**, 1627–1631.

21. J. I. Hedges, Global biogeochemical cycles—Progress and problems., *Mar. Chem.*, 1992, **39**, 67–93.

22. I. Kaplan and S. Rittenberg, Microbiological fractionation of sulphur isotopes., *J. Gen. Microbiol.*, 1964, **34**, 195–212.

23. B. Fry, J. Cox, H. Gest and J. Hayes, Discrimination between ^{34}S and ^{32}S during bacterial metabolism of inorganic sulfur compounds., *J. Bacteriol.*, 1986, **165**, 328–330.

24. J. R. Mossman, A. C. Aplin, C. D. Curtis and M. L. Coleman, Geochemistry of inorganic and organic sulphur in organic-rich sediments from the Peru Margin., *Geochim. Cosmochim. Acta*, 1991, **55**, 3581–3595.

25. N. Suits and M. Arthur, Sulfur diagenesis and partitioning in Holocene Peru shelf and upper slope sediments., *Chem. Geol.*, 2000, **163**, 219–234.

26. E. Henneke, G. Luther, G. de Lange and J. Hoefs, Sulfur speciation in anoxic hypersaline sediments from the eastern Mediterranean Sea., *Geochim. Cosmochim. Acta*, 1997, **61**, 307–321.

27. A. Bates, E. Spiker and C. Holmes, Speciation and isotopic composition of sedimentary sulfur in the Everglades, Florida, USA., *Chem. Geol.*, 1998, **146**, 155–170.

28. D. A. Zaback and L. M. Pratt, Isotopic composition and speciation of sulfur in the Miocene Monterey formation: reevaluation of sulfur reactions during early diagenesis in marine environments., *Geochim. Cosmochim. Acta*, 1992, **56**, 763–774.

29. L. A. Chambers, P. A. Trudinger, J. W. Smith and M. S. Burns, Sulfur isotope fractionation during sulfate reduction by dissimilatory sulfate-reducing bacteria., *Can. J. Microbiol.*, 1975, **21**, 1602–1607.

30. L. A. Chambers and P. A. Trudinger, Microbiological fractionation of stable sulfur isotopes: a review and critique., *Geomicrobiol. J.*, 1979, **1**, 249–293.

31. D. E. Canfield and B. Thamdrup, The production of ^{34}S-depleted sulfide during bacterial disproportionation of elemental sulfur., *Science*, 1994, **266**, 1973–1975.

32. H. Cypionka, A. M. Smock and M. E. Böttcher, A combined pathway of sulfur compound disproportionation in *Desulfovibrio desulfuricans.*, *FEMS Microbiol. Lett.*, 1998, **166**, 181–186.

33. D. E. Canfield, Biogeochemistry of sulfur isotopes., *Rev. Mineral. Geochem.*, 2001, **43**, 607–636.

34. M. S. Sim, S. Ono, K. Donovan, S. P. Templer and T. Bosak, Effect of electron donors on the fractionation of sulfur isotopes by a marine Desulfovibrio sp., *Geochim. Cosmochim. Acta*, 2011, **75**, 4244–4259.

35. U. Wortmann, S. Bernasconi and M. Bottcher, Hypersulfidic deep biosphere indicates extreme sulfur isotope fractionation during single-step microbial sulfate reduction., *Geology*, 2001, **29**, 647–650.

36. M. E. Böttcher, H.-J. Brumsack and C.-D. Dürselen, The isotopic composition of modern seawater sulfate: I. Coastal waters with special regard to the North Sea., *J. Mar. Syst.*, 2007, **67**, 73–82.

37. G. Paris, A. L. Sessions, A. V. Subhas and J. F. Adkins, MC-ICP-MS measurement of δ34S and Δ33S in small amounts of dissolved sulfate., *Chem. Geol.*, 2013, **345**, 1–12.

38. D. E. Canfield, B. Thamdrup and S. Fleischer, Isotope fractionation and sulfur metabolism by pure and enrichment cultures of elemental sulfur-disproportionating bacteria., *Limnol. Oceanogr.*, 1998, **43**, 253–264.

39. K. S. Habicht, D. E. Canfield and J. Rethmeier, Sulfur isotope fractionation during bacterial reduction and disproportionation of thiosulfate and sulfite., *Geochim. Cosmochim. Acta*, 1998, **62**, 2585–2595.

40. M. E. Böttcher, B. Thamdrup and T. W. Vennemann, Oxygen and sulfur isotope fractionation during anaerobic bacterial disproportionation of elemental sulfur., *Geochim. Cosmochim. Acta*, 2001, **65**, 1601–1609.

41. B. Tissot and D. Welte, *Petroleum Formation and Occurrence*, Springer, Heidelberg, 2nd edn, 1984.

42. D. E. Canfield, B. P. Boudreau, A. Mucci and J. K. Gundersen, The early diagenetic formation of organic sulfur in the sediments of Mangrove Lake, Bermuda., *Geochim. Cosmochim. Acta*, 1998, **62**, 767–781.

43. H. F. Passier, M. E. Böttcher and G. J. De Lange, Sulphur enrichment in organic matter of eastern Mediterranean sapropels: a study of sulphur isotope partitioning., *Aquatic Geochem.*, 1999, **5**, 99–118.

44. J. P. Werne, T. W. Lyons, D. J. Hollander, M. Formolo and J. S. Sinninghe-Damsté, Reduced sulfur in euxinic sediments of the Cariaco Basin: Sulfur isotope constraints on organic sulfur formation., *Chem. Geol.*, 2003, **195**, 159–179.

45. Z. Aizenshtat and A. Amrani, Significance of δ34S and evaluation of its imprint on sedimentary organic matter I. The role of reduced sulfur species in the diagenetic stage: a conceptual review, in *Geochemical Investigation in Earth and Space Science*, ed. R. J. Hill *et al.* Special publication 9, Geochemical Society, Saint Louis, MO, 2004, pp. 15–33.

46. E. M. Krupp, C. Pecheyran, S. Meffan-Main, O. F. X. Donard and J. Fresenius, Precise isotope-ratio measurements of lead species by

capillary gas chromatography hyphenated to hexapole Multicollector ICP-MS., *Anal. Chem.*, 2001, **370**, 573–580.

47. E. M. Krupp, C. Pecheyran, S. Meffan-Main and O. F. X. Donard, Precise isotope-ratio determination by CGC hyphenated to ICP–MCMS for speciation of trace amounts of gaseous sulfur, with SF6 as example compound., *Anal. Bioanal. Chem.*, 2004, **378**, 250–255.

48. M. Van Acker, A. Shahar, E. D. Young and M. L. Coleman, GC/multiple collector-ICPMS method for chlorine stable isotope analysis of chlorinated aliphatic hydrocarbons., *Anal. Chem.*, 2006, **78**, 4663–4667.

49. S. P. Sylva, L. A. Ball, R. K. Nelson and C. M. Reddy, Compound-specific 81Br/79Br analysis by capillary gas chromatography/multicollector inductively coupled plasma mass spectrometry., *Rapid Commun. Mass Spectrom.*, 2007, **21**, 3301–3305.

50. R. A. Werner and W. A. Brand, Referencing strategies and techniques in stable isotope ratio analysis., *Rapid Commun. Mass Spectrom.*, 2001, **15**, 501–519.

51. R. W. Page and I. P. Sweet, Geochronology of basin phases in the western Mt Isa Inlier, and correlation with the McArthur Basin., *Aust. J. Earth Sci.*, 1998, **45**, 219–232.

52. R. R. Large, S. W. Bull, D. R. Cooke and P. J. McGoldrick, A genetic model for the HYC deposit, Australia: Based on regional sedimentology, geochemistry, and sulfide-sediment relationships., *Econ. Geol.*, 1998, **93**, 1345–1368.

53. K. H. Williford, K. Grice, G. A. Logan, J. Chen and D. Huston, The molecular and isotopic effects of hydrothermal alteration of organic matter in the Paleoproterozoic McArthur River Pb/Zn/Ag ore deposit., *Earth Planet. Sci. Lett.*, 2011, **301**, 382–392.

54. C. S. Eldridge, N. Williams and J. L. Walshe, Sulfur isotope variability in sediment-hosted massive sulfide deposits as determined using the ion microprobe SHRIMP II: A study of the H. Y. C. deposit at McArthur River, Northern Territory, Australia., *Econ. Geol.*, 1993, **88**, 1–26.

55. T. G. Powell, M. J. Jackson, I. P. Sweet, I. J. Crick, C. J. Boreham and R. E. Summons, *Petroleum geology and geochemistry, Middle Proterozoic McArthur Basin*. Canberra, Bureau of Mineral Resources, 1987, p. 286.

56. G. A. Logan, M. C. Hinman, M. R. Walter and R. E. Summons, Biogeochemistry of the 1640 Ma McArthur River (HYC) lead-zinc ore and host sediments, Northern Territory, Australia., *Geochim. Cosmochim. Acta*, 2001, **65**, 2317–2336.

57. J. H. Chen, M. R. Walter, G. A. Logan, M. C. Hinman and R. E. Summons, The Paleoproterozoic McArthur River (HYC) Pb/Zn/Ag deposit of northern Australia: organic geochemistry and ore genesis., *Earth Planet. Sci. Lett.*, 2003, **210**, 467–479.

58. T. F. Anderson and L. M. Pratt, Isotope evidence for the origin of organic sulfur and elemental sulfur in marine sediments, in *Geochemical Transformations of Sedimentary Sulfur*, ed. M. A. Vairavamurthy and

M. A. A. Schoonen, Symposium Series 612, ACS, Washington, DC, 1995, pp. 378–396.

59. J. S. Sinninghe-Damsté, W. I. C. Rijpstra, J. W. De Leeuw and P. A. Schenck, The occurrence and identification of series of organic sulphur compounds in oils and sediment extracts: II. Their presence in samples from hypersaline and non-hypersaline palaeoenvironments and possible application as source, palaeoenvironmental and maturity indicators., *Geochim. Cosmochim. Acta*, 1989, **53**, 1323–1341.

60. A. Rieger, L. Schwark, M. E. Cisternas and H. Miller, Genesis and evolution of bitumen in Lower Cretaceous lavas and implications for stratabound copper deposits, north Chile., *Econ. Geol.*, 2008, **103**, 387–404.

61. A. Amrani, A. L. Sessions, Y. Tang, J. F. Adkins, R. J. Hills, M. J. Moldowan and Z. Wei, The sulfur-isotopic compositions of benzothiophenes and dibenzothiophenes as a proxy for thermochemical sulfate reduction., *Geochim. Cosmochim. Acta*, 2012, **84**, 152–164.

62. V. Brüchert and L. M. Pratt, Contemporaneous early diagenetic formation of organic and inorganic sulfur in estuarine sediments from St. Andrew Bay, Florida, USA., *Geochim. Cosmochim. Acta*, 1996, **60**, 2325–2332.

63. W. A. Hartgers, J. F. Lopez, J. S. Sinninghe-Damsté, C. Reiss, J. R. Maxwell and J. O. Grimalt, Sulfur-binding in recent environments: II. Speciation of sulfur and iron and implications for the occurrence of organo-sulfur compounds., *Geochim. Cosmochim. Acta*, 1997, **61**, 4769–4788.

64. A. I. Holman, K. Grice, P. F. Greenwood, M. E. Böttcher, J. L. Walshe and K. A. Evans, New aspects of sulfur biogeochemistry during ore deposition from $\delta^{34}S$ of elemental sulfur and organic sulfur from the 'Here's Your Chance' Pb-Zn-Ag deposit., *Chem. Geol.*, 2014. In Press.

65. H. Strauss, The sulfur isotopic record of Precambrian sulfates: new data and a critical evaluation of the existing record., *Precamb. Res.*, 1993, **63**, 225–246.

66. Y. A. Shen, R. Buick and D. E, Canfield, Isotopic evidence for microbial sulphate reduction in the early Archaean era., *Nature*, 2001, **410**, 77–81.

67. A. L. Zerkle, J. Farquhar, D. T. Johnston, R. P. Cox and D. E. Canfield, Fractionation of multiple sulfur isotopes during phototrophic oxidation of sulfide and elemental sulfur by a green sulfur bacterium., *Geochim. Cosmochim. Acta*, 2009, **73**, 291–306.

68. A. Amrani, A. Kamyshny Jr, O. Lev and Z. Aizenshtat, Sulfur stable isotope distribution of polysulfide anions in $(NH_4)_2Sn$ aqueous solution., *Inorg. Chem.*, 2006, **45**, 1427–1429.

69. A. Amrani and Z. Aizenshtat, Mechanisms of sulfur introduction chemically controlled: δ34S imprint., *Org. Geochem.*, 2004, **35**, 1319–1336.

70. M. Raven, A. Sessions, J. Adkins, J. Werne and T. Lyons, Sulfur-isotopic compositions of individual organic compounds from Cariaco Basin, under review.

71. H. Oduro, K. L. Van Alstyne and J. Farquhar, Sulfur isotope variability of oceanic DMSP generation and its contributions to marine biogenic sulfur emissions., *Proc. Natl Acad. Sci. U. S. A.*, 2012, **109**, 9012–9016.

72. A. Amrani, A. Sessions, J. Adkins, N. Dalleska, A. Dekas, S. John and V. Orphan, The δ34S of dimethyl sulfide in the surface ocean, in *22nd Goldschmidt Conference*, Montreal, Canada, 2012.

73. A. Amrani, W. Said-Ahmad, Y. Shaked and R. P. Kiene, Sulfur isotope homogeneity of oceanic DMSP and DMS., *Proc. Natl Acad. Sci. U. S. A.*, 2013, **110**, 18413–18418.

74. M. Vairavamurthy, W. Orr and B. Manowitz, Geochemical transformation of sedimentary sulfur: an introduction, in *Geochemical Transformations of Sedimentary Sulfur*, ed. M. A. Vairavamurthy and M. A. A. Schoonen, Symposium Series 612, ACS Washington, DC, 1995, pp. 1–14.

75. W. L. Orr, Sulfur in heavy oils, oil sands and oil shales, in *Oil Sand and Oil Shale Chemistry*, ed. O. P. Strausz and E. M. Lown, Verlag Chemie, New York, 1978, pp. 223–243.

76. K. E. Peters, C. C. Walters and J. M. Moldowan, *The Biomarker Guide*, Cambridge University Press, Cambridge, 2nd edn, 2005.

77. G. Faure and T. M. Mensing, *Isotopes, Principles and Applications*, John Wiley & Sons, Hoboken, NJ, 3rd edn, 2005.

78. J. S. Gaffney, E. T. Premuzic and B. Manowitz, On the usefulness of sulfur isotope ratios in crude oil correlations., *Geochim. Cosmochim. Acta*, 1980, **44**, 135–139.

79. H. G. Machel, Products and distinguishing criteria of bacterial and thermochemical sulfate reduction., *Appl. Geochem.*, 1995, **8**, 373–389.

80. H. G. Machel, Bacterial and thermochemical sulfate reduction in diagenetic settings—old and new insights., *Sediment. Geol.*, 2001, **140**, 143–175.

81. M. A. Rooney, Carbon isotopic evidence for the accelerated destruction of light hydrocarbons by thermochemical sulfate reduction. *1996 NSERC Thermochemical Sulfate Reduction (TSR) and Bacterial Sulfate Reduction (BSR) Workshop*, 25 April 1996, University of Calgary, Abstract.

82. B. K. Manzano, M. G. Fowler and H. G. Machel, The influence of thermochemical sulphate reduction on hydrocarbon composition in Nisku reservoirs, Brazeau River area, Alberta, Canada., *Org. Geochem.*, 1997, 27, 507–521.

83. M. M. Cross, D. A. C. Manning, S. H. Bottrell and R. H. Worden, Thermochemical sulfate reduction (TSR): experimental determination of reaction kinetics and implications of the observed reaction rates for petroleum reservoirs., *Org. Geochem.*, 2004, **35**, 393–404.

84. H. R. Krouse, C. A. Viau and L. S. Eliuk, Chemical and isotopic evidence of thermo chemical sulfate reduction by light hydrocarbon gases in deep carbonate reservoirs., *Nature*, 1988, **333**, 415–419.

85. R. Sassen, Geochemical and carbon isotopic studies of crude oil destruction, bitumen precipitation and sulfate reduction in the deep Smackover Formation., *Org. Geochem.*, 1988, **12**, 351–361.

86. W. L. Orr, Kerogen/asphaltene/sulfur relationships in sulfur-rich Monterey oils., *Org. Geochem.*, 1986, **10**, 499–516.
87. E. Idiz, E. Tannenbaum and I. Kaplan, Pyrolysis of high-sulfur Monterey kerogens: Stable isotopes of sulfur, carbon, and hydrogen, in *Geochemistry of Sulfur in Fossil Fuels*, ed. W. Orr and C. White, Symposium Series 429, ACS, Washington, DC, 1990, pp. 575–591.
88. A. Amrani, M. D. Lewan and Z. Aizenshtat, Stable sulfur isotope partitioning during simulated petroleum formation as determined by hydrous pyrolysis of Ghareb Limestone, Israel., *Geochim. Cosmochim. Acta*, 2005, **69**, 5317–5331.
89. A. Amrani, G. Dror, W. Said-Ahmad, S. Feinstein and I. J. Reznik, The distribution and sulfur isotope ratios of specific organic sulfur compounds during pyrolysis of thermally immature kerogen. *26th International Meeting of Organic Geochemistry*, Tenerife, Spain, 2013 *Book of Abstracts*, vol. 2, pp. 561–562.
90. S. Zhang, G. Zhu, J. Dai and Y. Liang, Comments by Worden and Cai (2006) on Zhang *et al.* (2005), *[Org. Geochem. 36, 1717–1730]*. *Org. Geochem.*, 2006, **37**, 512–515.
91. K. Grice, C. Cao, G. D. Love, M. E. Bottcher, R. J. Twitchett, E. Grosjean, R. E. Summons, S. Turgeon, W. J. Dunning and Y. Jin, Photic zone euxinia during the Permian-Triassic superanoxic event., *Science*, 2005, **307**, 706–709.
92. S. Fenton, K. Grice, R. T. Twitchett, M. Bottcher, C. V. Looy and B. Nabbefeld, Changes in biomarker abundances and sulfur isotopes of pyrite across the Permian-Triassic (P/Tr) Schuchert Dal section (East Greenland)., *Earth Planet. Sci. Lett.*, 2007, **262**, 230–239.
93. B. Nabbefeld, K. Grice, A. Schimmelmann, P. E. Sauer, M. E. Böttcher and R. J. Twitchett, Significance of δDkerogen, δ13Ckerogen and δ34Spyrite from several Permian/Triassic (P/Tr) sections., *Earth Planet. Sci. Lett.*, 2010, **295**, 21–29.
94. C. M. B. Jaraula, K. Grice, R. J. Twitchett, M. E. Bottcher, P. Le Metayer and A. G. Dastidar, Elevated pCO2 leading to Late Triassic extinction, persistent photic zone euxinia, and rising sea levels., *Geology*, 2013, **41**, 955–958.
95. Z. Wei, C. C. Walters, J. M. Moldowan, P. J. Mankiewicz, R. J. Pottorf, Y. Xiao, W. Maze, P. T. H. Nguyen, M. E. Madincea, N. T. Phan and K. E. Peters, Thiadiamondoids as proxies for the extent of thermochemical sulfate reduction., *Org. Geochem.*, 2012, **44**, 53–70.
96. S. Zhang and H. Huang, Geochemistry of Palaeozoic marine petroleum from the Tarim Basin, NW China: Part 1. Oil family classification., *Org. Geochem.*, 2005, **36**, 1204–1214.

CHAPTER 11

Applications of Liquid Chromatography–Isotope Ratio Mass Spectrometry in Geochemistry and Archaeological Science

ALISON J. BLYTH*[a] AND COLIN I. SMITH[b]

[a] Department of Chemistry, Curtin University, GPO Box U1987, Perth 6845, Western Australia, Australia; [b] Department of Archaeology, Environment and Community Planning, La Trobe University, Melbourne, Victoria, 3086, Australia
*Email: Alison.Blyth@curtin.edu.au

11.1 Introduction

Liquid chromatography–isotope ratio mass spectrometry (LC-IRMS) is a relatively recent analytical development in the field of compound-specific isotope analysis (CSIA). Its main advantage lies in allowing the analysis of compounds which are either not amenable to gas chromatography–isotope ratio mass spectrometry (GC-IRMS), or which require derivatisation to be resolved on a GC column.[1] A limitation of the technique is the restriction of analysis to compounds that are soluble in inorganic aqueous phases, as use of organic solvents in the mobile phase will increase the background signal and risk modification of the isotopic values. This means that while polar and

RSC Detection Science Series No. 4
Principles and Practice of Analytical Techniques in Geosciences
Edited by Kliti Grice
© The Royal Society of Chemistry 2015
Published by the Royal Society of Chemistry, www.rsc.org

non-volatile compounds such as carbohydrates, proteins, and amino acids can be easily measured, other LC-amenable compounds of importance in geochemistry, such as glycerol dialkyl glycerol tetraethers, cannot. A second advantage of an LC-IRMS system is the ability to use a flow injection mode to analyse bulk samples. As in compound-specific measurements, this function is restricted to samples amenable to an aqueous carrier, and so does not have quite the broad application of elemental analysis isotope ratio mass spectrometry (EA-IRMS). However, it does allow analysis of sample types, for example organic matter held in acid digests of calcite matrices, which cannot be analysed *via* other currently available techniques.[2]

Since the first commercial units became available in 2004, the technique has been applied in diverse fields including nutritional and biomedical research, food authentication, archaeology, and geo- and environmental sciences.[3-11] This review focuses on case studies of the methods and applications in the latter two fields.

11.2   Technology

A detailed discussion of the technology and development behind modern LC-IRMS is given by Godin and McCullagh.[1] In brief, current commercially available systems include the IsoLink system (Thermo Fisher Scientific GMBH, Ulm, Germany) and the Liquiface (Isoprime Ltd, Cheadle, UK). The original LC-Isolink system is outlined in Krummen *et al.*[12] and again summarised in Godin *et al.*,[13] and more details of the Liquiface can be found in Abaye *et al.*[14] Both instruments work in a similar fashion where the eluent of a liquid chromatograph carrying the analyte compounds flows into an interface with the mass spectrometer. The LC system can be used simply to deliver the liquid phase to the interface for bulk analysis or with a column prior to entry to the interface, for compound separation for compound-specific analysis. When the eluent enters the interface it is mixed with a continuous flow of oxidant and catalyst that mix and flow into a reactor (normally held at 99 °C) that will oxidise carbon-containing molecules to carbon dioxide (hence the restriction of using organic-free mobile phases). After oxidation the LC eluent is cooled and passes through a separating column where the analyte carbon dioxide is removed through a separating membrane *via* a counterflow of helium. The LC eluent (now containing no oxidisable carbon compounds) runs to waste, whereas the analyte carbon dioxide is carried in the helium flow through Nafion drying tubes (where the gas is dried using a helium counterflow across the Nafion membrane) and to the isotope ratio mass spectrometer.

The key technical challenge in the use of these systems has been developing chromatographic techniques for compound-specific work when no organic solvent can be used, as many modern chromatographic methods have some organic component. Moreover, low flow rates are used to be compatible with the interface, and complete resolution of analyte peaks is required for accurate analysis. Chromatographic separation has been a particular challenge for amino acid analysis (see below). Nevertheless, the performance of the instruments compares favourably to bulk and GC methods,[15] with the advantage of less sample preparation and often smaller sample sizes.

11.3   Compound-Specific Analysis

11.3.1   Amino Acids

Bone and tooth collagen plays an important role in archaeological science, as it is the substrate of choice for radiocarbon dating of skeletal material and for palaeodietary analysis. Bone collagen is often preserved in amounts suitable for isotopic analysis ($\delta^{13}C$, $\delta^{15}N$, and $\delta^{34}S$) and the isotopic values reflect a broad long-term dietary signal of protein intake (*i.e.* use of terrestrial C_3 or C_4 plant-based ecosystems or reliance on marine protein).[16–18] Bone collagen is preferred to bone mineral, as the bioapatite crystals of bone are prone to post-mortem alterations, whereas bone collagen is generally robust and can be extracted easily without significant diagenetic degradation[19,20] generally up to around 50 kya, although exceptionally more than 100 kya.[21] When the organisms have been living in certain environments and ecosystems, interpretations from bulk collagen isotope values can be difficult to make as there can be overlap between the isotope values of different dietary groups.[22] Investigating the collagen isotope values at the amino acid level has been one solution to this.[23] There is also interest in extending interpretations by investigating essential (amino acids that must come from dietary protein intake as they cannot be synthesised by the organism) and non-essential amino acids (that can come from dietary protein intake but can also be synthesised from other dietary carbon sources, *i.e.* carbohydrate and lipid). Differences in the two amino acid groups may reveal dietary information unavailable to bulk isotope analysis.

Amino acids have been investigated using GC-IRMS methods but these require derivatisation to make the amino acids more volatile, making carbon isotope analysis more difficult, and not all amino acids can be derivatised (and therefore analysed) (see Dunn *et al.*[15] for a useful overview). Since the advent of the LC-IRMS methods described above, amino acid analysis has been at the fore of the applications in modern[3,13,24] and ancient samples.[25] When analysing protein hydrolysates and trying to resolve as many as 12–18 amino acids, developing chromatographic methods has been an acute challenge.[1,26] Notably, amino acid analysis using the Thermo Scientific LC-Isolink has tended to be carried out using mixed-mode columns,[26] containing both cation-exchange and reverse-phase properties, whereas an anion-exchange method has been developed on the Isoprime Liquiface.[14]

Initial investigations on archaeological material by McCullagh *et al.*[25] demonstrated the potential of the technique applied to bone collagen. Smith *et al.*[27] improved on this initial method using a three-phase technique and several publications have followed using either variants of the McCullagh *et al.*[25] method[8,15,28] or the three-phase technique of Smith *et al.*[7,27] (It should also be noted that although not strictly an LC-IRMS application, the McCullagh *et al.*[25] chromatography technique has also been applied to purifying amino acid fractions for compound-specific radiocarbon analysis, in particular that of hydroxyproline.[29])

Various dietary markers have been investigated in these papers, based on the relationships between certain amino acids and certain dietary groups;[7,8] however, whether these are relationships are context specific or can be applied more universally requires further testing. The most promising approach to date appears to be that made by Honch *et al.*[28] who tested a number of archaeological populations with 'known diets' from contexts across the globe from different time periods. Despite the complexity of the data, it could be powerfully summarised in a bivariate relationship between $\delta^{13}C$ Phe *versus* $\delta^{13}C$ Val, which is able to distinguish terrestrial C_3 and C_4 consumers and high freshwater fish and high marine protein consumers (even in environments where normal bulk $\delta^{15}N$ and $\delta^{13}C$ values cannot). In addition to bone collagen, hair keratin amino acid $\delta^{13}C$ values have been analysed from archaeological material.[30] Hair survives in archaeological deposits less commonly than bone collagen, but remains an important source for palaeodietary data, especially as it records diet at a different time scale to bone collagen. Comparisons of bone and hair from the same individuals such as that made by Raghavan *et al.*[30] will no doubt prove useful in the future, revealing isotopic biographies of individuals from the past.

11.3.2 Volatile Fatty Acids

Understanding the carbon isotope composition of volatile fatty acids is important to the investigation of biogeochemical processes, particularly in extreme environments.[31,32] The analysis of acetate is especially important, as acetate fermentation is one of the two principal biochemical pathways by which methanogenesis occurs. Understanding changes in the $\delta^{13}C$ of acetate can therefore help elucidate the relative importance of different processes, sinks, and sources.[31,32] Previous methods existed for the isotopic analysis of acetate, but these were demanding in terms of preparatory steps, and required comparatively large sample sizes, rendering them unsuitable for the analysis of natural abundance samples in systems such as sedimentary porewaters. Heuer *et al.*[31] addressed this issue by developing an alternative protocol using reversed-phase high-performance liquid chromatrography (HPLC) linked to an isotope ratio mass spectrometer. This allowed porewater samples to be measured while still in their liquid phase, thus minimising or removing the need for prior preparation of the sample, and reducing the sample size required to 2–10 μM of compound. Analysis of model compounds showed good results for a wide range of volatile fatty acids, while natural samples allowed measurement of acetate, lactate, propionate, and butyrate, although isomers of the latter could not be resolved. A coeluting unidentified substance prevented measurement of formate. Analytical precision was around 0.7‰, with the mean values of dissolved model compounds varying from the EA values by around 0.6‰.[31] Further applications of this approach include the analysis of acetate and related carbon-bearing metabolites in subsurface sedimentary porewaters, in an attempt to enhance our understanding of carbon turnover by microbial communities in the deep biosphere[32] and the analysis of acetate in lake sediments to further understand biogeochemical processes and especially methanogenesis pathways in the freshwater environment.[33,34]

11.3.3 Plant Carbohydrates

Carbohydrates play a central role in organic matter pools in soils, sediments, and waterborne dissolved organic matter (DOM), are central to many food webs as substrates, and so are important in the study of carbon cycling. GC-IRMS methods exist for isotopic analysis[35] but involve extensive derivatisation, and are limited in the compound groups that can be measured.[9] Early protocols for LC-IRMS successfully measured isotopic compositions in some carbohydrates,[5] but were not able to separate the plant carbohydrates of interest in geochemistry and environmental science.[9] In order to address this, Boshker *et al.*[9] adapted an HPLC method which uses anion-exchange chromatography and pulsed amperometric detection (HPAEC-PAD) for use in LC-IRMS. The chromatography components of this method could not be directly transferred to isotope analyses, because it uses strongly alkaline eluents which can interfere with the wet oxidation, and produce high carbon backgrounds. The adaptations to the method lowered the backgrounds by changing to a lower-strength sodium hydroxide eluent, and changed the column type to a narrow bore in order to lower the flow rate. It was also necessary to investigate alternative pusher ions, in this case nitrate, as the sodium acetate used in conventional HPLC techniques is an organic reagent and therefore not suitable for isotopic work. The method proved able to resolve a number of plant carbohydrates of interest in environmental and sedimentary studies, with detection limits of 0.5–2 nmol. Proof-of-concept tests on natural and labelled sedimentary material found greater variability than in the analytical standards, attributed to small natural differences in the carbohydrate synthesis rates within the target core incubations, but indicated that the technique had considerable potential for work tracing carbohydrate dynamics in the natural environment.[9]

11.3.4 Amino Sugars

Microbial communities in soil play a vital role in carbon cycling and sequestration, and so understanding their structure, turnover, and response to different pressures is important.[36] Amino sugars in soils have been used as biomarkers for changes in the soil microbial biomass as the majority are of a microbial origin.[36–38] The introduction of stable isotope probing allows clearer resolution of the interactions and metabolic processes of the

different components of the soil microbial pool,[38] and a better understanding of the turnover of these compounds within a soil environment.[10,39]
By using labelling studies, or by exploiting known vegetation changes in
agricultural soils, it becomes possible to assess the contribution of different
age amino sugars to the soil pool,[10] and to assess the incorporation rates of
plant residue and the varying formation kinetics of different compounds.[39]
By combining isotopic labelling with selective inhibition of parts of the
microbial community, the interactions of the bacterial and fungal community components can also be studied.[38] GC-IRMS is possible for amino
sugars, but because of the need to derivatise the compounds, it is restricted
to samples reasonably abundant in carbon.[10] To enable such studies in
lower-abundance samples, and without the need for compound derivatisation, Bodé *et al.*[10] developed a new LC-IRMS technique, using an anion-exchange column to separate the compounds, and a guard column to provide
online purification to improve the baseline. Using separate methods to resolve basic and acidic amino sugars, this approach was able to resolve glu-
cosamine, galactosamine, and muramic acid, but not mannosamine, and
showed a variation of 0.3‰ for the $\delta^{13}C$ of spiked reference compounds.

11.4   Bulk Isotope Analysis

11.4.1   Water

By using the flow injection mode of LC-IRMS it is possible to measure the
$\delta^{13}C$ of both dissolved inorganic carbon (DIC) and dissolved organic carbon
(DOC) in (separate) liquid samples. This twin ability stems from the fact that
in normal use (*i.e.* to measure organic carbon), the liquid stream is mixed
with both an acid and oxidant before being passed through a reactor to
convert the organic carbon to carbon dioxide. By operating with only the
acid, milder oxidation conditions are used and the organic carbon is not
converted and DIC is measured.[40] This has been shown to allow rapid small
sample measurements at a high precision and with minimal preparation
required, although issues relating to sample storage and linearity remain.[40]
   The measurement of $\delta^{13}C$ in DOC in water samples is a more conventional
use for the instrumentation, and again, allows for rapid analysis of relatively
small samples, with minimal prior preparation.[41] In these cases, DIC must
be purged from the sample following acidification, either by vacuum[2] or by
individual purging with helium.[41] The type of acid used in acidifying the
sample is also of importance, with both phosphoric and nitric acids being
shown not to affect the isotopic values, provided analysis is completed
relatively soon after preparation of the solution,[2,41] while hydrochloric acid
significantly interfered with the oxidation of the organic matter, and therefore with the isotopic results. The conflict between the oxidation system and
halides also means that at present the technique has limited utility in salt-
water samples. It has been shown that the interference problem can be
overcome by increasing the quantity of persulfate in large volume wet

chemical oxidation systems,[42] but the problem remains an obstacle in small-volume systems working with low natural abundance levels of carbon.[41]

11.4.2 $\delta^{13}C$ of Organic Matter Preserved in Speleothems

Carbon isotope records in speleothems are useful to palaeoenvironmental research, as they can reflect both climatic changes, and alterations in vegetation and the soil ecosystem. Conventionally, analysis is undertaken on the carbon that forms the speleothem calcite, and which is derived from the carbon dioxide dissolved in the parent dripwater. However, although these types of records can be analysed at high resolution, and, in the right samples, are extremely informative about vegetation change or soil parameters,[43,44] they can also be subject to significant problems relating to inorganic fractionation during water transport and calcite precipitation.[45] Analysing the $\delta^{13}C$ of organic matter preserved in speleothems can provide an alternative or complimentary isotopic record directly linked to the overlying soil and vegetation. To date, little work has been done in this area, as researchers have been hampered by the very low organic content of most speleothems, and the need to decalcify the material in such a way that does not introduce organic contamination or concentrate acid salts that may be detrimental to instrumentation. These problems have been largely resolved by turning to LC-IRMS, and using the flow injection mode to undertake a bulk isotopic analysis of organic matter held in a phosphoric acid digest.[2] The technique requires only 0.2 g of calcite, allowing high-resolution records to be developed, and has been shown to have an analytical error of around 0.2‰.[2] Application of the approach to a stalagmite time-series from Scotland[46] showed that coherent records could be recovered, with the signal relating to climatic parameters *via* vegetation conditions and microbial activity. Crucially, pairing inorganic and organic records showed the potential to separate and identify plant and microbial influences on the record.[46]

11.5 Conclusion

The advent of LC-IRMS has opened up isotopic analysis to compounds which are not easily amenable to GC analysis, in particular water-soluble non-polar compounds. Development of specialised methods has allowed separation and reliable isotopic measurements of compounds of interest to the palaeosciences, such as amino acids crucial to understanding palaeodiet, and acetate from marine porewaters that provide information on pathways of methanogenesis, as well as plant carbohydrates and amino sugars relevant to understanding many aspects of carbon turnover in the geosphere. In addition to the chromatographic methods, the use of a flow injection mode also allows for bulk isotopic analysis of water samples and acid digests with minimal wet chemistry preparation. Much of the first decade of research with the commercially available instruments has been taken up with method development and proof-of-concept studies that will lay the solid foundations for further in depth work. With the field now entering its prime, there is scope for methodological advances to be made in improving separation and resolution of existing compounds, and for the techniques to be usefully transferred to new substrates and contexts, opening up novel applications within many geochemical and archaeological areas in the future.

Acknowledgements

CS would like to acknowledge the Australian Research Council for funding (Future Fellowship FT0992258). AJB is supported by an AINSE Fellowship.

References

1. J.-P. Godin and J. S. O. McCullagh, Review: Current applications and challenges for liquid chromatography coupled to isotope ratio mass spectrometry (LC/IRMS)., *Rapid Commun. Mass Spectrom.*, 2011, **25**, 3019–3028.
2. A. J. Blyth, Y. Shutova and C. I. Smith, $\delta^{13}C$ analysis of bulk organic matter in speleothems using liquid chromatography–isotope ratio mass spectrometry., *Org. Geochem.*, 2013, **55**, 22–25.
3. H. Schierbeek, T. F. Braake, J.-P. Godin, L.-B. Fay and J. B. Van Goudoever, Novel method for measurement of glutathione kinetics in neonates using liquid chromatography coupled to isotope ratio mass spectrometry., *Rapid Commun. Mass Spectrom.*, 2007, **21**, 2805–2812.
4. T. F. Braake, H. Schierbeek, K. de Groof and A. Vermes, Glutathione synthesis rates after amino acid administration directly after birth in preterm infants., *Am. J. Clin. Nutr.*, 2008, **88**, 333–339.
5. A. I. Cabañero, J. L. Recio and M. Rupérez, Liquid chromatography coupled to isotope ratio mass spectrometry: a new perspective on honey adulteration detection., *J. Agric. Food Chem.*, 2006, **54**, 9719–9727.
6. A. I. Cabañero, J. Recio and M. Rupérez, Isotope ratio mass spectrometry coupled to liquid and gas chromatography for wine ethanol characterization., *Rapid Commun. Mass Spectrom.*, 2008, **22**, 3111–3118.
7. K. Choy, C. I. Smith, B. T. Fuller and M. P. Richards, Investigation of amino acid δ13C signatures in bone collagen to reconstruct human palaeodiets using liquid chromatography–isotope ratio mass spectrometry., *Geochim. Cosmochim. Acta*, 2010, **74**, 6093–6111.
8. A. Pollard, P. Ditchfield, J. S. O. McCullagh, T. Allen, M. Gibson, C. Boston, S. Clough, N. Marquez-Grant and R. A. Nicholson, 'These boots were made for walking': The isotopic analysis of a C4 Roman inhumation from Gravesend, Kent, UK., *Am. J. Phys. Anthropol.*, 2011, **146**, 446–456.
9. H. T. S. Boschker, T. C. W. Moerdijk-Poortvliet, P. van Breugel, M. Houtekamer and J. J. Middelburg, A versatile method for stable carbon isotope analysis of carbohydrates by high performance liquid

chromatography/isotope ratio mass spectrometry., *Rapid Commun. Mass Spectrom.*, 2008, **22**, 3902–3908.

10. S. Bodé, K. Denef and P. Boeckx, Development and evaluation of a high-performance liquid chromatography/isotope ratio mass spectrometry methodology for $\delta^{13}C$ analyses of amino sugars in soils., *Rapid Commun. Mass Spectrom.*, 2009, **23**, 2519–2526.

11. T. C. W. Moerdijk-Poortvliet, L. J. Stal and H. T. S. Boschker, LC/IRMS analysis: A powerful technique to trace carbon flow in micro-phytobenthic communities in intertidal sediments., *J. Sea Res.*, 2013, DOI 10.1016/j.seares.2013.10.002.

12. M. Krummen, A. Hilkert, D. Juchelka, A. Duhr, H. Schlueter and R. Pesch, A new concept for isotope ratio monitoring liquid chroma-tography/mass spectrometry., *Rapid Commun. Mass Spectrom.*, 2004, **18**, 2260–2266.

13. J.-P. Godin, J. Hau, L.-B. Fay and G. Hopfgartner, Isotope ratio moni-toring of small molecules and macromolecules by liquid chroma-tography coupled to isotope ratio mass spectrometry., *Rapid Commun. Mass Spectrom.*, 2005, **19**, 2689–2698.

14. D. A. Abaye, D. J. Morrison and T. Preston, Strong anion exchange liquid chromatographic separation of protein amino acids for natural ^{13}C-abundance determination by isotope ratio mass spectrometry., *Rapid Commun. Mass Spectrom.*, 2011, **25**, 429–435.

15. P. J. H. Dunn, N. V. Honch and P. Evershed, Comparison of liquid chromatography–isotope ratio mass spectrometry (LC/IRMS) and gas chromatography–combustion–isotope ratio mass spectrometry (GC/C/IRMS) for the determination of collagen amino acid $\delta^{13}C$ values for palaeodietary and palaeoecological reconstruction., *Rapid Commun. Mass Spectrom.*, 2011, **25**, 2995–3011.

16. M. De, Niro, Postmortem preservation and alteration of *in vivo* bone collagen isotope ratios in relation to palaeodietary reconstruction., *Nature*, 1985, **317**, 806–809.

17. H. P. Schwarcz and M. J. Schoeninger, Stable isotope analyses in human nutritional ecology., *Yearbk Phys. Anthropol.*, 1991, **34**, 283–321.

18. O. Nehlich and M. P. Richards, Establishing collagen quality criteria for sulphur isotope analysis of archaeological bone collagen., *Archaeol. Anthropol. Sci.*, 2009, **1**, 59–75.

19. T. Brown, D. Nelson, J. Vogel and J. Southon, Improved collagen extraction by modified Longin method., *Radiocarbon*, 1988, **30**, 171–177.

20. G. J. van Klinken, Bone collagen quality indicators for palaeodietary and radiocarbon measurements., *J. Archaeol. Sci.*, 1999, **26**, 687–695.

21. K. Britton, S. Gaudzinski-Windheuser, W. Roebroeks, L. Kindler and M. P. Richards, Stable isotope analysis of well-preserved 120,000-year-old herbivore bone collagen from the Middle Palaeolithic site of Neumark-Nord 2, Germany reveals niche separation between bovids and equids., *Palaeogeog. Palaeoclimatol. Palaeoecol.*, 2012, **333–334**, 168–177.

22. S. H. Ambrose, Stable carbon and nitrogen isotope analysis of human and animal diet in Africa., *J. Hum. Evol.*, 1986, **15**, 707–731.

23. L. T. Corr, J. Sealy, M. Horton and R. Evershed, A novel marine dietary indicator utilising compound-specific bone collagen amino acid δ^{13}C values of ancient humans., *J. Archaeol. Sci.*, 2005, **32**, 321–330.

24. J.-P. Godin, D. Breuillé, C. Obled, I. Papet, H. Schierbeek, G. Hopfgartner and L. Fay, Liquid and gas chromatography coupled to isotope ratio mass spectrometry for the determination of ^{13}C–valine isotopic ratios in complex biological samples., *J. Mass Spectrom.*, 2008, **43**, 1334–1343.

25. J. S. O. McCullagh, D. Juchelka and R. E. M. Hedges, Analysis of amino acid ^{13}C abundance from human and faunal bone collagen using liquid chromatography/isotope ratio mass spectrometry., *Rapid Commun. Mass Spectrom.*, 2006, **20**, 2761–2768.

26. J. S. O. McCullagh, Mixed-mode chromatography/isotope ratio mass spectrometry., *Rapid Commun. Mass Spectrom.*, 2010, **24**, 483–494.

27. C. I. Smith, B. T. Fuller, K. Choy and M. P. Richards, A three-phase liquid chromatographic method for δ^{13}C analysis of amino acids from biological protein hydrolysates using liquid chromatography-isotope ratio mass spectrometry., *Anal. Biochem.*, 2009, **390**, 165–172.

28. N. V. Honch, J. S. O. McCullagh and R. E. M. Hedges, Variation of bone collagen amino acid δ^{13}C values in archaeological humans and fauna with different dietary regimes: Developing frameworks of dietary discrimination., *Am. J. Phys. Anthropol.*, 2012, **148**, 495–511.

29. J. S. O. McCullagh and A. Marom, Radiocarbon dating of individual amino acids from archaeological bone collagen., *Radiocarbon*, 2010, **52**, 620–634.

30. M. Raghavan, J. S. O. McCullagh, N. Lynnerup and R. E. M. Hedges, Amino acid δ^{13}C analysis of hair proteins and bone collagen using liquid chromatography/isotope ratio mass spectrometry: paleodietary implications from intra-individual comparisons., *Rapid Commun. Mass Spectrom.*, 2010, **24**, 541–548.

31. V. Heuer, M. Elvert, S. Tille, M. Krummen, X. Prieto Mollar, L. R. Hmelo and K.-U. Hinrichs, Online δ^{13}C analysis of volatile fatty acids in sediment/porewater systems by liquid chromatography – isotope ratio mass spectrometry., *Limnol. Oceanogr.*, 2006, **4**, 346–357.

32. V. Heuer, J. W. Pohlman, M. E. Torres, M. Elvert and K.-U. Hinrichs, The stable carbon isotoipe biogeochemistry of acetate and other dissolved carbon species in deep subseafloor sediments at the northern Cascadia Margin., *Geochim. Cosmochim. Acta*, 2009, **73**, 3323–3336.

33. V. B. Heuer, M. Krüger, M. Elvert and K.-U. Hinrichs, Experimental studies on the stable carbon isotope biogeochemistry of acetate in lake sediments., *Org. Geochem.*, 2010, **41**, 22–30.

34. Y. Liu, T. Yao, G. Gleixner, P. Claus and R. Conrad, Methanogenic pathways, ^{13}C isotope fractionation and archaeal community composition in lake sediments and wetland soils on the Tibetan Plateau., *J. Geophys. Res. Biogeosci.*, 2013, **118**, 650–664.

35. B. E. van Dongen, S. Schouten and J. S. Sinninghe Damsté, Gas chromatography/combustion/isotope-ratio-monitoring mass spectrometric analysis of methylboronic derivatives of monosaccharides: a new method for determining natural 13C abundances of carbohydrates., *Rapid Commun. Mass Spectrom.*, 2001, **15**, 496–500.

36. B. Glaser, M.-B. Turrión and K. Alef, Amino sugars and muramic acid – biomarkers for soil microbial community structure analysis., *Soil Biol. Biochem.*, 2004, **36**, 399–407.

37. R. T. Simpson, S. D. Frey, J. Six and R. K. Thiet, Preferential accumulation of microbial carbon in aggregate structures of no-tillage soils., *Soil Sci. Soc. Am. J.*, 2004, **68**, 1249–1255.

38. S. Bodé, R. Fancy and P. Boeckx, Stable isotope probing of amino sugars – a promising tool to assess microbial interactions in soils., *Rapid Commun. Mass Spectrom.*, 2013, **27**, 1367–1379.

39. Z. Bai, S. Bodé, D. Huygens, X. Zhang and P. Boeckx, Kinetics of amino sugar formation from organic residues of different quality., *Soil Biol. Biochem.*, 2013, **57**, 814–821.

40. J. A. Brandes, Rapid and precise $\delta^{13}C$ measurement of dissolved inorganic carbon in natural waters using liquid chromatography coupled to an isotope-ratio mass spectrometer., *Limnol. Oceanogr.*, 2009, 7, 730–739.

41. P. Albéric, Liquid chromatography/mass spectrometry stable isotope analysis of dissolved organic carbon in stream and soil waters., *Rapid Commun. Mass Spectrom.*, 2011, **25**, 3012–3018.

42. C. L. Osburn and G. St_Jean, The use of wet chemical oxidation with high-amplification isotope ratio mass spectrometry (WCO-IRMS) to measure stable isotope values of dissolved organic carbon in seawater., *Limnol. Oceanogr.*, 2007, 5, 296–308.

43. J. A. Dorale, L. A. González, M. K. Reagan, D. A. Pickett, M. T. Murrell and R. G. Baker, A high-resolution record of Holocene climate change in speleothem calcite from Cold Water Cave, northeast Iowa., *Science*, 1992, **258**, 1626–1630.

44. D. Genty, D. Blamart, B. Ghaleb, V. Plagnes, Ch. Causse, M. Bakalowicz, K. Zouari, N. Chkir, J. Hellstrom, K. Wainer and F. Bourges, Timing and dynamics of the last deglaciation from European and North African $\delta^{13}C$ stalagmite profiles—comparison with Chinese and South Hemisphere stalagmites., *Quat. Sci. Rev.*, 2006, **25**, 2118–2142.

45. V. E. Johnston, A. Borsato, C. Spötl, S. Frisia and R. Miorandi, Stable isotopes in caves over altitudinal gradients: fractionation behaviour and inferences for speleothem sensitivity to climate change., *Climate of the Past*, 2013, **9**, 99–118.

46. A. J. Blyth, C. I. Smith and R. N. Drysdale, A new perspective on the $\delta^{13}C$ signal preserved in speleothems using LC-IRMS analysis of bulk organic matter and compound specific stable isotope analysis., *Quat. Sci. Rev.*, 2013, **75**, 143–149.

CHAPTER 12

Advances in Comprehensive Two-Dimensional Gas Chromatography (GC×GC)

CHRISTIANE EISERBECK,*[a] ROBERT K. NELSON,[b] CHRISTOPHER M. REDDY[b] AND KLITI GRICE[a]

[a] Curtin University, WA Organic and Isotope Geochemistry Centre, Department of Chemistry, GPO Box U1987, Perth, WA 6845, Australia; [b] Department of Marine Chemistry and Geochemistry, Woods Hole Oceanographic Institution, Woods Hole, MA 02543, USA
*Email: Christiane.eiserbeck@gmail.com

12.1 Introduction

Comprehensive two-dimensional (2D) gas chromatography (GC×GC) started to attract analytical chemists about 20 years ago. Liu and Phillips[1] reported the first comprehensive 2D GC analysis of an oil sample and the 2D contour plot drew further attention by the scientific community.

A number of reviews published in the past summarise the development of GC×GC in the earlier years up to 2008.[2–9] More recent reviews published in 2012[10,11] capture the functionalities and provide comprehensive summary and discussion of recent developments in modulator and detector technology. This chapter extends these reviews, focusing on the application of GC×GC to a broad range of disciplines in the geosciences.

GC×GC is the next development step after 2D separation using heart-cutting. While heart-cutting can only accomplish 2D separation in a narrow,

RSC Detection Science Series No. 4
Principles and Practice of Analytical Techniques in Geosciences
Edited by Kliti Grice
Published by the Royal Society of Chemistry, www.rsc.org

predetermined time window, GC×GC transfers all effluent from the primary column through to the secondary column, maximising the sample resolution throughout the entire analysis. Hence, it is called comprehensive 2D GC as opposed to the heart-cutting technique, because the complete sample is separated in two dimensions. GC×GC uses two orthogonal stationary phases (such as non-polar and polar) within one analysis, thus adding a second dimension of chromatographic resolution. A chromatographic plane is created by these two orthogonal separation mechanisms. The obtained enhancement of chromatographic resolution (compared to conventional GC) is created by the additional peak capacity within that plane. A third dimension of separation can be added to GC×GC when it is coupled with a time-of-flight mass spectrometer (TOF-MS).

The placement within that plane provides invaluable insights into unknown compounds with regards to their elution order within a certain compound class (for details see section 12.2.6) as well as basic chemical and physical properties.[12]

The power of GC×GC rests in its capability of high-resolution separation of very complex mixtures such as biological and geological samples.[6,13–17] Applications range from the fields of metabolomics, medical and pharmaceutical separation problems right to the geosciences including petroleum geochemistry. GC×GC has also significant applications emerging in mineral exploration (Grice, personal communication).

Geological samples such as crude oils, condensates, and rock extracts are mixtures of thousands of compounds from highly abundant straight-chain n-alkanes to polycyclic saturated, unsaturated, and aromatic hydrocarbons, diamondoids, metalloporphyrins, and polar compounds containing heteroatoms. These compounds can vary significantly in concentration, challenging conventional separation techniques. The enhanced resolving power of GC×GC makes it an ideal choice for analysis of these complex mixtures.[4,6,14,18–27] The expanded peak capacity allows for simultaneous analysis of saturated and aromatic hydrocarbons as well as more polar compounds containing heteroatoms such as oxygen, nitrogen and sulfur that are commonly present in fluids and rocks from the subsurface.

This chapter introduces the principles of GC×GC, describes recent technical developments, and discusses a selection of applications to demonstrate the capabilities of GC×GC and its value for the geosciences. A summary of reviews and applications of GC×GC in the field of geosciences is listed in Table 12.1.

12.2   Technical Background

12.2.1   Basic Principle

In GC×GC, two capillary GC columns are fitted in the GC oven, connected *via* a transfer line. Compounds eluting from the first column within a predefined time window (modulation time) are cryogenically trapped and

Table 12.1   Reviews and applications of GC×GC in the geosciences.

- Reviews on GC×GC analysis[2–7,10,11,28]
- Whole fluid analysis of very complex samples such as crude oils[8,18,19,29–35]
- Separation and characterisation of unresolved complex mixtures[14,17,23,36]
- Fingerprinting[22,24,26,27,37,38]
- Statistical methods applied to data processing and interpretation[22,39–41]
- Monitoring of gradual changes in biomarker composition of oil in the environment (*e.g.* oil weathering)[42–51]
- Fluid-source-correlation studies[16,47,52]
- Detailed studies of biomarker distribution in geological samples including diamondoids,[53–55] specific biomarkers such as 18α(H)- and 18β(H)-oleanane, lupane[27,56]
- Paleoenvironmental proxies[34,35]
- Pyrolysis-GC×GC[25,57]
- Study of nitrogen-containing biomarkers[58–60] and sulfur-containing compounds[25] in crude oil fractions
- Technical studies (*e.g.* modulators, peak shape,[61] retention time reproducibility)[62,63]

---

focused before transfer onto the second column. A secondary oven housing the subsequent column is situated inside the main oven for independent temperature control of the second-dimension separation.

The continuous modulation of the complete sample in one analysis makes this technique a comprehensive 2D GC. In contrast, earlier 2D separation techniques that used a heart cut to transfer effluents of only one single time window for secondary separation were very limited to a specific separation problem.

12.2.2   Instrument Setup

The most commonly used GC×GC system is the LECO Pegasus 4D GC×GC (LECO Corporation, St. Joseph, MI, USA) which has a flame ionisation detector (GC×GC-FID) or can be coupled to a time-of-flight mass spectrometer (GC×GC-TOF-MS). The Zoex system (Zoex Corporation, Houston, TX, USA) is another system that is often applied in the petroleum field.

Figure 12.1 shows a schematic of a typical setup of a GC×GC-TOF-MS instrument. An exhaustive review of the setup and discussion of the individual components of the GC×GC system was presented by Adahchour *et al.*[2–6]

12.2.2.1   Columns

The primary column, along with the dual-stage cryogenic modulator, resides in the main oven, whereas the second-dimension column is fitted in a separate oven, allowing for independent temperature control.

Especially for applications in geosciences, the most common column combination includes a polar and a non-polar stationary phase. The

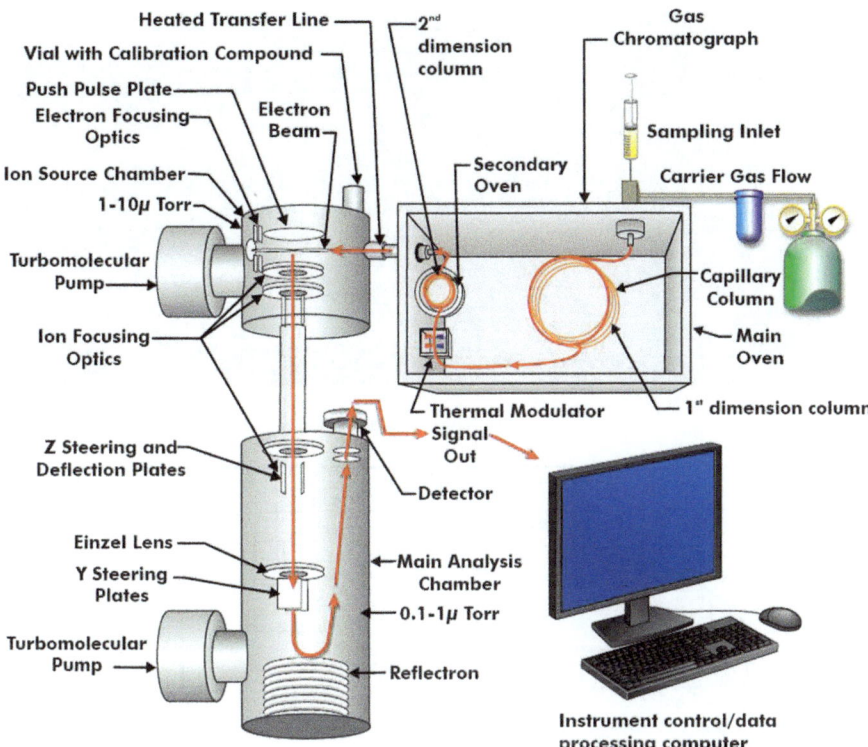

Figure 12.1   Schematic of a typical set-up of a GC×GC-TOFMS instrument (Source: LECO® website http://www.leco.com.au/files/sep_sci_pegasus_4d_gcxgc-tofms_209–183.pdf), reprinted with permission from Leco®).

first-dimension column is commonly a non-polar column (*e.g.* 100% dimethylpolysiloxane coated) of about 25–50 m length separating based on volatility. This setup allows for comparison with regular 1D GC analyses that are often performed on a DB-1 or equivalent column.

The subsequent second-dimension separation is commonly performed on a polar column (*e.g.* a 50% phenyl polysilphenylene-siloxane coated) of about 1.25–3 m in length. The fast second-dimension separation is practically isothermal with no boiling-point-contribution to this separation, thus the separation is orthogonal because retention is exclusively based on the specific interactions of each compound with the stationary phase.

A combination *vice versa* (polar/non-polar) is also often applied. This way the sample is separated based on polarity in the first separation step followed by a volatility separation. Examples of the effect of these column combinations on the elution order are shown in Figure 12.2.

The polar/non-polar configuration was reported to show better resolution for saturated hydrocarbons and might thus be better suited for samples with a higher concentration of these compound classes.[21] However, for a more holistic overview of a sample containing saturated, aromatic, and polar

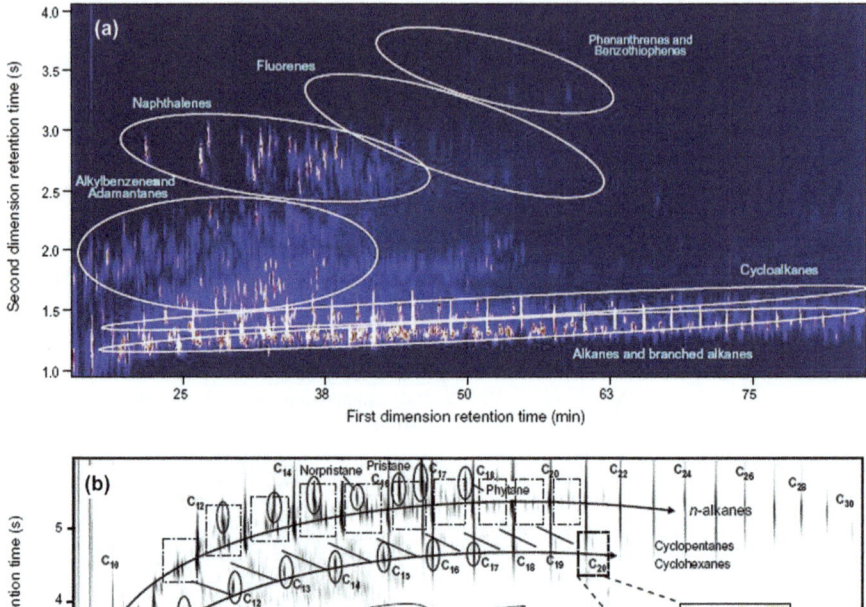

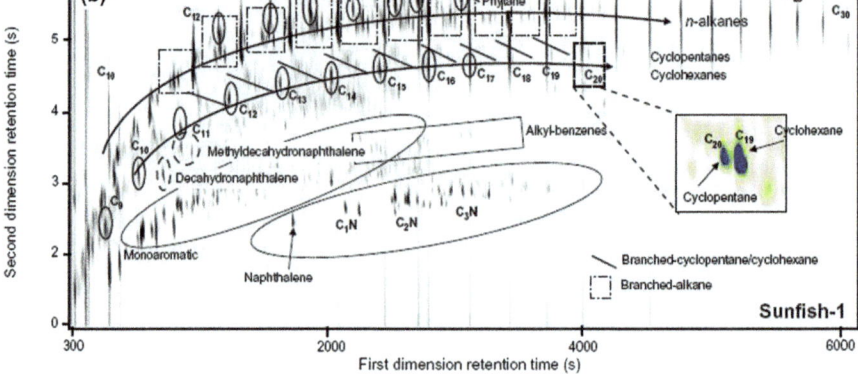

Figure 12.2   Exemplified elution pattern for the application of different column configurations: GC×GC analysis of land-plant oils. (a) GC×GC–FID plot of Sunfish-1 oil analysed using non-polar/polar column set. (b) GC×GC-TOFMS total ion chromatogram (TIC) of Sunfish-1 oil using polar/non-polar column set. b1, b2, and b3 display elution regions of alkanes/isoprenoids, alkylcyclopentanes/alkylcyclohexanes, and aromatics in 2D space, respectively.
Reprinted from Tran *et al.*,[14] Copyright (2010), with permission from Elsevier.

compounds, the non-polar/polar configuration is suggested and more applicable when comparing with 1D analyses.

Particularly in other disciplines such as medicine and metabolomics, non-orthogonal set-ups have been successfully applied. Enantioselective GC×GC of compounds in essential oils (*e.g.* tea tree oil, Chinese medicine) was achieved with both an orthogonal as well as a non-orthogonal approach.[6]

Depending on the separation problem at hand, a myriad of different combinations can be applied and the choice of the most suitable column combination is often the result of trial and error, particularly for very specific separation problems.

The length and diameter of the individual columns can also have an effect on the separation result, *e.g.* a wider internal diameter of the secondary column can extend the secondary retention time and thus increase the achieved separation. Alternatively, a secondary column with a reduced internal diameter can reduce the overall analysis time, which is usually long (*e.g.* 1.5–3 h) because of the slow temperature programme required to meet the modulation criteria.[6]

Overall, a large number of variables have to be taken into account when optimising the separation problem at hand such as the stationary phase, the column length, the column's internal diameter, the temperature programme, the modulation period. The separation success depends on the complex and sensitive interaction between these variables and often numerous attempts are required for optimised settings.

12.2.2.2 Modulator

The modulator is the crucial part in the separation system as it affects the separation and reproducibility. Modulators have developed significantly in the past decades. The most commonly used modulator is a thermal modulator consisting of a quad-jet system. It is placed between the two GC columns and creates two distinct trapping zones in which all effluents from the first column are cryogenically focused before thermal release onto the second column.

The dual-stage cryogenic modulator (*e.g.* LECO) operates with a cold and hot jet. The cold jet gas is dry nitrogen, chilled with liquid nitrogen. Alternatively, expanding liquid carbon dioxide is also used for cooling. The hot jet is operated with air that is heated to significantly higher temperatures than the column temperature, about 50 °C above the temperature of the main GC oven.

A more detailed summary of recent developments in modulator research was discussed by Marriott *et al.*[10] and Seeley.[11]

Modulation Period. The modulation period is the timespan during which eluents from the first separation step are trapped into bundles before they are released onto the second column.

The modulation period can affect the separation success significantly. The choice of the right modulation period for a particular separation depends on the nature of the hydrocarbon mix and the focus of the separation problem at hand. With longer modulation periods, more of the compounds that were separated on the first column are being mixed again in the modulator. For example, during a modulation period of 15 s all separated eluents from the first column are trapped and focused in a small band and injected together onto the second column as a mixture, losing the initial separation achieved in the first separation step. Therefore, the shortest possible modulation period is preferable.

However, this mixing effect needs to be balanced with the wrapping effect. Wrapping occurs when the modulation period is shorter than the time it takes all of the compounds that were transferred to the second column during the previous modulation phase to elute from the second column. If some compounds remain on the second column while the new set of compounds is being transferred onto the second column, these will overlap resulting in a false retention time as well as a misleading position within the chromatographic plane (see section 12.2.6). As an example, assume a modulation period of 5 s and a sample containing compounds which have a retention time of 7 s on the second column. In this case, these compounds will not elute during their modulation period of 5 s but will be carried over into the next modulation period and will elute after 2 s of that modulation period. They will therefore appear to have a retention time of 2 s and elute in a chromatographic space of less-retained compounds, *e.g.* high molecular weight aromatic compounds might elute together with n-alkanes and confuse the interpretation.

This is particularly relevant for unknown compounds, which in case of wrapping would elute in the chromatographic space of a different compound class from what they really are (see section 12.2.6). Furthermore, the separation of the following compounds will be compromised when there are compounds remaining on the column, and retention time shifts might be observed.

In summary, the preferred modulation time should be the minimum time (allowing for the most effective first-dimension separation) that is longer than the maximum secondary retention time of any compound in the sample. Commonly, modulation periods range from 5 to 10 s.

Another aspect that is affected by the modulation period is the peak width and shape. Compounds in high abundance have a larger peak with a broader peak width. If the peak width exceeds the modulation period, the compound will elute in two (or three) consecutive modulation periods, which can affect the peak shape as well as the automated interpretation, as the software might not recognise the two eluting parts as the same compound. In this case the peaks can be manually defined as the same compound.

12.2.2.3 Detector

The most commonly applied detector systems coupled to a GC×GC are the flame ionisation detector (FID) and the TOF-MS. Although double-focusing sector, quadrupole, and ion traps are popular MS detectors for GC, they have limited use in GC×GC because of their relatively slow scanning rates, compared to fast acquisition TOF-MS, better suited to characterise very narrow 2D peaks.

We do not go into detail about the technical background of these detectors here because they are widely used and a detailed description is not the focus of this chapter. A discussion of TOF-MS in general and in particular TOF-SIMS is given in Chapter 5 of this volume.

The FID is a very fast detector with high sensitivity. It allows for a precise quantification of individual compounds because it detects the compound and not just individual masses. Correct quantification requires a known retention time for the compound(s) of interest as well as baseline separation for each peak, which can often be challenging. The disadvantage of the FID is the lack of mass spectra and ion chromatograms to support compound identification.

TOF-MS, on the other hand, is equally fast and adds a third dimension of separation by obtaining mass spectra which allow for the use of ion chromatograms. Accurate identification of compounds, even unknowns, is possible *via* retention times and mass spectra. The study of series of *e.g.* diagenetic products is possible by using ion chromatograms.

TOF-MS is slightly less sensitive than FID and parameters such as the acquisition rate can have an effect on the separation result because it determines the signal-to-noise ratio. However, a high acquisition rate produces large data files which are challenging to handle afterwards. Again, this is a parameter that needs to be adjusted to the requirements of the separation problem.

TOF-MS operates at a push pulse rate of 5 kHz. This allows sufficient signal averaging time to ensure good signal-to-noise ratios while maintaining a data acquisition rate that is high enough to ensure accurate processing (signal average) of spectra for peaks eluting from the second-dimension column with second-dimension peak widths in the order of 50–200 ms at their base. Such narrow peaks require a high data acquisition rate. Hence, TOF-MS which can acquire data at up to 500 spectra per second, is the only mass spectrometer that can meet these requirements.

The sulfur chemiluminescence detector (SCD) has also successfully been used in combination with a GC×GC system to distinguish different sulfur-containing compound classes in the 2D space of a GC×GC, such as mercaptans, sulfides, thiophenes, benzothiophenes, dibenzothiophenes,[25] and thiodiamondoids.

FID *vs.* TOF-MS – Which to Choose?   GC×GC-TOF-MS and GC×GC-FID complement each other and the use of both systems is recommended. GC×GC-FID results in an improved peak shape, reproducibility, and quantitative peak areas, while GC×GC-TOF-MS analysis provides high-resolution separation with access to full mass spectra throughout the chromatogram. GC×GC-FID is better suited for quantification of compounds.

Peak identification is more challenging in FID due to the lack of mass spectra for confirmation. However, once the retention times (first and second order) are established for a given compound of interest, ideally by repeat analysis of standards, peak volumes are best determined using the FID detector. The advantage of GC×GC-FID compared to the TOF-MS is the reproducibility, increased sensitivity ($\sim$5 times), quantitative detection of peak abundances, and improved peak shape.[27,56] Similar response factors for all hydrocarbons in a GC×GC-FID chromatogram improve the comparability of obtained concentrations. Analysis of a whole oil results in a complete inventory of all GC-amendable compounds which is very powerful, due to the response factor of near unity.

A very simple advantage of GC×GC-FID is the lower maintenance cost, which allows for more affordable instrument settings with regard to separation problems that might require multiple analyses for an optimised result.

Furthermore, retention time stability is better in GC×GC-FID analyses. Retention time reproducibility was studied by Shellie *et al.*[62,63] Standard deviations of retention times (RT) in a run-to-run comparison based on 43 compounds in 6 consecutive runs were found to be on average 0.12% relative standard deviation (RSD) in the first dimension and 0.74% RSD in the second dimension. RT obtained on consecutive days were slightly less stable, although this was true mainly for very abundant compounds while RT of trace, minor, and intermediate concentration components were extremely reproducible. These RSD values represent ideal values obtained in very controlled conditions. Most applications of RT comparisons will require RT stability of compounds in different samples and in these cases the matter becomes more complex. Varying types of organic matter contain changing matrices for a particular compound that is to be compared which can result in slight changes in RT. Higher concentrations of one compound can also lead to elution in two separate modulation periods compared to low- to medium-abundance compounds which mostly elute within only one modulation period (depending on the peak width in relation to the chosen modulation period). Elution in multiple modulation periods can cause a slight shift in RT, depending on the point where the data processing software sets the peak maximum.

Eiserbeck *et al.*[27] determined the RT stability for multiple GC×GC-FID and GC×GC-TOF-MS analyses of $17\alpha,21\beta(H)\text{-}C_{30}$-hopane in seven different crude oil samples with varying $17\alpha,21\beta(H)\text{-}C_{30}$-hopane (hopane) concentrations. The GC×GC-FID measurements had a RSD of 0.05% and 0.54% for first and second RT compared to 0.16% and 1.82% for GC×GC-TOF-MS. The higher RSD compared to the results obtained by Shellie *et al.* are partly due to the varying matrices present in seven different oil samples compared to one single sample that was studied by Shellie *et al.*[62] Furthermore, Eiserbeck *et al.*[27] observed compounds coeluting with hopane in some of the analysed oil samples which were partially resolved in the second dimension, obscuring the true second dimension RT of $17\alpha,21\beta(H)\text{-}C_{30}$-hopane.

GC×GC-quadrupole MS and GC×GC-combustion isotope ratio mass spectrometry (GC×GC-irMS) have been employed in other disciplines such as drug screening or characterisation of essential oils; however, these detection techniques operate at lower acquisition rates ($\sim$20–25 Hz) and thus lose resolution capability of GC×GC (see Marriott *et al.*[10] and references therein). Nonetheless, GC×GC-quadrupole MS has been successfully applied to hydrocarbon mixtures in the past.[19] Developments towards more rapid acquisition rates up to 50 Hz for GC×GC-quadrupole MS are under way for scan and single-ion monitoring (SIM) mode (290 amu mass range, 40–330 m/z).[64,65]

GC×GC-TOF-MS is pre-eminent for detailed biomarker studies. The obtained mass spectra add a third dimension of separation which is required for identification of coeluting or unknown compounds. Selected ion chromatograms can provide invaluable information in areas of high peak density.

12.2.3 Sample Preparation

Because of the enhanced peak capacity of GC×GC, no sample separation is required before the analysis. The whole sample is diluted in hexane or similar solvents and analysed simultaneously. Separation of asphaltenes might be performed if required, using a centrifuge, to prevent blocking of the capillary column.

With whole fluid injection, saturated, aromatic, and polar compounds can be studied in one chromatogram without loss of resolution. Furthermore, potential losses of the more volatile, low-molecular-weight compounds during preparation of the sample as previously necessary (such as liquid chromatography) are reduced. Moreover, the risk of introducing contamination to a geological sample during sample preparation is also reduced. This can be particularly valuable in studies of early life in which the first appearance of compounds indicating living organisms, such as steranes, is the key question and the presence of such compounds in the sample due to contamination would be misleading and render the sample invalid.

Furthermore, some unsaturated compounds are thermally unstable or prone to rearrangement when treated with *e.g.* molecular sieves (see section 12.4.2.2). GC×GC separates these samples chromatographically without any prior wet chemical treatment and thus maintains the natural composition. Similarly, argentation chromatography of non-polar fractions obtained from liquid chromatography into saturated and unsaturated compounds can result in rearrangements of double bonds in olefins. None of these preseparation steps is necessary for GC×GC analysis, making it a valuable tool for analysis of unstable compounds.

However, for a specific, known separation problem that might be limited to only a certain number of compounds that are not at risk of rearrangement, fractionating the sample before analysis is recommended as this will simplify the chromatogram, allow more chromatographic space to separate particular compound classes without concern for 'wrapping' (see section 12.2.2.2)[17] and thus improve the interpretation as well as preserve the injector and capillary columns.

Pyrolysis-GC×GC is an alternative for sample injection of less accessible hydrocarbons in asphaltenes (see section 12.2.4). Furthermore, hydrous pyrolysis (HyPy, see Chapter 6) can be applied before GC×GC to obtain a GC-amendable sample.

12.2.4   Injector Methods Coupled with GC×GC

Wang *et al.*[25] first described the coupling of pyrolysis to GC×GC-FID and its application to the characterisation of petroleum source rocks. This study was extended by Payeur *et al.*[57] who evaluated the use of pyrolysis-GC×GC-TOF-MS to analyse the pyrolysis products of six whole-sediment samples from above, within, and below an organic carbon-rich Quaternary sapropel layer from the Mediterranean.

Valentine *et al.*[66] applied pyrolysis-GC×GC-TOF-MS to study the composition and source of asphalt volcanoes in the Santa Barbara Basin, off the coast of California, USA.

The combination of whole-sediment pyrolysis and GC×GC-TOF-MS is promising but the success is sensitive to the careful selection of the multiple analytical parameters, particularly the pyrolysis temperature and the operational temperatures and flow rates of the GC columns.

12.2.5   Data Processing

12.2.5.1   Signal Deconvolution in ChromaTOF® Software

The use of a TOF-MS ensures spectral continuity, thus allowing mass spectral deconvolution of coeluting peaks if these are characterised by different fragmentation patterns. The requirements for successful deconvolution are for the peak apexes of coeluting analytes to be at least separated by three scans and to differ somewhat in their mass spectra. This capability allows the identification of different coeluting compounds. The use of deconvoluted ion current (DIC) acts like a fourth dimension to the separation system. The ChromaTOF® GC software includes an automated signal deconvolution which can be added to the analysis sequence to produce extracted mass spectra free of interfering signals.

12.2.6   GC×GC Chromatogram—A Map for Biomarkers

GC×GC chromatograms are commonly displayed as contour plots (plan view, Figure 12.3), but can also be presented as mountain plots (see Figure 12.5c). In GC×GC chromatograms, compound classes are separated across the 2D plane spanned by the first and second dimension (Figure 12.3) according to their increasing volatility (non-polar column, first dimension) and polarity (polar column, second dimension).[24] With this sequence of capillary columns, the general elution order with increasing second-dimension retention time can be summarised as: n-alkanes < regular isoprenoids < irregular isoprenoids < alkylcyclopentanes < alkylcyclohexanes < bicyclic terpenoids < alkylbenzenes (Figure 12.3). Cyclic saturated compounds such as terpenoids elute further up in the first dimension and also slightly later in second dimension with each additional ring in the structure. Cyclic compounds (including mono- and diunsaturated compounds) elute

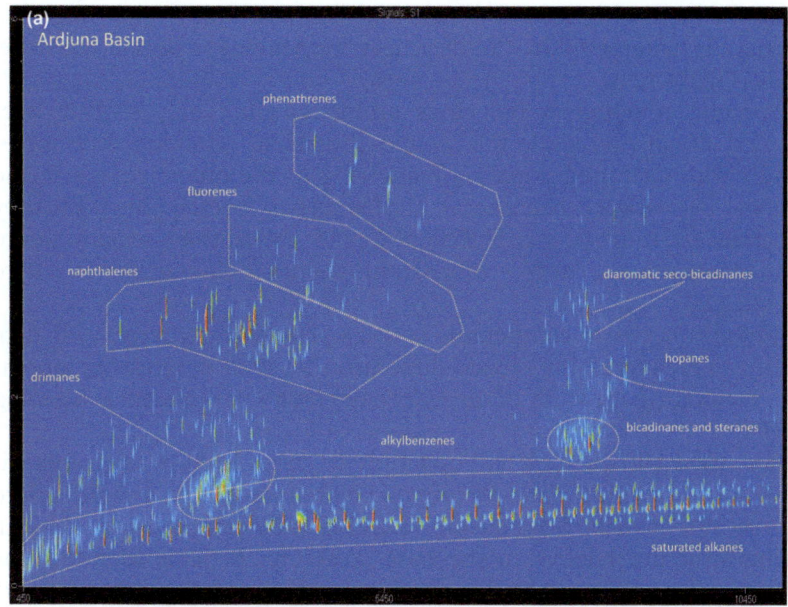

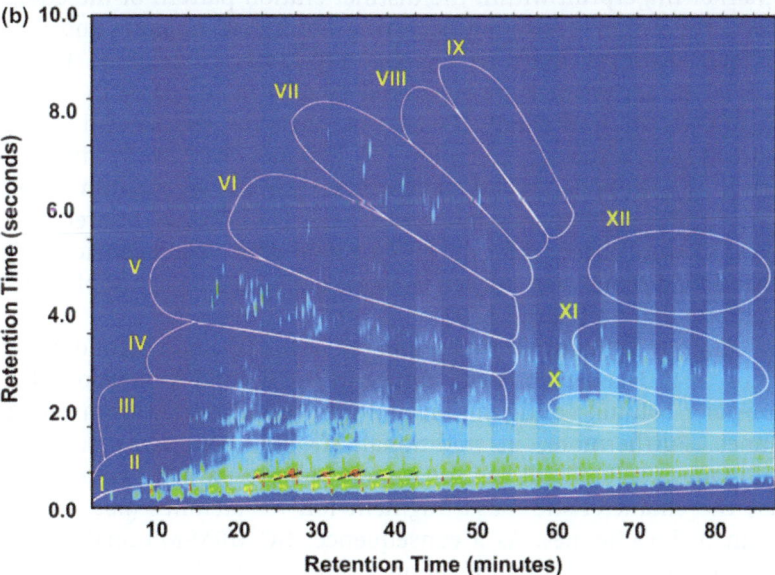

Figure 12.3   GC×GC-FID contour plots highlighting the compound class elution order and pattern with a non-polar/polar column configuration. a) Oil from the Ardjuna Basin (Indonesia), b) Oil P-2 described by Ventura *et al.*[24] I, normal and branched alkanes; II, cycloalkanes; III, alkyl benzenes and tricyclic terpanes; IV, indenes; V, naphthalenes and benzothiophenes; VI, fluorenes; VII, phenanthrenes and dibenzothiophenes; VIII, fluoranthenes; IX, pyrenes; X, steranes; XI, hopanes, and XII, benzohopanes.

Reprinted from Ventura *et al.*,[24] Copyright (2010), with permission from Elsevier.

along the increasing second dimension retention time in the general order of tricyclic triterpenoids < tetracyclic triterpenoids < pentacyclic triterpenoids (Figure 12.3).

Aromatic compounds (all rings aromatic) elute only slightly later in the first dimension while the second-dimension retention time increases rapidly with each additional aromatic ring in the order (groups including all alkyl derivatives): benzenes < alkylbenzenes (overlapping with tricyclic terpenanes < indenes < naphthalenes and benzothiophenes < fluorenes < phenanthrenes and dibenzothiophenes < fluoranthenes < pyrenes (Figure 12.3).[24]

Similarly, series of compounds such as pentacyclic triterpenoids demonstrate rapidly increasing second-dimension RT with progressing aromatisation, facilitating the study of diagenetic degradation products in a geological sample.

Biomarker classes such as hopanoids or steroids elute as groups within the chromatographic plane. Within one biomarker group, a slight difference in first- and/or second-dimension retention time can be related to changes in the structure relative to the base structure. Reliable identification of biomarkers is achieved based on the first- and second-dimension RT and the biomarker fingerprint within the distinct elution pattern of the compound classes. For example, naturally occurring 17β,21β(H)-hopanes and the series of altered epimers, such as C_{27}–C_{35} 17β,21α(H)-moretanes, 17α,21β(H)-hopanes, 25-norhopanes are separated by GC×GC not only in the first dimension, as in traditional 1D GC, but also in the second dimension (Figure 12.4), resolving coelution of hopanes with very similar mass spectra.[18] 25-Norhopanes elute relatively early in the second dimension, well before regular α,β-hopanes followed by the β,α-hopanes (moretanes). The longest RT were observed for the β,β-hopanes, which elute considerably later in the second dimension. All series create a distinct, recognisable pattern in the chromatogram (Figure 12.4).

All common steranes are well separated in the first and second dimension (Figure 12.4). Diasteranes elute earlier in the second dimension of a non-polar/polar column configuration and are well separated from the regular steranes. Furthermore, in traditional 1D chromatograms steranes and bicadinanes coelute with the separation being even more complicated because the characteristic mass fragment ion for steranes (m/z 217) is also present in bicadinanes. As a consequence, GC-MRM-MS analysis is often required for reliable quantification of these two biomarker groups. In GC×GC-TOF-MS chromatograms, both compound classes elute well separated in the second dimension (Figure 12.4).[27]

GC×GC analysis of whole oils facilitates identification of diagenetically related products. Saturated, mono-, di-, and triaromatic tetracyclic triterpenoids can easily be determined with the help of the two RT. Increasing secondary retention time responds to increasing polarity, *i.e.* aromaticity; mono-, di-, and triaromatic compounds elute along a line within the 2D chromatogram plane.

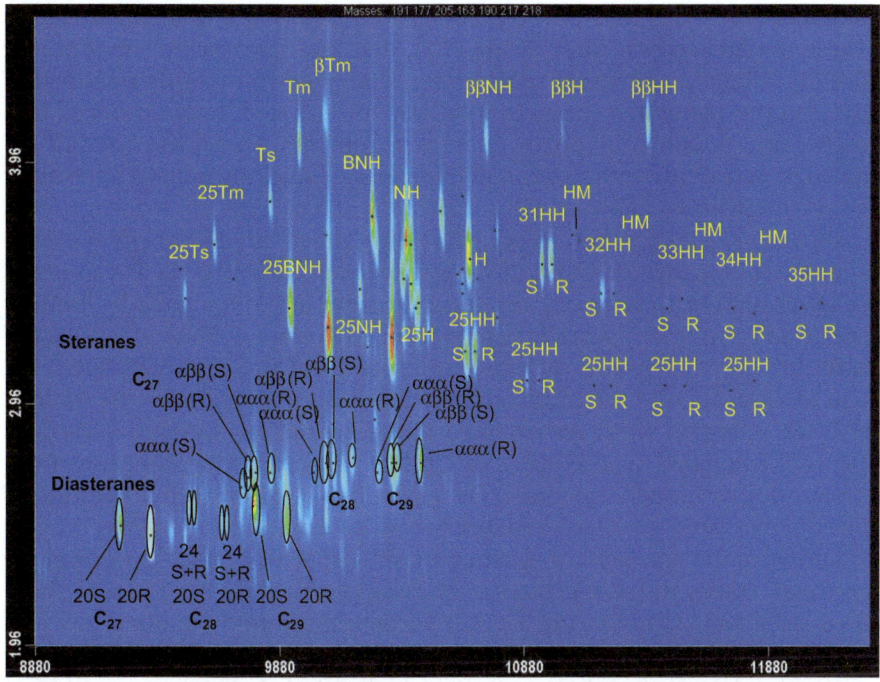

Figure 12.4 Separation of hopanes (yellow) and steranes (black) using GC×GC-TOF.
Shown is the combined EIC of the masses *m/z* 191, 177, 205, 163, 190,
217, and 218. H, hopane; 25, 25-norhopane; HH, homohopane; HM,
homomoretane; NH, norhopane; BNH, bisnorhopane.
Reprinted from Eiserbeck *et al.*,[27] Copyright (2012), with permission
from Elsevier.

12.3 Superiority of GC×GC Compared to Traditional 1D Techniques?

12.3.1 Comparison to GC-MS and GC-MRM-MS

The superiority of GC×GC over more traditional 1D gas chromatographic
techniques is evident in every GC×GC application presented in the literature. Eiserbeck *et al.*[27,67] demonstrated the separation and resolution improvement of GC×GC in comparison with traditional 1D GC as well as
metastable reaction monitoring techniques like GC-MRM-MS.

Most important to note is the improved resolution down to minor detailed
biomarker regions with only one analysis of the whole fluid. A similarly
detailed result using GC-MS or GC-MRM-MS required one to multiple separation steps prior to the analysis such as liquid chromatography, molecular
sieving, argentation chromatography, *etc.*, all of which can cause losses,
carry the risk of contamination of the sample, and are time consuming.
Subsequently, a series of different GC and MS methods (GC-MS and GC-
MRM-MS) have to be applied which are customised to the relevant fraction

or compound class in order to achieve the resolution and detail required. These methods can include scans, SIM, and multiple reaction monitoring (MRM) tailored to different questions and applied to multiple fractions of the sample (saturates, branched/cyclics, unsaturated hydrocarbons, aromatics and possibly the polars such as porphyrins. To achieve the full detailed picture of the composition of one sample can easily require up to 10 or even 20 individual analyses.

GC×GC separation can accomplish this detail in one single analysis without losing the detail of the mass spectra. GC-MRM-MS for example achieves high sensitivity and can separate coeluting compounds by limiting the acquisition to very specific parent-daughter-ion transitions. However, the resulting chromatograms lack the information of the full mass spectrum and identification of the peaks is limited to one or a handful of transitions and the retention time. An example of the spatial resolution achieved by GC×GC compared to coelution in GC-MS and separation resulting from monitoring of limited transitions as applied in GC-MRM is the coelution of bicadinanes with steranes as mentioned in section 12.2.6.

Furthermore, analysis of fractions using specified mass spectrometry methods masks the whole picture of a sample because each one of these analyses is only representing a small portion of the whole sample.

GC×GC-TOF-MS is the only gas chromatographic system that can acquire molecular data of the whole sample in high detail. A direct comparison of the relative abundances of saturated and aromatic compounds can be achieved, allowing for instant fingerprinting (see section 12.4.1) of oils, condensates and rock extracts.

12.3.1.1   Biodegradation

Gas chromatographic analysis of severely biodegraded samples with the traditional 1D technique often presents a challenge due to the extensive 'hump' of unresolved complex mixture (UCM). With GC-MS analysis, the chromatogram of the saturated hydrocarbon fraction of a severely biodegraded oil is dominated by UCM with only information retained of the remaining compounds. Mass spectra of the remaining compounds are often unclear due to low concentrations and can lead to loss of information of the molecular ion. Injection of higher concentrated sample could resolve this problem however the general resolution of the sample is disturbed by the increasing amount of UCM. Furthermore, injection of more UCM to the capillary column can cause damage to the column and significantly reduce its lifetime.

While the resolution and quantification of a limited number of compounds can be improved by applying a SIM, Figure 12.5a), the UCM still adds significant noise to the mass spectra and alters the baseline of the SIM chromatogram.

Application of GC-MRM-MS reduces the rise in the baseline caused by the UCM to near zero (Figure 12.5b) and increases the sensitivity for a few selected compounds (depending on the transitions chosen for a particular

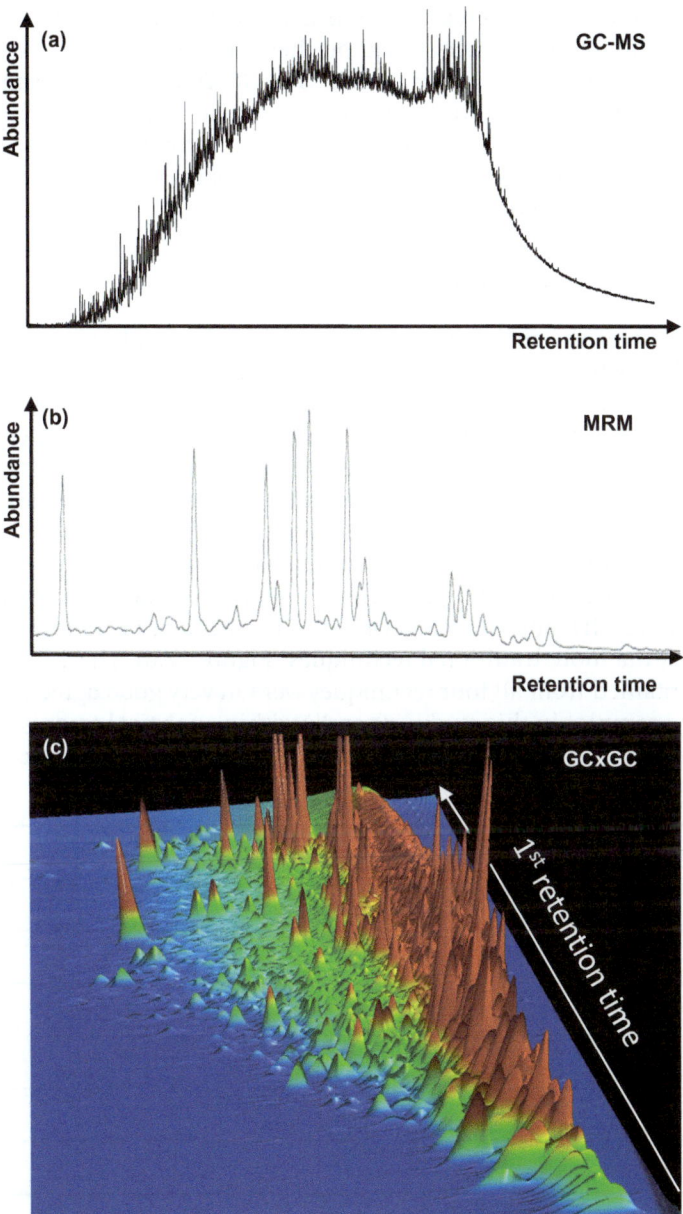

Figure 12.5   Comparison of the resolution of unresolved complex mixture (UCM) using (a) GC-MS, (b) GC-MRM-MS, and (c) Mountain plot of a GC×GC-FID analysis.
Reprinted from Eiserbeck *et al.*,[27] Copyright (2012), with permission from Elsevier.

method). However, screening of a biodegraded sample and retaining all mass spectral information needed for identification is not possible. GC-MRM-MS relies on only a few transitions per compound for

identification, which increases the sensitivity but prevents positive identification of many biomarkers based on their mass spectra.

The second-dimension separation in GC×GC-TOF-MS resolves most of the
compounds from the UCM in a whole-oil sample without significant losses
in sensitivity and retains full mass spectral information for identification
(see also section 12.4.3). Figure 12.5c illustrates the improvement in separation of a severely biodegraded oil.

12.3.1.2 Thermal Maturity

Assessment of thermal maturity of a geological sample using established
biomarker ratios[68] is an important part of the screening and interpretation
of a sample. Silva *et al.*[20] demonstrated the use of thermal maturity parameters from GC×GC-TOF-MS. But how do these parameters compare to
data obtained from 1D GC techniques? Are they comparable to the extensive
and valuable database of maturity parameters obtained in the past decades?
Eiserbeck *et al.*[27] evaluated this question by comparing results for the isomerisation ratio $S/(S + R)$ of C_{32} homohopanes, one of the most commonly
applied thermal maturity parameters,[68] obtained from GC-MS, GC-MRM-
MS, GC×GC-FID, and GC×GC-TOF-MS. GC×GC data proved to be comparable with the more traditional techniques (Figure 12.6). The $22S/(22S + 22R)$
ratios obtained from all four techniques were in very good agreement for the
analysed samples in this study, generally within analytical error (≤5%). A few
outliers were observed for severely biodegraded samples in which maturity

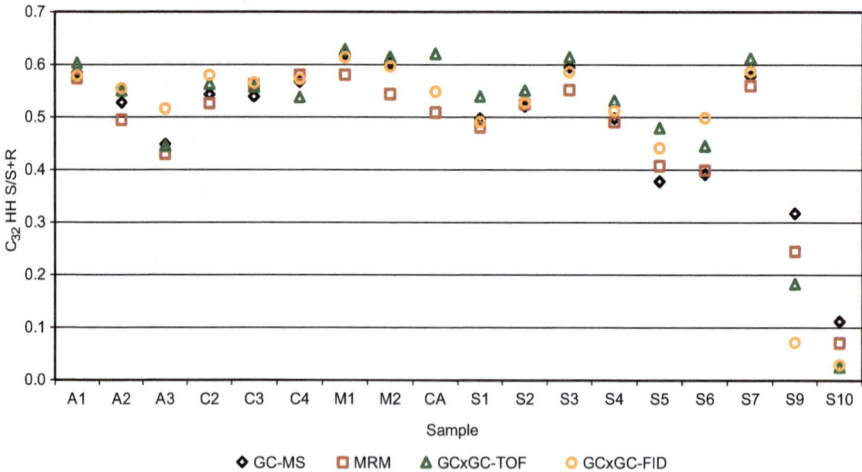

Figure 12.6 Comparison of thermal maturity parameters calculated from data derived from GC-MS, GC-MRM-MS, GC×GC-FID and GC×GC-TOF. The
C_{32} homohopane ratio C_{32} $S/(S + R)$ was chosen as a representative
parameter for thermal maturity.
Reprinted from Eiserbeck *et al.*,[27] Copyright (2012), with permission
from Elsevier.

parameters based on biomarker concentration are unreliable and alternative methods for the assessment of thermal maturity are recommended.

This result carries importance for the application of GC×GC for oil and rock screening, acquisition of much better resolved molecular data, and comparability to biomarker data reported in the past based on 1D analysis.

12.4   Applications of GC×GC in Geochemical Studies

The application of GC×GC in the geosciences is has increased significantly over the last 5–10 years. Summaries of less recent geochemical studies using GC×GC can be found in the reviews by Adahchour *et al.*[5,6,14]

This section focuses on recent applications and introduces concepts developed for tackling long-known separation problems such as the resolution of UCMs using GC×GC as well as approaches for the processing of the enormous amounts of data derived from GC×GC analyses.

12.4.1   Fingerprinting

GC×GC chromatograms of whole oils or sediment extracts provide an easy way for the first assessment of a fluid including dominant compound classes beyond the n-alkanes, which are mostly all that can be observed in a total ion chromatogram of traditional 1D GC. Geological samples can be compared in much more detail with just a quick glance at the 3D chromatogram (Figure 12.7).

Aromatic compounds elute further in the polar dimension than saturated compounds, and thus can be distinguished easily in whole-oil analysis. Most abundant compounds, compound classes, main features, and general similarities or differences between oils (Figure 12.7) appear clearly in much more detail than is possible using traditional 1D GC analysis.

Li *et al.*[26] used GC×GC in combination with sulfur isotope analysis to fingerprint petroleum fluids that are affected by thermochemical sulfur reduction (TSR). The authors clearly identified indicators for TSR in fluids from the Canadian Upper Devonian Nisku Formation and the Chinese north-eastern Jinxian Sag of Bohai Bay Basin. Molecular indicators of TSR in 'sour' condensates included lower saturated to aromatic hydrocarbon ratios (~ 2; for comparison 'sweet' oils commonly have a ratio of ~ 5), a high molar concentration of hydrogen sulfide in the associated gas ($\sim 20\%$ *vs.* <1% in sweet oils) and most importantly a strong predominance of sulfur-containing heterocyclics such as benzothiophenes and dibenzothiophenes over the common polycyclic aromatic hydrocarbons and diagenetically derived sulfur-bound biomarkers, clearly separated and identified in the GC×GC chromatogram (Figure 12.8).[26]

As described in section 12.2.6, chemical compound classes elute in a GC×GC chromatogram in clusters that can be grouped and also quantified as a group.

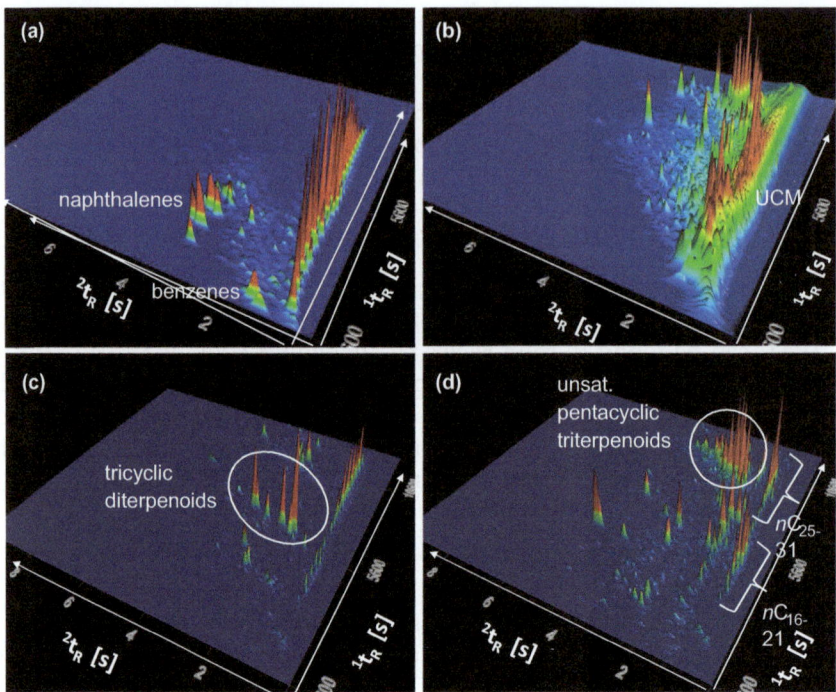

Figure 12.7   3D Mountain plots of GC×GC-FID chromatograms for two crude oils (a and b), and two sediment extracts (c and d). Highlighted compound classes group together and clearly indicate the main features of a sample in a spatial context. ${}^{1}t_R$ and ${}^{2}t_R$ are primary and secondary retention time, respectively and are given in seconds. Abundances are colour coded (blue = baseline, colour change from green to red with increasing abundance) and relative to the most abundant peak in each chromatogram. Reprinted from Eiserbeck et al.,[27] Copyright (2012), with permission from Elsevier.

The combination of high-resolution separation and quality quantification possibilities, and the grouping of compound classes can be applied to fingerprint fluids and compare them with regard to thermal maturity, source, degree of biodegradation, etc. Applying the full range of GC×GC capabilities, Ventura et al.[24] presented a method to identify compositional differences or similarities in oils, which is a defining question in oil reservoir characterisation. The method included a comparison of biomarkers that reflect thermal maturity and source of organic matter, a group comparison of compound classes, and comparison of compound classes across a specified retention index range. Quantitative assessment of oil similarities over a range of carbon numbers using traditional 1D GC techniques required a series of calibration standards and curves and was still limited to n-alkanes, iso-, cycloalkanes, and aromatic compounds.[69,70] With GC×GC the range of

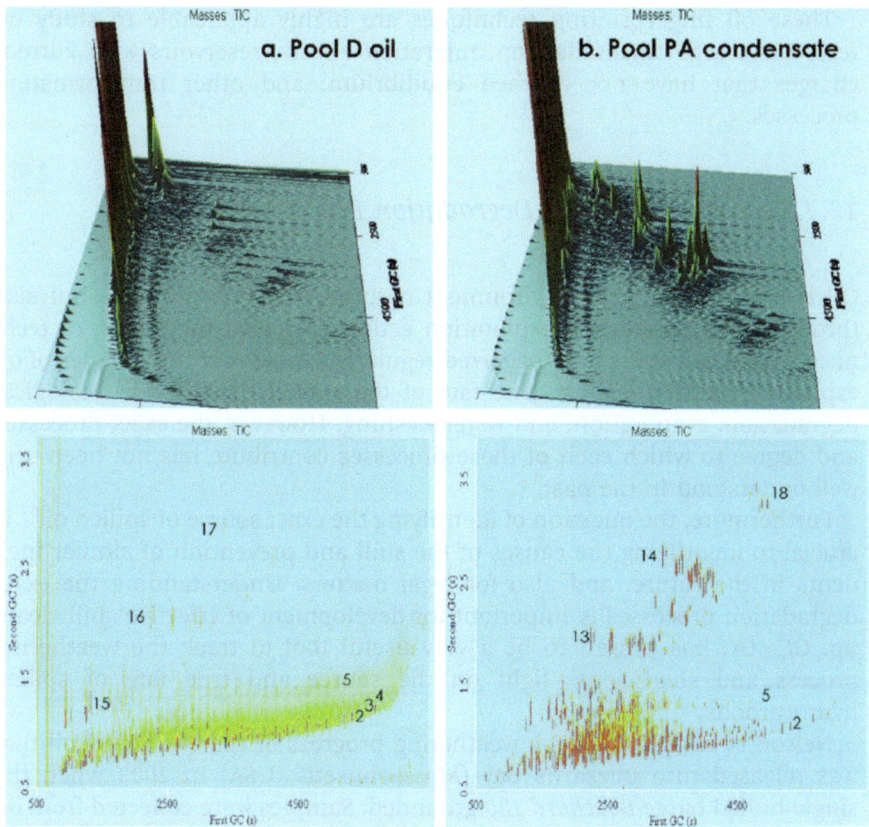

Figure 12.8 GC×GC chromatograms of fluids from the Brazeau River field, west-central Alberta (Canada). (a) Pool D oil, a 'sweet' oil without molecular signs of TSR. (b) Pool PA condensate with high abundance of benzothiophenes and dibenzithiophenes indicating TSR. (1) isoprenoid alkanes; (2) n-alkanes; (3, 4) alkylcyclopentanes and alkylcyclohexanes; (5) alkylbenzenes and alkyltoluenes including aryl isoprenoid alkanes; (6) steranes and methyl steranes; (7) pentacyclic terpanes; (8) hexahydrobenzohopanes; (9) thiolane and thiophene steroids; (10) thiolane and thiophene hopanoids; (11) monoaromatic steroids; (12) triaromatic steroids; (13) alkyl benzothiophenes; (14) alkyl dibenzothiophenes; (15) alkyl benzofurans; (16) alkyl naphthalenes; (17) alkyl phenanthrenes; (18) alkyl-9-thia-1,2-benzofluorenes.
Reprinted from Li *et al.*,[26] Copyright (2012), with permission from Elsevier.

comparable compound classes was extended to *e.g.* hopanes, steranes, mono- and dicyclic alkanes, naphthalenes and benzothiophenes, indenes, fluorenes, pyrenes, triaromatic steranes and benzohopanes, and more. Additionally, the resolving and data mining power of GC×GC could be utilised to identify minor chemical differences to a level that is not achievable with 1D GC.

These oil fingerprinting techniques are highly applicable to study oil reservoir compartmentalisation, migration effects, reservoirs with current charges that have not reached equilibrium, and other transformation processes.

12.4.1.1 Monitoring of Degradation Processes—Oil Spill Modelling

Oil is introduced to the environment naturally through oil seeps but also through incidents during exploration activities caused by human or technical failure. Oil spills have occurred regularly throughout the decades of oil exploration and transport. The fate of the expelled oil lies in biological degradation, evaporation, and water-washing. However, the exact processes and degree to which each of these processes contribute has not been very well understood in the past.

Furthermore, the question of identifying the exact source of spilled oil[52] is crucial to identifying the causes of the spill and prevention of similar incidents in the future, and also for legal matters. Understanding the exact degradation processes is important for development of effective spill clean-up. GC×GC has proven to be a very useful tool to track the weathering process and shed some light on the source and true fate of spilled hydrocarbons.

Nelson *et al.*[48] studied the weathering progress of the No. 6 fuel oil that was released into Buzzards Bay (Massachusetts, USA) in 2003 when the single-hulled barge *Bouchard 120* grounded. Samples were collected from oil covered rocks on days 12 and 179 after the accident. The GC×GC chromatograms of these two samples clearly showed the weathering (biodegradation, water-washing, and evaporation) and most affected compound classes could be identified as n-alkanes, alkylated naphthalenes, and phenanthrenes. A more sophisticated way to determine the changes between day 12 and day 179 was presented in form of difference, ratio, and addition chromatograms (Figure 12.9).

The difference chromatogram was obtained by normalising both chromatograms to $17\alpha(H),21\beta(H)$-hopane followed by the subtraction of the day 179 chromatogram from the day 12 chromatogram. Retention time stability was validated prior to this step. The resulting difference chromatograms (Figure 12.9 and Figure 12.10.a) show clearly all the compounds and compound classes that were reduced or eliminated between the two sampling events.

The ratio plot (Figure 12.10b) is a difference chromatogram resulting from the division of the slightly weathered sample (day 12) by the heavily weathered sample (day 179). Red peaks exhibit the largest relative loss from weathering. The addition chromatogram (Figure 12.10c) indicates compounds that are either completely removed from the slightly weathered sample (green) or were not present in the spilled oil sample at day 12 but

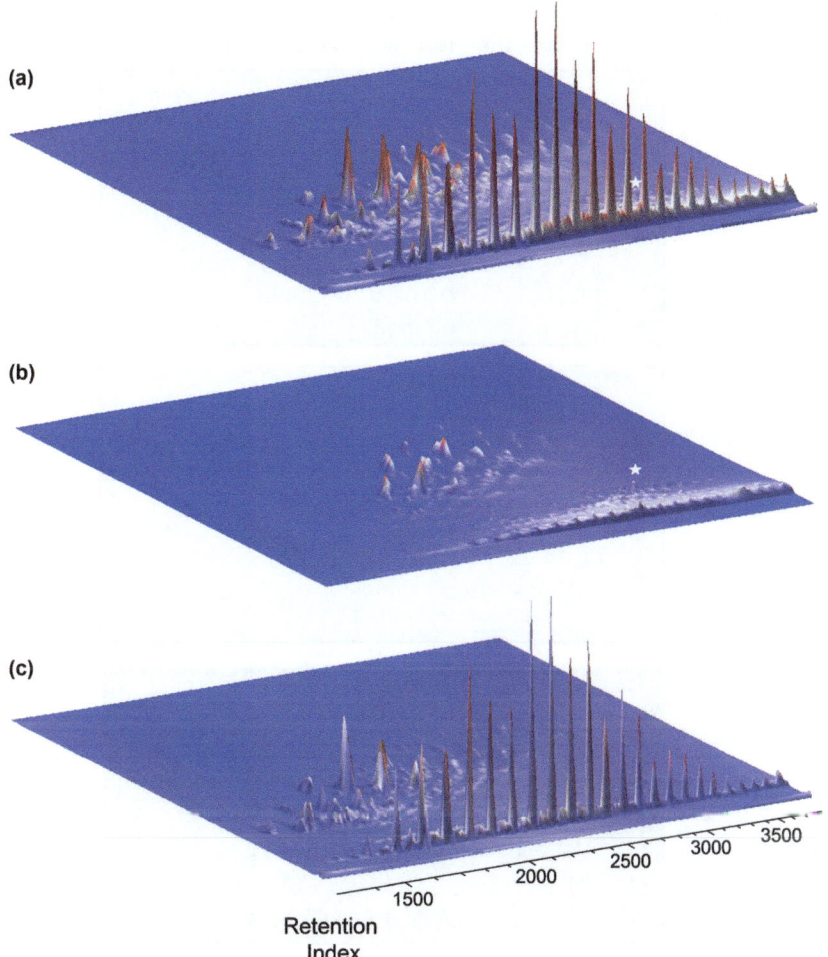

Figure 12.9 GC×GC mountain plot chromatogram of sample extracts from Bouchard 120 oil-covered rocks collected (a) 12 days after the spill (9 May
2003), and (b) 179 days after the spill (23 November 2003). The conserved standard 17α(H)-21β(H)-hopane is marked with a star.
Reprinted from Nelson *et al.*,[48] Copyright (2006), with permission of the
publisher Taylor & Francis Ltd (http://www.tandf.co.uk/journals).

were present on day 179 (red), representing compounds that have been
formed during the degradation process. The results showed a transition
from n-alkanes to polyaromatic hydrocarbon (PAH)-dominated hydrocarbons over the course of 6 months. During that time, 20–100% of the
n-alkanes were removed, while hopanes and steranes changed very little.
Image-based techniques like difference chromatograms identified weathering trends and facilitate source identification and monitoring of natural
and enhanced bioremediation of petroleum hydrocarbons.

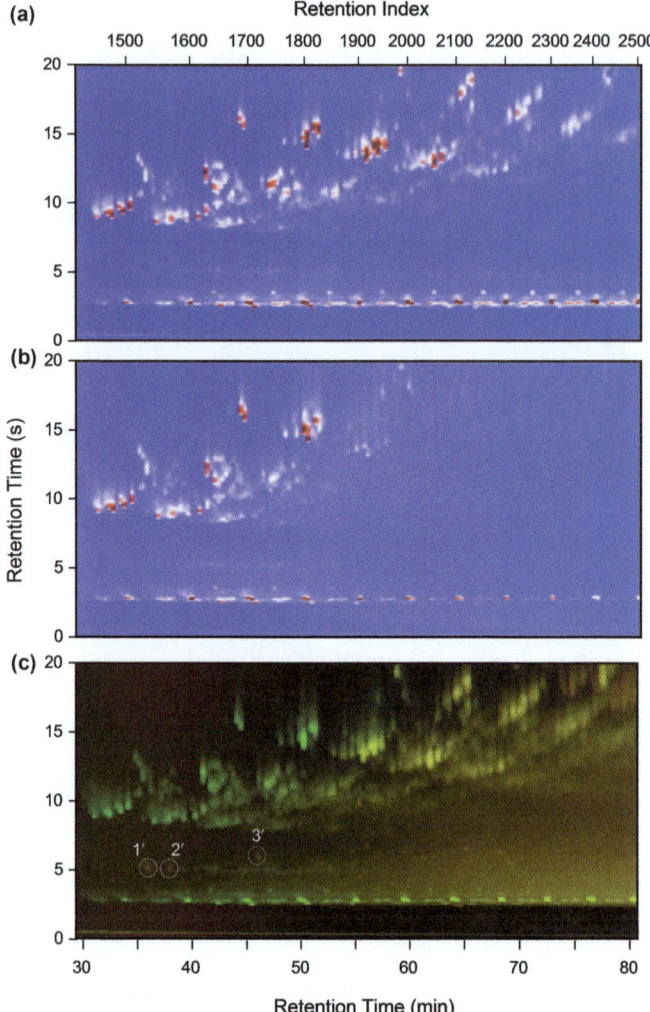

Figure 12.10   Enlarged GC×GC colour contour (a) difference chromatogram, (b) ratio chromatogram, and (c) addition chromatogram. Peak intensities in (a) and (b) are colour coded from white to red (most abundant). The addition chromatogram was produced using both colour and intensity values for each chromatogram. All the peaks in the day 12 (9 May 2003) chromatogram were assigned the colour green, and all the peaks from the day 179 (23 November 2003) chromatogram were assigned the colour red. Compounds that are present in one chromatogram but are absent in the other appear either green or red; compounds that are present in both chromatograms appear yellow. Red peaks at the left of the chromatogram (labelled 1, 2, and 3; c) indicate either that these compounds were not present in any appreciable amount in the day 12 chromatogram or that these compounds are preferentially conserved while proximal compounds are lost during oil weathering.
Reprinted from Nelson *et al.*,[48] Copyright (2006), with permission of the publisher Taylor & Francis Ltd (http://www.tandf.co.uk/journals).

Peacock *et al.*[49] characterised the residual petroleum hydrocarbons in salt marsh sediments in Winsor Cove (Buzzards Bay, Massachusetts), impacted from the 1974 spill of No. 2 fuel oil by the barge *Bouchard 65*. Significant amounts of PAHs, mainly C_4-naphthalenes and C_1-, C_2-, and C_3-phenanthrenes/anthracenes remained in the upper sediment layers whereas naphthalene and phenanthrene were removed. Biodegradation, evaporation, and water-washing as well as erosion were identified as removal processes.

Lemkau *et al.*[47] studied the weathering of heavy fuel oils from the MV *Cosco Busan* spill (November 2007; San Francisco Bay, CA, USA). They related the spilled oil to the exact source (ruptured tank 3 or 4) and characterised changes in the oil composition across location and time. Changes in n-C_{18}/phytane and benz[*a*]anthracene/chrysene ratios indicated biodegradation and photodegradation as causes of compound removal or decrease.

GC×GC has been applied extensively for studies of the fate of the oil after the Deepwater Horizon disaster.[42,50,71-74] On 20 April 2010 an explosion and fire aboard the drilling rig *Deepwater Horizon* in the Gulf of Mexico killed 11 crew members and marked the beginning of the largest offshore oil spill ever to occur in US territorial waters. The *Deepwater Horizon* capsized and sank on 21 April 2010, coming to rest approximately 400 m north-west of the well head at a depth of approximately 1500 m. Crude oil from the Macondo well flowed into the northern Gulf of Mexico between 20 April and 4 August 2010 (approximately 106 days) before the well was finally sealed with drilling mud and cement. Over the course of the spill an estimated 4.9 million barrels of oil flowed into the Gulf of Mexico, according to a US Federal On Scene Coordinators (FOSC) report.[75] Oil from the spill was found on beaches and in marshes in Texas, Louisiana, Mississippi, Alabama, and Florida.

Fingerprinting weathered oil and tar balls from this spill is challenging because (1) the crude oil was transported vertically through the water column from a depth of 1500 m to the surface; (2) the sea-surface water temperature was 30 °C and higher; (3) chemical dispersants were used both at the sea surface (initially) and at the well head; (4) *in situ* burning of oil at the sea surface took place; and (5) solar radiation is capable of transforming some of the chemical components in crude oil. The combined weathering processes of water-washing, evaporation, photooxidation, and bio-degradation altered the composition of spilled oil dramatically in many instances.

It became apparent shortly after the spill began that the most useful region of the 2D gas chromatogram for fingerprinting weathered oil samples in order to determine the source of spilled oil found on beaches of the northern Gulf of Mexico was the biomarker region and the hopanoids in particular appeared to be environmentally recalcitrant and thus very good molecules for fingerprinting (Figure 12.11a).[76] The advantage of using GC×GC for fingerprinting oil found on beaches and floating at the sea

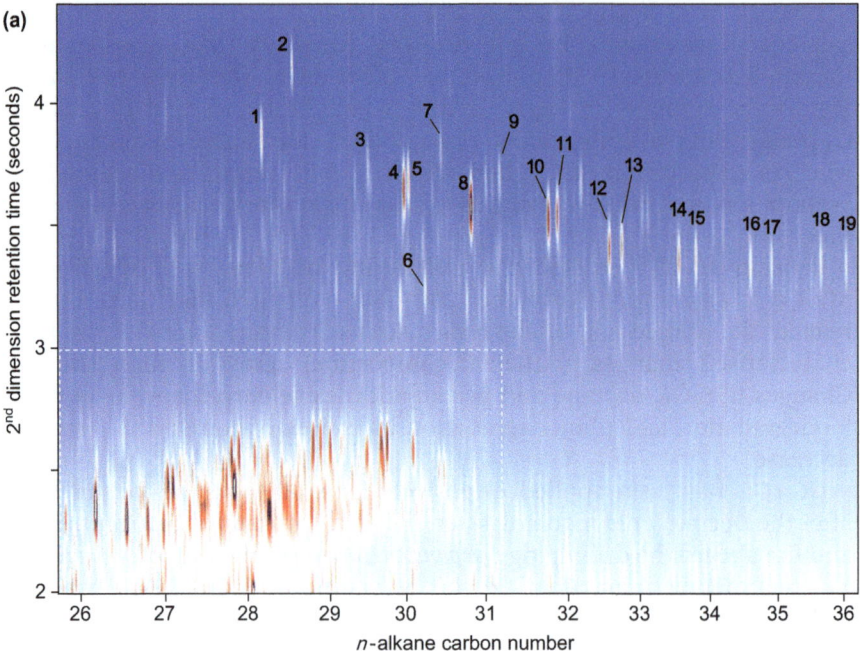

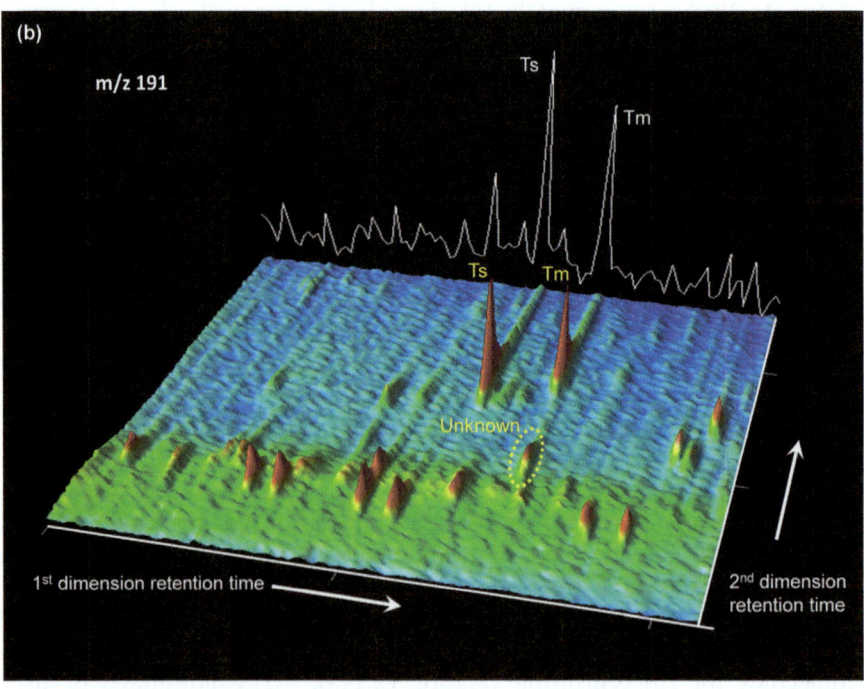

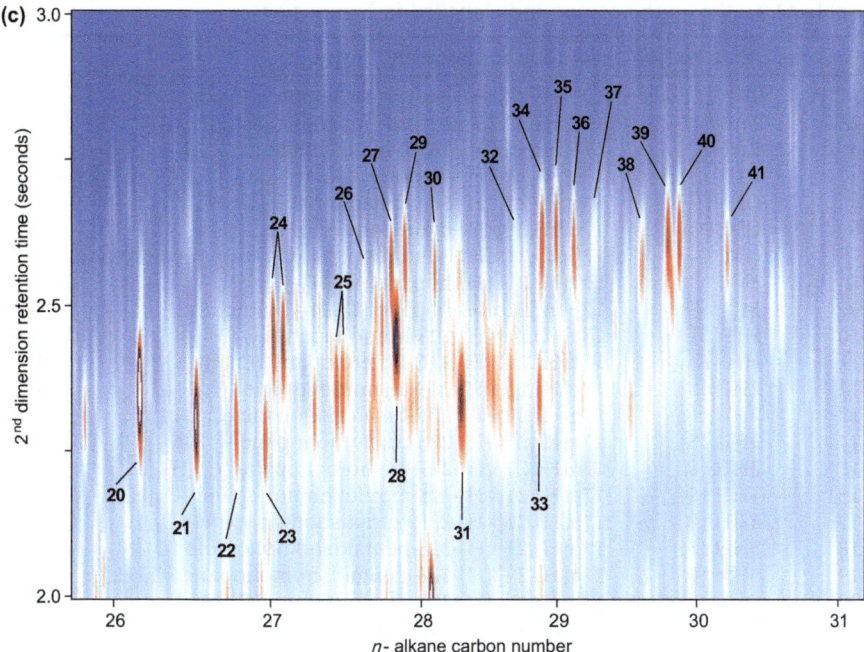

Figure 12.11   (a) Colour contour plot of the biomarker region of Macondo well crude oil. In this figure the hopanoid components are numbered and the chemical names, molecular formulas, and masses are shown in Table 12.2. The area highlighted inside the yellow box (diasteranes and steranes) is expanded in Figure 12.11c. (b) Example of a coelution that is readily apparent using GC×GC but frequently overlooked by laboratories using standard GC-MS. When laboratories integrate the unknown and the Tm peak together as one peak the Ts/Tm ratio underestimates the thermal maturity of the crude oil being studied. (c) This figure highlights the complexity of the sterane/diasterane region of Macondo well crude oil.

surface is the very high resolving power of the GC×GC technique compared to standard GC techniques. 2D GC affords approximately 10× more resolution compared with traditional GC. There are numerous examples of biomarker components that coelute in 1D GC which are resolved using GC×GC.

Figure 12.11b highlights the coelution of an unknown component and 17α-22,29,30-trisnorhopane (Tm). The unknown component eluting with Tm on GC-MS instruments produces an *m/z* 191 ion and this component elutes directly below the Tm molecule in 2D chromatographic space. Laboratories that utilise the Ts/Tm thermal maturity indicator ratio (Ts: 18α-22,29,30-trisnorhopane) using standard GC-MS techniques underestimate the thermal maturity of crude oils when this unknown component is present and integrated together with the Tm peak (as in the case of Macondo oil).

Table 12.2   Key to the compounds labelled in Figure 12.11.

Peak no.	Abbreviation	Compound name and formula	Mass
1	Ts	18α(H)-22,29,30-trinorneohopane ($C_{27}H_{46}$)	370
2	Tm	17α(H)-22,29,30-trinorhopane ($C_{27}H_{46}$)	370
3	BNH	17α(H),21β(H)-28,30-bisnorhopane ($C_{28}H_{48}$)	384
4	NH	17α(H),21β(H)-30-norhopane ($C_{29}H_{50}$)	398
5	C29-Ts	18α(H),21β(H)-30-norneohopane ($C_{29}H_{50}$)	398
6	C30-Dia	17α(H),21β(H)-diahopane ($C_{30}H_{52}$)	412
7	NM	17β(H),21α(H)-30-norhopane ($C_{29}H_{50}$) normoretane	398
8	H	17α(H),21β(H)-hopane ($C_{30}H_{52}$)	412
9	M	17β(H),21α(H)-hopane ($C_{30}H_{52}$) moretane	412
10	HH (S)	17α(H),21β(H)-22S-homohopane ($C_{31}H_{54}$)	426
11	HH (R)	17α(H),21β(H)-22R-homohopane ($C_{31}H_{54}$)	426
12	2HH (S)	17α(H),21β(H)-22S-bishomohopane ($C_{32}H_{56}$)	440
13	2HH (R)	17α(H),21β(H)-22R-bishomohopane ($C_{32}H_{56}$)	440
14	3HH (S)	17α(H),21β(H)-22S-trishomohopane ($C_{33}H_{58}$)	454
15	3HH (R)	17α(H),21β(H)-22R-trishomohopane ($C_{33}H_{58}$)	454
16	4HH (S)	17α(H),21β(H)-22S-tetrakishomohopane ($C_{34}H_{60}$)	468
17	4HH (R)	17α(H),21β(H)-22R-tetrakishomohopane ($C_{34}H_{60}$)	468
18	5HH (S)	17α(H),21β(H)-22S-pentakishomohopane ($C_{35}H_{62}$)	482
19	5HH (R)	17α(H),21β(H)-22R-pentakishomohopane ($C_{35}H_{62}$)	482
20	DiaC$_{27}$βα-20S	13β(H),17α(H)-20S-diacholestane ($C_{27}H_{48}$)	372
21	DiaC$_{27}$βα-20R	13β(H),17α(H)-20R-diacholestane ($C_{27}H_{48}$)	372
22	DiaC$_{27}$αβ-20S	13α(H),17β(H)-20S-diacholestane ($C_{27}H_{48}$)	372
23	DiaC$_{27}$αβ-20R	13α(H),17β(H)-20R-diacholestane ($C_{27}H_{48}$)	372
24	DiaC$_{28}$βα-20S	24-methyl(S&R)-13β(H),17α(H)-20S-diacholestane ($C_{28}H_{50}$)	386
25	DiaC$_{28}$βα-20R	24-methyl(S&R)-13β(H),17α(H)-20R-diacholestane ($C_{28}H_{50}$)	386
26	C$_{27}$ααα-20S	5α(H),14α(H),17α(H)-20S-cholestane ($C_{27}H_{48}$)	372
27	C$_{27}$αββ-20R	5α(H),14β(H),17β(H)-20R-cholestane ($C_{27}H_{48}$)	372
28	DiaC$_{29}$βα-20S	24-ethyl(S&R)-13β(H),17α(H)-20S-diacholestane ($C_{29}H_{52}$)	400
29	C$_{27}$αββ-20S	5α(H),14β(H),17β(H)-20S-cholestane ($C_{27}H_{48}$)	372
30	C$_{27}$ααα-20R	5α(H),14α(H),17α(H)-20R-cholestane ($C_{27}H_{48}$)	372
31	DiaC$_{29}$βα-20R	24-ethyl(S&R)-13β(H),17α(H)-20R-diacholestane ($C_{29}H_{52}$)	400
32	C$_{28}$ααα-20S	24-methyl-5α(H),14α(H),17α(H)-20S-cholestane ($C_{28}H_{50}$)	386
33	DiaC$_{29}$αβ-20R	24-ethyl(S&R)-13α(H),17β(H)-20R-diacholestane ($C_{29}H_{52}$)	400
34	C$_{28}$αββ-20R	24-methyl-5α(H),14β(H),17β(H)-20R-cholestane ($C_{28}H_{50}$)	386
35	C$_{28}$αββ-20S	24-methyl-5α(H),14β(H),17β(H)-20S-cholestane ($C_{28}H_{50}$)	386
36	Unknown	Unknown ($C_{29}H_{52}$)	400
37	C$_{28}$ααα-20R	24-methyl-5α(H),14α(H),17α(H)-20R-cholestane ($C_{28}H_{50}$)	386
38	C$_{29}$ααα-20S	24-ethyl-5α(H),14α(H),17α(H)-20S-cholestane ($C_{29}H_{52}$)	400
39	C$_{29}$αββ-20R	24-ethyl-5α(H),14β(H),17β(H)-20R-cholestane ($C_{29}H_{52}$)	400
40	C$_{29}$αββ-20S	24-ethyl-5α(H),14β(H),17β(H)-20S-cholestane ($C_{29}H_{52}$)	400
41	C$_{29}$ααα-20R	24-ethyl-5α(H),14α(H),17α(H)-20R-cholestane ($C_{29}H_{52}$)	400

An expanded view of the region of the GC×GC chromatogram that contains diasteranes and steranes is presented in Figure 12.11c (this is the area highlighted by a yellow box in Figure 12.11a). This figure illustrates the complexity of the diasterane/sterane region as well as the resolving power of the GC×GC technique. This level of resolution is unattainable by conventional GC techniques, and coeluting peaks (that are frequently misinterpreted) are resolved from one another here. Examples of components that coelute on standard GC instruments but are resolved using GC×GC are seen as peaks 27 and 28 and peaks 33 and 34 in Figure 12.11c. Peaks 27 and 28 are 5α(H),14β(H),17β(H)-20R-cholestane and 24-ethyl(S&R)-13β(H),17α(H)-20S-diacholestane respectively and commonly coelute on 1D instruments. Using GC×GC, these two coeluting components are very nearly completely resolved. Likewise, peaks 33 and 34, (24-ethyl(S&R)-13α(H),17β(H)-20R-diacholestane and 24-methyl-5α(H),14β(H),17β(H)-20R-cholestane), coelute on 1D instruments but are completely resolved using GC×GC. Close inspection of this chromatogram highlights many more components that coelute on 1D instruments that are resolved on GC×GC instruments (see the small coeluting peak to the lower right of peak 39, for example).

The detailed characterisation by GC×GC of the Macondo oil and gas end member sampled directly from the well head was applied by Reddy *et al.*[50] to determine the gas-to-oil ratio (GOR), as well as the trajectories and fates of the hydrocarbons in the deep water marine environment. In contrast to oil spills at the sea surface, where evaporation and dissolution are equally important factors in the removal of hydrocarbons, deep water spills such as the *Deepwater Horizon* disaster at 1.5 km depth lack the evaporative component in the fate of the hydrocarbons. Compositional comparison of the released Macondo fluid and samples taken from the water column within a deep water plume that was defined by Camilli *et al.*[77] make it possible to study the partitioning of hydrocarbons into the aqueous phase in the absence of atmospheric evaporation.

Detailed compositional analysis of the whole fluid showed that the total C_1–C_5 hydrocarbon fraction accounted for half of the mass of hydrocarbons released from the Macondo well. These light hydrocarbons are commonly not detected by traditional oil characterisation methods on a molecular level, which can lead to significant underestimation of the total hydrocarbon mass released during a spill.

Compositional analyses of the hydrocarbon content in the deep water plume at 1100 m depth indicated that the hydrocarbon levels of low-molecular-weight n-alkanes and aromatic compounds were below their saturation levels. Furthermore, evidence was found that methane was quantitatively trapped in the deep water plume and was nearly absent in shallower depths of the water column. With increasing chain length (C_2, C_3, *etc.*), less n-alkanes were dissolved in the deep water plume due to their decreasing solubility in water. With increasing molecular mass an increasing proportion of the total n-alkane content was retained in the buoyant liquid oil phase ascending to the sea surface. Water-soluble aromatic compounds were also monitored in their coexistence to n-alkanes and their decrease in abundance over 27 km of trajectory of the plume. Data by Reddy *et al.*[50] indicate no significant biodegradation progress in aromatic compounds over 27 km (4 days).

Overall, it was demonstrated that most of the C_1–C_3 hydrocarbons and water-soluble aromatic compounds were retained in the deep water column while less soluble components were transported to the surface or sank to the sea floor.

Natural seeps contribute almost half of the oil entering the coastal ocean. Nonetheless, the rates and relevance of physical, chemical, and biological weathering processes are poorly understood. Wardlaw *et al.*[51] studied natural oil seeps offshore Santa Barbara, California, USA with comprehensive 2D GC to investigate the specific changes in petroleum composition and hydrocarbon abundance between subsurface reservoirs, a proximal sea floor seep, and the sea surface overlying the seep. The group developed methods to account for hydrocarbon mass losses to the different processes including biodegradation, dissolution, and evaporation differentiating biological and physical weathering processes of complex mixtures at a molecular level using GC×GC.

Similarly, Farwell *et al.* studied the long-term fate and fallout plume of heavy oil from strong natural petroleum seeps near Coal Oil Point, California.[46]

12.4.1.2   Statistical Analysis of GC×GC Data

Ventura *et al.*[22] extended their fingerprinting approach and applied statistical methods to effectively data-mine the large molecular data sets obtained by a high-resolution technique such as GC×GC. To utilise the full spectrum of molecular information contained in a complex crude oil and resolved by GC×GC for the accurate establishment of oil similarities instead of relying on a relatively small number of biomarkers, Ventura and co-workers used multiway principal component analysis (MPCA). MPCA in this application quantitatively compares every compound eluting within the same first- and second-dimension RT with eluents in that same GC×GC space in other fluids. The comparison based on GC×GC chromatograms included 3500 and more quantified components. Furthermore, three individual regions were identified by Ventura *et al.* that are suitable to analyse, resolve, and evaluate compositional changes related to the source organic matter of the hydrocarbons and their exposure to biological and physical weathering.

MPCA was demonstrated to be an effective tool to resolve multimolecular differences, even between very similar fluids, as well as to evaluate the degree of molecular relatedness and group oils into families. It was possible to quantitatively determine compositional and instrumental artefacts in order to exclude their impact on the establishment of similarities and differences.

Contaminants (*e.g.* alkenes introduced from drilling fluids) could be identified using factor loading plots and did not disturb the evaluation of the degree of the fluids' relatedness.

In order to identify minute differences in otherwise very similar oils, high-resolution techniques like GC×GC are required. Powerful data mining methods such as MPCA are needed to extract the unique molecular differences from the large data sets obtained from GC×GC, making GC×GC-MPCA a valuable combination to study reservoir connectivity, or discriminate between unique molecular compositions derived from source-dependent or weathering-related processes.

12.4.1.3 Identification and Quantification of Contaminants in Crude Oil

Drilling fluids play an essential role in the drilling process. However, they are dreaded by geochemists because they can significantly contaminate valuable samples, both fluid samples and sediment samples, essentially rendering them useless. Companies lose valuable information required to understand the reservoir, thermal maturity, the source of the petroleum hydrocarbons, and source rock quality in the drilled section which in the long term cause significant financial losses. Oil-based drilling fluids contain compounds that are equally present in the reservoir fluid, thus contamination with synthetic oil-based muds such as linear α-olefins (LAOs) or internal olefins (IOs) irreversibly alter the sample and prevent analysis of the indigenous oil composition.

Because of the complexity of crude oils, recognition and moreover quantification of IOs and LAOs can be challenging using 1D GC techniques. Reddy *et al.*[78] applied GC×GC-FID to identify and quantify alkene-based drilling fluids in crude oils, allowing them to eliminate the compounds that are derived from drilling fluids from the molecular composition of the crude oil.

As described in section 12.2.6, alkenes are separated from the range of n-alkanes in second dimension on a non-polar/polar column combination. Hence, the presence of artificially introduced alkenes can be detected and quantified in a GC×GC analysis down to an estimated 1% of contamination in crude oil.[78] This method is valuable to the industry as it enables detailed and accurate biomarker analysis of contaminated fluids and can also be used to monitor areas near wells for possible leaks or discharge into the environment.

Recently, Aeppli *et al.*[42] reported the use of olefins commonly found in synthetic drilling fluids to determine the source of oil sheens sampled in 2012 near the *Deepwater Horizon* site. Application of GC×GC-FID and GC×GC-TOF-MS allowed clear identification of three olefins in these oil sheen samples which clearly indicated that the oil sheens are sourced from leaking oil that was in contact or mixed with drilling fluid before release into the water. This observation ruled out the source of the sheens being the coffer-dam oil, as was proposed by BP, because this oil does not contain any drilling fluid residues. Instead, presence of the olefins together with biomarker ratios and alkene ratios obtained from the highly resolved GC×GC-TOF-MS chromatograms suggest that the observed sheens originated from tanks and pits on the *Deepwater Horizon* wreckage.

These findings introduce a method to track drilling fluids released to the environment and further provide a framework for assessment of the fate of drilling fluids commonly released during exploration activities around the world.

12.4.2   Detailed Biomarker Analysis

12.4.2.1   Diamondoids

Diamondoids are very resistant to biodegradation[79] and are of particular interest to petroleum geochemists for thermal maturity assessment of highly mature oils.

With the use of GC×GC-TOF-MS, diamondoids could be effectively separated from normal and branched alkanes and other saturated hydrocarbons.[53–55] A large number of additional diamondoids could be separated and identified in addition to the diamondoids that were known to date using 1D GC. Wang *et al.*[54] recognised and identified over 100 individual substitutions of adamantane, diamantine, and triamantane. Silva *et al.*[55] identified tetramantanes and propyltriamantanes. These very recent studies present the start of a fruitful in depth discovery of diamondoids and their formation mechanisms.

12.4.2.2   Plant Markers

The GC×GC separation of angiosperm-derived saturated pentacyclic triterpenoids 18α(H)-, 18β(H)-oleanane and lupane was described by Eiserbeck *et al.*[56] The resolution of 18α(H)-oleanane and lupane using a similar column combination (non-polar/polar) was also reported by Silva *et al.*[20] The oleanane isomers and lupane coelute on a 1D GC system. Separation of all three of these compounds in one analysis could not be achieved previously. These compounds are very important angiosperm biomarkers for the recognition of terrigenous organic matter. However, the presence and abundance of individual plant biomarkers has not been well studied and understood in their information content, because of the challenges of separation. This applies not only to the three most commonly applied and well known angiosperm biomarkers—the oleanane isomers and lupine—but in particular to the amount of information that could be obtained from the detailed study of the distribution of the various diagenetic products of the biological precursors amyrin and betulin. GC×GC provides the tools to investigate the presence and abundances of the various compound classes present in one single sample that are derived from the same precursors and can be studied in one single analysis, and direct comparison allows conclusions about the diagenetic processes within this biomarker class.

Eiserbeck *et al.*[27] extended the separation of plant biomarkers using GC×GC to these various compound classes. Only one whole-oil/source rock extract analysis was required to achieve baseline separation of commonly low abundant higher plant biomarkers such as the saturated C_{24} des-A- and C_{23} nor-des-A-compounds series including lupanoids, oleanoids, and ursanoids and their mono- and diunsaturated homologues.[27] A clear elution pattern for each series could be displayed (see section 12.2.6) with the series eluting parallel to each other, shifted in first- and second-dimension retention time (Figure 12.12).

Furthermore, components with identical molecular masses and fragmentation patterns were resolved in the second GC dimension, for instance tentatively identified des-A-lupane and nor-des-A-ursane/taraxastane (Figure 12.12a), or the coelution of des-A-lupene and des-A-lupadiene with two isomers, respectively (Figure 12.12b). These compounds could not be identified or quantified using 1D GC techniques.

GC×GC-TOF-MS is particularly valuable for separation problems of coeluting components with identical molecular masses and fragmentation pattern because the mass spectral similarity prevents separation based on extracted ion chromatograms or even metastable transition monitoring 1D GC systems.

Additional critical 1D GC separation problems in studies of angiosperm biomarkers that were resolved by GC×GC in second-dimension separation include the coelution of 24-norlupane and norhopane as well as 24,28-bisnorlupane and bisnorhopane (Figure 12.13), monoaromatic des-A-compounds ($M^+ = 310$), and triaromatic pentacyclic triterpenoids ($M^+ = 342$) (Figure 12.12c).

Taraxastane, another biomarker indicating land-plant input to the source organic matter,[80] was reported to elute later in the first and second dimension than α,β-hopane, moretane and oleanane.[27] The 25-norhopane series in biodegraded samples elutes earlier in the second dimension and thus does not coelute with higher-plant-derived saturated C_{28}- and C_{29}-compounds.

Many more compounds could be identified due to the sensitivity and resolution of GC×GC than were detected by traditional 1D GC, including isomers of known compounds as well as potentially new biomarkers. Some unsaturated triterpenoids are thermally unstable and prone to rearrangement. Taraxer-14-ene, for instance, coelutes in traditional GC-MS analysis with n-C_{30} alkane. Molecular sieving procedures[81] to separate these compounds before the analysis activate rearrangement of taraxer-14-ene to a mixture of oleanene isomers.[82] The increased resolution and sensitivity of GC×GC-TOF-MS enables the separation of such unstable compounds in a whole-oil analysis without loss in resolution maintaining the natural distribution of biomarkers in the sample.

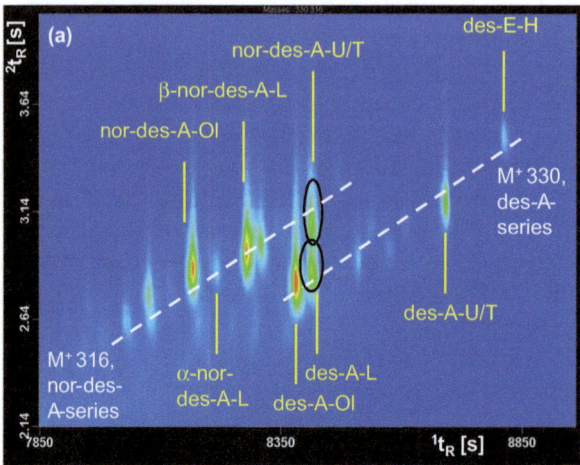

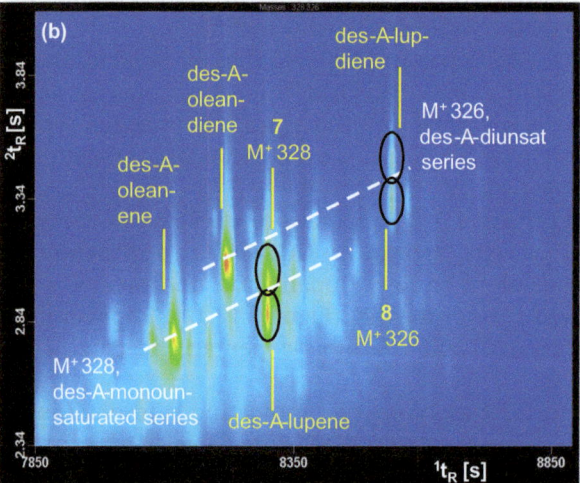

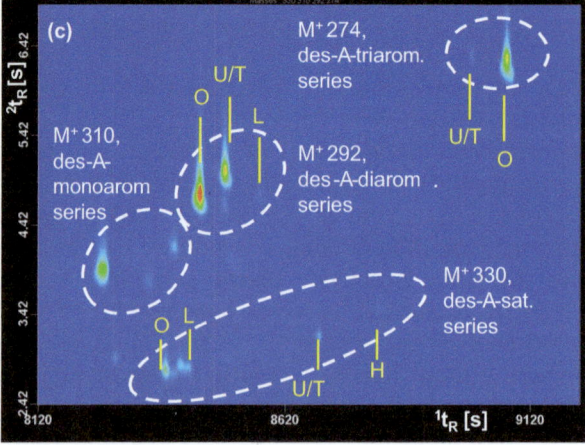

12.4.3 Unresolved Complex Mixtures

UCMs present challenges for the identification and quantification of resolved compounds using traditional GC-MS. Furthermore, the thousands of unresolved compounds within the UCM contain valuable information about the organic matter and also can contribute to the understanding of biodegradation processes. The high resolving power of GC×GC was applied to study non-degraded compounds in biodegraded samples as well as the composition of the UCM itself.

Many studies have been published on the successful application of GC×GC to questions related to degraded fluids.[14,17,23,27,83] Tran *et al.*[14] investigated the chemical composition of UCMs in three sets of oils from Australia, each set representing one oil family, and each sample within one set was exposed to a different level of biodegradation. The three sets were from marine, land-plant, and mixed marine and land-plant-derived organic matter, respectively. Chemical changes with progressing biodegradation were observed for each type of organic matter. A polar/non-polar column configuration was applied based on the suggestion that UCM is dominated by aliphatic, non-polar compounds,[17,84] which was confirmed by the position of the bulk of the UCM within the chromatographic plane. Tran *et al.*[14] showed that the different sources of the oils had little impact on the composition of the UCM. In all three oil families, the main compound class resolved from the UCM with GC×GC was C_1–C_7 alkyl-decahydronaphthalenes (alkyl-decalin). The dominating alkyl substitutions shift with increasing degree of biodegradation, from C_1–C_3-substituted isomers dominating oils exhibiting lower levels of degradation to C_4–C_7 isomers, which are more abundant in highly degraded oils. The longer the alkyl chain, the more isomers were observed in the GC×GC chromatogram, eluting in a parallel bands across the chromatographic plane. It was proposed that the coelution of both the isomers within one band as well as the bands with each other accounts for the UCM 'hump' in traditional 1D GC chromatograms. Alkyl-decalins were observed as constituents of UCM derived from biodegradation before, using GC×GC analysis,[17,23,49,52] *e.g.* in intertidal sediments contaminated with residual hydrocarbons after oil spills,[49] although less obvious due to the use of the more common

---

Figure 12.12 GC×GC-TOFMS ion chromatograms illustrating elution order and separation results for tetracyclic triterpenoids in oil C2. (a) Extracted ion chromatogram m/z $330 + 316$. (b) Extracted ion chromatogram m/z $328 + 326$. (c) Extracted ion chromatogram m/z $330 + 310 + 292 + 274$ representing saturated, mono-, di- and triaromatic tetracyclic triterpenoids. O, oleanane; U/T, ursane/taraxastanes; L, lupine; H, hopane. Dashed lines represent the location of the series. Des-A-compounds were tentatively identified *via* comparison of GC-MS, GC-MRM-MS, and GC×GC-TOF chromatograms and mass spectra with elution orders and mass spectra reported in the literature.
Reprinted from Eiserbeck *et al.*,[27] Copyright (2012), with permission from Elsevier.

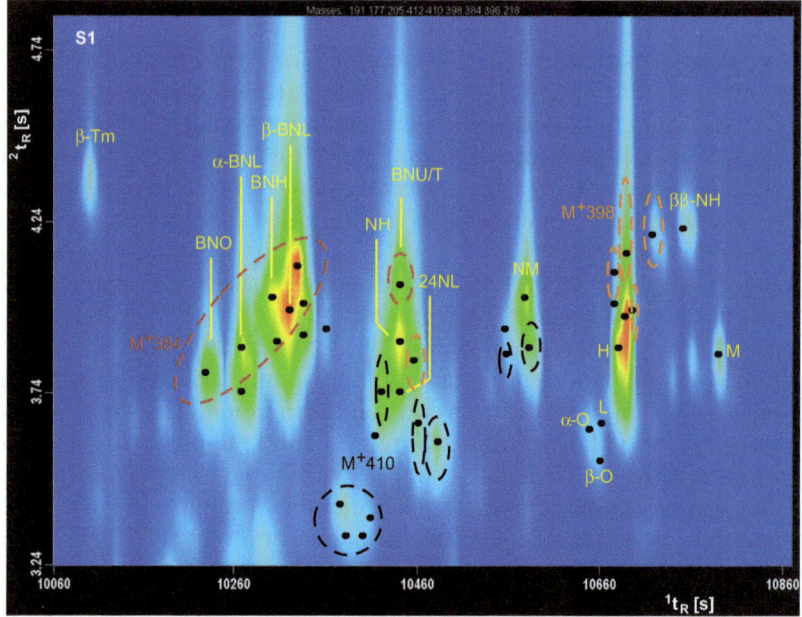

Figure 12.13   Section of the combined GC×GC-TOFMS EIC of the masses *m/z* 191, 177, 205, 412, 410, 398, 396, 384, 218 of rock extract S1. Key: dashed circles indicate groups of compounds (especially unidentified compounds based on the same molecular mass. Red: $M^+ = 384$, orange; $M^+ = 398$, black; $M^+ = 410$. β-Tm, 22,29,30-trinor-17β-hopane; α-BNL, 17α-24,28-bisnorlupane; β-BNL, 17β-24,28-bisnorlupane; BNH, 28,30-bisnorhopane; BNO, tentatively identified as 24,28-bisnoroleanane; NH, 30-norhopane; BNU/T, tentatively identified as 24,28-bisnorursane/taraxastane; 24NL, 24-norlupane; NM, 30-normoretane; H, 17α,21β(H)-hopane; α-O, 18α(H)-oleanane; β-O, 18β(H)-oleanane; L, lupine; ββ-NH, 17β,21β(H)-30-norhopane; M, moretane.
Reprinted from Eiserbeck *et al.*,[27] Copyright (2012), with permission from Elsevier.

non-polar/polar column configuration in these earlier studies. The polar/non-polar column configuration used by Tran *et al.*[14] was considered preferable to the more commonly used non-polar/polar combination to study the composition of UCM.

Although microbial biodegradation as the source of UCMs is more commonly studied and better understood, UCMs can also occur in systems that have been hydrothermally altered.[85–88]

Attempts to identify the constituents of UCMs in the past have been very involved and had varying success. Even extensive application of chemical degradation methods and use of capillary GC columns could not resolve some complex mixtures. The nature of the compounds that form the UCM remains unclear.

Ventura *et al.*[23] analysed samples from the Archaean which exposed varying degrees and topologies of UCMs that were irresolvable by traditional GC-MS. The researchers applied the increased separation power of GC×GC to facilitate tentative identification of a range of compound classes that were embedded within these UCM types. Based on the topology of the UCMs derived from 1D GC and the chemical composition derived from GC×GC analysis, they were able to characterise three different types of UCM in their sample set.

Type I UCM had the majority of the UCM in the lower molecular weight region and contained a range of mono- to hexacycloalkanes of unknown origin. Type II was characterised by a UCM of higher molecular weight and contained mostly C_{35}–C_{40} archaeal lipids. Type III represents a mix of type I and type II UCMs. This study demonstrated that GC×GC analysis of UCM in archaeal samples can provide valuable information about the biological source and diagenetic history of the organic matter. Furthermore, the analysis revealed that the samples from the Porcupine Gold Camp (northern Ontario, Canada)were unlikely to be derived from selective preservation of bioresistant compounds as it occurs during biodegradation because easily biodegradable compounds such as n-alkanes and acyclic isoprenoids were abundant in all samples. Very low concentrations of 25-norhopanes and the presence of highly isomerised archaeal lipids observed in type II UCM and the high abundance of mono- to hexacycloalkanes in type I UCM cannot be explained by biodegradation as source of the UCM, but rather suggest hydrothermal alteration as the likely degradation process. Type II UCM was proposed to result from the diagenetic alteration of biogenic material *via* a coupled process of isomerisation caused by hydrothermal stress in presence of high hydrogen pressure (hydrothermal conditions) and hydrocarbon cracking.

These insights support the study of the effects of hydrothermal alteration on the composition and preservation of Archaean organic matter and thus can help to determine the diversity of geological localities bearing extractable hydrocarbons.

In 2012, Ventura *et al.*[83] further investigated the genesis and evolution of hydrothermal petroleum, again utilising the power of GC×GC coupled with statistical partial least squares (PLS) linear regression analysis. Two simple metrics were introduced to quantitatively determine the alteration of maltene fractions in simple and complex mixtures and UCMs within GC×GC data. A molecular complexity metric quantifies the chromatographic complexity of a GC-amenable hydrocarbon fluid. The second metric is PLS linear regression of RT of individual compounds in petroleum. These two proxies were found to constrain the geochemical parameters that lead to structural ordering of the UCM in a GC×GC chromatogram. The authors compared samples collected from different stratigraphic or geographic hydrothermal environments employing the above metrics, as well as subtracted GC×GC chromatograms, which identified a characteristic multimolecular pattern of high molecular weight PAH migration.

The results showed evidence that, unlike the UCM derived from bio-degradation which is characterised by an intensified baseline due to the degradation of n-alkanes, the UCMs in hydrothermal petroleum are a unique assemblage of biomarkers dominating the hydrocarbon composition. They take shape in a continuous series of compounds across compound classes. The structural ordering of the UCM that resulted from the composition of the source organic matter coupled with the degree of thermal alteration of the oil was only visible in a GC×GC chromatogram. The study concluded that UCMs from hydrothermal alteration/oxidative weathering appear to represent the early stages of hydrothermal petroleum formation. Continued pyrolysis and oxidation reduces hydrocarbon complexity as a result of extensive dehydrogenation and dealkylation.

12.4.4 Depositional Environments

Oliveira *et al.*[34,35] recognised the presence of onoceranes as a possible additional proxy for depositional paleoenvironments. With the use of GC×GC-TOF-MS, the researchers identified a series of onocerane isomers in representative oil samples from the Potiguar Basin and Cumuruxatiba Basin in Brazil. Both oil families are the result of multiple charge events, combining characteristics of an older charge which was subsequently biodegraded, and a younger non-biodegraded charge. Based on biomarker parameters the depositional environments of these oils could be classified as lacustrine and marine. With the use of GC×GC-TOF-MS, it was identified that the lacustrine samples contained two distinct series of 3β-methylhopane and onocerane isomers, respectively, whereas the marine samples exhibited a predominance of the 2α-methylhopane series and lack of onocerane isomers. The results suggest that the presence of an onocerane series might be a proxy for lacustrine depositional paleoenvironments.

12.5 Limitations

A critical part in the comprehensive 2D GC is the modulator which traps, focuses, and releases preseparated compound packages onto the secondary column. Improvements in the modulation times could increase the resolution power of GC×GC, as indicated by Blumberg *et al.*[89]

Further expansion in the application of GC×GC could be achieved by technical developments in acquisition rates of detectors such as quadrupole MS or irMS in order to couple the full separation power of GC×GC with a broader variety of detectors. A combination of the resolution of GC×GC coupled to irMS is particularly desirable because compound-specific isotope analysis (CSIA) requires baseline separation of the compounds, which is often a very challenging or even unachievable task and currently requires numerous wet chemical separation steps.

However, the biggest challenge in the application of GC×GC to date is the handling of the overwhelming amount of data produced by this technique.

More advanced data processing and data mining software and techniques are required to fully utilise the data provided in a GC×GC analysis. To date, information reduction in form of extracted ion chromatograms is often applied to reduce the available amount of data to a level we humans can comprehend, still neglecting the full picture that is presented to us in a GC×GC chromatogram.

12.6 Summary and Future of GC×GC

GC×GC-TOF-MS will become a platform technology dedicated specifically to energy, environment, and earth science research. Its various applications will not be limited to petroleum, but will also be extended to mineral exploration, evolution of our planet's history, and tracing contaminants in our environment. GC×GC-TOF-MS will support research from a very broad range of disciplines in the areas of oil and mineral exploration, human and coastal impacts on our environment, potable water supplies, climate, and paleo-climate change.

Acknowledgements

CMR would like to thank the GOMRI DEEP-C consortium for support.

References

1. Z. Liu and J. B. Phillips, *J. Chromatogr. Sci.*, 1991, **29**, 227–231.
2. M. Adahchour, J. Beens, R. J. J. Vreuls and U. A. T. Brinkman, *Trends Anal. Chem.*, 2006, **25**, 438–454.
3. M. Adahchour, J. Beens, R. J. J. Vreuls and U. A. T. Brinkman, *Trends Anal. Chem.*, 2006, **25**, 540–553.
4. M. Adahchour, J. Beens, R. J. J. Vreuls and U. A. T. Brinkman, *Trends Anal. Chem.*, 2006, **25**, 821–840.
5. M. Adahchour, J. Beens, R. J. J. Vreuls and U. A. T. Brinkman, *Trends Anal. Chem.*, 2006, **25**, 726–741.
6. M. Adahchour, J. Beens and U. A. T. Brinkman, *J. Chromatogr. A*, 2008, **1186**, 67–108.
7. J. B. Phillips and J. Beens, *J. Chromatogr. A*, 1999, **856**, 331–347.
8. J. Beens and U. A. T. Brinkman, *Trends Anal. Chem.*, 2000, **19**, 260–275.
9. E. B. Ledford, C. A. Billesbach and Q. Y. Zhu, *J. High Resolut. Chromatogr.*, 2000, **23**, 205–207.
10. P. J. Marriott, S.-T. Chin, B. Maikhunthod, H.-G. Schmarr and S. Bieri, *Trends Anal. Chem.*, 2012, **34**, 1–21.
11. J. V. Seeley, in *Gas Chromatography*, Elsevier, Amsterdam, 2012, pp. 161–185.
12. J. S. Arey, R. K. Nelson, L. Xu and C. M. Reddy, *Anal. Chem.*, 2005, **77**, 7172–7182.
13. J. M. D. Dimandja, *Anal. Chem.*, 2004, **76**, 167A–174A.

14. T. C. Tran, G. A. Logan, E. Grosjean, D. Ryan and P. J. Marriott, *Geochim. Cosmochim. Acta*, 2010, **74**, 6468–6484.
15. G. T. Ventura, F. Kenig, C. M. Reddy, J. Schieber, G. S. Frysinger, R. K. Nelson, E. Dinel, R. B. Gaines and P. Schaeffer, *Proc. Natl Acad. Sci. U. S. A.*, 2007, **104**, 14260–14265.
16. R. B. Gaines, G. S. Frysinger, M. S. Hendrick-Smith and J. D. Stuart, *Environ. Sci. Technol.*, 1999, **33**, 2106–2112.
17. G. S. Frysinger, R. B. Gaines, L. Xu and C. M. Reddy, *Environ. Sci. Technol.*, 2003, **37**, 1653–1662.
18. A. Aguiar, A. I. Silva Júnior, D. A. Azevedo and F. R. Aquino Neto, *Fuel*, 2010, **89**, 2760–2768.
19. G. S. Frysinger and R. B. Gaines, *J. Sep. Sci.*, 2001, **24**, 87–96.
20. R. S. F. Silva, H. G. M. Aguiar, M. D. Rangel, D. A. Azevedo and F. R. Aquino Neto, *Fuel*, 2011, **90**, 2694–2699.
21. T. C. Tran, G. A. Logan, E. Grosjean, J. Harynuk, D. Ryan and P. Marriott, *Org. Geochem.*, 2006, **37**, 1190–1194.
22. G. T. Ventura, G. J. Hall, R. K. Nelson, G. S. Frysinger, B. Raghuraman, A. E. Pomerantz, O. C. Mullins and C. M. Reddy, *J. Chromatogr. A*, 2011, **1218**, 2584–2592.
23. G. T. Ventura, F. Kenig, C. M. Reddy, G. S. Frysinger, R. K. Nelson, B. V. Mooy and R. B. Gaines, *Org. Geochem.*, 2008, **39**, 846–867.
24. G. T. Ventura, B. Raghuraman, R. K. Nelson, O. C. Mullins and C. M. Reddy, *Org. Geochem.*, 2010, **41**, 1026–1035.
25. F. C. Y. Wang and C. C. Walters, *Anal. Chem.*, 2007, **79**, 5642–5650.
26. M. Li, S. Zhang, C. Jiang, G. Zhu, M. Fowler, S. Achal, M. Milovic, R. Robinson and S. Larter, *Org. Geochem.*, 2008, **39**, 1144–1149.
27. C. Eiserbeck, R. K. Nelson, K. Grice, J. Curiale and C. M. Reddy, *Geochim. Cosmochim. Acta*, 2012, **87**, 299–322.
28. J. B. Phillips and J. Xu, *J. Chromatogr. A*, 1995, **703**, 327–334.
29. B. M. F. Ávila, A. Aguiar, A. O. Gomes and D. A. Azevedo, *Org. Geochem.*, 2010, **41**, 863–866.
30. B. M. F. Ávila, R. Pereira, A. O. Gomes and D. A. Azevedo, *J. Chromatogr. A*, 2011, **1218**, 3208–3216.
31. T. Dutriez, M. Courtiade, J. Ponthus, D. Thiébaut, H. Dulot and M.-C. Hennion, *Fuel*, 2012, **96**, 108–119.
32. T. Dutriez, D. Thiébaut, M. Courtiade, H. Dulot, F. Bertoncini and M.-C. Hennion, *Fuel*, 2013, **104**, 583–592.
33. L. S. Freitas, C. Von Mühlen, J. H. Bortoluzzi, C. A. Zini, M. Fortuny, C. Dariva, R. C. C. Coutinho, A. F. Santos and E. B. Caramão, *J. Chromatogr. A*, 2009, **1216**, 2860–2865.
34. C. R. Oliveira, A. A. Ferreira, C. J. F. Oliveira, D. A. Azevedo, E. V. Santos Neto and F. R. Aquino Neto, *Org. Geochem.*, 2012, **46**, 154–164.
35. C. R. Oliveira, C. J. F. Oliveira, A. A. Ferreira, D. A. Azevedo and F. R. Aquino Neto, *Org. Geochem.*, 2012, **53**, 131–136.
36. A. M. Booth, A. G. Scarlett, C. A. Lewis, S. T. Belt and S. J. Rowland, *Environ. Sci. Technol.*, 2008, **42**, 8122–8126.

37. M. F. Almstetter, I. J. Appel, M. A. Gruber, C. Lottaz, B. Timischl, R. Spang, K. Dettmer and P. J. Oefner, *Anal. Chem.*, 2009.
38. G. Saravanabhavan, A. Helferty, P. V. Hodson and R. S. Brown, *J. Chromatogr. A*, 2007, **1156**, 124–133.
39. C. A. Bruckner, B. J. Prazen and R. E. Synovec, *Anal. Chem.*, 1998, **70**, 2796–2804.
40. L. Xie, P. J. Marriott and M. Adams, *Anal. Chim. Acta*, 2003, **500**, 211–222.
41. T. Gröger, M. Schäffer, M. Pütz, B. Ahrens, K. Drew, M. Eschner and R. Zimmermann, *J. Chromatogr. A*, 2008, **1200**, 8–16.
42. C. Aeppli, C. A. Carmichael, R. K. Nelson, K. L. Lemkau, W. M. Graham, M. C. Redmond, D. L. Valentine and C. M. Reddy, *Environ. Sci. Technol.*, 2012, **46**, 8799–8807.
43. J. S. Arey, R. K. Nelson, D. L. Plata and C. M. Reddy, *Environ. Sci. Technol.*, 2007, **41**, 5747–5755.
44. J. S. Arey, R. K. Nelson and C. M. Reddy, *Environ. Sci. Technol.*, 2007, **41**, 5738–5746.
45. J. A. DeMello, C. A. Carmichael, E. E. Peacock, R. K. Nelson, J. Samuel Arey and C. M. Reddy, *Marine Pollution Bulletin*, 2007, **54**, 894–904.
46. C. Farwell, C. M. Reddy, E. Peacock, R. K. Nelson, L. Washburn and D. L. Valentine, *Environ. Sci. Technol.*, 2009, **43**, 3542–3548.
47. K. L. Lemkau, E. E. Peacock, R. K. Nelson, G. T. Ventura, J. L. Kovecses and C. M. Reddy, *Mar. Pollut. Bull.*, 2010, **60**, 2123–2129.
48. R. K. Nelson, B. M. Kile, D. L. Plata, S. P. Sylva, L. Xu, C. M. Reddy, R. B. Gaines, G. S. Frysinger and S. E. Reichenbach, *Environ. Forensics*, 2006, 7, 33–44.
49. E. E. Peacock, G. R. Hampson, R. K. Nelson, L. Xu, G. S. Frysinger, R. B. Gaines, J. W. Farrington, B. W. Tripp and C. M. Reddy, *Mar. Pollut. Bull.*, 2007, **54**, 214–225.
50. C. M. Reddy, J. S. Arey, J. S. Seewald, S. P. Sylva, K. L. Lemkau, R. K. Nelson, C. A. Carmichael, C. P. McIntyre, J. Fenwick, G. T. Ventura, B. A. S. Van Mooy and R. Camilli, *Proc. Natl Acad. Sci. U. S. A.*, 2012, **109**, 20229–20234.
51. G. D. Wardlaw, J. S. Arey, C. M. Reddy, R. K. Nelson, G. T. Ventura and D. L. Valentine, *Environ. Sci. Technol.*, 2008, **42**, 7166–7173.
52. R. B. Gaines, G. S. Frysinger, C. M. Reddy and R. K. Nelson, in *Spill Oil Fingerprinting and Source Identification*, ed. Z. Wang and S. Stout, Academic Press, New York, 2006, pp. 169–202.
53. S. Li, S. Hu, J. Cao, M. Wu and D. Zhang, *Int. J. Mol. Sci.*, 2012, **13**, 11399–11410.
54. G. Wang, S. Shi, P. Wang and T. G. Wang, *Fuel*, 2013, **107**, 706–714.
55. R. C. Silva, R. S. F. Silva, E. V. R. de Castro, K. E. Peters and D. A. Azevedo, *Fuel*, 2013, **112**, 125–133.
56. C. Eiserbeck, R. K. Nelson, K. Grice, J. Curiale, C. M. Reddy and P. Raiteri, *J. Chromatogr. A*, 2011, **1218**, 5549–5553.
57. A. L. Payeur, P. A. Meyers and R. D. Sacks, *Org. Geochem.*, 2011, **42**, 1263–1270.

58. C. von Mühlen, E. C. de Oliveira, P. D. Morrison, C. A. Zini, E. B. Caramao and P. J. Marriott, *J. Sep. Sci.*, 2007, **30**, 3223–3232.
59. C. von Mühlen, E. C. de Oliveira, C. A. Zini, E. B. Caramão and P. J. Marriott, *Energy Fuels*, 2010, **24**, 3572–3580.
60. C. Flego and C. Zannoni, *Fuel*, 2011, **90**, 2863–2869.
61. P. M. Harvey and R. A. Shellie, *J. Chromatogr. A*, 2011, **1218**, 3153–3158.
62. R. Shellie, P. Marriott, M. Leus, J.-P. Dufour, L. Mondello, G. Dugo, K. Sun, B. Winniford, J. Griffith and J. Luong, *J. Chromatogr. A*, 2003, **1019**, 273–278.
63. R. A. Shellie, L.-L. Xie and P. J. Marriott, *J. Chromatogr. A*, 2002, **968**, 161–170.
64. G. Purcaro, P. Quinto Tranchida, L. Conte, A. Obiedzińska, P. Dugo, G. Dugo and L. Mondello, *J. Sep. Sci.*, 2011, **34**, 2411–2417.
65. G. Purcaro, P. Q. Tranchida, C. Ragonese, L. Conte, P. Dugo, G. Dugo and L. Mondello, *Anal. Chem.*, 2010, **82**, 8583–8590.
66. D. Valentine, C. Reddy, C. Farwell, T. Hill, O. Pizarro, D. Yoerger, R. Camilli, R. Nelson, E. Peacock, S. Bagby, B. Clarke, C. Roman and M. Soloway, *Nat. Geosci.*, 2010, **3**, 345–348.
67. C. Eiserbeck, PhD, Curtin University, 2011.
68. K. E. Peters, C. C. Walters and J. M. Moldowan, *The Biomarker Guide, Volume 2: Biomarkers and Isotopes in Petroleum Exploration and Earth History*, Cambridge University Press, Cambridge, 2005.
69. N. Varotsis and P. Guieze, *J. Petrol. Sci. Eng.*, 1996, **15**, 81–89.
70. H. P. Roenningsen, I. Skjevrak and E. Osjord, *Energy Fuels*, 1989, **3**, 744–755.
71. C. A. Carmichael, J. S. Arey, W. M. Graham, L. J. Linn, K. L. Lemkau, R. K. Nelson and C. M. Reddy, *Environ. Res. Lett.*, 2012, 7.
72. R. P. Rodgers, A. M. McKenna, R. K. Nelson, W. K. Robbins, C. S. Hsu, C. M. Reddy and A. G. Marshall, *Abstr. Papers Am. Chem. Soc.*, 2011, **241**.
73. H. K. White, P.-Y. Hsing, W. Cho, T. M. Shank, E. E. Cordes, A. M. Quattrini, R. K. Nelson, R. Camilli, A. W. J. Demopoulos, C. R. German, J. M. Brooks, H. H. Roberts, W. Shedd, C. M. Reddy and C. R. Fisher, *Proc. Nat.l Acad. Sci. U. S. A.*, 2012, **109**, 20303–20308.
74. J. P. Ryan, Y. Zhang, H. Thomas, E. V. Rienecker, R. K. Nelson and S. R. Cummings in *Monitoring and Modeling the Deepwater Horizon Oil Spill: A Record-Breaking Enterprise*, ed. Y. Liu, A. MacFadyen, Z. G. Ji and R. H. Weisberg, 2011, vol. 195, 63–75.
75. *On Scene Coordinator Report, Deepwater Horizon Oil Spill*, U.S. Coast Guard, 2011.
76. R. C. Prince, D. L. Elmendorf, J. R. Lute, C. S. Hsu, C. E. Haith, J. D. Senius, G. J. Dechert, G. S. Douglas and E. L. Butler, *Environ. Sci. Technol.*, 1994, **38**, 142–145.
77. R. Camilli, C. M. Reddy, D. R. Yoerger, B. A. S. Van Mooy, M. V. Jakuba, J. C. Kinsey, C. P. McIntyre, S. P. Sylva and J. V. Maloney, *Science*, 2010, **330**, 201–204.

78. C. M. Reddy, R. K. Nelson, S. P. Sylva, L. Xu, E. A. Peacock, B. Raghuraman and O. C. Mullins, *J. Chromatogr. A*, 2007, **1148**, 100–107.
79. K. Grice, R. Alexander and R. I. Kagi, *Org. Geochem.*, 2000, **31**, 67–73.
80. H. P. Nytoft, G. Kildahl-Andersen and O. J. Samuel, *Org. Geochem.*, 2010, **41**, 1104–1118.
81. K. Grice, R. d. Mesmay, A. Glucina and S. Wang, *Org. Geochem.*, 2008, **39**, 284–288.
82. J. Rullkötter, T. M. Peakman and H. L. Ten Haven, *Org. Geochem.*, 1994, **21**, 215–233.
83. G. T. Ventura, B. R. T. Simoneit, R. K. Nelson and C. M. Reddy, *Org. Geochem.*, 2012, **45**, 48–65.
84. S. D. Killops and M. A. H. A. Al-Juboori, *Org. Geochem.*, 1990, **15**, 147–160.
85. A. I. Rushdi and B. R. T. Simoneit, *Appl. Geochem.*, 2002, **17**, 1401–1428.
86. A. I. Rushdi and B. R. T. Simoneit, *Appl. Geochem.*, 2002, **17**, 1467–1494.
87. B. R. T. Simoneit, A. Y. Lein, V. I. Peresypkin and G. A. Osipov, *Geochim. Cosmochim. Acta*, 2004, **68**, 2275–2294.
88. P. F. Z.-d. Valle and B. R. T. Simoneit, *Appl. Geochem.*, 2005, **20**, 2343–2350.
89. L. M. Blumberg, F. David, M. S. Klee and P. Sandra, *J. Chromatogr. A*, 2008, **1188**, 2–16.

Subject Index

Locators in **bold** refer to figures and tables.